Symmetry in Science & Nature

" The Cosmic Balance "

Edited by Paul F. Kisak

Contents

56.2 Fish 459

Chapter 1

Symmetry

For other uses, see Symmetry (disambiguation).

Symmetry (from Greek συμμετρία *symmetria* "agreement in dimensions, due proportion, arrangement")[1] in everyday

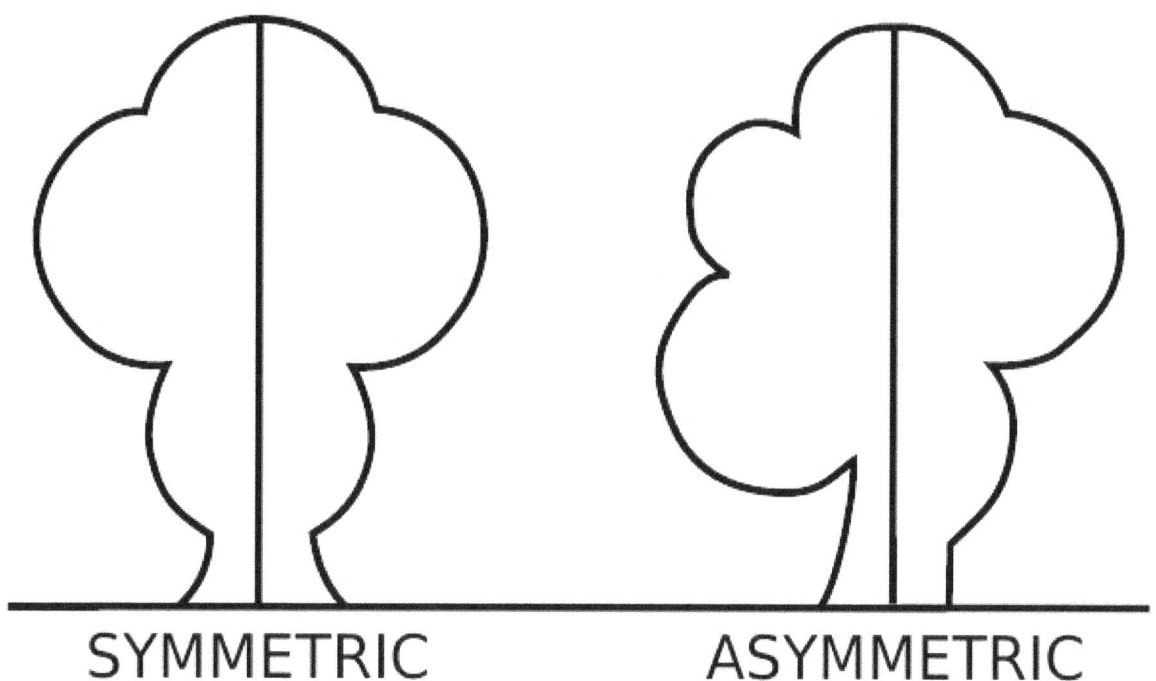

language refers to a sense of harmonious and beautiful proportion and balance.[2][lower-alpha 1] In mathematics, "symmetry" has a more precise definition, that an object is invariant to a transformation, such as reflection but including other transforms too. Although these two meanings of "symmetry" can sometimes be told apart, they are related, so they are here discussed together.

Mathematical symmetry may be observed with respect to the passage of time; as a spatial relationship; through geometric transformations such as scaling, reflection, and rotation; through other kinds of functional transformations; and as an aspect of abstract objects, theoretic models, language, music and even knowledge itself.[3][lower-alpha 2]

This article describes symmetry from three perspectives: in mathematics, including geometry, the most familiar type of symmetry for many people; in science and nature; and in the arts, covering architecture, art and music.

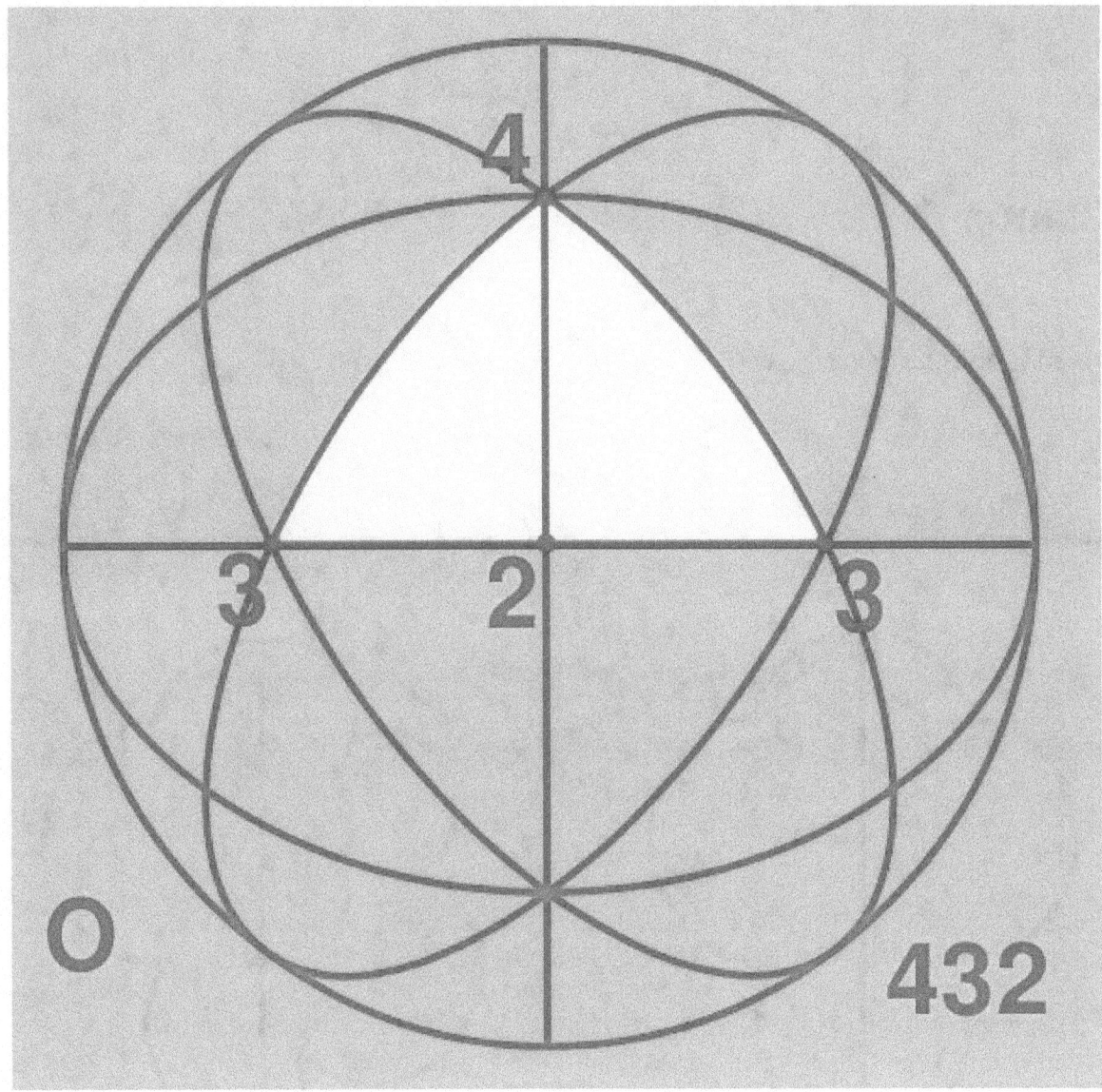

Sphere symmetrical group o representing an octahedral rotational symmetry. The yellow region shows the fundamental domain.

The opposite of symmetry is asymmetry.

1.1   In mathematics

1.1.1   In geometry

Main article: Symmetry (geometry)

A geometric shape or object is symmetric if it can be divided into two or more identical pieces that are arranged in an organized fashion.[4] This means that an object is symmetric if there is a transformation that moves individual pieces of the object but doesn't change the overall shape. The type of symmetry is determined by the way the pieces are organized, or by the type of transformation:

- An object has reflectional symmetry if there is a line of symmetry going through it which divides it into two pieces which are mirror images of each other.[5]

- An object has rotational symmetry if the object can be rotated about a fixed point without changing the overall shape.[6]

- An object has translational symmetry if it can be translated without changing its overall shape.[7]

- An object has helical symmetry if it can be simultaneously translated and rotated in three-dimensional space along a line known as a screw axis.[8]

- An object has scale symmetry if it does not change shape when it is expanded or contracted.[9] Fractals also exhibit a form of scale symmetry, where small portions of the fractal are similar in shape to large portions.[10]

- Other symmetries include glide reflection symmetry and rotoreflection symmetry.

1.1.2 In logic

A dyadic relation R is symmetric if and only if, whenever it's true that Rab, it's true that Rba.[11] Thus, "is the same age as" is symmetrical, for if Paul is the same age as Mary, then Mary is the same age as Paul.

Symmetric binary logical connectives are *and* (∧, or &), *or* (∨, or |), *biconditional* (if and only if) (↔), *nand* (not-and, or $\overline{\wedge}$), *xor* (not-biconditional, or ⊻), and *nor* (not-or, or $\overline{\vee}$).

1.1.3 Other areas of mathematics

Main article: Symmetry (mathematics)

Generalizing from geometrical symmetry in the previous section, we say that a mathematical object is *symmetric* with respect to a given mathematical operation, if, when applied to the object, this operation preserves some property of the object.[12] The set of operations that preserve a given property of the object form a group.

In general, every kind of structure in mathematics will have its own kind of symmetry. Examples include even and odd functions in calculus; the symmetric group in abstract algebra; symmetric matrices in linear algebra; and the Galois group in Galois theory. In statistics, it appears as symmetric probability distributions, and as skewness, asymmetry of distributions.

1.2 In science and nature

Further information: Patterns in nature

1.2.1 In physics

Main article: Symmetry in physics

Symmetry in physics has been generalized to mean invariance—that is, lack of change—under any kind of transformation, for example arbitrary coordinate transformations.[13] This concept has become one of the most powerful tools of theoretical physics, as it has become evident that practically all laws of nature originate in symmetries. In fact, this role inspired the Nobel laureate PW Anderson to write in his widely read 1972 article *More is Different* that "it is only slightly overstating the case to say that physics is the study of symmetry."[14] See Noether's theorem (which, in greatly simplified form, states that for every continuous mathematical symmetry, there is a corresponding conserved quantity; a conserved current, in Noether's original language);[15] and also, Wigner's classification, which says that the symmetries of the laws of physics determine the properties of the particles found in nature.[16]

Important symmetries in physics include continuous symmetries and discrete symmetries of spacetime; internal symmetries of particles; and supersymmetry of physical theories.

1.2.2   In biology

Further information: symmetry in biology, facial symmetry and patterns in nature

Bilateral animals, including humans, are more or less symmetric with respect to the sagittal plane which divides the body into left and right halves.[17] Animals that move in one direction necessarily have upper and lower sides, head and tail ends, and therefore a left and a right. The head becomes specialized with a mouth and sense organs, and the body becomes bilaterally symmetric for the purpose of movement, with symmetrical pairs of muscles and skeletal elements, though internal organs often remain asymmetric.[18]

Plants and sessile (attached) animals such as sea anemones often have radial or rotational symmetry, which suits them because food or threats may arrive from any direction. Fivefold symmetry is found in the echinoderms, the group that includes starfish, sea urchins, and sea lilies.[19]

1.2.3   In chemistry

Main article: molecular symmetry

Symmetry is important to chemistry because it undergirds essentially all *specific* interactions between molecules in nature (i.e., via the interaction of natural and human-made chiral molecules with inherently chiral biological systems). The control of the symmetry of molecules produced in modern chemical synthesis contributes to the ability of scientists to offer therapeutic interventions with minimal side effects. A rigorous understanding of symmetry explains fundamental observations in quantum chemistry, and in the applied areas of spectroscopy and crystallography. The theory and application of symmetry to these areas of physical science draws heavily on the mathematical area of group theory.[20]

1.3   In social interactions

People observe the symmetrical nature, often including asymmetrical balance, of social interactions in a variety of contexts. These include assessments of Reciprocity, empathy, sympathy, apology, dialog, respect, justice, and revenge. Reflective equilibrium is the balance that may be attained through deliberative mutual adjustment among general principles and specific judgments.[21] Symmetrical interactions send the moral message "we are all the same" while asymmetrical interactions may send the message "I am special; better than you." Peer relationships, such as can be governed by the golden rule, are based on symmetry, whereas power relationships are based on asymmetry.[22] Symmetrical relationships can to some degree be maintained by simple (game theory) strategies seen in symmetric games such as tit for tat.[23]

1.4   In the arts

Further information: Mathematics and art

1.4.1   In architecture

Further information: Mathematics and architecture

Symmetry finds its ways into architecture at every scale, from the overall external views of buildings such as Gothic cathedrals and The White House, through the layout of the individual floor plans, and down to the design of individual building elements such as tile mosaics. Islamic buildings such as the Taj Mahal and the Lotfollah mosque make elaborate use of symmetry both in their structure and in their ornamentation.[24][25] Moorish buildings like the Alhambra are ornamented with complex patterns made using translational and reflection symmetries as well as rotations.[26]

It has been said that only bad architects rely on a "symmetrical layout of blocks, masses and structures";[27] Modernist architecture, starting with International style, relies instead on "wings and balance of masses".[27]

1.4.2 In pottery and metal vessels

Since the earliest uses of pottery wheels to help shape clay vessels, pottery has had a strong relationship to symmetry. Pottery created using a wheel acquires full rotational symmetry in its cross-section, while allowing substantial freedom of shape in the vertical direction. Upon this inherently symmetrical starting point, potters from ancient times onwards have added patterns that modify the rotational symmetry to achieve visual objectives.

Cast metal vessels lacked the inherent rotational symmetry of wheel-made pottery, but otherwise provided a similar opportunity to decorate their surfaces with patterns pleasing to those who used them. The ancient Chinese, for example, used symmetrical patterns in their bronze castings as early as the 17th century BC. Bronze vessels exhibited both a bilateral main motif and a repetitive translated border design.[28]

1.4.3 In quilts

As quilts are made from square blocks (usually 9, 16, or 25 pieces to a block) with each smaller piece usually consisting of fabric triangles, the craft lends itself readily to the application of symmetry.[29]

1.4.4 In carpets and rugs

A long tradition of the use of symmetry in carpet and rug patterns spans a variety of cultures. American Navajo Indians used bold diagonals and rectangular motifs. Many Oriental rugs have intricate reflected centers and borders that translate a pattern. Not surprisingly, rectangular rugs typically use quadrilateral symmetry—that is, motifs that are reflected across both the horizontal and vertical axes.[30][31]

1.4.5 In music

Major and minor triads on the white piano keys are symmetrical to the D. (compare article) (file)

Symmetry is not restricted to the visual arts. Its role in the history of music touches many aspects of the creation and perception of music.

Musical form

Symmetry has been used as a formal constraint by many composers, such as the arch (swell) form (ABCBA) used by Steve Reich, Béla Bartók, and James Tenney. In classical music, Bach used the symmetry concepts of permutation and invariance.[32]

Pitch structures

Symmetry is also an important consideration in the formation of scales and chords, traditional or tonal music being made up of non-symmetrical groups of pitches, such as the diatonic scale or the major chord. Symmetrical scales or chords, such as the whole tone scale, augmented chord, or diminished seventh chord (diminished-diminished seventh), are said to lack direction or a sense of forward motion, are ambiguous as to the key or tonal center, and have a less specific diatonic functionality. However, composers such as Alban Berg, Béla Bartók, and George Perle have used axes of symmetry and/or interval cycles in an analogous way to keys or non-tonal tonal centers.

Perle (1992)[33] explains "C–E, D–F♯, [and] Eb–G, are different instances of the same interval ... the other kind of identity. ... has to do with axes of symmetry. C–E belongs to a family of symmetrically related dyads as follows:"

Thus in addition to being part of the interval-4 family, C–E is also a part of the sum-4 family (with C equal to 0).

Interval cycles are symmetrical and thus non-diatonic. However, a seven pitch segment of C5 (the cycle of fifths, which are enharmonic with the cycle of fourths) will produce the diatonic major scale. Cyclic tonal progressions in the works of Romantic composers such as Gustav Mahler and Richard Wagner form a link with the cyclic pitch successions in the atonal music of Modernists such as Bartók, Alexander Scriabin, Edgard Varèse, and the Vienna school. At the same time, these progressions signal the end of tonality.

The first extended composition consistently based on symmetrical pitch relations was probably Alban Berg's *Quartet*, Op. 3 (1910).[34]

Equivalency

Tone rows or pitch class sets which are invariant under retrograde are horizontally symmetrical, under inversion vertically. See also Asymmetric rhythm.

1.4.6   In other arts and crafts

Symmetries appear in the design of objects of all kinds. Examples include beadwork, furniture, sand paintings, knotwork, masks, and musical instruments. Symmetries are central to the art of M.C. Escher and the many applications of tessellation in art and craft forms such as wallpaper, ceramic tilework, batik, ikat, carpet-making, and many kinds of textile and embroidery patterns.[35]

1.4.7   In aesthetics

Main article: Symmetry (physical attractiveness)

The relationship of symmetry to aesthetics is complex. Humans find bilateral symmetry in faces physically attractive;[36] it indicates health and genetic fitness.[37][38] Opposed to this is the tendency for excessive symmetry to be perceived as boring or uninteresting. People prefer shapes that have some symmetry, but enough complexity to make them interesting.[39]

1.4.8   In literature

Symmetry can be found in various forms in literature, a simple example being the palindrome where a brief text reads the same forwards or backwards. Stories may have a symmetrical structure, as in the rise:fall pattern of *Beowulf*.

1.5   See also

- Burnside's lemma

- Chirality

- Even and odd functions

- Fixed points of isometry groups in Euclidean space – center of symmetry

- Spacetime symmetries

- Spontaneous symmetry breaking

- Symmetry-breaking constraints

- Symmetric relation

- Symmetries of polyiamonds

- Symmetries of polyominoes

- Symmetry group

- Time symmetry

- Wallpaper group

1.6 Notes

[1] For example, Aristotle ascribed spherical shape to the heavenly bodies, attributing this formally defined geometric measure of symmetry to the natural order and perfection of the cosmos.

[2] Symmetric objects can be material, such as a person, crystal, quilt, floor tiles, or molecule, or it can be an abstract structure such as a mathematical equation or a series of tones (music).

1.7 References

[1] "symmetry". Online Etymology Dictionary.

[2] Zee, A. (2007). *Fearful Symmetry*. Princeton, N.J.: Princeton University Press. ISBN 978-0-691-13482-6.

[3] Mainzer, Klaus (2005). *Symmetry And Complexity: The Spirit and Beauty of Nonlinear Science*. World Scientific. ISBN 981-256-192-7.

[4] E. H. Lockwood, R. H. Macmillan, *Geometric Symmetry*, London: Cambridge Press, 1978

[5] Weyl, Hermann (1982) [1952]. *Symmetry*. Princeton: Princeton University Press. ISBN 0-691-02374-3.

[6] Singer, David A. (1998). *Geometry: Plane and Fancy*. Springer Science & Business Media.

[7] Stenger, Victor J. (2000) and Mahou Shiro (2007). *Timeless Reality*. Prometheus Books. Especially chapter 12. Nontechnical.

[8] Bottema, O. and B. Roth, *Theoretical Kinematics*, Dover Publications (September 1990)

[9] Tian Yu Cao *Conceptual Foundations of Quantum Field Theory* Cambridge University Press p.154-155

[10] Gouyet, Jean-François (1996). *Physics and fractal structures*. Paris/New York: Masson Springer. ISBN 978-0-387-94153-0.

[11] Josiah Royce, Ignas K. Skrupskelis (2005) *The Basic Writings of Josiah Royce: Logic, loyalty, and community (Google eBook)* Fordham Univ Press, p. 790

[12] Christopher G. Morris (1992) *Academic Press Dictionary of Science and Technology* Gulf Professional Publishing

[13] Costa, Giovanni; Fogli, Gianluigi (2012). *Symmetries and Group Theory in Particle Physics: An Introduction to Space-Time and Internal Symmetries*. Springer Science & Business Media. p. 112.

[14] Anderson, P.W. (1972). "More is Different" (PDF). *Science* **177** (4047): 393–396. Bibcode:1972Sci...177..393A. doi:.393. PMID 17796623.

[15] Kosmann-Schwarzbach, Yvette (2010). *The Noether theorems: Invariance and conservation laws in the twentieth century*. Sources and Studies in the History of Mathematics and Physical Sciences. Springer-Verlag. ISBN 978-0-387-87867-6.

[16] Wigner, E. P. (1939). "On unitary representations of the inhomogeneous Lorentz group", *Annals of Mathematics* **40** (1): 149–204, doi:10.2307/1968551, MR 1503456.

[17] Valentine, James W. "Bilateria". AccessScience. Retrieved 29 May 2013.

[18] Hickman, Cleveland P.; Roberts, Larry S.; Larson, Allan (2002). "Animal Diversity (Third Edition)" (PDF). *Chapter 8: Acoelomate Bilateral Animals*. McGraw-Hill. p. 139. Retrieved October 25, 2012.

[19] Stewart, Ian (2001). *What Shape is a Snowflake? Magical Numbers in Nature*. Weidenfeld & Nicolson. pp. 64–65.

[20] Lowe, John P; Peterson, Kirk (2005). *Quantum Chemistry* (Third ed.). Academic Press. ISBN 0-12-457551-X.

[21] Reflective Equilibrium entry by Norman Daniels in the *Stanford Encyclopedia of Philosophy*, 2003-04-28

[22] Emotional Competency: Symmetry

[23] Lutus, P. (2008). "The Symmetry Principle". Retrieved 28 September 2015.

[24] Williams: Symmetry in Architecture. Members.tripod.com (1998-12-31). Retrieved on 2013-04-16.

[25] Aslaksen: Mathematics in Art and Architecture. Math.nus.edu.sg. Retrieved on 2013-04-16.

[26] Derry, Gregory N. (2002). *What Science Is and How It Works*. Princeton University Press. pp. 269–. ISBN 978-1-4008-2311-6.

[27] Dunlap, David W. (31 July 2009). "Behind the Scenes: Edgar Martins Speaks". New York Times. Retrieved 11 November 2014. "My starting point for this construction was a simple statement which I once read (and which does not necessarily reflect my personal views): 'Only a bad architect relies on symmetry; instead of symmetrical layout of blocks, masses and structures, Modernist architecture relies on wings and balance of masses.'

[28] The Art of Chinese Bronzes. Chinavoc (2007-11-19). Retrieved on 2013-04-16.

[29] Quate: Exploring Geometry Through Quilts. Its.guilford.k12.nc.us. Retrieved on 2013-04-16.

[30] Marla Mallett Textiles & Tribal Oriental Rugs. The Metropolitan Museum of Art, New York.

[31] Dilucchio: Navajo Rugs. Navajocentral.org (2003-10-26). Retrieved on 2013-04-16.

[32] see ("Fugue No. 21," pdf or Shockwave)

[33] Perle,George(1992). "Symmetry,the twelve-tone scale,and tonality".*Contemporary Music Review***6**(2): 81–96. doi:10151.

[34] Perle, George (1990). *The Listening Composer*. University of California Press.

[35] Cucker, Felix (2013). *Manifold Mirrors: The Crossing Paths of the Arts and Mathematics*. Cambridge University Press. pp. 77–78, 83, 89, 103. ISBN 978-0-521-72876-8.

[36] Grammer, K., & Thornhill, R. (1994). Human (Homo sapiens) facial attractiveness and sexual selection: the role of symmetry and averageness. Journal of comparative psychology (Washington, D.C. : 1983), 108(3), 233–42.

[37] Rhodes, Gillian; Zebrowitz, Leslie, A. (2002). *Facial Attractiveness - Evolutionary, Cognitive, and Social Perspectives*. Ablex. ISBN 1-56750-636-4.

[38] Jones, B. C., Little, A. C., Tiddeman, B. P., Burt, D. M., & Perrett, D. I. (2001). Facial symmetry and judgements of apparent health Support for a '' good genes '' explanation of the attractiveness – symmetry relationship, 22, 417–429.

[39] Arnheim, Rudolf (1969). *Visual Thinking*. University of California Press.

1.8   Further reading

- *The Equation That Couldn't Be Solved: How Mathematical Genius Discovered the Language of Symmetry*, Mario Livio, Souvenir Press 2006, ISBN 0-285-63743-6

1.9 External links

- Dutch: Symmetry Around a Point in the Plane

- Chapman: Aesthetics of Symmetry

- ISIS Symmetry

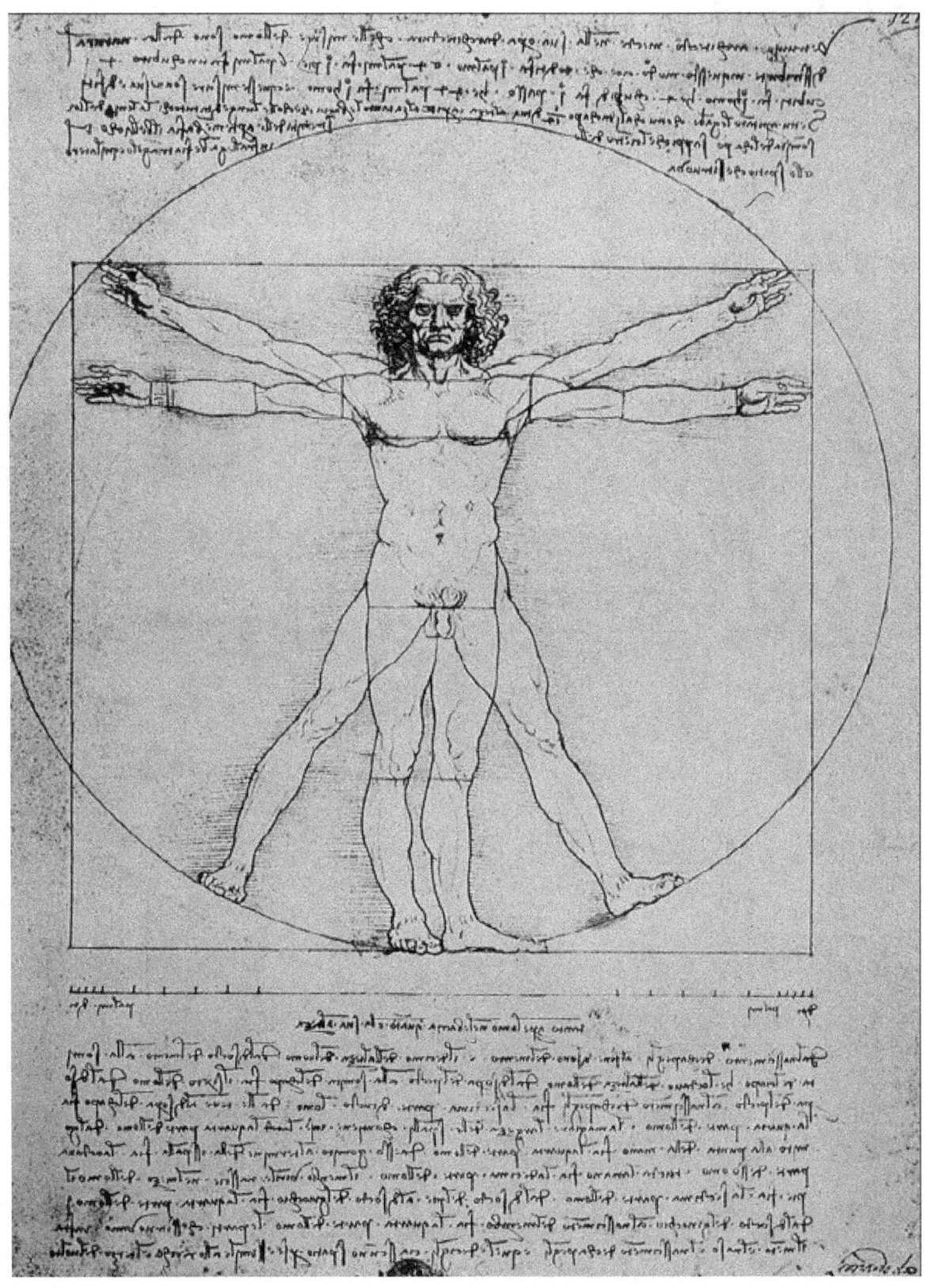

Leonardo da Vinci's 'Vitruvian Man' (ca. 1487) is often used as a representation of symmetry in the human body and, by extension, the natural universe.

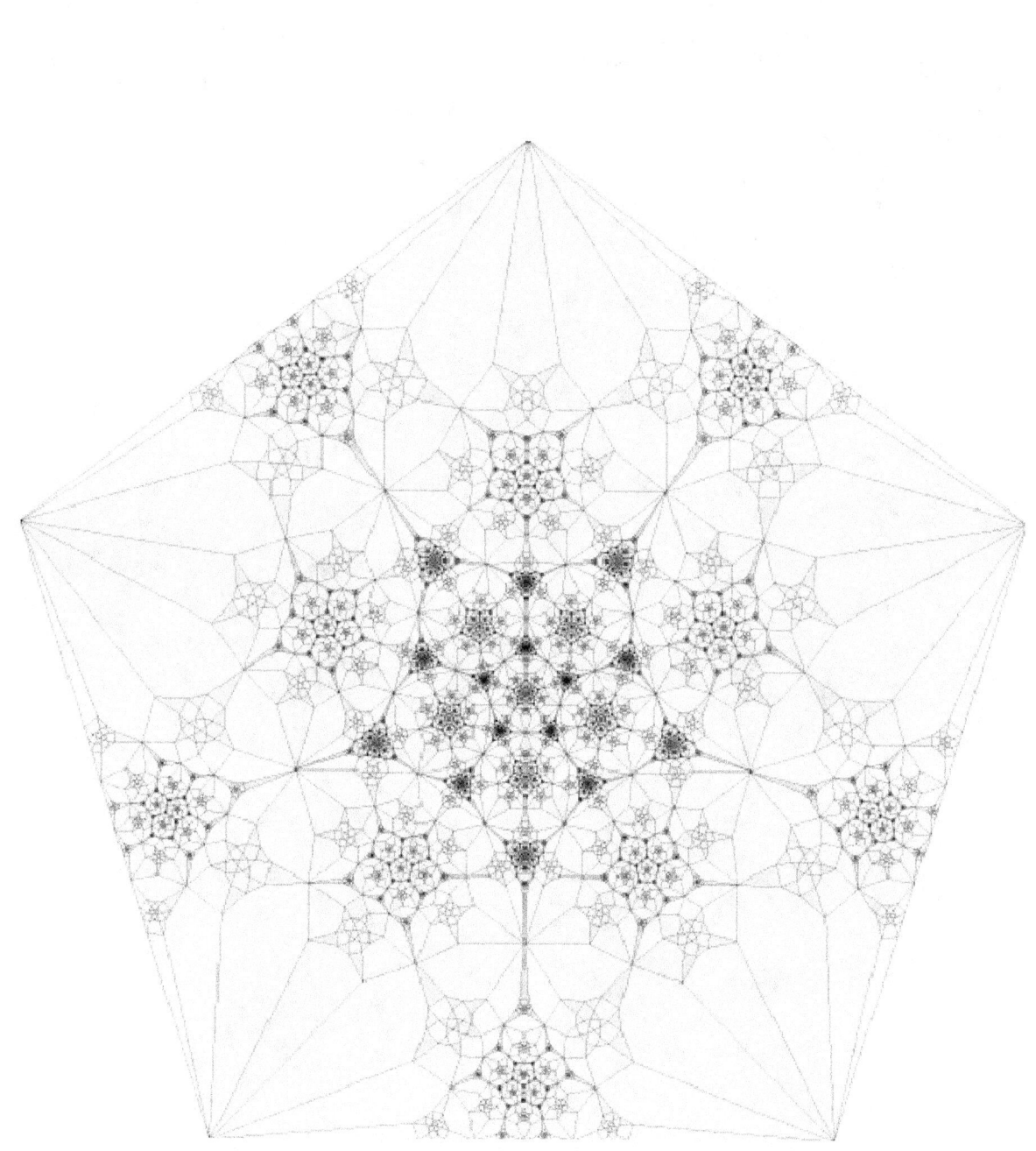

A fractal-like shape that has reflectional symmetry, rotational symmetry and self-similarity, three forms of symmetry. This shape is obtained by a finite subdivision rule.

Symmetric arcades of a portico in the Great Mosque of Kairouan also called the Mosque of Uqba, in Tunisia.

The triskelion has 3-fold rotational symmetry.

Many animals are approximately mirror-symmetric, though internal organs are often arranged asymmetrically.

The ceiling of Lotfollah mosque, Isfahan, Iran has 8-fold symmetries.

Seen from the side, the Taj Mahal has bilateral symmetry; from the top (in plan), it has fourfold symmetry.

Clay pots thrown on a pottery wheel acquire rotational symmetry.

Kitchen Kaleidoscope Block

Persian rug.

Celtic knotwork

Chapter 2

Symmetry (physics)

For other uses, see Symmetry (disambiguation).

In physics, a **symmetry** of a physical system is a physical or mathematical feature of the system (observed or intrinsic) that is preserved or remains unchanged under some transformation.

A family of particular transformations may be *continuous* (such as rotation of a circle) or *discrete* (e.g., reflection of a bilaterally symmetric figure, or rotation of a regular polygon). Continuous and discrete transformations give rise to corresponding types of symmetries. Continuous symmetries can be described by Lie groups while discrete symmetries are described by finite groups (see Symmetry group).

These two concepts, Lie and finite groups, are the foundation for the fundamental theories of modern physics. Symmetries are frequently amenable to mathematical formulations such as group representations and can, in addition, be exploited to simplify many problems.

Arguably the most important example of a symmetry in physics is that the speed of light has the same value in all frames of reference, which is known in mathematical terms as Poincare group, the symmetry group of special relativity. Another important example is the invariance of the form of physical laws under arbitrary differentiable coordinate transformations, which is an important idea in general relativity.

2.1 Symmetry as invariance

Invariance is specified mathematically by transformations that leave some quantity unchanged. This idea can apply to basic real-world observations. For example, temperature may be constant throughout a room. Since the temperature is independent of position within the room, the temperature is *invariant* under a shift in the measurer's position.

Similarly, a uniform sphere rotated about its center will appear exactly as it did before the rotation. The sphere is said to exhibit spherical symmetry. A rotation about any axis of the sphere will preserve how the sphere "looks".

2.1.1 Invariance in force

The above ideas lead to the useful idea of *invariance* when discussing observed physical symmetry; this can be applied to symmetries in forces as well.

For example, an electric field due to a wire is said to exhibit cylindrical symmetry, because the electric field strength at a given distance r from the electrically charged wire of infinite length will have the same magnitude at each point on the surface of a cylinder (whose axis is the wire) with radius r. Rotating the wire about its own axis does not change its position or charge density, hence it will preserve the field. The field strength at a rotated position is the same. Suppose some configuration of charges (may be non-stationary) produce an electric field in some direction, then rotating the configuration of the charges (without disturbing the internal dynamics that produces the particular field) will lead to a net

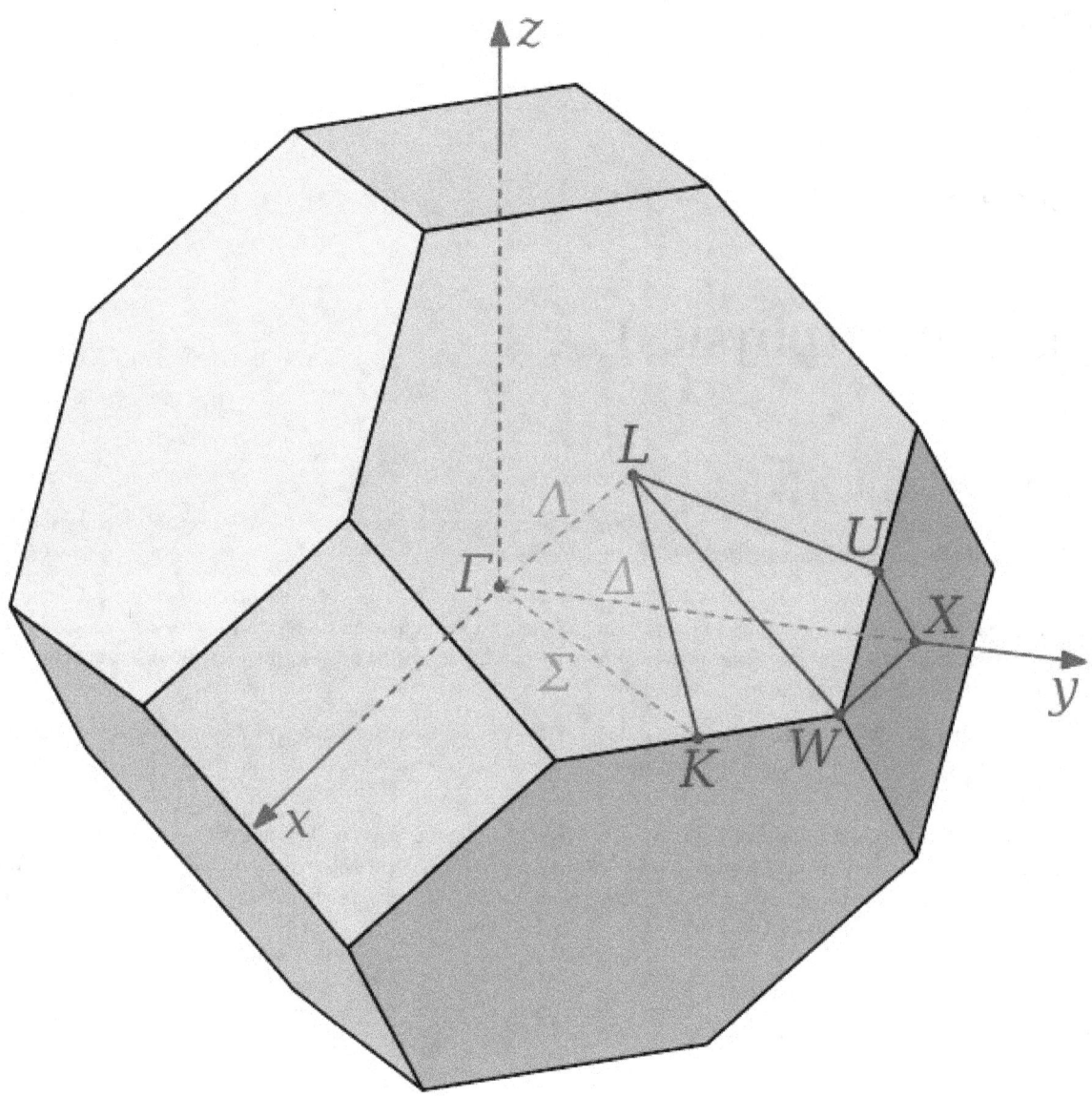

First Brillouin zone of FCC lattice showing symmetry labels

rotation of the direction of the electric field. These two properties are interconnected through the more general property that rotating *any* system of charges causes a corresponding rotation of the electric field.

In Newton's theory of mechanics, given two bodies, each with mass m, starting from rest at the origin and moving along the x-axis in opposite directions, one with speed v_1 and the other with speed v_2 the total kinetic energy of the system (as calculated from an observer at the origin) is $\frac{1}{2}m(v_1{}^2 + v_2{}^2)$ and remains the same if the velocities are interchanged. The total kinetic energy is preserved under a reflection in the y-axis.

The last example above illustrates another way of expressing symmetries, namely through the equations that describe some aspect of the physical system. The above example shows that the total kinetic energy will be the same if v_1 and v_2 are interchanged.

2.2 Local and global symmetries

Main articles: Global symmetry and Local symmetry

Symmetries may be broadly classified as *global* or *local*. A *global symmetry* is one that holds at all points of spacetime, whereas a *local symmetry* is one that has a different symmetry transformation at different points of spacetime; specifically a local symmetry transformation is parameterised by the spacetime co-ordinates. Local symmetries play an important role in physics as they form the basis for gauge theories.

2.3 Continuous symmetries

The two examples of rotational symmetry described above - spherical and cylindrical - are each instances of continuous symmetry. These are characterised by invariance following a continuous change in the geometry of the system. For example, the wire may be rotated through any angle about its axis and the field strength will be the same on a given cylinder. Mathematically, continuous symmetries are described by continuous or smooth functions. An important subclass of continuous symmetries in physics are spacetime symmetries.

2.3.1 Spacetime symmetries

Main article: Spacetime symmetries

Continuous *spacetime symmetries* are symmetries involving transformations of space and time. These may be further classified as *spatial symmetries*, involving only the spatial geometry associated with a physical system; *temporal symmetries*, involving only changes in time; or *spatio-temporal symmetries*, involving changes in both space and time.

- **Time translation**: A physical system may have the same features over a certain interval of time δt ; this is expressed mathematically as invariance under the transformation $t \rightarrow t + a$ for any real numbers t and a in the interval. For example, in classical mechanics, a particle solely acted upon by gravity will have gravitational potential energy mgh when suspended from a height h above the Earth's surface. Assuming no change in the height of the particle, this will be the total gravitational potential energy of the particle at all times. In other words, by considering the state of the particle at some time (in seconds) t_0 and also at $t_0 + 3$, say, the particle's total gravitational potential energy will be preserved.

- **Spatial translation**: These spatial symmetries are represented by transformations of the form $\vec{r} \rightarrow \vec{r} + \vec{a}$ and describe those situations where a property of the system does not change with a continuous change in location. For example, the temperature in a room may be independent of where the thermometer is located in the room.

- **Spatial rotation**: These spatial symmetries are classified as proper rotations and improper rotations. The former are just the 'ordinary' rotations; mathematically, they are represented by square matrices with unit determinant. The latter are represented by square matrices with determinant -1 and consist of a proper rotation combined with a spatial reflection (inversion). For example, a sphere has proper rotational symmetry. Other types of spatial rotations are described in the article *Rotation symmetry*.

- **Poincaré transformations**: These are spatio-temporal symmetries which preserve distances in Minkowski space-time, i.e. they are isometries of Minkowski space. They are studied primarily in special relativity. Those isometries that leave the origin fixed are called Lorentz transformations and give rise to the symmetry known as Lorentz co-variance.

- **Projective symmetries**: These are spatio-temporal symmetries which preserve the geodesic structure of spacetime. They may be defined on any smooth manifold, but find many applications in the study of exact solutions in general relativity.

- *Inversion transformations*: These are spatio-temporal symmetries which generalise Poincaré transformations to include other conformal one-to-one transformations on the space-time coordinates. Lengths are not invariant under inversion transformations but there is a cross-ratio on four points that is invariant.

Mathematically, spacetime symmetries are usually described by smooth vector fields on a smooth manifold. The underlying local diffeomorphisms associated with the vector fields correspond more directly to the physical symmetries, but the vector fields themselves are more often used when classifying the symmetries of the physical system.

Some of the most important vector fields are Killing vector fields which are those spacetime symmetries that preserve the underlying metric structure of a manifold. In rough terms, Killing vector fields preserve the distance between any two points of the manifold and often go by the name of isometries.

2.4   Discrete symmetries

Main article: Discrete symmetry

A **discrete symmetry** is a symmetry that describes non-continuous changes in a system. For example, a square possesses discrete rotational symmetry, as only rotations by multiples of right angles will preserve the square's original appearance. Discrete symmetries sometimes involve some type of 'swapping', these swaps usually being called *reflections* or *interchanges*.

- *Time reversal*: Many laws of physics describe real phenomena when the direction of time is reversed. Mathematically, this is represented by the transformation, $t \rightarrow -t$. For example, Newton's second law of motion still holds if, in the equation $F = m\ddot{r}$, t is replaced by $-t$. This may be illustrated by recording the motion of an object thrown up vertically (neglecting air resistance) and then playing it back. The object will follow the same parabolic trajectory through the air, whether the recording is played normally or in reverse. Thus, position is symmetric with respect to the instant that the object is at its maximum height.

- *Spatial inversion*: These are represented by transformations of the form $\vec{r} \rightarrow -\vec{r}$ and indicate an invariance property of a system when the coordinates are 'inverted'. Said another way, these are symmetries between a certain object and its mirror image.

- *Glide reflection*: These are represented by a composition of a translation and a reflection. These symmetries occur in some crystals and in some planar symmetries, known as wallpaper symmetries.

2.4.1   C, P, and T symmetries

The Standard model of particle physics has three related natural near-symmetries. These state that the actual universe about us is indistinguishable from one where:

- Every particle is replaced with its antiparticle. This is C-symmetry (charge symmetry);

- Everything appears as if reflected in a mirror. This is P-symmetry (parity symmetry);

- The direction of time is reversed. This is T-symmetry (time symmetry).

T-symmetry is counterintuitive (surely the future and the past are not symmetrical) but explained by the fact that the Standard model describes local properties, not global ones like entropy. To properly reverse the direction of time, one would have to put the big bang and the resulting low-entropy state in the "future." Since we perceive the "past" ("future") as having lower (higher) entropy than the present (see perception of time), the inhabitants of this hypothetical time-reversed universe would perceive the future in the same way as we perceive the past.

These symmetries are near-symmetries because each is broken in the present-day universe. However, the Standard Model predicts that the combination of the three (that is, the simultaneous application of all three transformations) must be a symmetry, called CPT symmetry. In <ref name=*qm*>G. Kalmbach H.E.: *Quantum Mathematics: WIGRIS*. RGN Publications, Delhi, 2014.</ref> the 4 dimensional matrix description of P,T is through a diagonal matrix, the negative identity, as well as C. Hence CPT is the identity operator. CP violation, the violation of the combination of C- and P-symmetry, is necessary for the presence of significant amounts of baryonic matter in the universe. CP violation is a fruitful area of current research in particle physics.

2.4.2 Supersymmetry

Main article: Supersymmetry

A type of symmetry known as supersymmetry has been used to try to make theoretical advances in the standard model. Supersymmetry is based on the idea that there is another physical symmetry beyond those already developed in the standard model, specifically a symmetry between bosons and fermions. Supersymmetry asserts that each type of boson has, as a supersymmetric partner, a fermion, called a superpartner, and vice versa. Supersymmetry has not yet been experimentally verified: no known particle has the correct properties to be a superpartner of any other known particle. If superpartners exist they must have masses greater than current particle accelerators can generate.

2.5 Mathematics of physical symmetry

Main article: Symmetry group
See also: Symmetry in quantum mechanics and Symmetries in general relativity

The transformations describing physical symmetries typically form a mathematical group. Group theory is an important area of mathematics for physicists.

Continuous symmetries are specified mathematically by *continuous groups* (called Lie groups). Many physical symmetries are isometries and are specified by symmetry groups. Sometimes this term is used for more general types of symmetries. The set of all proper rotations (about any angle) through any axis of a sphere form a Lie group called the special orthogonal group $SO(3)$. (The *3* refers to the three-dimensional space of an ordinary sphere.) Thus, the symmetry group of the sphere with proper rotations is $SO(3)$. Any rotation preserves distances on the surface of the ball. The set of all Lorentz transformations form a group called the Lorentz group (this may be generalised to the Poincaré group).

Discrete symmetries are described by discrete groups. For example, the symmetries of an equilateral triangle are described by the symmetric group S_3.

An important type of physical theory based on *local* symmetries is called a *gauge* theory and the symmetries natural to such a theory are called gauge symmetries. Gauge symmetries in the Standard model, used to describe three of the fundamental interactions, are based on the SU(3) × SU(2) × U(1) group. (Roughly speaking, the symmetries of the SU(3) group describe the strong force, the SU(2) group describes the weak interaction and the U(1) group describes the electromagnetic force.)

Also, the reduction by symmetry of the energy functional under the action by a group and spontaneous symmetry breaking of transformations of symmetric groups appear to elucidate topics in particle physics (for example, the unification of electromagnetism and the weak force in physical cosmology).

2.5.1 Conservation laws and symmetry

Main article: Noether's theorem

The symmetry properties of a physical system are intimately related to the conservation laws characterizing that system.

Noether's theorem gives a precise description of this relation. The theorem states that each continuous symmetry of a physical system implies that some physical property of that system is conserved. Conversely, each conserved quantity has a corresponding symmetry. For example, the isometry of space gives rise to conservation of (linear) momentum, and isometry of time gives rise to conservation of energy.

The following table summarizes some fundamental symmetries and the associated conserved quantity.

2.6   Mathematics

Continuous symmetries in physics preserve transformations. One can specify a symmetry by showing how a very small transformation affects various particle fields. The commutator of two of these infinitessimal transformations are equivalent to a third infinitessimal transformation of the same kind hence they form a Lie algebra.

A general coordinate transformation (also known as a diffeomorphism) has the infinitessimal effect on a scalar, spinor and vector field for example:

$$\delta\phi(x) = h^\mu(x)\partial_\mu\phi(x)$$

$$\delta\psi^\alpha(x) = h^\mu(x)\partial_\mu\psi^\alpha(x) + \partial_\mu h_\nu(x)\sigma^{\alpha\beta}_{\mu\nu}\psi^\beta(x)$$

$$\delta A_\mu(x) = h^\nu(x)\partial_\nu A_\mu(x) + A_\nu(x)\partial_\mu h^\nu(x)$$

for a general field, $h(x)$. Without gravity only the Poincaré symmetries are preserved which restricts $h(x)$ to be of the form:

$$h^\mu(x) = M^{\mu\nu}x_\nu + P^\mu$$

where **M** is an antisymmetric matrix (giving the Lorentz and rotational symmetries) and **P** is a general vector (giving the translational symmetries). Other symmetries affect multiple fields simultaneously. For example local gauge transformations apply to both a vector and spinor field:

$$\delta\psi^\alpha(x) = \lambda(x).\tau^{\alpha\beta}\psi^\beta(x)$$

$$\delta A_\mu(x) = \partial_\mu\lambda(x)$$

where τ are generators of a particular Lie group. So far the transformations on the right have only included fields of the same type. Supersymmetries are defined according to how the mix fields of *different* types.

Another symmetry which is part of some theories of physics and not in others is scale invariance which involve Weyl transformations of the following kind:

$$\delta\phi(x) = \Omega(x)\phi(x)$$

If the fields have this symmetry then it can be shown that the field theory is almost certainly conformally invariant also. This means that in the absence of gravity h(x) would restricted to the form:

$$h^\mu(x) = M^{\mu\nu}x_\nu + P^\mu + Dx_\mu + K^\mu|x|^2 - 2K^\nu x_\nu x_\mu$$

with **D** generating scale transformations and **K** generating special conformal transformations. For example N=4 super-Yang-Mills theory has this symmetry while General Relativity doesn't although other theories of gravity such as conformal gravity do. The 'action' of a field theory is an invariant under all the symmetries of the theory. Much of modern theoretical physics is to do with speculating on the various symmetries the Universe may have and finding the invariants to construct field theories as models.

In string theories, since a string can be decomposed into an infinite number of particle fields, the symmetries on the string world sheet is equivalent to special transformations which mix an infinite number of fields.

2.7   See also

- Conservation law
- Conserved current

- Coordinate-free

- Covariance and contravariance

- Diffeomorphism

- Fictitious force

- Galilean invariance

- Gauge theory

- General covariance

- Harmonic coordinate condition

- Inertial frame of reference

- Lie group

- List of mathematical topics in relativity

- Lorentz covariance

- Noether's theorem

- Poincaré group

- Special relativity

- Spontaneous symmetry breaking

- Standard model

- Standard model (mathematical formulation)

- Symmetry breaking

- Wheeler–Feynman Time-Symmetric Theory

2.8 References

2.8.1 General readers

- Leon Lederman and Christopher T. Hill (2005) *Symmetry and the Beautiful Universe.* Amherst NY: Prometheus Books.

- Schumm, Bruce (2004) *Deep Down Things.* Johns Hopkins Univ. Press.

- Victor J. Stenger (2000) *Timeless Reality: Symmetry, Simplicity, and Multiple Universes.* Buffalo NY: Prometheus Books. Chpt. 12 is a gentle introduction to symmetry, invariance, and conservation laws.

- Anthony Zee (2007) *Fearful Symmetry: The search for beauty in modern physics.* 2nd ed. Princeton University Press. ISBN 978-0-691-00946-9. 1986 1st ed. published by Macmillan.

2.8.2 Technical readers

- Brading, K., and Castellani, E., eds. (2003) *Symmetries in Physics: Philosophical Reflections.* Cambridge Univ. Press.

- -------- (2007) "Symmetries and Invariances in Classical Physics" in Butterfield, J., and John Earman, eds., *Philosophy of Physic Part B.* North Holland: 1331-68.

- Debs, T. and Redhead, M. (2007) *Objectivity, Invariance, and Convention: Symmetry in Physical Science.* Harvard Univ. Press.

- John Earman (2002) "Laws, Symmetry, and Symmetry Breaking: Invariance, Conservations Principles, and Objectivity." Address to the 2002 meeting of the Philosophy of Science Association.

- G. Kalmbach H.E.: *Quantum Mathematics: WIGRIS.* RGN Publications, Delhi, 2014

- Mainzer, K. (1996) *Symmetries of nature.* Berlin: De Gruyter.

- Mouchet, A. "Reflections on the four facets of symmetry: how physics exemplifies rational thinking". European Physical Journal H 38 (2013) 661 hal.archives-ouvertes.fr:hal-00637572

- Thompson, William J. (1994) *Angular Momentum: An Illustrated Guide to Rotational Symmetries for Physical Systems.* Wiley. ISBN 0-471-55264-X.

- Bas Van Fraassen (1989) *Laws and symmetry.* Oxford Univ. Press.

- Eugene Wigner (1967) *Symmetries and Reflections.* Indiana Univ. Press.

2.9 External links

- Stanford Encyclopedia of Philosophy: "Symmetry"—by K. Brading and E. Castellani.

- Pedagogic Aids to Quantum Field Theory Click on link to Chapter 6: Symmetry, Invariance, and Conservation for a simplified, step-by-step introduction to symmetry in physics.

Chapter 3

Physical system

In physics, a **physical system** is a portion of the physical universe chosen for analysis. Everything outside the system is known as the environment. The environment is ignored except for its effects on itself. In a physical system, a lower probability states that the vector is equivalent to a higher complexity.

The split between system and environment is the analyst's choice, generally made to simplify the analysis. For example, the water in a lake, the water in half of a lake, or an individual molecule of water in the lake can each be considered a physical system. An isolated system is one that has negligible interaction with its environment. Often a system in this sense is chosen to correspond to the more usual meaning of system, such as a particular machine.

In the study of quantum coherence the "system" may refer to the microscopic properties of an object (e.g. the mean of a pendulum bob), while the relevant "environment" may be the internal degrees of freedom, described classically by the pendulum's thermal vibrations.

3.1 See also

- Conceptual systems

- Phase space

- Physical phenomenon

- Thermodynamic system

- Physical ontology

- Signal-flow graph

3.2 External links

- Conceptual vs Physical Systems

- Research in simulation and modeling of various physical systems

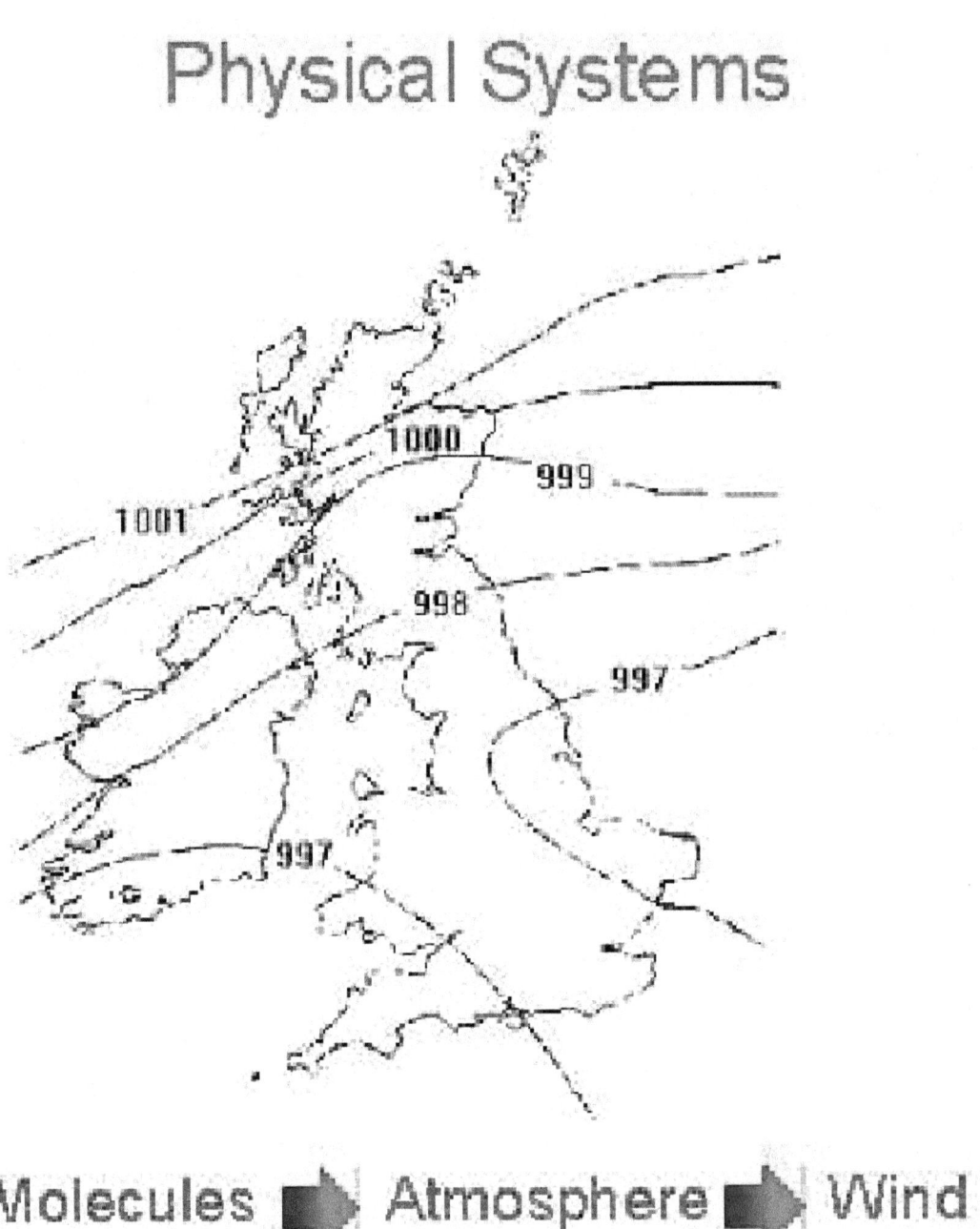

Weather map as an example of a physical system

Chapter 4

Symmetry group

Not to be confused with Symmetric group.

This article is about the abstract algebraic structures. For other meanings, see Symmetry group (disambiguation).

 In abstract algebra, the **symmetry group** of an object (image, signal, etc.) is the group of all transformations under which the object is invariant with composition as the group operation. For a space with a metric, it is a subgroup of the isometry group of the space concerned. If not stated otherwise, this article considers symmetry groups in Euclidean geometry, but the concept may also be studied in more general contexts as expanded below.

4.1   Introduction

The "objects" may be geometric figures, images, and patterns, such as a wallpaper pattern. The definition can be made more precise by specifying what is meant by image or pattern, e.g., a function of position with values in a set of colors. For symmetry of physical objects, one may also want to take their physical composition into account. The group of isometries of space induces a group action on objects in it.

The symmetry group is sometimes also called **full symmetry group** in order to emphasize that it includes the orientation-reversing isometries (like reflections, glide reflections and improper rotations) under which the figure is invariant. The subgroup of orientation-preserving isometries (i.e. translations, rotations, and compositions of these) that leave the figure invariant is called its **proper symmetry group**. The proper symmetry group of an object is equal to its full symmetry group if and only if the object is chiral (and thus there are no orientation-reversing isometries under which it is invariant).

Any symmetry group whose elements have a common fixed point, which is true for all finite symmetry groups and also for the symmetry groups of bounded figures, can be represented as a subgroup of the orthogonal group $O(n)$ by choosing the origin to be a fixed point. The proper symmetry group is then a subgroup of the special orthogonal group $SO(n)$, and is therefore also called **rotation group** of the figure.

A **discrete symmetry group** is a symmetry group such that for every point of the space the set of images of the point under the isometries in the symmetry group is a discrete set.

Discrete symmetry groups come in three types: (1) finite **point groups**, which include only rotations, reflections, inversion and rotoinversion -- they are just the finite subgroups of $O(n)$, (2) infinite **lattice groups**, which include only translations, and (3) infinite **space groups** which combines elements of both previous types, and may also include extra transformations like screw axis and glide reflection. There are also *continuous* symmetry groups, which contain rotations of arbitrarily small angles or translations of arbitrarily small distances. The group of all symmetries of a sphere $O(3)$ is an example of this, and in general such continuous symmetry groups are studied as Lie groups. With a categorization of subgroups of the Euclidean group corresponds a categorization of symmetry groups.

Two geometric figures are considered to be of the same symmetry type if their symmetry groups are conjugate subgroups of the Euclidean group $E(n)$ (the isometry group of $\mathbf{R}^n$), where two subgroups H_1, H_2 of a group G are *conjugate*, if there exists $g \in G$ such that $H_1 = g^{-1} H_2 g$. For example:

- two 3D figures have mirror symmetry, but with respect to different mirror planes.

- two 3D figures have 3-fold rotational symmetry, but with respect to different axes.

- two 2D patterns have translational symmetry, each in one direction; the two translation vectors have the same length but a different direction.

When considering isometry groups, one may restrict oneself to those where for all points the set of images under the isometries is topologically closed. This includes all discrete isometry groups and also those involved in continuous symmetries, but excludes for example in 1D the group of translations by a rational number. A "figure" with this symmetry group is non-drawable and up to arbitrarily fine detail homogeneous, without being really homogeneous.

4.2   One dimension

The isometry groups in one dimension where for all points the set of images under the isometries is topologically closed are:

- the trivial group C_1

- the groups of two elements generated by a reflection in a point; they are isomorphic with C_2

- the infinite discrete groups generated by a translation; they are isomorphic with Z, the additive group of the integers

- the infinite discrete groups generated by a translation and a reflection in a point; they are isomorphic with the generalized dihedral group of Z, Dih(Z), also denoted by D∞ (which is a semidirect product of Z and C_2).

- the group generated by all translations (isomorphic with the additive group of the real numbers **R**); this group cannot be the symmetry group of a "pattern": it would be homogeneous, hence could also be reflected. However, a uniform one-dimensional vector field has this symmetry group.

- the group generated by all translations and reflections in points; they are isomorphic with the generalized dihedral group of **R**, Dih(**R**).

See also symmetry groups in one dimension.

4.3   Two dimensions

Up to conjugacy the discrete point groups in two-dimensional space are the following classes:

- cyclic groups C_1, C_2, C_3, C_4, ... where Cn consists of all rotations about a fixed point by multiples of the angle $360°/n$

- dihedral groups D_1, D_2, D_3, D_4, ..., where Dn (of order $2n$) consists of the rotations in Cn together with reflections in n axes that pass through the fixed point.

C_1 is the trivial group containing only the identity operation, which occurs when the figure has no symmetry at all, for example the letter **F**. C_2 is the symmetry group of the letter **Z**, C_3 that of a triskelion, C_4 of a swastika, and C_5, C_6, etc. are the symmetry groups of similar swastika-like figures with five, six, etc. arms instead of four.

D_1 is the 2-element group containing the identity operation and a single reflection, which occurs when the figure has only a single axis of bilateral symmetry, for example the letter **A**. D_2, which is isomorphic to the Klein four-group, is the symmetry group of a non-equilateral isosceles triangle, and D_3, D_4 etc. are the symmetry groups of the regular polygons.

The actual symmetry groups in each of these cases have two degrees of freedom for the center of rotation, and in the case of the dihedral groups, one more for the positions of the mirrors.

The remaining isometry groups in two dimensions with a fixed point, where for all points the set of images under the isometries is topologically closed are:

- the special orthogonal group SO(2) consisting of all rotations about a fixed point; it is also called the circle group S^1, the multiplicative group of complex numbers of absolute value 1. It is the *proper* symmetry group of a circle and the continuous equivalent of Cn. There is no geometric figure that has as *full* symmetry group the circle group, but for a vector field it may apply (see the three-dimensional case below).

- the orthogonal group O(2) consisting of all rotations about a fixed point and reflections in any axis through that fixed point. This is the symmetry group of a circle. It is also called Dih(S^1) as it is the generalized dihedral group of S^1.

For non-bounded figures, the additional isometry groups can include translations; the closed ones are:

- the 7 frieze groups

- the 17 wallpaper groups

- for each of the symmetry groups in one dimension, the combination of all symmetries in that group in one direction, and the group of all translations in the perpendicular direction

- ditto with also reflections in a line in the first direction

4.4 Three dimensions

See also: Point groups in three dimensions

Up to conjugacy the set of three-dimensional point groups consists of 7 infinite series, and 7 separate ones. In crystallography they are restricted to be compatible with the discrete translation symmetries of a crystal lattice. This crystallographic restriction of the infinite families of general point groups results in 32 crystallographic point groups (27 from the 7 infinite series, and 5 of the 7 others).

The continuous symmetry groups with a fixed point include those of:

- cylindrical symmetry without a symmetry plane perpendicular to the axis, this applies for example often for a bottle

- cylindrical symmetry with a symmetry plane perpendicular to the axis

- spherical symmetry

For objects and scalar fields the cylindrical symmetry implies vertical planes of reflection. However, for vector fields it does not: in cylindrical coordinates with respect to some axis, $\mathbf{A} = A_\rho \hat{\rho} + A_\phi \hat{\phi} + A_z \hat{z}$ has cylindrical symmetry with respect to the axis if and only if A_ρ, A_ϕ, and A_z have this symmetry, i.e., they do not depend on φ. Additionally there is reflectional symmetry if and only if $A_\phi = 0$.

For spherical symmetry there is no such distinction, it implies planes of reflection.

The continuous symmetry groups without a fixed point include those with a screw axis, such as an infinite helix. See also subgroups of the Euclidean group.

4.5   Symmetry groups in general

See also: Automorphism

In wider contexts, a **symmetry group** may be any kind of **transformation group**, or automorphism group. Once we know what kind of mathematical structure we are concerned with, we should be able to pinpoint what mappings preserve the structure. Conversely, specifying the symmetry can define the structure, or at least clarify what we mean by an invariant, geometric language in which to discuss it; this is one way of looking at the Erlangen programme.

For example, automorphism groups of certain models of finite geometries are not "symmetry groups" in the usual sense, although they preserve symmetry. They do this by preserving *families* of point-sets rather than point-sets (or "objects") themselves.

Like above, the group of automorphisms of space induces a group action on objects in it.

For a given geometric figure in a given geometric space, consider the following equivalence relation: two automorphisms of space are equivalent if and only if the two images of the figure are the same (here "the same" does not mean something like e.g. "the same up to translation and rotation", but it means "exactly the same"). Then the equivalence class of the identity is the symmetry group of the figure, and every equivalence class corresponds to one isomorphic version of the figure.

There is a bijection between every pair of equivalence classes: the inverse of a representative of the first equivalence class, composed with a representative of the second.

In the case of a finite automorphism group of the whole space, its order is the order of the symmetry group of the figure multiplied by the number of isomorphic versions of the figure.

Examples:

- Isometries of the Euclidean plane, the figure is a rectangle: there are infinitely many equivalence classes; each contains 4 isometries.

- The space is a cube with Euclidean metric; the figures include cubes of the same size as the space, with colors or patterns on the faces; the automorphisms of the space are the 48 isometries; the figure is a cube of which one face has a different color; the figure has a symmetry group of 8 isometries, there are 6 equivalence classes of 8 isometries, for 6 isomorphic versions of the figure.

Compare Lagrange's theorem (group theory) and its proof.

4.6   See also

- Crystallography

- Crystal system

- Euclidean plane isometry

- Fixed points of isometry groups in Euclidean space

- Group action

- Permutation group

- Point group

- Space group

- Symmetric group

- Symmetry

- Symmetry in quantum mechanics

4.7 Further reading

- Burns, G.; Glazer, A. M. (1990). *Space Groups for Scientists and Engineers* (2nd ed.). Boston: Academic Press, Inc. ISBN 0-12-145761-3.

- Clegg, W (1998). *Crystal Structure Determination (Oxford Chemistry Primer)*. Oxford: Oxford University Press. ISBN 0-19-855901-1.

- O'Keeffe, M.; Hyde, B. G. (1996). *Crystal Structures; I. Patterns and Symmetry*. Washington, DC: Mineralogical Society of America, *Monograph Series*. ISBN 0-939950-40-5.

- Miller, Willard Jr. (1972). *Symmetry Groups and Their Applications*. New York: Academic Press. OCLC 589081. Retrieved 2009-09-28.

4.8 External links

- Weisstein, Eric W., "Symmetry Group", *MathWorld*.

- Weisstein, Eric W., "Tetrahedral Group", *MathWorld*.

- Overview of the 32 crystallographic point groups - form the first parts (apart from skipping $n=5$) of the 7 infinite series and 5 of the 7 separate 3D point groups

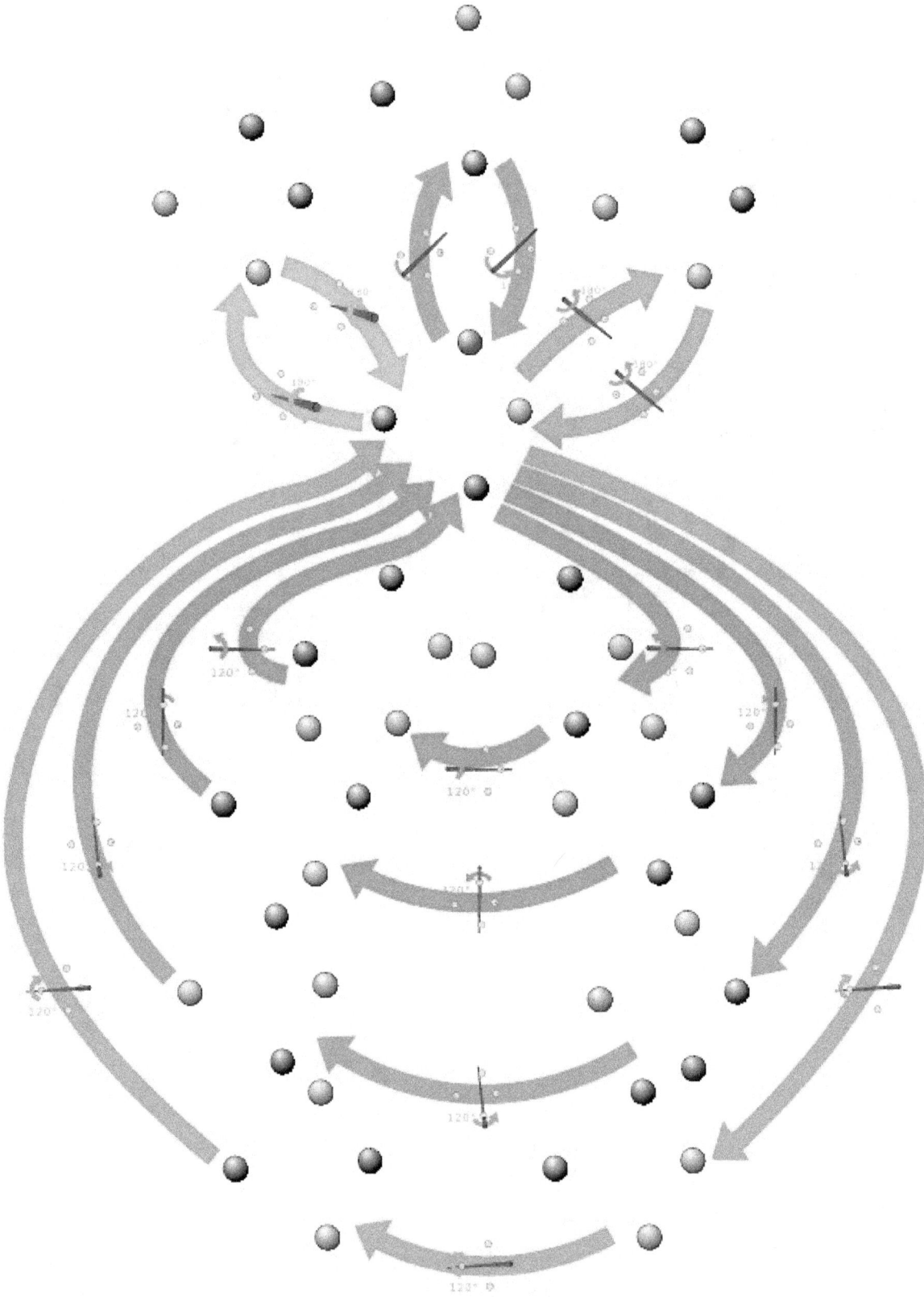

A tetrahedron is invariant under 12 distinct rotations, reflections excluded. These are illustrated here in the cycle graph format, along with the 180° edge (blue arrows) and 120° vertex (reddish arrows) rotations that permute the tetrahedron through the positions. The 12 rotations form the **rotation (symmetry) group** of the figure.

Chapter 5

Circular symmetry

Circular symmetry in mathematical physics applies to a 2-dimensional field which can be expressed as a function of distance from a central point only. This means that all points on each circle take the same value.

An example would be magnetic field intensity in a plane perpendicular to a current-carrying wire. A pattern with circular symmetry would consist of concentric circles.

The 3-dimensional equivalent term is **spherical symmetry**. A scalar field has spherical symmetry if it depends on the distance to the origin only, such as the potential of a central force. A vector field has spherical symmetry if it is in radially inward or outward direction with a magnitude and orientation (inward/outward) depending on the distance to the origin only, such as a central force.

5.1 See also

- Rotational symmetry

- Particle in a spherically symmetric potential

- Gauss's theorem

Chapter 6

Rotational symmetry

The triskelion appearing on the Isle of Man flag.

Generally, an object with 'rotational symmetry' also known in biological contexts as 'radial symmetry', is an object that looks the same after a certain amount of rotation. An object may have more than one rotational symmetry; for instance, if reflections or turning it over are not counted. The degree of rotational symmetry is how many degrees the shape has to be turned to look the same on a different side or vertex. It cannot be the same side or vertex.

6.1 Formal treatment

See also: Rotational invariance

Formally the rotational symmetry is symmetry with respect to some or all rotations in m-dimensional Euclidean space. Rotations are direct isometries, i.e., isometries preserving orientation. Therefore a symmetry group of rotational symmetry is a subgroup of $E^+(m)$ (see Euclidean group).

Symmetry with respect to all rotations about all points implies translational symmetry with respect to all translations, so space is homogeneous, and the symmetry group is the whole $E(m)$. With the modified notion of symmetry for vector fields the symmetry group can also be $E^+(m)$.

For symmetry with respect to rotations about a point we can take that point as origin. These rotations form the special orthogonal group $SO(m)$, the group of $m \times m$ orthogonal matrices with determinant 1. For $m = 3$ this is the rotation group $SO(3)$.

In another meaning of the word, the rotation group *of an object* is the symmetry group within $E^+(n)$, the group of direct isometries; in other words, the intersection of the full symmetry group and the group of direct isometries. For chiral objects it is the same as the full symmetry group.

Laws of physics are $SO(3)$-invariant if they do not distinguish different directions in space. Because of Noether's theorem, rotational symmetry of a physical system is equivalent to the angular momentum conservation law.

6.1.1   n-fold rotational symmetry

Rotational symmetry of order n, also called **n-fold rotational symmetry**, or **discrete rotational symmetry of the nth order**, with respect to a particular point (in 2D) or axis (in 3D) means that rotation by an angle of $360°/n$ ($180°$, $120°$, $90°$, $72°$, $60°$, $51\,{}^3/_7°$, etc.) does not change the object. Note that "1-fold" symmetry is no symmetry, and "2-fold" is the simplest symmetry, so it does not mean "more than basic".

The notation for n-fold symmetry is Cn or simply "n". The actual symmetry group is specified by the point or axis of symmetry, together with the n. For each point or axis of symmetry the abstract group type is cyclic group Zn of order n. Although for the latter also the notation Cn is used, the geometric and abstract Cn should be distinguished: there are other symmetry groups of the same abstract group type which are geometrically different, see cyclic symmetry groups in 3D.

The fundamental domain is a sector of $360°/n$.

Examples without additional reflection symmetry:

- $n = 2$, $180°$: the *dyad*, quadrilaterals with this symmetry are the parallelograms; other examples: letters Z, N, S; apart from the colors: yin and yang

- $n = 3$, $120°$: *triad*, triskelion, Borromean rings; sometimes the term *trilateral symmetry* is used;

- $n = 4$, $90°$: *tetrad*, swastika

- $n = 6$, $60°$: *hexad*, Star of David

- $n = 8$, $45°$: *octad*, Octagonal muqarnas, computer-generated (CG), ceiling

Cn is the rotation group of a regular n-sided polygon in 2D and of a regular n-sided pyramid in 3D.

If there is e.g. rotational symmetry with respect to an angle of $100°$, then also with respect to one of $20°$, the greatest common divisor of $100°$ and $360°$.

A typical 3D object with rotational symmetry (possibly also with perpendicular axes) but no mirror symmetry is a propeller.

6.1.2   Examples

6.1.3   Multiple symmetry axes through the same point

For discrete symmetry with multiple symmetry axes through the same point, there are the following possibilities:

- In addition to an n-fold axis, n perpendicular 2-fold axes: the dihedral groups D_n of order $2n$ ($n \geq 2$). This is the rotation group of a regular prism, or regular bipyramid. Although the same notation is used, the geometric and abstract D_n should be distinguished: there are other symmetry groups of the same abstract group type which are geometrically different, see dihedral symmetry groups in 3D.

- 4×3-fold and 3×2-fold axes: the rotation group T of order 12 of a regular tetrahedron. The group is isomorphic to alternating group A_4.

- 3×4-fold, 4×3-fold, and 6×2-fold axes: the rotation group O of order 24 of a cube and a regular octahedron. The group is isomorphic to symmetric group S_4.

- 6×5-fold, 10×3-fold, and 15×2-fold axes: the rotation group I of order 60 of a dodecahedron and an icosahedron. The group is isomorphic to alternating group A_5. The group contains 10 versions of D_3 and 6 versions of D_5 (rotational symmetries like prisms and antiprisms).

In the case of the Platonic solids, the 2-fold axes are through the midpoints of opposite edges, the number of them is half the number of edges. The other axes are through opposite vertices and through centers of opposite faces, except in the case of the tetrahedron, where the 3-fold axes are each through one vertex and the center of one face.

6.1.4   Rotational symmetry with respect to any angle

Rotational symmetry with respect to any angle is, in two dimensions, circular symmetry. The fundamental domain is a half-line.

In three dimensions we can distinguish **cylindrical symmetry** and **spherical symmetry** (no change when rotating about one axis, or for any rotation). That is, no dependence on the angle using cylindrical coordinates and no dependence on either angle using spherical coordinates. The fundamental domain is a half-plane through the axis, and a radial half-line, respectively. **Axisymmetric** or **axisymmetrical** are adjectives which refer to an object having cylindrical symmetry, or **axisymmetry**. An example of approximate spherical symmetry is the Earth (with respect to density and other physical and chemical properties).

In 4D, continuous or discrete rotational symmetry about a plane corresponds to corresponding 2D rotational symmetry in every perpendicular plane, about the point of intersection. An object can also have rotational symmetry about two perpendicular planes, e.g. if it is the Cartesian product of two rotationally symmetry 2D figures, as in the case of e.g. the duocylinder and various regular duoprisms.

6.1.5   Rotational symmetry with translational symmetry

2-fold rotational symmetry together with single translational symmetry is one of the Frieze groups. There are two roto-centers per primitive cell.

Together with double translational symmetry the rotation groups are the following wallpaper groups, with axes per primitive cell:

- p2 (2222): 4×2-fold; rotation group of a parallelogrammic, rectangular, and rhombic lattice.

- p3 (333): 3×3-fold; *not* the rotation group of any lattice (every lattice is upside-down the same, but that does not apply for this symmetry); it is e.g. the rotation group of the regular triangular tiling with the equilateral triangles alternatingly colored.

- p4 (442): 2×4-fold, 2×2-fold; rotation group of a square lattice.

- p6 (632): 1×6-fold, 2×3-fold, 3×2-fold; rotation group of a hexagonal lattice.

- 2-fold rotocenters (including possible 4-fold and 6-fold), if present at all, form the translate of a lattice equal to the translational lattice, scaled by a factor 1/2. In the case translational symmetry in one dimension, a similar property applies, though the term "lattice" does not apply.

- 3-fold rotocenters (including possible 6-fold), if present at all, form a regular hexagonal lattice equal to the translational lattice, rotated by $30°$ (or equivalently $90°$), and scaled by a factor $\frac{1}{3}\sqrt{3}$

- 4-fold rotocenters, if present at all, form a regular square lattice equal to the translational lattice, rotated by $45°$, and scaled by a factor $\frac{1}{2}\sqrt{2}$

- 6-fold rotocenters, if present at all, form a regular hexagonal lattice which is the translate of the translational lattice.

Scaling of a lattice divides the number of points per unit area by the square of the scale factor. Therefore the number of 2-, 3-, 4-, and 6-fold rotocenters per primitive cell is 4, 3, 2, and 1, respectively, again including 4-fold as a special case of 2-fold, etc.

3-fold rotational symmetry at one point and 2-fold at another one (or ditto in 3D with respect to parallel axes) implies rotation group p6, i.e. double translational symmetry and 6-fold rotational symmetry at some point (or, in 3D, parallel axis). The translation distance for the symmetry generated by one such pair of rotocenters is 2√3 times their distance.

6.2 See also

- Ambigram

- Axial symmetry

- Crystallographic restriction theorem

- Lorentz symmetry

- Point groups in three dimensions

- Recycling symbol

- Screw axis

- Space group

- Three hares

6.3 References

- Weyl, Hermann (1982) [1952]. *Symmetry*. Princeton: Princeton University Press. ISBN 0-691-02374-3.

6.4 External links

- Media related to Rotational symmetry by order at Wikimedia Commons

- Rotational Symmetry Examples from Math Is Fun

Chapter 7

Global symmetry

In physics, a **global symmetry** is a symmetry that holds at all points in the spacetime under consideration, as opposed to a local symmetry which varies from point to point.

Global symmetries require conservation laws, but not forces, in physics.

An example of a global symmetry is the action of the $U(1) = e^{iq\theta}$ (for θ a constant - making it a global transformation) group on the Dirac Lagrangian:

$$\mathcal{L}_D = \bar{\psi}\left(i\gamma^\mu \partial_\mu - m\right)\psi$$

Under this transformation the wavefunction changes as $\psi \to e^{iq\theta}\psi$ and $\bar{\psi} \to e^{-iq\theta}\bar{\psi}$ and so:

$$\mathcal{L} \to \bar{\mathcal{L}} = e^{-iq\theta}\bar{\psi}\left(i\gamma^\mu\partial_\mu - m\right)e^{iq\theta}\psi = e^{-iq\theta}e^{iq\theta}\bar{\psi}\left(i\gamma^\mu\partial_\mu - m\right)\psi = \mathcal{L}$$

7.1 See also

- Field (physics)
- Global spacetime structure
- Local spacetime structure

Chapter 8

Local symmetry

In physics, a **local symmetry** is symmetry of some physical quantity, which smoothly depends on the point of the base manifold. Such quantities can be for example an observable, a tensor or the Lagrangian of a theory. If a symmetry is local in this sense, then one can apply a local transformation (resp. local gauge transformation), which means that the representation of the symmetry group is a function of the manifold and can thus be taken to act differently on different points of spacetime.

The diffeomorphism group is a local symmetry and thus every geometrical or generally covariant theory (i.e. a theory whose equations are tensor equations, for example general relativity) has local symmetries.

Often the term local symmetry is specifically associated with local gauge symmetries in Yang–Mills theory (see also standard model) where the Lagrangian is locally symmetric under some compact Lie group. Local gauge symmetries always come together with some bosonic gauge fields, like the photon or gluon field, which induce a force in addition to requiring conservation laws.[1]

8.1 Examples

- General relativity has a local symmetry (general covariance, diffeomorphisms) which can be seen as generating the gravitational force.[2] Special relativity only has a global symmetry (Lorentz symmetry or more generally Poincaré symmetry)

- There are many global symmetries (such as SU(2) of isospin symmetry) and local symmetries (like SU(2) of weak interactions) in particle physics. The standard model of particle physics consists of Yang-Mills Theories

- The symmetry group of Supergravity is a local symmetry, whereas supersymmetry is a global symmetry.

8.2 See also

- Field (physics)

- Global spacetime structure

- Local spacetime structure

- Gauge theory

- Gravitation (book)

8.3   References

[1] Kaku, Michio (1993). *Quantum Field Theory: A Modern Introduction*. New York: Oxford University Press. ISBN 0-19-507652-4.

[2] Misner, Charles W.; Thorne, Kip S.; Wheeler, John Archibald (1973-09-15). "Gravitation". San Francisco: W. H. Freeman. ISBN 978-0-7167-0344-0.

Chapter 9

Gauge theory

For a more accessible and less technical introduction to this topic, see Introduction to gauge theory.

In physics, a **gauge theory** is a type of field theory in which the Lagrangian is invariant under a continuous group of local transformations.

The term *gauge* refers to redundant degrees of freedom in the Lagrangian. The transformations between possible gauges, called *gauge transformations*, form a Lie group—referred to as the *symmetry group* or the *gauge group* of the theory. Associated with any Lie group is the Lie algebra of group generators. For each group generator there necessarily arises a corresponding vector field called the *gauge field*. Gauge fields are included in the Lagrangian to ensure its invariance under the local group transformations (called *gauge invariance*). When such a theory is quantized, the quanta of the gauge fields are called *gauge bosons*. If the symmetry group is non-commutative, the gauge theory is referred to as *non-abelian*, the usual example being the Yang–Mills theory.

Many powerful theories in physics are described by Lagrangians that are invariant under some symmetry transformation groups. When they are invariant under a transformation identically performed at *every* point in the space in which the physical processes occur, they are said to have a global symmetry. The requirement of local symmetry, the cornerstone of gauge theories, is a stricter constraint. In fact, a global symmetry is just a local symmetry whose group's parameters are fixed in space-time.

Gauge theories are important as the successful field theories explaining the dynamics of elementary particles. Quantum electrodynamics is an abelian gauge theory with the symmetry group U(1) and has one gauge field, the electromagnetic four-potential, with the photon being the gauge boson. The Standard Model is a non-abelian gauge theory with the symmetry group U(1)×SU(2)×SU(3) and has a total of twelve gauge bosons: the photon, three weak bosons and eight gluons.

Gauge theories are also important in explaining gravitation in the theory of general relativity. Its case is somewhat unique in that the gauge field is a tensor, the Lanczos tensor. Theories of quantum gravity, beginning with gauge gravitation theory, also postulate the existence of a gauge boson known as the graviton. Gauge symmetries can be viewed as analogues of the principle of general covariance of general relativity in which the coordinate system can be chosen freely under arbitrary diffeomorphisms of spacetime. Both gauge invariance and diffeomorphism invariance reflect a redundancy in the description of the system. An alternative theory of gravitation, gauge theory gravity, replaces the principle of general covariance with a true gauge principle with new gauge fields.

Historically, these ideas were first stated in the context of classical electromagnetism and later in general relativity. However, the modern importance of gauge symmetries appeared first in the relativistic quantum mechanics of electrons – quantum electrodynamics, elaborated on below. Today, gauge theories are useful in condensed matter, nuclear and high energy physics among other subfields.

9.1   History and importance

The earliest field theory having a gauge symmetry was Maxwell's formulation of electrodynamics in 1864. The importance of this symmetry remained unnoticed in the earliest formulations. Similarly unnoticed, Hilbert had derived the Einstein field equations by postulating the invariance of the action under a general coordinate transformation. Later Hermann Weyl, in an attempt to unify general relativity and electromagnetism, conjectured that *Eichinvarianz* or invariance under the change of scale (or "gauge") might also be a local symmetry of general relativity. After the development of quantum mechanics, Weyl, Vladimir Fock and Fritz London modified gauge by replacing the scale factor with a complex quantity and turned the scale transformation into a change of phase, which is a U(1) gauge symmetry. This explained the electromagnetic field effect on the wave function of a charged quantum mechanical particle. This was the first widely recognised gauge theory, popularised by Pauli in the 1940s.[1]

In 1954, attempting to resolve some of the great confusion in elementary particle physics, Chen Ning Yang and Robert Mills introduced **non-abelian gauge theories** as models to understand the strong interaction holding together nucleons in atomic nuclei. (Ronald Shaw, working under Abdus Salam, independently introduced the same notion in his doctoral thesis.) Generalizing the gauge invariance of electromagnetism, they attempted to construct a theory based on the action of the (non-abelian) SU(2) symmetry group on the isospin doublet of protons and neutrons. This is similar to the action of the U(1) group on the spinor fields of quantum electrodynamics. In particle physics the emphasis was on using **quantized gauge theories**.

This idea later found application in the quantum field theory of the weak force, and its unification with electromagnetism in the electroweak theory. Gauge theories became even more attractive when it was realized that non-abelian gauge theories reproduced a feature called asymptotic freedom. Asymptotic freedom was believed to be an important characteristic of strong interactions. This motivated searching for a strong force gauge theory. This theory, now known as quantum chromodynamics, is a gauge theory with the action of the SU(3) group on the color triplet of quarks. The Standard Model unifies the description of electromagnetism, weak interactions and strong interactions in the language of gauge theory.

In the 1970s, Sir Michael Atiyah began studying the mathematics of solutions to the classical Yang–Mills equations. In 1983, Atiyah's student Simon Donaldson built on this work to show that the differentiable classification of smooth 4-manifolds is very different from their classification up to homeomorphism. Michael Freedman used Donaldson's work to exhibit exotic $\mathbf{R}^4$s, that is, exotic differentiable structures on Euclidean 4-dimensional space. This led to an increasing interest in gauge theory for its own sake, independent of its successes in fundamental physics. In 1994, Edward Witten and Nathan Seiberg invented gauge-theoretic techniques based on supersymmetry that enabled the calculation of certain topological invariants (the Seiberg–Witten invariants). These contributions to mathematics from gauge theory have led to a renewed interest in this area.

The importance of gauge theories in physics is exemplified in the tremendous success of the mathematical formalism in providing a unified framework to describe the quantum field theories of electromagnetism, the weak force and the strong force. This theory, known as the Standard Model, accurately describes experimental predictions regarding three of the four fundamental forces of nature, and is a gauge theory with the gauge group SU(3) × SU(2) × U(1). Modern theories like string theory, as well as general relativity, are, in one way or another, gauge theories.

See Pickering[2] for more about the history of gauge and quantum field theories.

9.2   Description

9.2.1   Global and local symmetries

In physics, the mathematical description of any physical situation usually contains excess degrees of freedom; the same physical situation is equally well described by many equivalent mathematical configurations. For instance, in Newtonian dynamics, if two configurations are related by a Galilean transformation (an inertial change of reference frame) they represent the same physical situation. These transformations form a group of "symmetries" of the theory, and a physical situation corresponds not to an individual mathematical configuration but to a class of configurations related to one another by this symmetry group.

This idea can be generalized to include local as well as global symmetries, analogous to much more abstract "changes of coordinates" in a situation where there is no preferred "inertial" coordinate system that covers the entire physical system. A gauge theory is a mathematical model that has symmetries of this kind, together with a set of techniques for making physical predictions consistent with the symmetries of the model.

9.2.2   Example of global symmetry

When a quantity occurring in the mathematical configuration is not just a number but has some geometrical significance, such as a velocity or an axis of rotation, its representation as numbers arranged in a vector or matrix is also changed by a coordinate transformation. For instance, if one description of a pattern of fluid flow states that the fluid velocity in the neighborhood of $(x=1, y=0)$ is 1 m/s in the positive x direction, then a description of the same situation in which the coordinate system has been rotated clockwise by 90 degrees states that the fluid velocity in the neighborhood of $(x=0, y=1)$ is 1 m/s in the positive y direction. The coordinate transformation has affected both the coordinate system used to identify the *location* of the measurement and the basis in which its *value* is expressed. As long as this transformation is performed globally (affecting the coordinate basis in the same way at every point), the effect on values that represent the *rate of change* of some quantity along some path in space and time as it passes through point P is the same as the effect on values that are truly local to P.

9.2.3   Use of fiber bundles to describe local symmetries

In order to adequately describe physical situations in more complex theories, it is often necessary to introduce a "coordinate basis" for some of the objects of the theory that do not have this simple relationship to the coordinates used to label points in space and time. (In mathematical terms, the theory involves a fiber bundle in which the fiber at each point of the base space consists of possible coordinate bases for use when describing the values of objects at that point.) In order to spell out a mathematical configuration, one must choose a particular coordinate basis at each point (a *local section* of the fiber bundle) and express the values of the objects of the theory (usually "fields" in the physicist's sense) using this basis. Two such mathematical configurations are equivalent (describe the same physical situation) if they are related by a transformation of this abstract coordinate basis (a change of local section, or *gauge transformation*).

In most gauge theories, the set of possible transformations of the abstract gauge basis at an individual point in space and time is a finite-dimensional Lie group. The simplest such group is U(1), which appears in the modern formulation of quantum electrodynamics (QED) via its use of complex numbers. QED is generally regarded as the first, and simplest, physical gauge theory. The set of possible gauge transformations of the entire configuration of a given gauge theory also forms a group, the *gauge group* of the theory. An element of the gauge group can be parameterized by a smoothly varying function from the points of spacetime to the (finite-dimensional) Lie group, such that the value of the function and its derivatives at each point represents the action of the gauge transformation on the fiber over that point.

A gauge transformation with constant parameter at every point in space and time is analogous to a rigid rotation of the geometric coordinate system; it represents a global symmetry of the gauge representation. As in the case of a rigid rotation, this gauge transformation affects expressions that represent the rate of change along a path of some gauge-dependent quantity in the same way as those that represent a truly local quantity. A gauge transformation whose parameter is *not* a constant function is referred to as a local symmetry; its effect on expressions that involve a derivative is qualitatively different from that on expressions that don't. (This is analogous to a non-inertial change of reference frame, which can produce a Coriolis effect.)

9.2.4   Gauge fields

The "gauge covariant" version of a gauge theory accounts for this effect by introducing a gauge field (in mathematical language, an Ehresmann connection) and formulating all rates of change in terms of the covariant derivative with respect to this connection. The gauge field becomes an essential part of the description of a mathematical configuration. A configuration in which the gauge field can be eliminated by a gauge transformation has the property that its field strength (in mathematical language, its curvature) is zero everywhere; a gauge theory is *not* limited to these configurations. In other words, the distinguishing characteristic of a gauge theory is that the gauge field does not merely compensate for a poor choice of coordinate system; there is generally no gauge transformation that makes the gauge field vanish.

When analyzing the dynamics of a gauge theory, the gauge field must be treated as a dynamical variable, similarly to other objects in the description of a physical situation. In addition to its interaction with other objects via the covariant derivative, the gauge field typically contributes energy in the form of a "self-energy" term. One can obtain the equations for the gauge theory by:

- starting from a naïve ansatz without the gauge field (in which the derivatives appear in a "bare" form);

- listing those global symmetries of the theory that can be characterized by a continuous parameter (generally an abstract equivalent of a rotation angle);

- computing the correction terms that result from allowing the symmetry parameter to vary from place to place; and

- reinterpreting these correction terms as couplings to one or more gauge fields, and giving these fields appropriate self-energy terms and dynamical behavior.

This is the sense in which a gauge theory "extends" a global symmetry to a local symmetry, and closely resembles the historical development of the gauge theory of gravity known as general relativity.

9.2.5   Physical experiments

Gauge theories are used to model the results of physical experiments, essentially by:

- limiting the universe of possible configurations to those consistent with the information used to set up the experiment, and then

- computing the probability distribution of the possible outcomes that the experiment is designed to measure.

The mathematical descriptions of the "setup information" and the "possible measurement outcomes" (loosely speaking, the "boundary conditions" of the experiment) are generally not expressible without reference to a particular coordinate system, including a choice of gauge. (If nothing else, one assumes that the experiment has been adequately isolated from "external" influence, which is itself a gauge-dependent statement.) Mishandling gauge dependence in boundary conditions is a frequent source of anomalies in gauge theory calculations, and gauge theories can be broadly classified by their approaches to anomaly avoidance.

9.2.6   Continuum theories

The two gauge theories mentioned above (continuum electrodynamics and general relativity) are examples of continuum field theories. The techniques of calculation in a continuum theory implicitly assume that:

- given a completely fixed choice of gauge, the boundary conditions of an individual configuration can in principle be completely described;

- given a completely fixed gauge and a complete set of boundary conditions, the principle of least action determines a unique mathematical configuration (and therefore a unique physical situation) consistent with these bounds;

- the likelihood of possible measurement outcomes can be determined by:

  - establishing a probability distribution over all physical situations determined by boundary conditions that are consistent with the setup information,

  - establishing a probability distribution of measurement outcomes for each possible physical situation, and

- convolving these two probability distributions to get a distribution of possible measurement outcomes consistent with the setup information; and

- fixing the gauge introduces no anomalies in the calculation, due either to gauge dependence in describing partial information about boundary conditions or to incompleteness of the theory.

These assumptions are close enough to be valid across a wide range of energy scales and experimental conditions, to allow these theories to make accurate predictions about almost all of the phenomena encountered in daily life, from light, heat, and electricity to eclipses and spaceflight. They fail only at the smallest and largest scales (due to omissions in the theories themselves) and when the mathematical techniques themselves break down (most notably in the case of turbulence and other chaotic phenomena).

9.2.7 Quantum field theories

Other than these classical continuum field theories, the most widely known gauge theories are quantum field theories, including quantum electrodynamics and the Standard Model of elementary particle physics. The starting point of a quantum field theory is much like that of its continuum analog: a gauge-covariant action integral that characterizes "allowable" physical situations according to the principle of least action. However, continuum and quantum theories differ significantly in how they handle the excess degrees of freedom represented by gauge transformations. Continuum theories, and most pedagogical treatments of the simplest quantum field theories, use a gauge fixing prescription to reduce the orbit of mathematical configurations that represent a given physical situation to a smaller orbit related by a smaller gauge group (the global symmetry group, or perhaps even the trivial group).

More sophisticated quantum field theories, in particular those that involve a non-abelian gauge group, break the gauge symmetry within the techniques of perturbation theory by introducing additional fields (the Faddeev–Popov ghosts) and counterterms motivated by anomaly cancellation, in an approach known as BRST quantization. While these concerns are in one sense highly technical, they are also closely related to the nature of measurement, the limits on knowledge of a physical situation, and the interactions between incompletely specified experimental conditions and incompletely understood physical theory. The mathematical techniques that have been developed in order to make gauge theories tractable have found many other applications, from solid-state physics and crystallography to low-dimensional topology.

9.3 Classical gauge theory

9.3.1 Classical electromagnetism

Historically, the first example of gauge symmetry discovered was classical electromagnetism. In electrostatics, one can either discuss the electric field, $\mathbf{E}$, or its corresponding electric potential, V. Knowledge of one makes it possible to find the other, except that potentials differing by a constant, $V \to V + C$, correspond to the same electric field. This is because the electric field relates to *changes* in the potential from one point in space to another, and the constant C would cancel out when subtracting to find the change in potential. In terms of vector calculus, the electric field is the gradient of the potential, $\mathbf{E} = -\nabla V$. Generalizing from static electricity to electromagnetism, we have a second potential, the vector potential $\mathbf{A}$, with

$$\mathbf{E} = -\nabla V - \frac{\partial \mathbf{A}}{\partial t}$$
$$\mathbf{B} = \nabla \times \mathbf{A}$$

The general gauge transformations now become not just $V \to V + C$ but

$$\mathbf{A} \to \mathbf{A} + \nabla f$$
$$V \to V - \frac{\partial f}{\partial t}$$

where f is any function that depends on position and time. The fields remain the same under the gauge transformation, and therefore Maxwell's equations are still satisfied. That is, Maxwell's equations have a gauge symmetry.

9.3.2   An example: Scalar O(n) gauge theory

The remainder of this section requires some familiarity with classical or quantum field theory, and the use of Lagrangians.

Definitions in this section: gauge group, gauge field, interaction Lagrangian, gauge boson.

The following illustrates how local gauge invariance can be "motivated" heuristically starting from global symmetry properties, and how it leads to an interaction between originally non-interacting fields.

Consider a set of n non-interacting real scalar fields, with equal masses m. This system is described by an action that is the sum of the (usual) action for each scalar field φ_i

$$S = \int d^4x \sum_{i=1}^{n} \left[\frac{1}{2} \partial_\mu \varphi_i \partial^\mu \varphi_i - \frac{1}{2} m^2 \varphi_i^2 \right]$$

The Lagrangian (density) can be compactly written as

$$\mathcal{L} = \frac{1}{2} (\partial_\mu \Phi)^T \partial^\mu \Phi - \frac{1}{2} m^2 \Phi^T \Phi$$

by introducing a vector of fields

$$\Phi = (\varphi_1, \varphi_2, \ldots, \varphi_n)^T$$

The term ∂_μ is Einstein notation for the partial derivative of Φ in each of the four dimensions.

It is now transparent that the Lagrangian is invariant under the transformation

$$\Phi \mapsto \Phi' = G\Phi$$

whenever G is a *constant* matrix belonging to the n-by-n orthogonal group O(n). This is seen to preserve the Lagrangian, since the derivative of Φ transforms identically to Φ and both quantities appear inside dot products in the Lagrangian (orthogonal transformations preserve the dot product).

$$(\partial_\mu \Phi) \mapsto (\partial_\mu \Phi)' = G\partial_\mu \Phi$$

This characterizes the *global* symmetry of this particular Lagrangian, and the symmetry group is often called the **gauge group**; the mathematical term is **structure group**, especially in the theory of G-structures. Incidentally, Noether's theorem implies that invariance under this group of transformations leads to the conservation of the *currents*

$$J_\mu^a = i\partial_\mu \Phi^T T^a \Phi$$

where the T^a matrices are generators of the SO(n) group. There is one conserved current for every generator.

Now, demanding that this Lagrangian should have *local* O(n)-invariance requires that the G matrices (which were earlier constant) should be allowed to become functions of the space-time coordinates x.

In this case, the G matrices do not "pass through" the derivatives, when $G = G(x)$.

$$\partial_\mu (G\Phi) \neq G(\partial_\mu \Phi)$$

The failure of the derivative to commute with "G" introduces an additional term (in keeping with the product rule), which spoils the invariance of the Lagrangian. In order to rectify this we define a new derivative operator such that the derivative of Φ again transforms identically with Φ

$$(D_\mu \Phi)' = G D_\mu \Phi$$

This new "derivative" is called a (gauge) covariant derivative and takes the form

$$D_\mu = \partial_\mu + ig A_\mu$$

Where g is called the coupling constant; a quantity defining the strength of an interaction. After a simple calculation we can see that the **gauge field** $A(x)$ must transform as follows

$$A'_\mu = G A_\mu G^{-1} + \frac{i}{g}(\partial_\mu G)G^{-1}$$

The gauge field is an element of the Lie algebra, and can therefore be expanded as

$$A_\mu = \sum_a A^a_\mu T^a$$

There are therefore as many gauge fields as there are generators of the Lie algebra.

Finally, we now have a *locally gauge invariant* Lagrangian

$$\mathcal{L}_{\text{loc}} = \frac{1}{2}(D_\mu \Phi)^T D^\mu \Phi - \frac{1}{2}m^2 \Phi^T \Phi$$

Pauli uses the term *gauge transformation of the first type* to mean the transformation of Φ, while the compensating transformation in A is called a *gauge transformation of the second type*.

The difference between this Lagrangian and the original *globally gauge-invariant* Lagrangian is seen to be the **interaction Lagrangian**

$$\mathcal{L}_{\text{int}} = i\frac{g}{2}\Phi^T A^T_\mu \partial^\mu \Phi + i\frac{g}{2}(\partial_\mu \Phi)^T A^\mu \Phi - \frac{g^2}{2}(A_\mu \Phi)^T A^\mu \Phi$$

This term introduces interactions between the n scalar fields just as a consequence of the demand for local gauge invariance. However, to make this interaction physical and not completely arbitrary, the mediator $A(x)$ needs to propagate in space. That is dealt with in the next section by adding yet another term, $\mathcal{L}_{\text{gf}}$, to the Lagrangian. In the quantized version of the obtained classical field theory, the quanta of the gauge field $A(x)$ are called gauge bosons. The interpretation of the interaction Lagrangian in quantum field theory is of scalar bosons interacting by the exchange of these gauge bosons.

9.3.3 The Yang–Mills Lagrangian for the gauge field

Main article: Yang–Mills theory

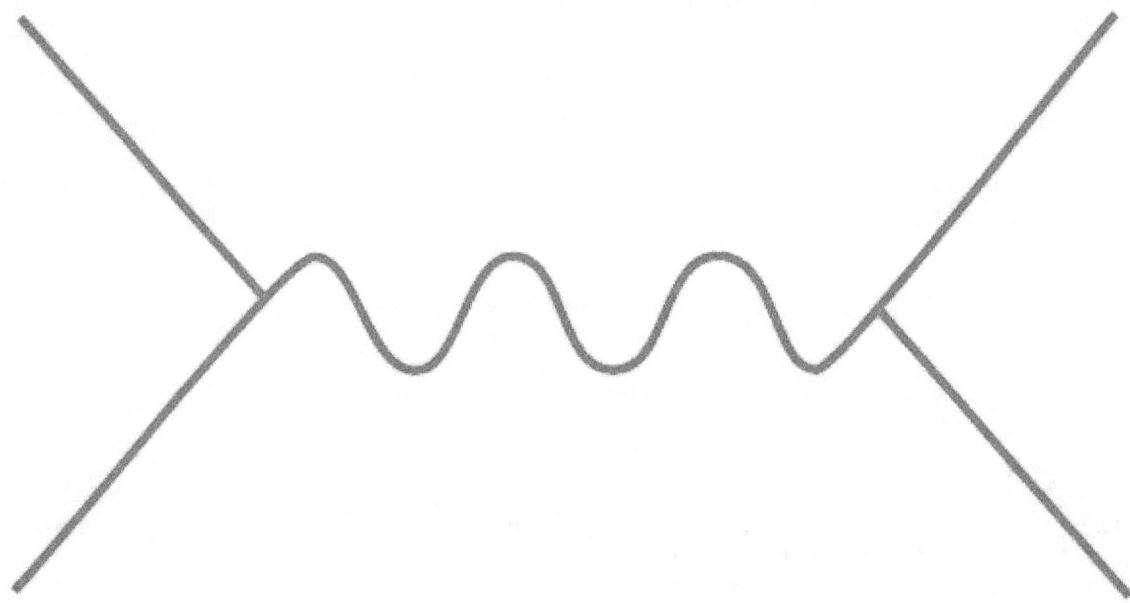

Feynman diagram of scalar bosons interacting via a gauge boson

The picture of a classical gauge theory developed in the previous section is almost complete, except for the fact that to define the covariant derivatives D, one needs to know the value of the gauge field $A(x)$ at all space-time points. Instead of manually specifying the values of this field, it can be given as the solution to a field equation. Further requiring that the Lagrangian that generates this field equation is locally gauge invariant as well, one possible form for the gauge field Lagrangian is (conventionally) written as

$$\mathcal{L}_{\text{gf}} = -\frac{1}{2} \operatorname{Tr}(F^{\mu\nu} F_{\mu\nu})$$

with

$$F_{\mu\nu} = \frac{1}{ig}[D_\mu, D_\nu]$$

and the trace being taken over the vector space of the fields. This is called the **Yang–Mills action**. Other gauge invariant actions also exist (e.g., nonlinear electrodynamics, Born–Infeld action, Chern–Simons model, theta term, etc.).

Note that in this Lagrangian term there is no field whose transformation counterweighs the one of A . Invariance of this term under gauge transformations is a particular case of *a priori* classical (geometrical) symmetry. This symmetry must be restricted in order to perform quantization, the procedure being denominated gauge fixing, but even after restriction, gauge transformations may be possible.[3]

The complete Lagrangian for the gauge theory is now

$$\mathcal{L} = \mathcal{L}_{\text{loc}} + \mathcal{L}_{\text{gf}} = \mathcal{L}_{\text{global}} + \mathcal{L}_{\text{int}} + \mathcal{L}_{\text{gf}}$$

9.3.4   An example: Electrodynamics

As a simple application of the formalism developed in the previous sections, consider the case of electrodynamics, with only the electron field. The bare-bones action that generates the electron field's Dirac equation is

$$S = \int \bar{\psi}(i\hbar c\,\gamma^{\mu}\partial_{\mu} - mc^2)\psi\,\mathrm{d}^4x$$

The global symmetry for this system is

$$\psi \mapsto e^{i\theta}\psi$$

The gauge group here is U(1), just rotations of the phase angle of the field, with the particular rotation determined by the constant θ.

"Localising" this symmetry implies the replacement of θ by $\theta(x)$. An appropriate covariant derivative is then

$$D_{\mu} = \partial_{\mu} - i\frac{e}{\hbar}A_{\mu}$$

Identifying the "charge" e (not to be confused with the mathematical constant e in the symmetry description) with the usual electric charge (this is the origin of the usage of the term in gauge theories), and the gauge field $A(x)$ with the four-vector potential of electromagnetic field results in an interaction Lagrangian

$$\mathcal{L}_{\mathrm{int}} = \frac{e}{\hbar}\bar{\psi}(x)\gamma^{\mu}\psi(x)A_{\mu}(x) = J^{\mu}(x)A_{\mu}(x)$$

where $J^{\mu}(x)$ is the usual four vector electric current density. The gauge principle is therefore seen to naturally introduce the so-called minimal coupling of the electromagnetic field to the electron field.

Adding a Lagrangian for the gauge field $A_{\mu}(x)$ in terms of the field strength tensor exactly as in electrodynamics, one obtains the Lagrangian used as the starting point in quantum electrodynamics.

$$\mathcal{L}_{\mathrm{QED}} = \bar{\psi}(i\hbar c\,\gamma^{\mu}D_{\mu} - mc^2)\psi - \frac{1}{4\mu_0}F_{\mu\nu}F^{\mu\nu}$$

See also: Dirac equation, Maxwell's equations, Quantum electrodynamics

9.4 Mathematical formalism

Gauge theories are usually discussed in the language of differential geometry. Mathematically, a *gauge* is just a choice of a (local) section of some principal bundle. A **gauge transformation** is just a transformation between two such sections.

Although gauge theory is dominated by the study of connections (primarily because it's mainly studied by high-energy physicists), the idea of a connection is not central to gauge theory in general. In fact, a result in general gauge theory shows that affine representations (i.e., affine modules) of the gauge transformations can be classified as sections of a jet bundle satisfying certain properties. There are representations that transform covariantly pointwise (called by physicists gauge transformations of the first kind), representations that transform as a connection form (called by physicists gauge transformations of the second kind, an affine representation)—and other more general representations, such as the B field in BF theory. There are more general nonlinear representations (realizations), but these are extremely complicated. Still, nonlinear sigma models transform nonlinearly, so there are applications.

If there is a principal bundle P whose base space is space or spacetime and structure group is a Lie group, then the sections of P form a principal homogeneous space of the group of gauge transformations.

Connections (gauge connection) define this principal bundle, yielding a covariant derivative ∇ in each associated vector bundle. If a local frame is chosen (a local basis of sections), then this covariant derivative is represented by the connection form A, a Lie algebra-valued 1-form, which is called the **gauge potential** in physics. This is evidently not an intrinsic but a frame-dependent quantity. The curvature form F, a Lie algebra-valued 2-form that is an intrinsic quantity, is constructed from a connection form by

$$\mathbf{F} = d\mathbf{A} + \mathbf{A} \wedge \mathbf{A}$$

where d stands for the exterior derivative and $\wedge$ stands for the wedge product. ($\mathbf{A}$ is an element of the vector space spanned by the generators T^a, and so the components of $\mathbf{A}$ do not commute with one another. Hence the wedge product $\mathbf{A} \wedge \mathbf{A}$ does not vanish.)

Infinitesimal gauge transformations form a Lie algebra, which is characterized by a smooth Lie-algebra-valued scalar, ε. Under such an infinitesimal gauge transformation,

$$\delta_\varepsilon \mathbf{A} = [\varepsilon, \mathbf{A}] - d\varepsilon$$

where $[\cdot, \cdot]$ is the Lie bracket.

One nice thing is that if $\delta_\varepsilon X = \varepsilon X$, then $\delta_\varepsilon DX = \varepsilon DX$ where D is the covariant derivative

$$DX \overset{\text{def}}{=} dX + \mathbf{A}X$$

Also, $\delta_\varepsilon \mathbf{F} = \varepsilon \mathbf{F}$, which means $\mathbf{F}$ transforms covariantly.

Not all gauge transformations can be generated by infinitesimal gauge transformations in general. An example is when the base manifold is a compact manifold without boundary such that the homotopy class of mappings from that manifold to the Lie group is nontrivial. See instanton for an example.

The *Yang–Mills action* is now given by

$$\frac{1}{4g^2} \int \text{Tr}[*F \wedge F]$$

where * stands for the Hodge dual and the integral is defined as in differential geometry.

A quantity which is **gauge-invariant** (i.e., invariant under gauge transformations) is the Wilson loop, which is defined over any closed path, γ, as follows:

$$\chi^{(\rho)}\left(\mathcal{P}\left\{e^{\int_\gamma A}\right\}\right)$$

where χ is the character of a complex representation ρ and $\mathcal{P}$ represents the path-ordered operator.

9.5 Quantization of gauge theories

Main article: Quantum gauge theory

Gauge theories may be quantized by specialization of methods which are applicable to any quantum field theory. However, because of the subtleties imposed by the gauge constraints (see section on Mathematical formalism, above) there are many technical problems to be solved which do not arise in other field theories. At the same time, the richer structure of gauge theories allows simplification of some computations: for example Ward identities connect different renormalization constants.

9.5.1 Methods and aims

The first gauge theory quantized was quantum electrodynamics (QED). The first methods developed for this involved gauge fixing and then applying canonical quantization. The Gupta–Bleuler method was also developed to handle this problem. Non-abelian gauge theories are now handled by a variety of means. Methods for quantization are covered in the article on quantization.

The main point to quantization is to be able to compute quantum amplitudes for various processes allowed by the theory. Technically, they reduce to the computations of certain correlation functions in the vacuum state. This involves a renormalization of the theory.

When the running coupling of the theory is small enough, then all required quantities may be computed in perturbation theory. Quantization schemes intended to simplify such computations (such as canonical quantization) may be called **perturbative quantization schemes**. At present some of these methods lead to the most precise experimental tests of gauge theories.

However, in most gauge theories, there are many interesting questions which are non-perturbative. Quantization schemes suited to these problems (such as lattice gauge theory) may be called **non-perturbative quantization schemes**. Precise computations in such schemes often require supercomputing, and are therefore less well-developed currently than other schemes.

9.5.2 Anomalies

Some of the symmetries of the classical theory are then seen not to hold in the quantum theory; a phenomenon called an **anomaly**. Among the most well known are:

- The scale anomaly, which gives rise to a *running coupling constant*. In QED this gives rise to the phenomenon of the Landau pole. In Quantum Chromodynamics (QCD) this leads to asymptotic freedom.

- The chiral anomaly in either chiral or vector field theories with fermions. This has close connection with topology through the notion of instantons. In QCD this anomaly causes the decay of a pion to two photons.

- The gauge anomaly, which must cancel in any consistent physical theory. In the electroweak theory this cancellation requires an equal number of quarks and leptons.

9.6 Pure gauge

A pure gauge is the set of field configurations obtained by a gauge transformation on the null-field configuration, i.e., a gauge-transform of zero. So it is a particular "gauge orbit" in the field configuration's space.

Thus, in the abelian case, where $A_\mu(x) \to A'_\mu(x) = A_\mu(x) + \partial_\mu f(x)$, the pure gauge is just the set of field configurations $A'_\mu(x) = \partial_\mu f(x)$ for all $f(x)$.

9.7 See also

9.8 References

[1] Wolfgang Pauli (1941) "Relativistic Field Theories of Elementary Particles," *Rev. Mod. Phys.* **13**: 203–32.

[2] Pickering, A. (1984). *Constructing Quarks*. University of Chicago Press. ISBN 0-226-66799-5.

[3] Sakurai, *Advanced Quantum Mechanics*, sect 1–4

9.9   Bibliography

General readers

- Schumm, Bruce (2004) *Deep Down Things*. Johns Hopkins University Press. Esp. chpt. 8. A serious attempt by a physicist to explain gauge theory and the Standard Model with little formal mathematics.

Texts

- Bromley, D.A. (2000). *Gauge Theory of Weak Interactions*. Springer. ISBN 3-540-67672-4.

- Cheng, T.-P.; Li, L.-F. (1983). *Gauge Theory of Elementary Particle Physics*. Oxford University Press. ISBN 0-19-851961-3.

- Frampton, P. (2008). *Gauge Field Theories* (3rd ed.). Wiley-VCH.

- Kane, G.L. (1987). *Modern Elementary Particle Physics*. Perseus Books. ISBN 0-201-11749-5.

Articles

- Becchi,C. (1997). "Introduction to Gauge Theories". p. 5211. arXiv:hep-ph/9705211. Bibcode:1997hep.pB.

- Gross, D. (1992). "Gauge theory – Past, Present and Future" (PDF). Retrieved 2009-04-23.

- Jackson, J.D. (2002). "From Lorenz to Coulomb and other explicit gauge transformations". *Am.J.Phys* **70** (9): 917–928. arXiv:physics/0204034. Bibcode:2002AmJPh..70..917J. doi:10.1119/1.1491265.

- Svetlichny,George(1999). "Preparation for Gauge Theory". p. 2027. arXiv:math-ph/9902027. Bibcode:1999.

9.10   External links

- Hazewinkel, Michiel, ed. (2001), "Gauge transformation", *Encyclopedia of Mathematics*, Springer, ISBN 978-1-55608-010-4

- Yang–Mills equations on DispersiveWiki

- Gauge theories on Scholarpedia

Chapter 10

Continuous symmetry

In mathematics, **continuous symmetry** is an intuitive idea corresponding to the concept of viewing some symmetries as motions, as opposed to e.g. reflection symmetry, which is invariance under a kind of flip from one state to another.

10.1 Formalization

The notion of continuous symmetry has largely and successfully been formalised in the mathematical notions of topological group, Lie group and group action. For most practical purposes continuous symmetry is modelled by a *group action* of a topological group.

10.1.1 One-parameter subgroups

The simplest motions follow a one-parameter subgroup of a Lie group, such as the Euclidean group of three-dimensional space. For example translation parallel to the x-axis by u units, as u varies, is a one-parameter group of motions. Rotation around the z-axis is also a one-parameter group.

10.2 Noether's theorem

Continuous symmetry has a basic role in Noether's theorem in theoretical physics, in the derivation of conservation laws from symmetry principles, specifically for continuous symmetries. The search for continuous symmetries only intensified with the further developments of quantum field theory.

10.3 See also

- Goldstone's theorem

- Infinitesimal transformation

- Noether's theorem

- Sophus Lie

10.4   References

- William H. Barker, Roger Howe (2007), *Continuous Symmetry: from Euclid to Klein*

Chapter 11

Spacetime symmetries

For the notation, see Ricci calculus.

Spacetime symmetries are features of spacetime that can be described as exhibiting some form of symmetry. The role of symmetry in physics is important in simplifying solutions to many problems, spacetime symmetries are used in the study of exact solutions of Einstein's field equations of general relativity.

11.1 Physical motivation

Physical problems are often investigated and solved by noticing features which have some form of symmetry. For example, in the Schwarzschild solution, the role of spherical symmetry is important in deriving the Schwarzschild solution and deducing the physical consequences of this symmetry (such as the non-existence of gravitational radiation in a spherically pulsating star). In cosmological problems, symmetry finds a role to play in the cosmological principle which restricts the type of universes that are consistent with large-scale observations (e.g. the Friedmann-Lemaître-Robertson-Walker (FLRW) metric). Symmetries usually require some form of preserving property, the most important of which in general relativity include the following:

- preserving geodesics of the spacetime

- preserving the metric tensor

- preserving the curvature tensor

These and other symmetries will be discussed in more detail later. This preservation feature can be used to motivate a useful definition of symmetries.

11.2 Mathematical definition

A rigorous definition of symmetries in general relativity has been given by Hall (2004). In this approach, the idea is to use (smooth) vector fields whose local flow diffeomorphisms preserve some property of the spacetime. This preserving property of the diffeomorphisms is made precise as follows. A smooth vector field X on a spacetime M is said to *preserve* a smooth tensor T on M (or T is **invariant** under X) if, for each smooth local flow diffeomorphism ϕt associated with X, the tensors T and $\phi t^*(T)$ are equal on the domain of ϕt. This statement is equivalent to the more usable condition that the Lie derivative of the tensor under the vector field vanishes:

$$\mathcal{L}_X T = 0$$

on M. This has the consequence that, given any two points p and q on M, the coordinates of T in a coordinate system around p are equal to the coordinates of T in a coordinate system around q. A *symmetry on the spacetime* is a smooth vector field whose local flow diffeomorphisms preserve some (usually geometrical) feature of the spacetime. The (geometrical) feature may refer to specific tensors (such as the metric, or the energy-momentum tensor) or to other aspects of the spacetime such as its geodesic structure. The vector fields are sometimes referred to as *collineations, symmetry vector fields* or just *symmetries*. The set of all symmetry vector fields on M forms a Lie algebra under the Lie bracket operation as can be seen from the identity:

$$\mathcal{L}_{[X,Y]}T = \mathcal{L}_X(\mathcal{L}_Y T) - \mathcal{L}_Y(\mathcal{L}_X T)$$

the term on the right usually being written, with an abuse of notation, as $[\mathcal{L}_X, \mathcal{L}_Y]T$.

11.3   Killing symmetry

Main article: Killing vector field

A Killing vector field is one of the most important types of symmetries and is defined to be a smooth vector field that preserves the metric tensor:

$$\mathcal{L}_X g_{ab} = 0$$

This is usually written in the expanded form as:

$$X_{a;b} + X_{b;a} = 0$$

Killing vector fields find extensive applications (including in classical mechanics) and are related to conservation laws.

11.4   Homothetic symmetry

Main article: Homothetic vector field

A homothetic vector field is one which satisfies:

$$\mathcal{L}_X g_{ab} = 2c g_{ab}$$

where c is a real constant. Homothetic vector fields find application in the study of singularities in general relativity.

11.5   Affine symmetry

Main article: Affine vector field

An affine vector field is one that satisfies:

$$(\mathcal{L}_X g_{ab})_{;c} = 0$$

An affine vector field preserves geodesics and preserves the affine parameter.

The above three vector field types are special cases of projective vector fields which preserve geodesics without necessarily preserving the affine parameter.

11.6 Conformal symmetry

Main article: Conformal vector field

A conformal vector field is one which satisfies:

$$\mathcal{L}_X g_{ab} = \phi g_{ab}$$

where ϕ is a smooth real-valued function on M.

11.7 Curvature symmetry

Main article: Curvature collineation

A curvature collineation is a vector field which preserves the Riemann tensor:

$$\mathcal{L}_X R^a{}_{bcd} = 0$$

where $R^a bcd$ are the components of the Riemann tensor. The set of all smooth curvature collineations forms a Lie algebra under the Lie bracket operation (if the smoothness condition is dropped, the set of all curvature collineations need not form a Lie algebra). The Lie algebra is denoted by $CC(M)$ and may be infinite-dimensional. Every affine vector field is a curvature collineation.

11.8 Matter symmetry

Main article: Matter collineation

A less well-known form of symmetry concerns vector fields that preserve the energy-momentum tensor. These are variously referred to as matter collineations or matter symmetries and are defined by:

$$\mathcal{L}_X T_{ab} = 0$$

where Tab are the energy-momentum tensor components. The intimate relation between geometry and physics may be highlighted here, as the vector field X is regarded as preserving certain physical quantities along the flow lines of X, this being true for any two observers. In connection with this, it may be shown that *every Killing vector field is a matter collineation* (by the Einstein field equations, with or without cosmological constant). Thus, given a solution of the EFE, *a vector field that preserves the metric necessarily preserves the corresponding energy-momentum tensor*. When the energy-momentum tensor represents a perfect fluid, every Killing vector field preserves the energy density, pressure and the fluid flow vector field. When the energy-momentum tensor represents an electromagnetic field, a Killing vector field does *not necessarily* preserve the electric and magnetic fields.

11.9   Local and global symmetries

Main articles: Local symmetry and Global symmetry

11.10   Applications

As mentioned at the start of this article, the main application of these symmetries occur in general relativity, where solutions of Einstein's equations may be classified by imposing some certain symmetries on the spacetime.

11.10.1   Spacetime classifications

Classifying solutions of the EFE constitutes a large part of general relativity research. Various approaches to classifying spacetimes, including using the Segre classification of the energy-momentum tensor or the Petrov classification of the Weyl tensor have been studied extensively by many researchers, most notably Stephani et al. (2003). They also classify spacetimes using symmetry vector fields (especially Killing and homothetic symmetries). For example, Killing vector fields may be used to classify spacetimes, as there is a limit to the number of global, smooth Killing vector fields that a spacetime may possess (the maximum being 10 for 4-dimensional spacetimes). Generally speaking, the higher the dimension of the algebra of symmetry vector fields on a spacetime, the more symmetry the spacetime admits. For example, the Schwarzschild solution has a Killing algebra of dimension 4 (3 spatial rotational vector fields and a time translation), whereas the Friedmann-Lemaître-Robertson-Walker (FLRW) metric (excluding the Einstein static subcase) has a Killing algebra of dimension 6 (3 translations and 3 rotations). The Einstein static metric has a Killing algebra of dimension 7 (the previous 6 plus a time translation).

The assumption of a spacetime admitting a certain symmetry vector field can place restrictions on the spacetime.

11.11   See also

- Field (physics)

- Killing tensor

- Lie groups

- Noether's theorem

- Ricci decomposition

- Symmetry in physics

- Symmetry in quantum mechanics

- Derivations of the Lorentz transformations

11.12   References

- Hall, Graham (2004). *Symmetries and Curvature Structure in General Relativity (World Scientific Lecture Notes in Physics)*. Singapore: World Scientific Pub. Co. ISBN 981-02-1051-5. See *Section 10.1* for a definition of symmetries.

- Stephani, Hans; Kramer, Dietrich; MacCallum, Malcolm; Hoenselaers, Cornelius & Herlt, Eduard (2003). *Exact Solutions of Einstein's Field Equations*. Cambridge: Cambridge University Press. ISBN 0-521-46136-7.

- Schutz, Bernard (1980). *Geometrical Methods of Mathematical Physics*. Cambridge: Cambridge University Press. ISBN 0-521-29887-3. See *Chapter 3* for properties of the Lie derivative and *Section 3.10* for a definition of invariance.

Chapter 12

Continuous function

In mathematics, a **continuous function** is, roughly speaking, a function for which small changes in the input result in small changes in the output. Otherwise, a function is said to be a *discontinuous* function. A continuous function with a continuous inverse function is called a homeomorphism.

Continuity of functions is one of the core concepts of topology, which is treated in full generality below. The introductory portion of this article focuses on the special case where the inputs and outputs of functions are real numbers. In addition, this article discusses the definition for the more general case of functions between two metric spaces. In order theory, especially in domain theory, one considers a notion of continuity known as Scott continuity. Other forms of continuity do exist but they are not discussed in this article.

As an example, consider the function $h(t)$, which describes the height of a growing flower at time t. This function is continuous. By contrast, if $M(t)$ denotes the amount of money in a bank account at time t, then the function jumps whenever money is deposited or withdrawn, so the function $M(t)$ is discontinuous.

12.1 History

A form of this epsilon-delta definition of continuity was first given by Bernard Bolzano in 1817. Augustin-Louis Cauchy defined continuity of $y = f(x)$ as follows: an infinitely small increment α of the independent variable x always produces an infinitely small change $f(x + \alpha) - f(x)$ of the dependent variable y (see e.g., *Cours d'Analyse*, p. 34). Cauchy defined infinitely small quantities in terms of variable quantities, and his definition of continuity closely parallels the infinitesimal definition used today (see microcontinuity). The formal definition and the distinction between pointwise continuity and uniform continuity were first given by Bolzano in the 1830s but the work wasn't published until the 1930s. Eduard Heine provided the first published definition of uniform continuity in 1872, but based these ideas on lectures given by Peter Gustav Lejeune Dirichlet in 1854.[1]

12.2 Real-valued continuous functions

12.2.1 Definition

A function from the set of real numbers to the real numbers can be represented by a graph in the Cartesian plane; such a function is continuous if, roughly speaking, the graph is a single unbroken curve with no "holes" or "jumps".

A function is *continuous at a point* if it does not have a hole or jump. A "hole" or "jump" in the graph of a function occurs if the value of the function at a point c differs from its limiting value along points that are nearby. Such a point is called a *discontinuity*. A function is then *continuous* if it has no holes or jumps: that is, if it is continuous at every point of its domain. Otherwise, a function is *discontinuous*, at the points where the value of the function differs from its limiting value (if any).

There are several ways to make this definition mathematically rigorous. These definitions are equivalent to one another, so the most convenient definition can be used to determine whether a given function is continuous or not. In the definitions below,

$$f : I \to \mathbf{R}.$$

is a function defined on a subset I of the set $\mathbf{R}$ of real numbers. This subset I is referred to as the domain of f. Some possible choices include $I=\mathbf{R}$, the whole set of real numbers, an open interval

$$I = (a, b) = \{x \in \mathbf{R} \mid a < x < b\}.$$

or a closed interval

$$I = [a, b] = \{x \in \mathbf{R} \mid a \leq x \leq b\}.$$

Here, a and b are real numbers.

Definition in terms of limits of functions

The function f is *continuous at some point* c of its domain if the limit of $f(x)$ as x approaches c through the domain of f exists and is equal to $f(c)$.[2] In mathematical notation, this is written as

$$\lim_{x \to c} f(x) = f(c).$$

In detail this means three conditions: first, f has to be defined at c. Second, the limit on the left hand side of that equation has to exist. Third, the value of this limit must equal $f(c)$.

(We have here assumed that the domain of f does not have any isolated points. For example, an interval or union of intervals has no isolated points.)

Definition in terms of neighborhoods

A neighborhood of a point c is a set that contains all points of the domain within some fixed distance of c. Intuitively, a function is continuous at a point c if the range of the restriction of f to a neighborhood of c shrinks to a single point $f(c)$ as the width of the neighborhood shrinks to zero. More precisely, a function f is continuous at a point c of its domain if, for any neighborhood $N_1(f(c))$ there is a neighborhood $N_2(c)$ such that $f(x) \in N_1(f(c))$ whenever $x \in N_2(c)$.

This definition does not require any assumption on the nature of the domain. For instance, the function f is automatically continuous at every isolated point of its domain. As a specific example, every real valued function on the set of integers is continuous.

Definition in terms of limits of sequences

One can instead require that for any sequence $(x_n)_{n \in \mathbb{N}}$ of points in the domain which converges to c, the corresponding sequence $(f(x_n))_{n \in \mathbb{N}}$ converges to $f(c)$. In mathematical notation, $\forall (x_n)_{n \in \mathbb{N}} \subset I : \lim_{n \to \infty} x_n = c \Rightarrow \lim_{n \to \infty} f(x_n) = f(c)$.

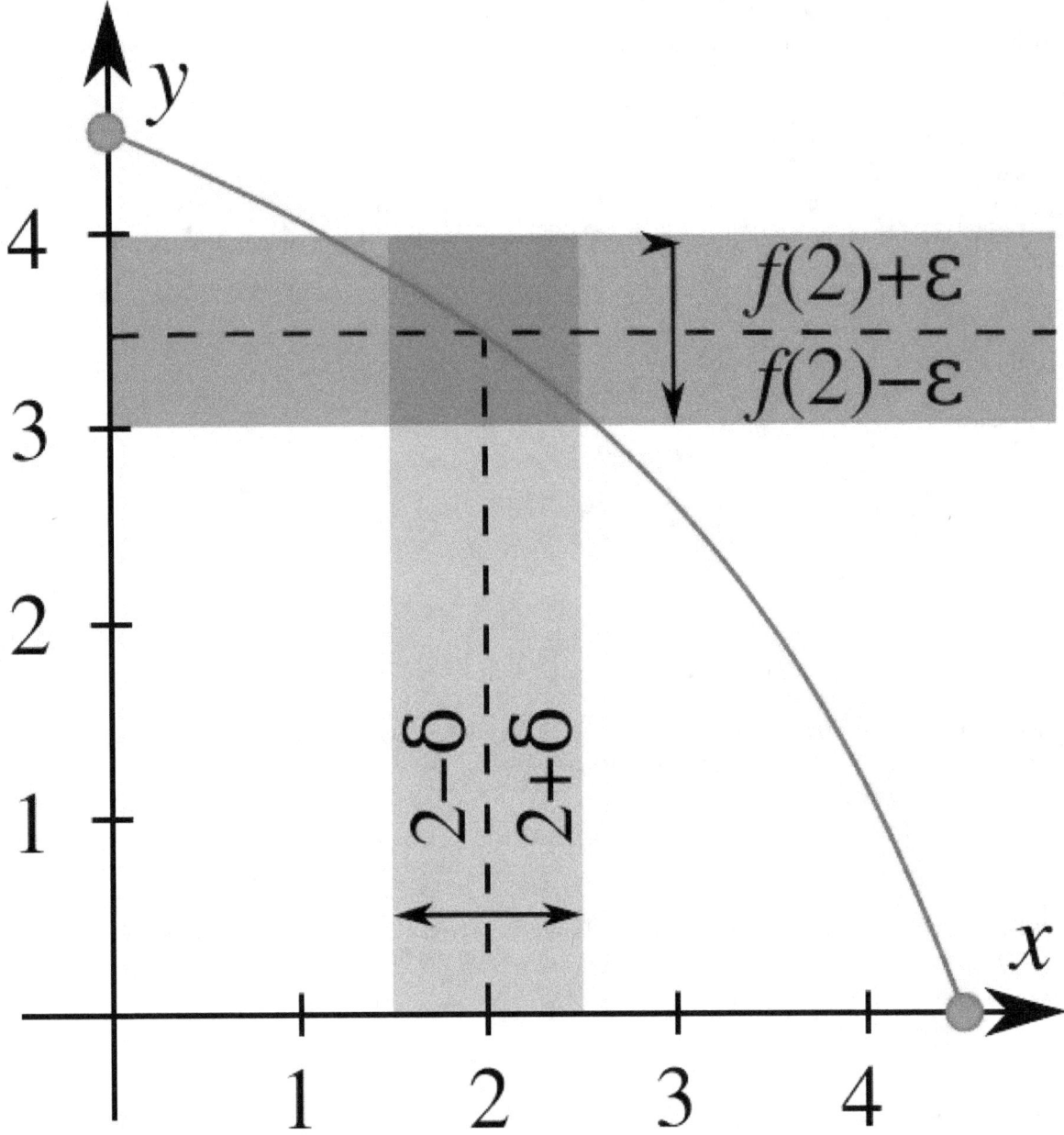

Illustration of the ε-δ-definition: for ε=0.5, c=2, the value δ=0.5 satisfies the condition of the definition.

Weierstrass definition (epsilon–delta) of continuous functions

Explicitly including the definition of the limit of a function, we obtain a self-contained definition: Given a function f as above and an element c of the domain I, f is said to be continuous at the point c if the following holds: For any number ε > 0, however small, there exists some number δ > 0 such that for all x in the domain of f with $c - \delta < x < c + \delta$, the value of $f(x)$ satisfies

$$f(c) - \varepsilon < f(x) < f(c) + \varepsilon.$$

Alternatively written, continuity of $f : I \to R$ at $c \in I$ means that for every $\varepsilon > 0$ there exists a $\delta > 0$ such that for all $x \in I$,:

$|x - c| < \delta \Rightarrow |f(x) - f(c)| < \varepsilon.$

More intuitively, we can say that if we want to get all the $f(x)$ values to stay in some small neighborhood around $f(c)$, we simply need to choose a small enough neighborhood for the x values around c, and we can do that no matter how small the $f(x)$ neighborhood is; f is then continuous at c.

In modern terms, this is generalized by the definition of continuity of a function with respect to a basis for the topology, here the metric topology.

Definition using oscillation

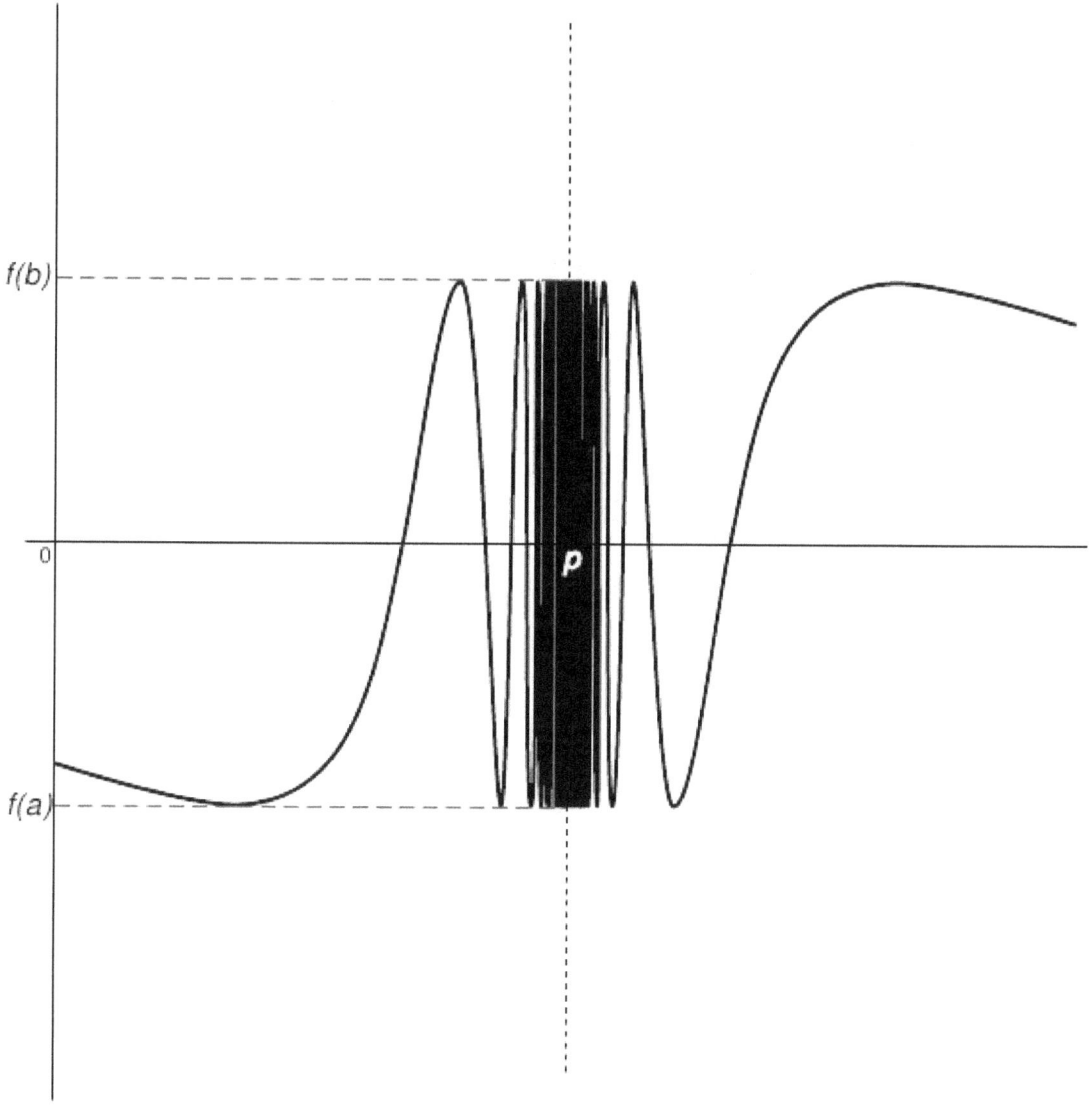

The failure of a function to be continuous at a point is quantified by its oscillation.

Continuity can also be defined in terms of oscillation: a function f is continuous at a point x_0 if and only if its oscillation at that point is zero:[3] in symbols, $\omega_f(x_0) = 0$. A benefit of this definition is that it *quantifies* discontinuity: the oscillation gives how *much* the function is discontinuous at a point.

This definition is useful in descriptive set theory to study the set of discontinuities and continuous points – the continuous points are the intersection of the sets where the oscillation is less than ε (hence a Gδ set) – and gives a very quick proof of one direction of the Lebesgue integrability condition.[4]

The oscillation is equivalent to the ε-δ definition by a simple re-arrangement, and by using a limit (lim sup, lim inf) to define oscillation: if (at a given point) for a given ε_0 there is no δ that satisfies the ε-δ definition, then the oscillation is at least ε_0, and conversely if for every ε there is a desired δ, the oscillation is 0. The oscillation definition can be naturally generalized to maps from a topological space to a metric space.

Definition using the hyperreals

Cauchy defined continuity of a function in the following intuitive terms: an infinitesimal change in the independent variable corresponds to an infinitesimal change of the dependent variable (see *Cours d'analyse*, page 34). Non-standard analysis is a way of making this mathematically rigorous. The real line is augmented by the addition of infinite and infinitesimal numbers to form the hyperreal numbers. In nonstandard analysis, continuity can be defined as follows.

> A real-valued function f is continuous at x if its natural extension to the hyperreals has the property that for all infinitesimal dx, $f(x+dx) - f(x)$ is infinitesimal[5]

(see microcontinuity). In other words, an infinitesimal increment of the independent variable always produces to an infinitesimal change of the dependent variable, giving a modern expression to Augustin-Louis Cauchy's definition of continuity.

12.2.2   Examples

All polynomial functions, such as $f(x) = x^3 + x^2 - 5x + 3$ (pictured), are continuous. This is a consequence of the fact that, given two continuous functions

$$f, g: I \to \mathbf{R}$$

defined on the same domain I, then the sum $f + g$ and the product fg of the two functions are continuous (on the same domain I). Moreover, the function

$$\frac{f}{g}: \{x \in I | g(x) \neq 0\} \to \mathbf{R}, x \mapsto \frac{f(x)}{g(x)}$$

is continuous. (The points where $g(x)$ is zero are discarded, as they are not in the domain of f/g.) For example, the function (pictured)

$$f(x) = \frac{2x - 1}{x + 2}$$

is defined for all real numbers $x \neq -2$ and is continuous at every such point. Thus it is a continuous function. The question of continuity at $x = -2$ does not arise, since $x = -2$ is not in the domain of f. There is no continuous function $F: \mathbf{R} \to \mathbf{R}$ that agrees with $f(x)$ for all $x \neq -2$. The sinc function $g(x) = (\sin x)/x$, defined for all $x \neq 0$ is continuous at these points. Thus it is a continuous function, too. However, unlike the one of the previous example, this one *can* be extended to a continuous function on all real numbers, namely

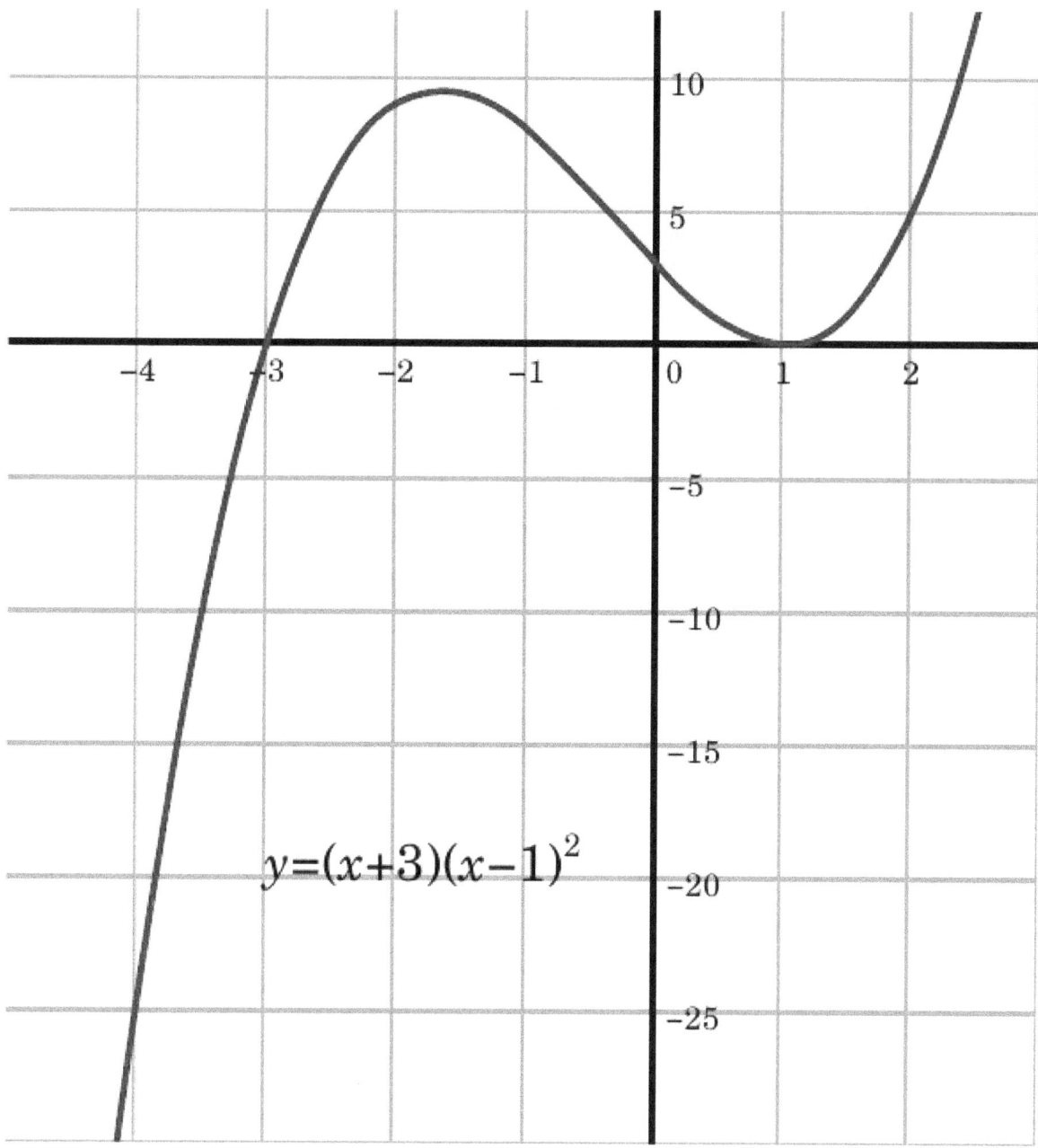

$$y=(x+3)(x-1)^2$$

The graph of a cubic function has no jumps or holes. The function is continuous.

$$G(x) = \begin{cases} \frac{\sin(x)}{x} & \text{if } x \neq 0 \\ 1 & \text{if } x = 0. \end{cases}$$

since the limit of $g(x)$, when x approaches 0, is 1. Therefore, the point $x=0$ is called a removable singularity of g.

Given two continuous functions

$$f : I \to J(\subset \mathbf{R}), g : J \to \mathbf{R}.$$

the composition

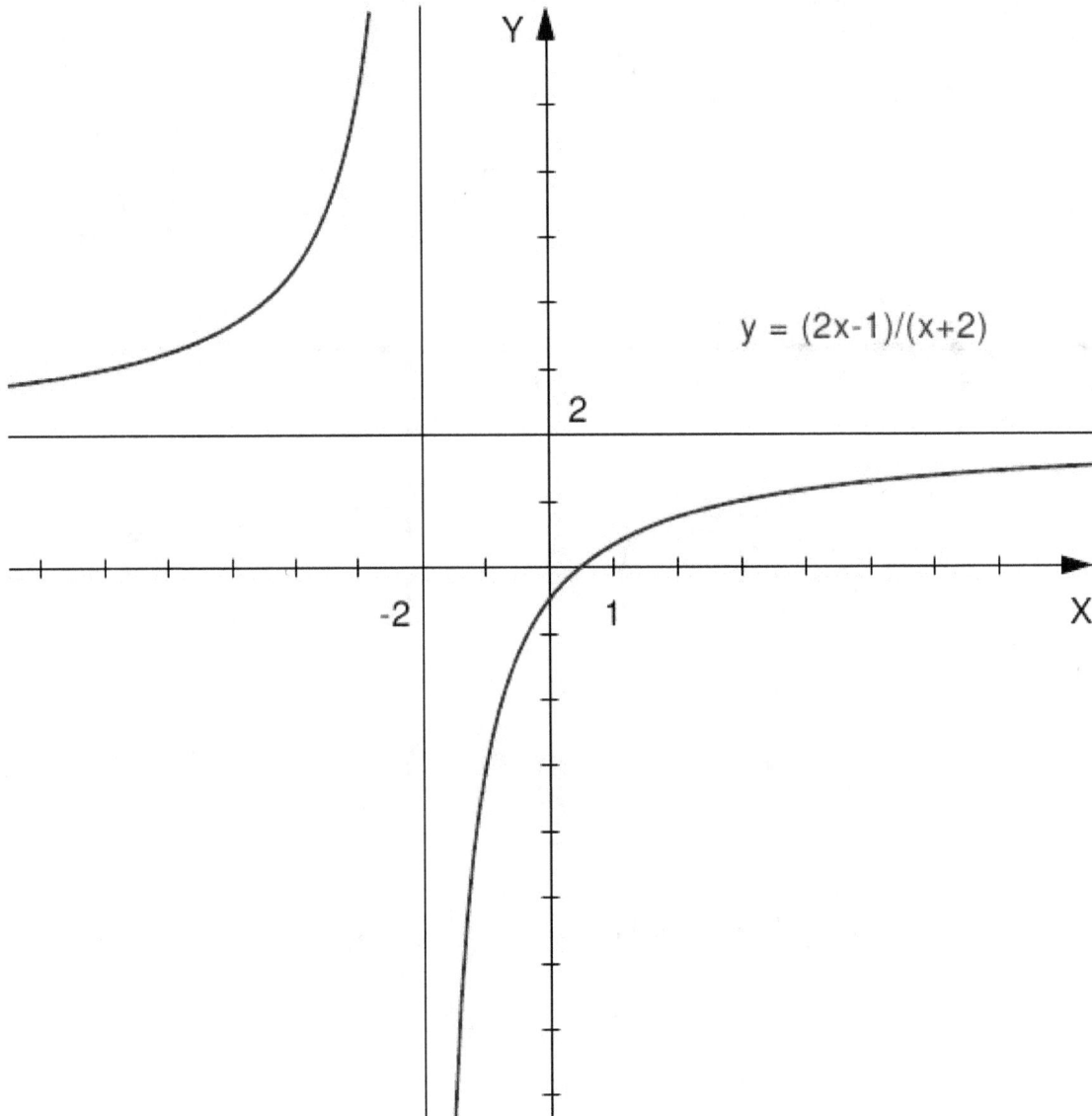

The graph of a continuous rational function. The function is not defined for x = −2. The vertical and horizontal lines are asymptotes.

$$g \circ f : I \to \mathbf{R}. x \mapsto g(f(x))$$

is continuous.

12.2.3 Non-examples

An example of a discontinuous function is the function f defined by $f(x) = 1$ if $x > 0$, $f(x) = 0$ if $x \le 0$. Pick for instance $\varepsilon = \frac{1}{2}$. There is no δ-neighborhood around $x = 0$ that will force all the $f(x)$ values to be within ε of $f(0)$. Intuitively we can think of this type of discontinuity as a sudden jump in function values. Similarly, the signum or sign function

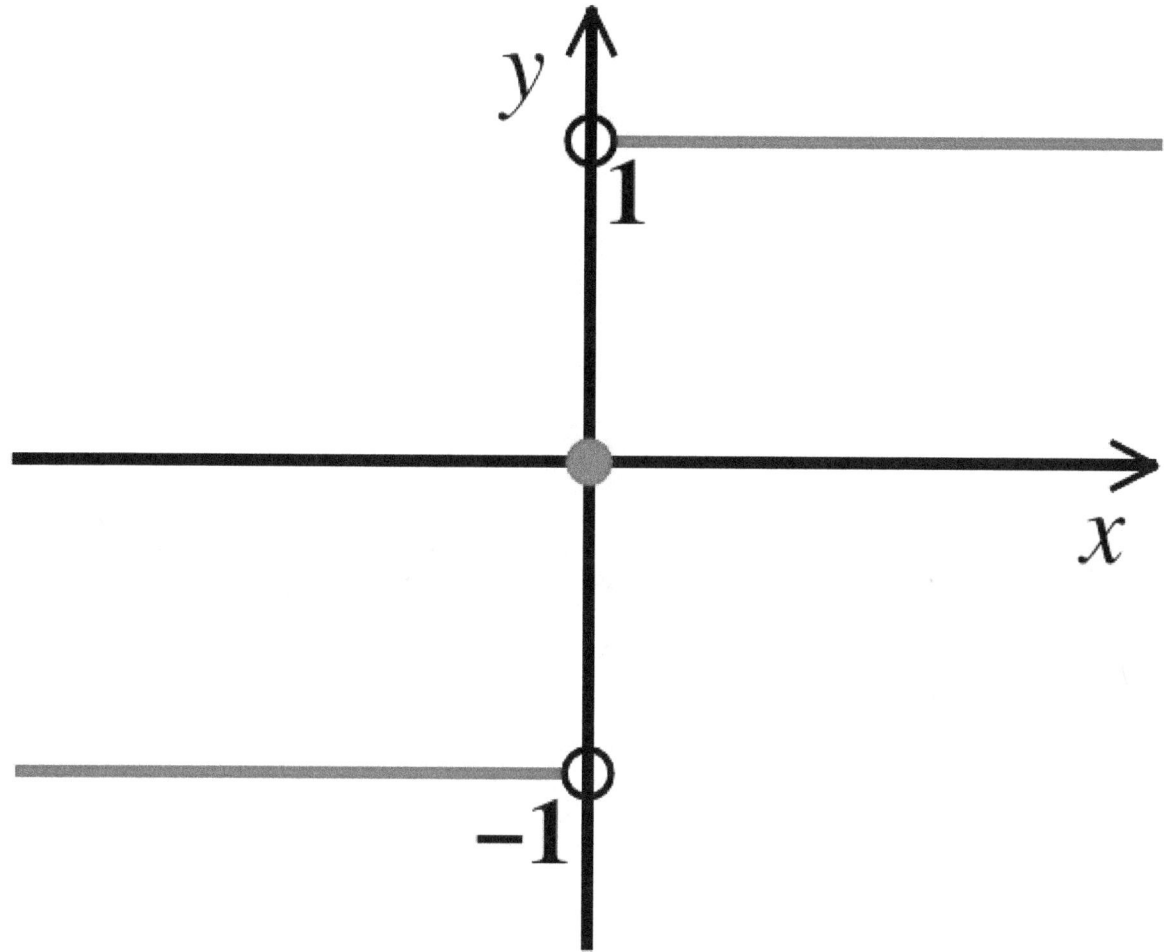

Plot of the signum function. The hollow dots indicate that sgn(x) is 1 for all x>0 and −1 for all x<0.

$$sgn(x) = \begin{cases} 1 & \text{if } x > 0 \\ 0 & \text{if } x = 0 \\ -1 & \text{if } x < 0 \end{cases}$$

is discontinuous at $x = 0$ but continuous everywhere else. Yet another example: the function

$$f(x) = \begin{cases} \sin\left(\frac{1}{x^2}\right) & \text{if } x \neq 0 \\ 0 & \text{if } x = 0 \end{cases}$$

is continuous everywhere apart from $x = 0$.

Thomae's function,

$$f(x) = \begin{cases} 1 & \text{if } x = 0 \\ \frac{1}{q} & \text{if } x = \frac{p}{q} \text{number rational a is terms) lowest (in} \\ 0 & \text{if } x \text{irrational is .} \end{cases}$$

is continuous at all irrational numbers and discontinuous at all rational numbers. In a similar vein, Dirichlet's function

Plot of Thomae's function for the domain 0<x<1.

$$D(x) = \begin{cases} 0 \text{ if } x \text{ irrational is } (\in \mathbb{R} \setminus \mathbb{Q}) \\ 1 \text{ if } x \text{ rational is } (\in \mathbb{Q}) \end{cases}$$

is nowhere continuous.

12.2.4   Properties

Intermediate value theorem

The intermediate value theorem is an existence theorem, based on the real number property of completeness, and states:

> If the real-valued function f is continuous on the closed interval $[a, b]$ and k is some number between $f(a)$ and $f(b)$, then there is some number c in $[a, b]$ such that $f(c) = k$.

For example, if a child grows from 1 m to 1.5 m between the ages of two and six years, then, at some time between two and six years of age, the child's height must have been 1.25 m.

As a consequence, if f is continuous on $[a, b]$ and $f(a)$ and $f(b)$ differ in sign, then, at some point c in $[a, b]$, $f(c)$ must equal zero.

Extreme value theorem

The extreme value theorem states that if a function f is defined on a closed interval $[a,b]$ (or any closed and bounded set) and is continuous there, then the function attains its maximum, i.e. there exists $c \in [a,b]$ with $f(c) \geq f(x)$ for all $x \in [a,b]$. The same is true of the minimum of f. These statements are not, in general, true if the function is defined on an open interval (a,b) (or any set that is not both closed and bounded), as, for example, the continuous function $f(x) = 1/x$, defined on the open interval $(0,1)$, does not attain a maximum, being unbounded above.

Relation to differentiability and integrability

Every differentiable function

$$f : (a, b) \to \mathbf{R}$$

is continuous, as can be shown. The converse does not hold: for example, the absolute value function

$$f(x) = |x| = \begin{cases} x \text{ if } x \geq 0 \\ -x \text{ if } x < 0 \end{cases}$$

is everywhere continuous. However, it is not differentiable at $x = 0$ (but is so everywhere else). Weierstrass's function is also everywhere continuous but nowhere differentiable.

The derivative $f'(x)$ of a differentiable function $f(x)$ need not be continuous. If $f'(x)$ is continuous, $f(x)$ is said to be continuously differentiable. The set of such functions is denoted $C^1((a, b))$. More generally, the set of functions

$$f : \Omega \to \mathbf{R}$$

(from an open interval (or open subset of $\mathbf{R}$) Ω to the reals) such that f is n times differentiable and such that the n-th derivative of f is continuous is denoted $C^n(\Omega)$. See differentiability class. In the field of computer graphics, these three levels are sometimes called G^0 (continuity of position), G^1 (continuity of tangency), and G^2 (continuity of curvature).

Every continuous function

$$f : [a, b] \to \mathbf{R}$$

is integrable (for example in the sense of the Riemann integral). The converse does not hold, as the (integrable, but discontinuous) sign function shows.

Pointwise and uniform limits

Given a sequence

$$f_1, f_2, \ldots : I \to \mathbf{R}$$

of functions such that the limit

$$f(x) := \lim_{n \to \infty} f_n(x)$$

exists for all x in I, the resulting function $f(x)$ is referred to as the pointwise limit of the sequence of functions $(fn)n{\in}\mathbf{N}$. The pointwise limit function need not be continuous, even if all functions fn are continuous, as the animation at the right shows. However, f is continuous when the sequence converges uniformly, by the uniform convergence theorem. This theorem can be used to show that the exponential functions, logarithms, square root function, trigonometric functions are continuous.

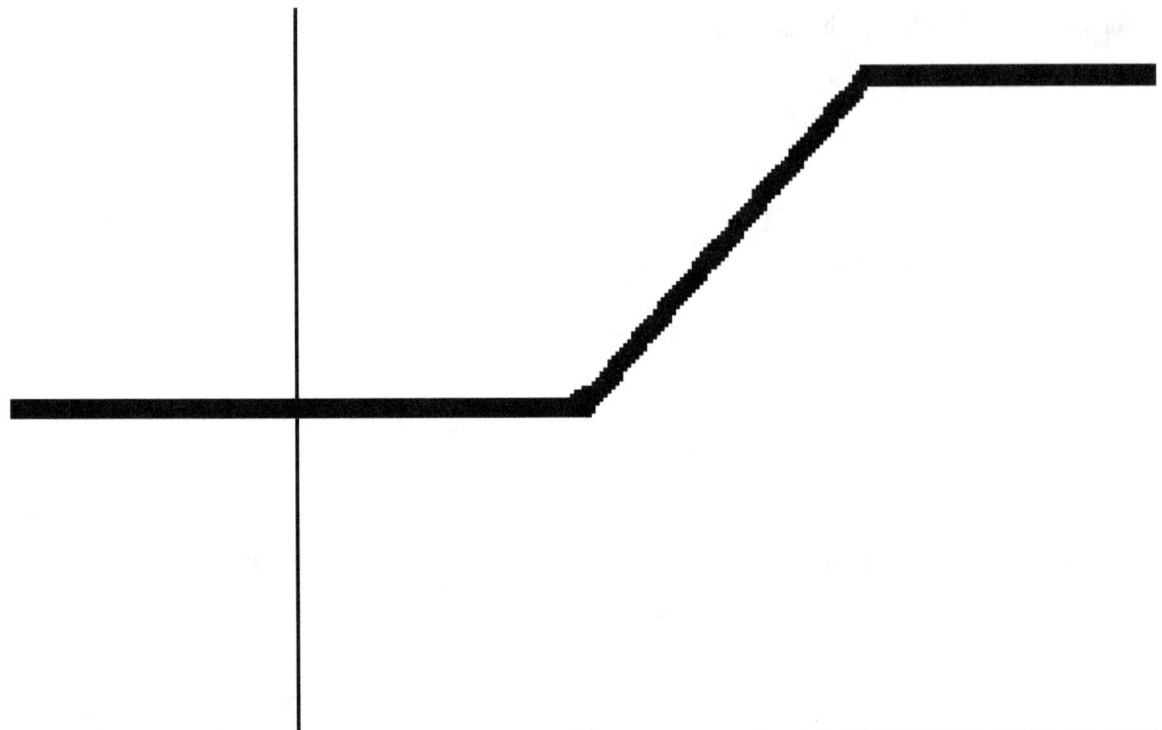

A sequence of continuous functions $f_n(x)$ *whose (pointwise) limit function* $f(x)$ *is discontinuous. The convergence is not uniform.*

12.2.5   Directional and semi-continuity

- A right-continuous function

- A left-continuous function

Discontinuous functions may be discontinuous in a restricted way, giving rise to the concept of directional continuity (or right and left continuous functions) and semi-continuity. Roughly speaking, a function is *right-continuous* if no jump occurs when the limit point is approached from the right. More formally, f is said to be right-continuous at the point c if the following holds: For any number $\varepsilon > 0$ however small, there exists some number $\delta > 0$ such that for all x in the domain with $c < x < c + \delta$, the value of $f(x)$ will satisfy

$$|f(x) - f(c)| < \varepsilon.$$

This is the same condition as for continuous functions, except that it is required to hold for x strictly larger than c only. Requiring it instead for all x with $c - \delta < x < c$ yields the notion of *left-continuous* functions. A function is continuous if and only if it is both right-continuous and left-continuous.

A function f is *lower semi-continuous* if, roughly, any jumps that might occur only go down, but not up. That is, for any $\varepsilon > 0$, there exists some number $\delta > 0$ such that for all x in the domain with $|x - c| < \delta$, the value of $f(x)$ satisfies

$$f(x) \geq f(c) - \epsilon.$$

The reverse condition is *upper semi-continuity*.

12.3 Continuous functions between metric spaces

The concept of continuous real-valued functions can be generalized to functions between metric spaces. A metric space is a set X equipped with a function (called metric) dX, that can be thought of as a measurement of the distance of any two elements in X. Formally, the metric is a function

$$d_X : X \times X \to \mathbf{R}$$

that satisfies a number of requirements, notably the triangle inequality. Given two metric spaces (X, dX) and (Y, dY) and a function

$$f : X \to Y$$

then f is continuous at the point c in X (with respect to the given metrics) if for any positive real number ε, there exists a positive real number δ such that all x in X satisfying $dX(x, c) < \delta$ will also satisfy $dY(f(x), f(c)) < \varepsilon$. As in the case of real functions above, this is equivalent to the condition that for every sequence (xn) in X with limit $\lim xn = c$, we have $\lim f(xn) = f(c)$. The latter condition can be weakened as follows: f is continuous at the point c if and only if for every convergent sequence (xn) in X with limit c, the sequence $(f(xn))$ is a Cauchy sequence, and c is in the domain of f.

The set of points at which a function between metric spaces is continuous is a Gδ set – this follows from the ε-δ definition of continuity.

This notion of continuity is applied, for example, in functional analysis. A key statement in this area says that a linear operator

$$T : V \to W$$

between normed vector spaces V and W (which are vector spaces equipped with a compatible norm, denoted $\|x\|$) is continuous if and only if it is bounded, that is, there is a constant K such that

$$\|T(x)\| \leq K \|x\|$$

for all x in V.

12.3.1 Uniform, Hölder and Lipschitz continuity

The concept of continuity for functions between metric spaces can be strengthened in various ways by limiting the way δ depends on ε and c in the definition above. Intuitively, a function f as above is uniformly continuous if the δ does not depend on the point c. More precisely, it is required that for every real number $\varepsilon > 0$ there exists $\delta > 0$ such that for every $c, b \in X$ with $dX(b, c) < \delta$, we have that $dY(f(b), f(c)) < \varepsilon$. Thus, any uniformly continuous function is continuous. The converse does not hold in general, but holds when the domain space X is compact. Uniformly continuous maps can be defined in the more general situation of uniform spaces.[6]

A function is Hölder continuous with exponent α (a real number) if there is a constant K such that for all b and c in X, the inequality

$$d_Y(f(b), f(c)) \leq K \cdot (d_X(b, c))^{\alpha}$$

holds. Any Hölder continuous function is uniformly continuous. The particular case $\alpha = 1$ is referred to as Lipschitz continuity. That is, a function is Lipschitz continuous if there is a constant K such that the inequality

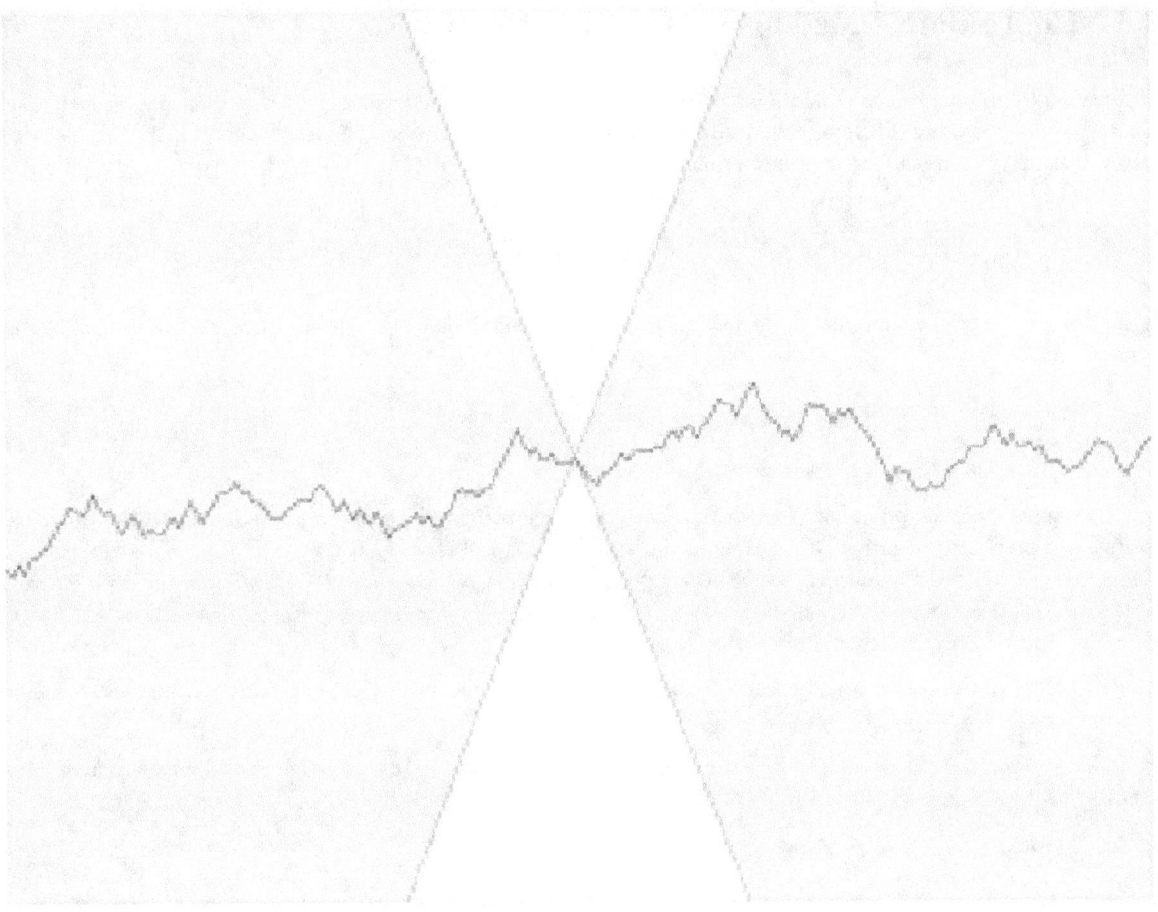

For a Lipschitz continuous function, there is a double cone (shown in white) whose vertex can be translated along the graph, so that the graph always remains entirely outside the cone.

$$d_Y(f(b), f(c)) \leq K \cdot d_X(b, c)$$

holds any b, c in X.[7] The Lipschitz condition occurs, for example, in the Picard–Lindelöf theorem concerning the solutions of ordinary differential equations.

12.4   Continuous functions between topological spaces

Another, more abstract, notion of continuity is continuity of functions between topological spaces in which there generally is no formal notion of distance, as there is in the case of metric spaces. A topological space is a set X together with a topology on X, which is a set of subsets of X satisfying a few requirements with respect to their unions and intersections that generalize the properties of the open balls in metric spaces while still allowing to talk about the neighbourhoods of a given point. The elements of a topology are called open subsets of X (with respect to the topology).

A function

$$f : X \to Y$$

between two topological spaces X and Y is continuous if for every open set $V \subseteq Y$, the inverse image

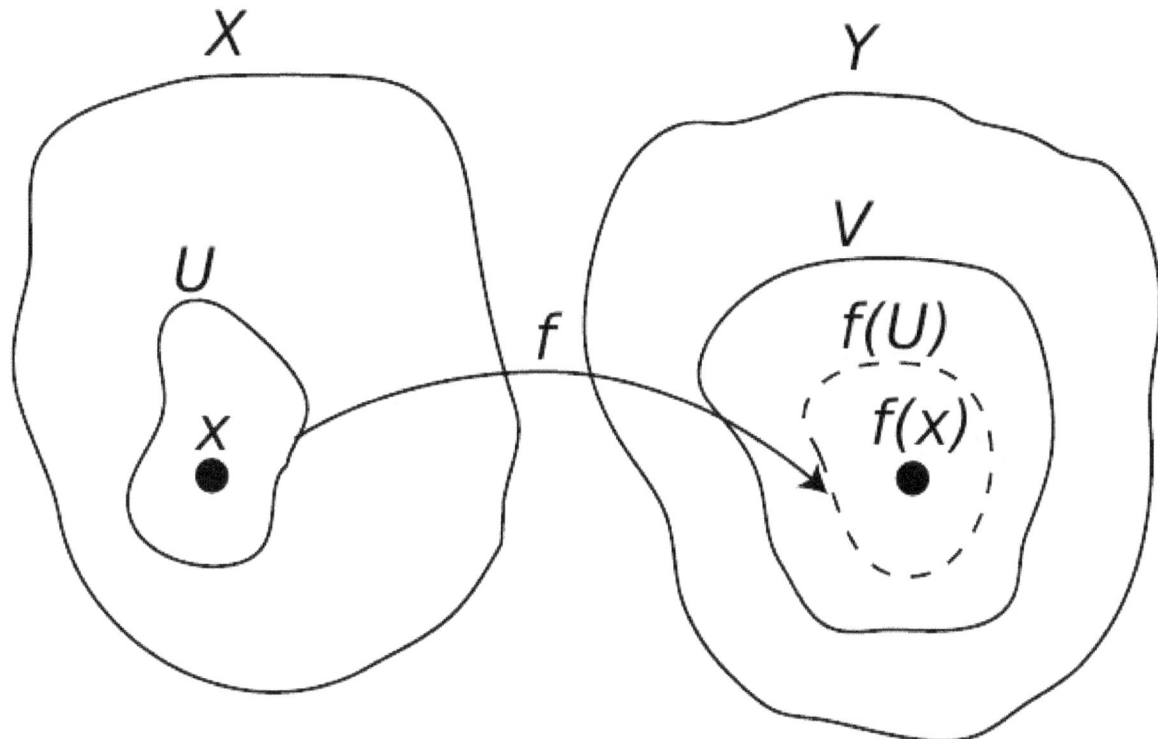

Continuity of a function at a point.

$$f^{-1}(V) = \{x \in X \mid f(x) \in V\}$$

is an open subset of X. That is, f is a function between the sets X and Y (not on the elements of the topology TX), but the continuity of f depends on the topologies used on X and Y.

This is equivalent to the condition that the preimages of the closed sets (which are the complements of the open subsets) in Y are closed in X.

An extreme example: if a set X is given the discrete topology (in which every subset is open), all functions

$$f: X \rightarrow T$$

to any topological space T are continuous. On the other hand, if X is equipped with the indiscrete topology (in which the only open subsets are the empty set and X) and the space T set is at least T_0, then the only continuous functions are the constant functions. Conversely, any function whose range is indiscrete is continuous.

12.4.1 Alternative definitions

Several equivalent definitions for a topological structure exist and thus there are several equivalent ways to define a continuous function.

Neighborhood definition

Neighborhoods continuity for functions between topological spaces $(X, \mathcal{T}_X)$ and $(Y, \mathcal{T}_Y)$ at a point may be defined:
A function $f: X \rightarrow Y$ is continuous at a point $x \in X$ iff for any neighborhood of its image $f(x) \in Y$ the preimage is again a neighborhood of that point: $\forall N \in \mathcal{N}_{f(x)} : f^{-1}(N) \in \mathcal{M}_x$

According to the property that neighborhood systems being upper sets this can be restated as follows:
$$\forall N \in \mathcal{N}_{f(x)} \exists M \in \mathcal{M}_x : M \subseteq f^{-1}(N)$$
$$\forall N \in \mathcal{N}_{f(x)} \exists M \in \mathcal{M}_x : f(M) \subseteq N$$
The second one being a restatement involving the image rather than the preimage.

Literally, this means no matter how small the neighborhood is chosen one can always find a neighborhood mapped into it.

Besides, there's a simplification involving only open neighborhoods. In fact, they're equivalent:
$$\forall V \in \mathcal{T}_Y, f(x) \in V \exists U \in \mathcal{T}_X, x \in U : U \subseteq f^{-1}(V)$$
$$\forall V \in \mathcal{T}_Y, f(x) \in V \exists U \in \mathcal{T}_X, x \in U : f(U) \subseteq V$$
The second one again being a restatement using images rather than preimages.

If X and Y are metric spaces, it is equivalent to consider the neighborhood system of open balls centered at x and $f(x)$ instead of all neighborhoods. This gives back the above δ-ε definition of continuity in the context of metric spaces. However, in general topological spaces, there is no notion of nearness or distance.

Note, however, that if the target space is Hausdorff, it is still true that f is continuous at a if and only if the limit of f as x approaches a is $f(a)$. At an isolated point, every function is continuous.

Sequences and nets

In several contexts, the topology of a space is conveniently specified in terms of limit points. In many instances, this is accomplished by specifying when a point is the limit of a sequence, but for some spaces that are too large in some sense, one specifies also when a point is the limit of more general sets of points indexed by a directed set, known as nets. A function is (Heine-)continuous only if it takes limits of sequences to limits of sequences. In the former case, preservation of limits is also sufficient; in the latter, a function may preserve all limits of sequences yet still fail to be continuous, and preservation of nets is a necessary and sufficient condition.

In detail, a function $f \colon X \to Y$ is **sequentially continuous** if whenever a sequence (x_n) in X converges to a limit x, the sequence $(f(x_n))$ converges to $f(x)$. Thus sequentially continuous functions "preserve sequential limits". Every continuous function is sequentially continuous. If X is a first-countable space and countable choice holds, then the converse also holds: any function preserving sequential limits is continuous. In particular, if X is a metric space, sequential continuity and continuity are equivalent. For non first-countable spaces, sequential continuity might be strictly weaker than continuity. (The spaces for which the two properties are equivalent are called sequential spaces.) This motivates the consideration of nets instead of sequences in general topological spaces. Continuous functions preserve limits of nets, and in fact this property characterizes continuous functions.

Closure operator definition

Instead of specifying the open subsets of a topological space, the topology can also be determined by a closure operator (denoted cl) which assigns to any subset $A \subseteq X$ its closure, or an interior operator (denoted int), which assigns to any subset A of X its interior. In these terms, a function

$$f \colon (X, \mathrm{cl}) \to (X', \mathrm{cl}')$$

between topological spaces is continuous in the sense above if and only if for all subsets A of X

$$f(\mathrm{cl}(A)) \subseteq \mathrm{cl}'(f(A)).$$

That is to say, given any element x of X that is in the closure of any subset A, $f(x)$ belongs to the closure of $f(A)$. This is equivalent to the requirement that for all subsets A' of X'

$$f^{-1}(\mathrm{cl}'(A')) \supseteq \mathrm{cl}(f^{-1}(A')).$$

Moreover,

$$f: (X, \text{int}) \to (X', \text{int}')$$

is continuous if and only if

$$f^{-1}(\text{int}'(A')) \subseteq \text{int}(f^{-1}(A'))$$

for any subset A' of Y.

12.4.2 Properties

If $f: X \to Y$ and $g: Y \to Z$ are continuous, then so is the composition $g \circ f: X \to Z$. If $f: X \to Y$ is continuous and

- X is compact, then $f(X)$ is compact.
- X is connected, then $f(X)$ is connected.
- X is path-connected, then $f(X)$ is path-connected.
- X is Lindelöf, then $f(X)$ is Lindelöf.
- X is separable, then $f(X)$ is separable.

The possible topologies on a fixed set X are partially ordered: a topology τ_1 is said to be coarser than another topology τ_2 (notation: $\tau_1 \subseteq \tau_2$) if every open subset with respect to τ_1 is also open with respect to τ_2. Then, the identity map

$$\text{id}X: (X, \tau_2) \to (X, \tau_1)$$

is continuous if and only if $\tau_1 \subseteq \tau_2$ (see also comparison of topologies). More generally, a continuous function

$$(X, \tau_X) \to (Y, \tau_Y)$$

stays continuous if the topology τY is replaced by a coarser topology and/or τX is replaced by a finer topology.

12.4.3 Homeomorphisms

Symmetric to the concept of a continuous map is an open map, for which *images* of open sets are open. In fact, if an open map f has an inverse function, that inverse is continuous, and if a continuous map g has an inverse, that inverse is open. Given a bijective function f between two topological spaces, the inverse function f^{-1} need not be continuous. A bijective continuous function with continuous inverse function is called a *homeomorphism*.

If a continuous bijection has as its domain a compact space and its codomain is Hausdorff, then it is a homeomorphism.

12.4.4 Defining topologies via continuous functions

Given a function

$$f: X \to S,$$

where X is a topological space and S is a set (without a specified topology), the final topology on S is defined by letting the open sets of S be those subsets A of S for which $f^{-1}(A)$ is open in X. If S has an existing topology, f is continuous with respect to this topology if and only if the existing topology is coarser than the final topology on S. Thus the final topology can be characterized as the finest topology on S that makes f continuous. If f is surjective, this topology is canonically identified with the quotient topology under the equivalence relation defined by f.

Dually, for a function f from a set S to a topological space, the initial topology on S has as open subsets A of S those subsets for which $f(A)$ is open in X. If S has an existing topology, f is continuous with respect to this topology if and only if the existing topology is finer than the initial topology on S. Thus the initial topology can be characterized as the coarsest topology on S that makes f continuous. If f is injective, this topology is canonically identified with the subspace topology of S, viewed as a subset of X.

More generally, given a set S, specifying the set of continuous functions

$$S \to X$$

into all topological spaces X defines a topology. Dually, a similar idea can be applied to maps

$$X \to S.$$

This is an instance of a universal property.

12.5   Related notions

Various other mathematical domains use the concept of continuity in different, but related meanings. For example, in order theory, an order-preserving function $f: X \to Y$ between particular types of partially ordered sets X and Y is continuous if for each directed subset A of X, we have $\sup(f(A)) = f(\sup(A))$. Here sup is the supremum with respect to the orderings in X and Y, respectively. This notion of continuity is the same as topological continuity when the partially ordered sets are given the Scott topology.[8][9]

In category theory, a functor

$$F : C \to D$$

between two categories is called *continuous*, if it commutes with small limits. That is to say,

$$\varprojlim_{i \in I} F(C_i) \cong F(\varprojlim_{i \in I} C_i)$$

for any small (i.e., indexed by a set I, as opposed to a class) diagram of objects in C .

A *continuity space* is a generalization of metric spaces and posets,[10][11] which uses the concept of quantales, and that can be used to unify the notions of metric spaces and domains.[12]

12.6   See also

- Absolute continuity

- Classification of discontinuities

- Coarse function

- Continuous stochastic process

- Dini continuity

- Discrete function

- Equicontinuity

- Normal function

- Piecewise

- Symmetrically continuous function

12.7 Notes

[1] Rusnock,P.;Kerr-Lawson,A. (2005), "Bolzano and uniform continuity",*Historia Mathematica***32**(3): 303–311,doi:10.10163

[2] Lang, Serge (1997), *Undergraduate analysis*, Undergraduate Texts in Mathematics (2nd ed.), Berlin, New York: Springer-Verlag, ISBN 978-0-387-94841-6, section II.4

[3] *Introduction to Real Analysis*, updated April 2010, William F. Trench, Theorem 3.5.2, p. 172

[4] *Introduction to Real Analysis*, updated April 2010, William F. Trench, 3.5 "A More Advanced Look at the Existence of the Proper Riemann Integral", pp. 171–177

[5] "Elementary Calculus", *wisc.edu*.

[6] Gaal, Steven A. (2009), *Point set topology*, New York: Dover Publications, ISBN 978-0-486-47222-5, section IV.10

[7] Searcóid, Mícheál Ó (2006), *Metric spaces*, Springer undergraduate mathematics series, Berlin, New York: Springer-Verlag, ISBN 978-1-84628-369-7, section 9.4

[8] Goubault-Larrecq, Jean (2013). *Non-Hausdorff Topology and Domain Theory: Selected Topics in Point-Set Topology*. Cambridge University Press. ISBN 1107034132.

[9] Gierz, G.; Hofmann, K. H.; Keimel, K.; Lawson, J. D.; Mislove, M. W.; Scott, D. S. (2003). *Continuous Lattices and Domains*. Encyclopedia of Mathematics and its Applications **93**. Cambridge University Press. ISBN 0521803381.

[10] Flagg, R. C. (1997). "Quantales and continuity spaces". *Algebra Universalis*. CiteSeerX: 10.1.1.48.851.

[11] Kopperman, R. (1988). "All topologies come from generalized metrics". *American Mathematical Monthly* **95** (2): 89–97. doi:10.2307/2323060.

[12] Flagg, B.; Kopperman, R. (1997). "Continuity spaces: Reconciling domains and metric spaces". *Theoretical Computer Science* **177** (1): 111–138. doi:10.1016/S0304-3975(97)00236-3.

12.8 References

- Hazewinkel, Michiel, ed. (2001), "Continuous function", *Encyclopedia of Mathematics*, Springer, ISBN 978-1-55608-010-4

- Visual Calculus by Lawrence S. Husch, University of Tennessee (2001).

Chapter 13

Smoothness

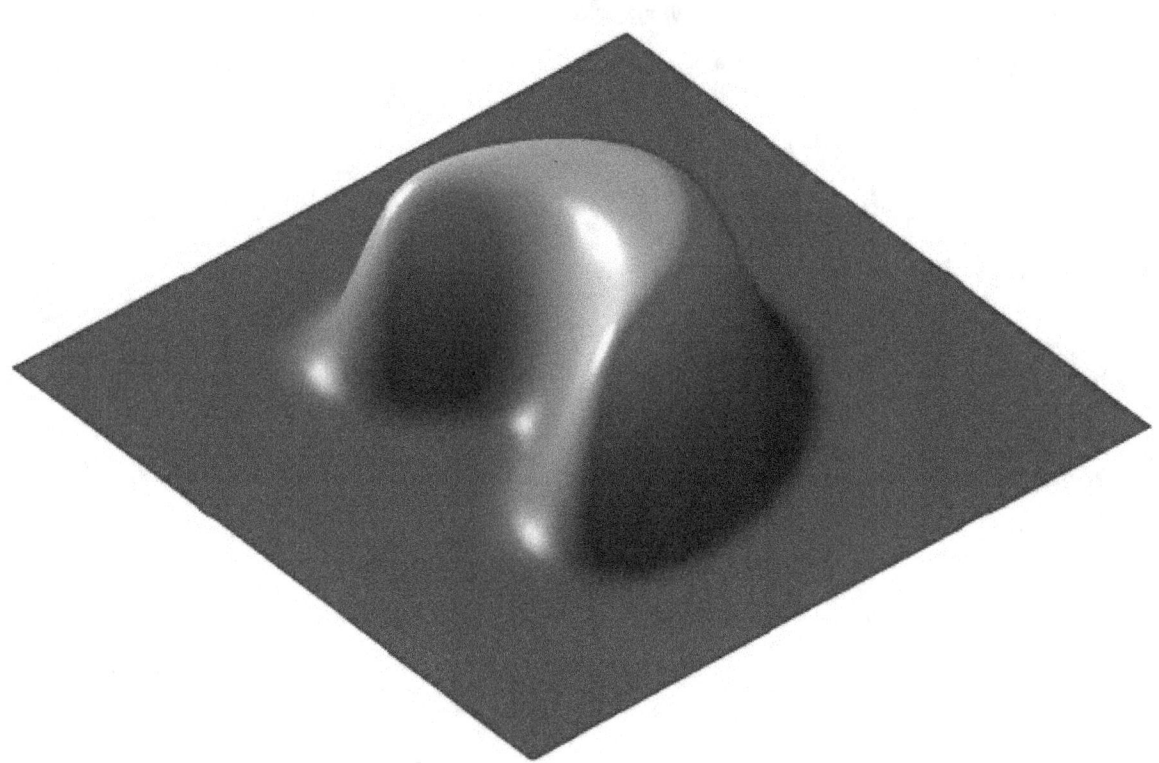

A bump function is a smooth function with compact support.

In mathematical analysis, **smoothness** has to do with how many derivatives of a function exist and are continuous. The term **smooth function** is often used technically to mean a function that has derivatives of all orders everywhere in its domain.

13.1 Differentiability classes

Differentiability class is a classification of functions according to the properties of their derivatives. Higher order differentiability classes correspond to the existence of more derivatives.

Consider an open set on the real line and a function f defined on that set with real values. Let k be a non-negative integer. The function f is said to be of (differentiability) **class C^k** if the derivatives f', f'', ..., $f^{(k)}$ exist and are continuous (the continuity is implied by differentiability for all the derivatives except for $f^{(k)}$). The function f is said to be of **class C^∞**, or **smooth**, if it has derivatives of all orders.[1] The function f is said to be of **class C^ω**, or **analytic**, if f is smooth *and* if it equals its Taylor series expansion around any point in its domain. C^ω is thus strictly contained in C^∞. Bump functions are examples of functions in C^∞ but *not* in C^ω.

To put it differently, the class C^0 consists of all continuous functions. The class C^1 consists of all differentiable functions whose derivative is continuous; such functions are called **continuously differentiable**. Thus, a C^1 function is exactly a function whose derivative exists and is of class C^0. In general, the classes C^k can be defined recursively by declaring C^0 to be the set of all continuous functions and declaring C^k for any positive integer k to be the set of all differentiable functions whose derivative is in C^{k-1}. In particular, C^k is contained in C^{k-1} for every k, and there are examples to show that this containment is strict. C^∞, the class of **infinitely differentiable** functions, is the intersection of the sets C^k as k varies over the non-negative integers (i.e. from 0 to ∞).

13.1.1 Examples

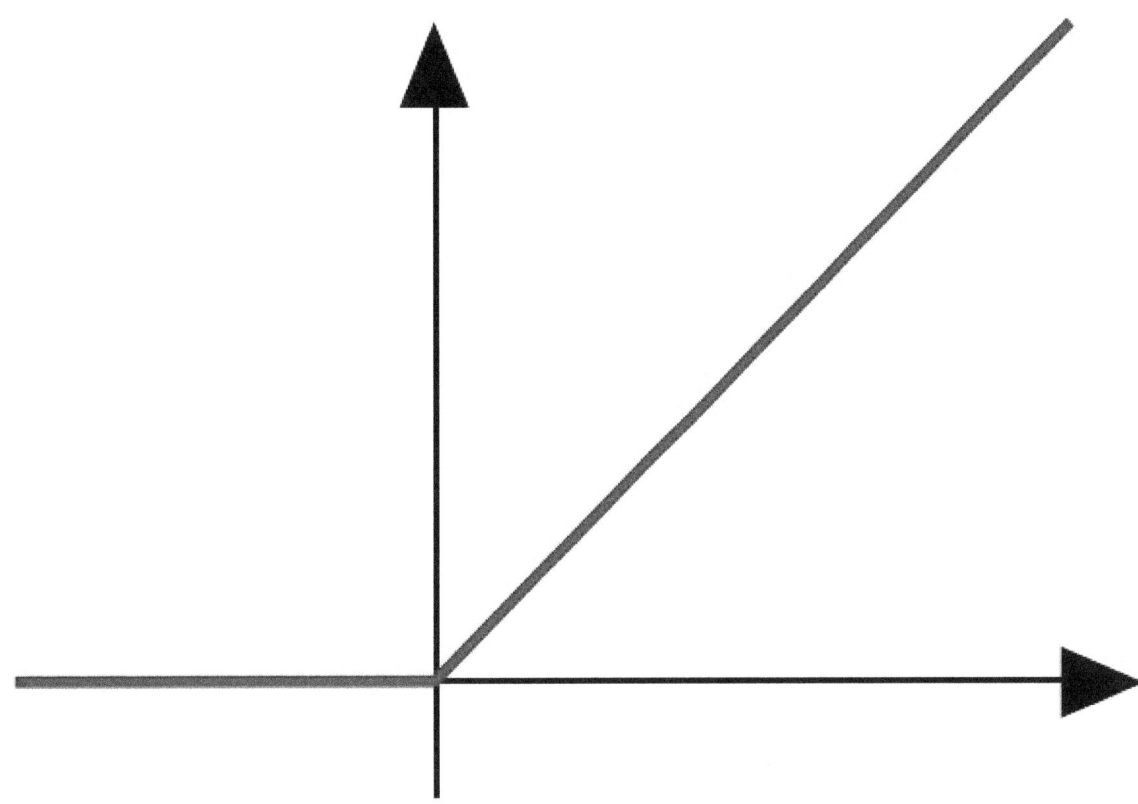

The C^0 function f(x)=x for x≥0 and 0 otherwise.

The function

$$f(x) = \begin{cases} x & \text{if } x \geq 0, \\ 0 & \text{if } x < 0 \end{cases}$$

is continuous, but not differentiable at $x = 0$, so it is of class C^0 but not of class C^1.

The function

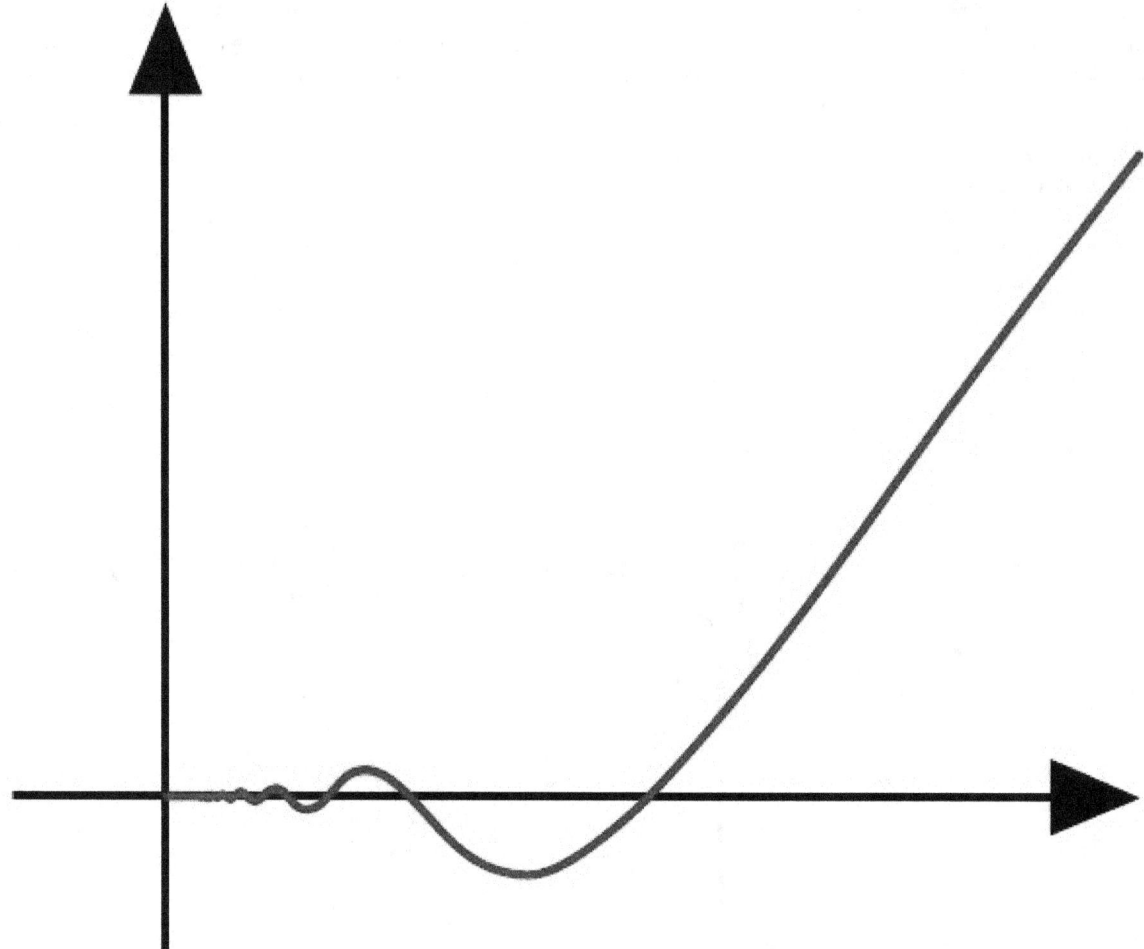

The function f(x)=x² *sin(1/x) for x>0.*

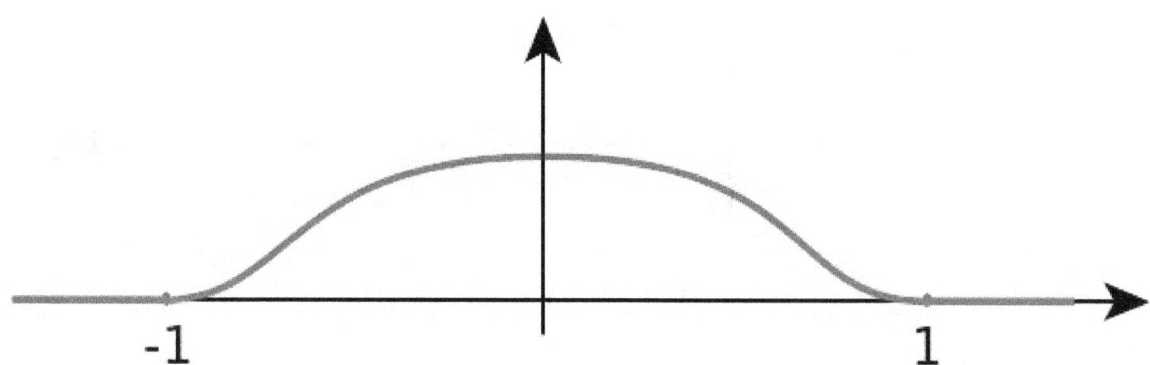

A smooth function that is not analytic.

$$f(x) = \begin{cases} x^2 \sin\left(\frac{1}{x}\right) & \text{if } x \neq 0, \\ 0 & \text{if } x = 0 \end{cases}$$

is differentiable, with derivative

$$f'(x) = \begin{cases} -\cos(\frac{1}{x}) + 2x\sin(\frac{1}{x}) & \text{if } x \neq 0, \\ 0 & \text{if } x = 0. \end{cases}$$

Because $\cos(1/x)$ oscillates as $x \to 0$, $f'(x)$ is not continuous at zero. Therefore, this function is differentiable but not of class C^1. Moreover, if one takes $f(x) = x^{4/3}\sin(1/x)$ $(x \neq 0)$ in this example, it can be used to show that the derivative function of a differentiable function can be unbounded on a compact set and, therefore, that a differentiable function on a compact set may not be locally Lipschitz continuous.

The functions

$$f(x) = |x|^{k+1}$$

where k is even, are continuous and k times differentiable at all x. But at $x = 0$ they are not $(k+1)$ times differentiable, so they are of class C^k but not of class C^j where $j > k$.

The exponential function is analytic, so, of class C^∞. The trigonometric functions are also analytic wherever they are defined.

The function

$$f(x) = \begin{cases} e^{-\frac{1}{1-x^2}} & \text{if } |x| < 1, \\ 0 & \text{otherwise} \end{cases}$$

is smooth, so of class C^∞, but it is not analytic at $x = \pm 1$, so it is not of class C^ω. The function f is an example of a smooth function with compact support.

13.1.2 Multivariate differentiability classes

Let n and m be some positive integers. If f is a function from an open subset of $\mathbf{R}^n$ with values in $\mathbf{R}^m$, then f has component functions $f_1, ..., fm$. Each of these may or may not have partial derivatives. For a non-negative integer ℓ, we say that f is of **class C^ℓ** if all of the partial derivatives $\frac{\partial^\ell f_i}{\partial x_{i_1}^{\ell_1} \partial x_{i_2}^{\ell_2} \cdots \partial x_{i_k}^{\ell_k}}$ exist and are continuous, where k is a non-negative integer, i is an integer between 1 and m, each of $i_1, i_2, \ldots, i_k$ is an integer between 1 and n, each of $\ell, \ell_1, \ell_2, \ldots, \ell_k$ is an integer between 0 and ℓ, and $\ell_1 + \ell_2 + \cdots + \ell_k = \ell$.[1] The classes C^∞ and C^ω are defined as before.[1]

These criteria of differentiability can be applied to the transition functions of a differential structure. The resulting space is called a C^k manifold.

If one wishes to start with a coordinate-independent definition of the **class C^k**, one may start by considering maps between Banach spaces. A map from one Banach space to another is differentiable at a point if there is an affine map which approximates it at that point. The derivative of the map assigns to the point x the linear part of the affine approximation to the map at x. Since the space of linear maps from one Banach space to another is again a Banach space, we may continue this procedure to define higher order derivatives. A map f is of **class C^k** if it has continuous derivatives up to order k, as before.

Note that $\mathbf{R}^n$ is a Banach space for any value of n, so the coordinate-free approach is applicable in this instance. It can be shown that the definition in terms of partial derivatives and the coordinate-free approach are equivalent; that is, a function f is of **class C^k** by one definition iff it is so by the other definition.

13.1.3 The space of C^k functions

Let D be an open subset of the real line. The set of all C^k functions defined on D and taking real values is a Fréchet vector space with the countable family of seminorms

$$p_{K,m} = \sup_{x \in K} \left| f^{(m)}(x) \right|$$

where K varies over an increasing sequence of compact sets whose union is D, and $m = 0, 1, \ldots, k$.

The set of C^∞ functions over D also forms a Fréchet space. One uses the same seminorms as above, except that m is allowed to range over all non-negative integer values.

The above spaces occur naturally in applications where functions having derivatives of certain orders are necessary; however, particularly in the study of partial differential equations, it can sometimes be more fruitful to work instead with the Sobolev spaces.

13.2   Parametric continuity

Parametric continuity is a concept applied to parametric curves describing the smoothness of the parameter's value with distance along the curve.

13.2.1   Definition

A curve can be said to have C^n continuity if $\dfrac{d^n s}{dt^n}$ is continuous of value throughout the curve.

As an example of a practical application of this concept, a curve describing the motion of an object with a parameter of time, must have C^1 continuity for the object to have finite acceleration. For smoother motion, such as that of a camera's path while making a film, higher orders of parametric continuity are required.

13.2.2   Order of continuity

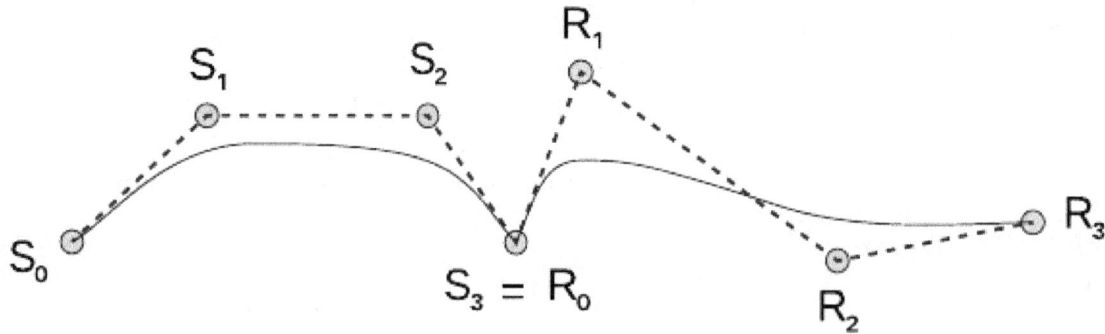

Two Bézier curve segments attached that is only C^0 continuous.

The various order of parametric continuity can be described as follows:[2]

- C^{-1}: curves include discontinuities

- C^0: curves are joined

- C^1: first derivatives are continuous

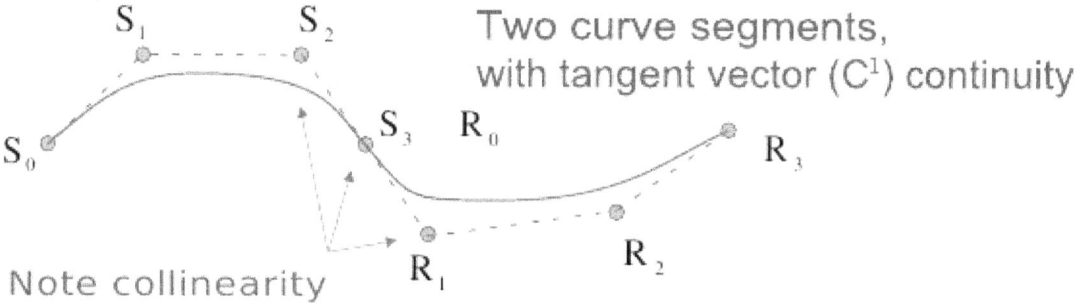

Two curve segments,
with tangent vector (C^1) continuity

Note collinearity

Two Bézier curve segments attached in such a way that they are C^1 continuous.

- C^2: first and second derivatives are continuous

- C^n: first through n^{th} derivatives are continuous

The term *parametric continuity* was introduced to distinguish it from *geometric continuity* (G^n) which removes restrictions on the speed with which the parameter traces out the curve.[3]

13.3 Geometric continuity

Geometric continuity is the continuity of the implicit function.

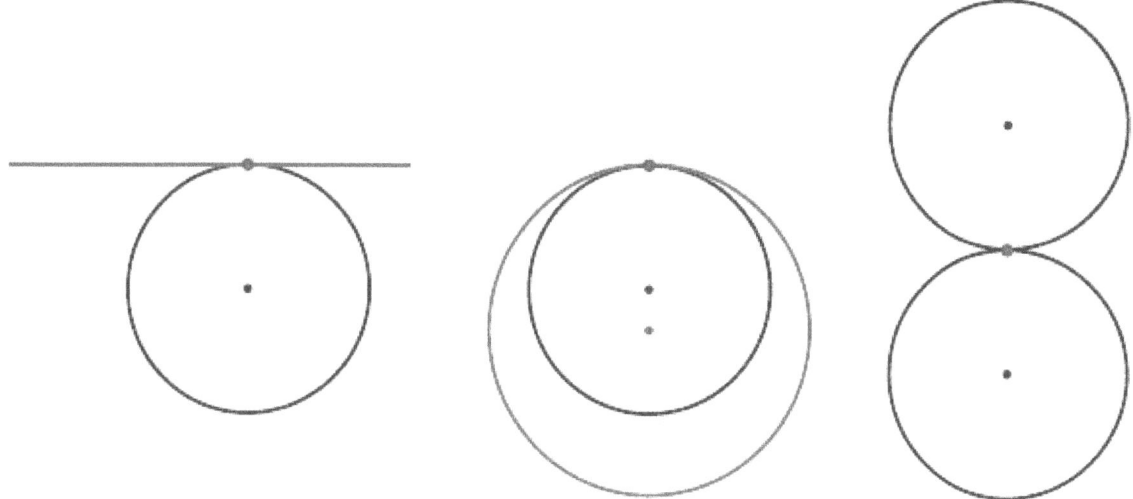

Curves with G^1-contact (circles,line)

The concept of **geometrical** or **geometric continuity** was primarily applied to the conic sections and related shapes by mathematicians such as Leibniz, Kepler, and Poncelet. The concept was an early attempt at describing, through geometry rather than algebra, the concept of continuity as expressed through a parametric function.

The basic idea behind geometric continuity was that the five conic sections were really five different versions of the same shape. An ellipse tends to a circle as the eccentricity approaches zero, or to a parabola as it approaches one; and a hyperbola tends to a parabola as the eccentricity drops toward one; it can also tend to intersecting lines. Thus, there was *continuity* between the conic sections. These ideas led to other concepts of continuity. For instance, if a circle and a straight line were two expressions of the same shape, perhaps a line could be thought of as a circle of infinite radius. For

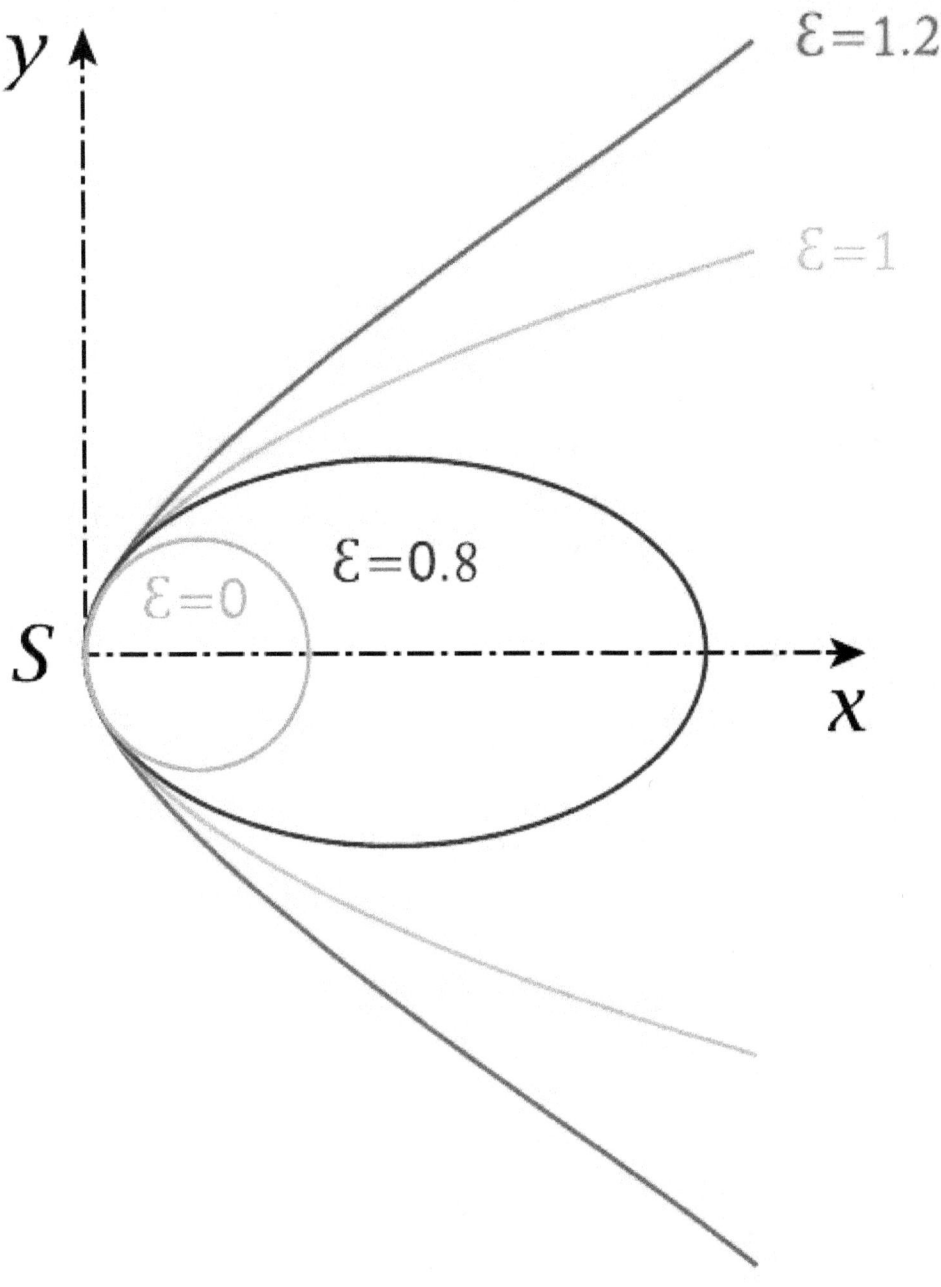

$(1 - \varepsilon^2)x^2 - 2px + y^2 = 0, \ p > 0, \varepsilon \geq 0$
pencil of conic sections with G^2-*contact: p fix, ε variable*
($\varepsilon = 0$: circle, $\varepsilon = 0.8$: ellipse, $\varepsilon = 1$: parabola, $\varepsilon = 1.2$: hyperbola)

such to be the case, one would have to make the line closed by allowing the point $x = \infty$ to be a point on the circle, and for $x = +\infty$ and $x = -\infty$ to be identical. Such ideas were useful in crafting the modern, algebraically defined, idea of the continuity of a function and of ∞.

13.3.1 Smoothness of curves and surfaces

A curve or surface can be described as having G^n continuity, n being the increasing measure of smoothness. Consider the segments either side of a point on a curve:

- G^0: The curves touch at the join point.

- G^1: The curves also share a common tangent direction at the join point.

- G^2: The curves also share a common center of curvature at the join point.

In general, G^n continuity exists if the curves can be reparameterized to have C^n (parametric) continuity.[4][5] A reparametrtion of the curve is geometrically identical to the original; only the parameter is affected.

Equivalently, two vector functions $f(t)$ and $g(t)$ have G^n continuity if $f^{(n)}(t) \neq 0$ and $f^{(n)}(t) \equiv k g^{(n)}(t)$, for a scalar $k > 0$ (i.e., if the direction, but not necessarily the magnitude, of the two vectors is equal).

While it may be obvious that a curve would require G^1 continuity to appear smooth, for good aesthetics, such as those aspired to in architecture and sports car design, higher levels of geometric continuity are required. For example, reflections in a car body will not appear smooth unless the body has G^2 continuity.

A *rounded rectangle* (with ninety degree circular arcs at the four corners) has G^1 continuity, but does not have G^2 continuity. The same is true for a *rounded cube*, with octants of a sphere at its corners and quarter-cylinders along its edges. If an editable curve with G^2 continuity is required, then cubic splines are typically chosen; these curves are frequently used in industrial design.

13.3.2 Smoothness of piecewise defined curves and surfaces

13.4 Smoothness

13.4.1 Relation to analyticity

While all analytic functions are "smooth" (i.e. have all derivatives continuous) on the set on which they are analytic, examples such as bump functions (mentioned above) show that the converse is not true for functions on the reals: there exist smooth real functions which are not analytic. Simple examples of functions which are smooth but not analytic at any point can be made by means of Fourier series; another example is the Fabius function. Although it might seem that such functions are the exception rather than the rule, it turns out that the analytic functions are scattered very thinly among the smooth ones; more rigorously, the analytic functions form a meagre subset of the smooth functions. Furthermore, for every open subset A of the real line, there exist smooth functions which are analytic on A and nowhere else.

It is useful to compare the situation to that of the ubiquity of transcendental numbers on the real line. Both on the real line and the set of smooth functions, the examples we come up with at first thought (algebraic/rational numbers and analytic functions) are far better behaved than the majority of cases: the transcendental numbers and nowhere analytic functions have full measure (their complements are meagre).

The situation thus described is in marked contrast to complex differentiable functions. If a complex function is differentiable just once on an open set it is both infinitely differentiable and analytic on that set.

13.4.2 Smooth partitions of unity

Smooth functions with given closed support are used in the construction of **smooth partitions of unity** (see *partition of unity* and topology glossary); these are essential in the study of smooth manifolds, for example to show that Riemannian metrics can be defined globally starting from their local existence. A simple case is that of a **bump function** on the real line, that is, a smooth function f that takes the value 0 outside an interval $[a,b]$ and such that

$f(x) > 0 \quad \text{for} \quad a < x < b.$

Given a number of overlapping intervals on the line, bump functions can be constructed on each of them, and on semi-infinite intervals $(-\infty, c]$ and $[d,+\infty)$ to cover the whole line, such that the sum of the functions is always 1.

From what has just been said, partitions of unity don't apply to holomorphic functions; their different behavior relative to existence and analytic continuation is one of the roots of sheaf theory. In contrast, sheaves of smooth functions tend not to carry much topological information.

13.4.3   Smooth functions between manifolds

Smooth maps between smooth manifolds may be defined by means of charts, since the idea of smoothness of function is independent of the particular chart used. If F is a map from an m-manifold M to an n-manifold N, then F is smooth if, for every $p \in M$, there is a chart (U, φ) in M containing p and a chart (V, ψ) in N containing $F(p)$ with $F(U) \subset V$, such that $\psi \circ F \circ \varphi^{-1}$ is smooth from $\varphi(U)$ to $\psi(V)$ as a function from $\mathbf{R}^m$ to $\mathbf{R}^n$.

Such a map has a first derivative defined on tangent vectors; it gives a fibre-wise linear mapping on the level of tangent bundles.

13.4.4   Smooth functions between subsets of manifolds

There is a corresponding notion of **smooth map** for arbitrary subsets of manifolds. If $f : X \to Y$ is a function whose domain and range are subsets of manifolds $X \subset M$ and $Y \subset N$ respectively. f is said to be **smooth** if for all $x \in X$ there is an open set $U \subset M$ with $x \in U$ and a smooth function $F : U \to N$ such that $F(p) = f(p)$ for all $p \in U \cap X$.

13.5   See also

- Non-analytic smooth function

- Quasi-analytic function

- Spline

- Smooth number (number theory)

- Sinuosity

13.6   References

[1] Warner (1983), p. 5, Definition 1.2.

[2] Parametric Curves

[3] Richard H. Bartels; John C. Beatty; Brian A. Barsky (1987). *An Introduction to Splines for Use in Computer Graphics and Geometric Modeling*. Morgan Kaufmann. Chapter 13. Parametric vs. Geometric Continuity. ISBN 978-1-55860-400-1.

[4] Brian A. Barsky and Tony D. DeRose, "Geometric Continuity of Parametric Curves: Three Equivalent Characterizations," IEEE Computer Graphics and Applications, 9(6), Nov. 1989, pp. 60–68.

[5] Erich Hartmann:*Geometry and Algorithms for COMPUTER AIDED DESIGN*, page 55

- This article incorporates text from a publication now in the public domain: Chisholm, Hugh, ed. (1911). *Encyclopædia Britannica* (11th ed.). Cambridge University Press.

- Guillemin, Victor; Pollack, Alan (1974). *Differential Topology*. Englewood Cliffs: Prentice-Hall. ISBN 0-13-212605-2.

- Warner, Frank Wilson (1983). *Foundations of differentiable manifolds and Lie groups*. Springer. ISBN 978-0-387-90894-6.

Chapter 14

Discrete symmetry

In mathematics, a **discrete symmetry** is a symmetry that describes non-continuous changes in a system. For example, a square possesses discrete rotational symmetry, as only rotations by multiples of right angles will preserve the square's original appearance. Discrete symmetries sometimes involve some type of 'swapping', these swaps usually being called *reflections* or *interchanges*. In mathematics and theoretical physics, a **discrete symmetry** is a symmetry under the transformations of a discrete group—e.g. a topological group with a discrete topology whose elements form a finite or a countable set.

One of the most prominent discrete symmetries in physics is parity symmetry. It manifests itself in various elementary physical quantum systems, such as quantum harmonic oscillator, electron orbitals of Hydrogen-like atoms by forcing wavefunctions to be even or odd. This in turn gives rise to selection rules that determine which transition lines are visible in atomic absorption spectra.

14.1   References

- Slavik V. Jablan, *Symmetry, Ornament and Modularity*, Volume 30 of K & E Series on Knots and Everything, World Scientific, 2002. ISBN 9812380809

Chapter 15

T-symmetry

In theoretical physics, **T-symmetry** is the theoretical symmetry of physical laws under a **time reversal** transformation:

$$T : t \mapsto -t.$$

Although in restricted contexts one may find this symmetry, the observable universe itself does not show symmetry under time reversal, primarily due to the second law of thermodynamics. Hence time is said to be non-symmetric, or asymmetric, except for equilibrium states when the second law of thermodynamics predicts the time symmetry to hold. However, quantum noninvasive measurements are predicted to violate time symmetry even in equilibrium,[1] contrary to their classical counterparts, although it has not yet been experimentally confirmed.

Time *asymmetries* are generally distinguished as between those intrinsic to the dynamic physical laws, those due to the initial conditions of our universe, and due to measurements

1. The T-asymmetry of the weak force is of the first kind.

2. The T-asymmetry of the second law of thermodynamics is of the second kind, while

3. The T-asymmetry of the noninvasive measurements is of the third kind.

15.1 Invariance

Physicists also discuss the time-reversal invariance of local and/or macroscopic descriptions of physical systems, independent of the invariance of the underlying microscopic physical laws. For example, Maxwell's equations with material absorption or Newtonian mechanics with friction are not time-reversal invariant at the macroscopic level where they are normally applied, even if they are invariant at the microscopic level; when one includes the atomic motions, the "lost" energy is translated into heat.

15.2 Macroscopic phenomena: the second law of thermodynamics

Our daily experience shows that T-symmetry does not hold for the behavior of bulk materials. Of these macroscopic laws, most notable is the second law of thermodynamics. Many other phenomena, such as the relative motion of bodies with friction, or viscous motion of fluids, reduce to this, because the underlying mechanism is the dissipation of usable energy (for example, kinetic energy) into heat.

The question of whether this time-asymmetric dissipation is really inevitable has been considered by many physicists, often in the context of **Maxwell's demon**. The name comes from a thought experiment described by James Clerk Maxwell in which a microscopic demon guards a gate between two halves of a room. It only lets slow molecules into one half, only fast ones into the other. By eventually making one side of the room cooler than before and the other hotter, it seems to reduce the entropy of the room, and reverse the arrow of time. Many analyses have been made of this; all show that when the entropy of room and demon are taken together, this total entropy does increase. Modern analyses of this problem have taken into account Claude E. Shannon's relation between entropy and information. Many interesting results in modern computing are closely related to this problem — reversible computing, quantum computing and physical limits to computing, are examples. These seemingly metaphysical questions are today, in these ways, slowly being converted to the stuff of the physical sciences.

The current consensus hinges upon the Boltzmann-Shannon identification of the logarithm of phase space volume with the negative of Shannon information, and hence to entropy. In this notion, a fixed initial state of a macroscopic system corresponds to relatively low entropy because the coordinates of the molecules of the body are constrained. As the system evolves in the presence of dissipation, the molecular coordinates can move into larger volumes of phase space, becoming more uncertain, and thus leading to increase in entropy.

One can, however, equally well imagine a state of the universe in which the motions of all of the particles at one instant were the reverse (strictly, the CPT reverse). Such a state would then evolve in reverse, so presumably entropy would decrease (Loschmidt's paradox). Why is 'our' state preferred over the other?

One position is to say that the constant increase of entropy we observe happens *only* because of the initial state of our universe. Other possible states of the universe (for example, a universe at heat death equilibrium) would actually result in no increase of entropy. In this view, the apparent T-asymmetry of our universe is a problem in cosmology: why did the universe start with a low entropy? This view, if it remains viable in the light of future cosmological observation, would connect this problem to one of the big open questions beyond the reach of today's physics — the question of *initial conditions* of the universe.

15.3 Macroscopic phenomena: black holes

An object can cross through the event horizon of a black hole from the outside, and then fall rapidly to the central region where our understanding of physics breaks down. Since within a black hole the forward light-cone is directed towards the center and the backward light-cone is directed outward, it is not even possible to define time-reversal in the usual manner. The only way anything can escape from a black hole is as Hawking radiation.

The time reversal of a black hole would be a hypothetical object known as a white hole. From the outside they appear similar. While a black hole has a beginning and is inescapable, a white hole has an ending and cannot be entered. The forward light-cones of a white hole are directed outward; and its backward light-cones are directed towards the center.

The event horizon of a black hole may be thought of as a surface moving outward at the local speed of light and is just on the edge between escaping and falling back. The event horizon of a white hole is a surface moving inward at the local speed of light and is just on the edge between being swept outward and succeeding in reaching the center. They are two different kinds of horizons—the horizon of a white hole is like the horizon of a black hole turned inside-out.

The modern view of black hole irreversibility is to relate it to the second law of thermodynamics, since black holes are viewed as thermodynamic objects. Indeed, according to the Gauge–gravity duality conjecture, all microscopic processes in a black hole are reversible, and only the collective behavior is irreversible, as in any other macroscopic, thermal system.

15.4 Kinetic consequences: detailed balance and Onsager reciprocal relations

In physical and chemical kinetics, T-symmetry of the mechanical microscopic equations implies two important laws: the principle of detailed balance and the Onsager reciprocal relations. T-symmetry of the microscopic description together with its kinetic consequences are called microscopic reversibility.

15.5 Effect of time reversal on some variables of classical physics

15.5.1 Even

Classical variables that do not change upon time reversal include:

$\vec{r}$, Position of a particle in three-space

$\vec{a}$, Acceleration of the particle

$\vec{F}$, Force on the particle

E, Energy of the particle

ϕ, Electric potential (voltage)

$\vec{E}$, Electric field

$\vec{D}$, Electric displacement

ρ, Density of electric charge

$\vec{P}$, Electric polarization

Energy density of the electromagnetic field

Maxwell stress tensor

All masses, charges, coupling constants, and other physical constants, except those associated with the weak force.

15.5.2 Odd

Classical variables that time reversal negates include:

t, The time when an event occurs

$\vec{v}$, Velocity of a particle

$\vec{p}$, Linear momentum of a particle

$\vec{l}$, Angular momentum of a particle (both orbital and spin)

$\vec{A}$, Electromagnetic vector potential

$\vec{B}$, Magnetic induction

$\vec{H}$, Magnetic field

$\vec{j}$, Density of electric current

$\vec{M}$, Magnetization

$\vec{S}$, Poynting vector

Power (rate of work done).

15.6 Microscopic phenomena: time reversal invariance

Since most systems are asymmetric under time reversal, it is interesting to ask whether there are phenomena that do have this symmetry. In classical mechanics, a velocity v reverses under the operation of T, but an acceleration does not. Therefore, one models dissipative phenomena through terms that are odd in v. However, delicate experiments in which known sources of dissipation are removed reveal that the laws of mechanics are time reversal invariant. Dissipation itself is originated in the second law of thermodynamics.

The motion of a charged body in a magnetic field, B involves the velocity through the Lorentz force term $v \times B$, and might seem at first to be asymmetric under T. A closer look assures us that B also changes sign under time reversal. This happens because a magnetic field is produced by an electric current, J, which reverses sign under T. Thus, the motion of classical charged particles in electromagnetic fields is also time reversal invariant. (Despite this, it is still useful to consider the time-reversal non-invariance in a *local* sense when the external field is held fixed, as when the magneto-optic effect is analyzed. This allows one to analyze the conditions under which optical phenomena that locally break time-reversal, such as Faraday isolators and directional dichroism, can occur.) The laws of gravity also seem to be time reversal invariant in classical mechanics.

In physics one separates the laws of motion, called kinematics, from the laws of force, called dynamics. Following the classical kinematics of Newton's laws of motion, the kinematics of quantum mechanics is built in such a way that it presupposes nothing about the time reversal symmetry of the dynamics. In other words, if the dynamics are invariant, then the kinematics will allow it to remain invariant; if the dynamics is not, then the kinematics will also show this. The structure of the quantum laws of motion are richer, and we examine these next.

15.6.1   Time reversal in quantum mechanics

This section contains a discussion of the three most important properties of time reversal in quantum mechanics; chiefly,

1. that it must be represented as an anti-unitary operator,

2. that it protects non-degenerate quantum states from having an electric dipole moment,

3. that it has two-dimensional representations with the property $T^2 = -1$.

The strangeness of this result is clear if one compares it with parity. If parity transforms a pair of quantum states into each other, then the sum and difference of these two basis states are states of good parity. Time reversal does not behave like this. It seems to violate the theorem that all abelian groups be represented by one-dimensional irreducible representations. The reason it does this is that it is represented by an anti-unitary operator. It thus opens the way to spinors in quantum mechanics.

15.6.2   Anti-unitary representation of time reversal

Eugene Wigner showed that a symmetry operation S of a Hamiltonian is represented, in quantum mechanics either by a **unitary** operator, $S = U$, or an **antiunitary** one, $S = UK$ where U is unitary, and K denotes complex conjugation. These are the only operations that act on Hilbert space so as to preserve the *length* of the projection of any one state-vector onto another state-vector.

Consider the parity operator. Acting on the position, it reverses the directions of space, so that $P^{-1}xP = -x$. Similarly, it reverses the direction of *momentum*, so that $PpP^{-1} = -p$, where x and p are the position and momentum operators. This preserves the canonical commutator $[x, p] = i\hbar$, where $\hbar$ is the reduced Planck constant, only if P is chosen to be unitary, $PiP^{-1} = i$.

On the other hand, for time reversal, the time-component of the momentum is the energy. If time reversal were implemented as a unitary operator, it would reverse the sign of the energy just as space-reversal reverses the sign of the momentum. This is not possible, because, unlike momentum, energy is always positive. Since energy in quantum mechanics is defined as the phase factor $\exp(-iEt)$ that one gets when one moves forward in time, the way to reverse time while preserving the sign of the energy is to reverse the sense of "i", so that the sense of phases is reversed.

Similarly, any operation that reverses the sense of phase, which changes the sign of i, will turn positive energies into negative energies unless it also changes the direction of time. So every antiunitary symmetry in a theory with positive energy must reverse the direction of time. The only antiunitary symmetry is time reversal, together with a unitary symmetry that does not reverse time.

Given the *time reversal* operator T, it does nothing to the x-operator, $TxT^{-1} = x$, but it reverses the direction of p, so that $TpT^{-1} = -p$. The canonical commutator is invariant only if T is chosen to be anti-unitary, i.e., $TiT^{-1} = -i$. For a particle with spin J, one can use the representation

$$T = e^{-i\pi J_y/\hbar} K,$$

where J_y is the y-component of the spin, and use of $TJT^{-1} = -J$ has been made.

15.6.3 Electric dipole moments

This has an interesting consequence on the electric dipole moment (EDM) of any particle. The EDM is defined through the shift in the energy of a state when it is put in an external electric field: $\Delta e = d \cdot E + E \cdot \delta \cdot E$, where d is called the EDM and δ, the induced dipole moment. One important property of an EDM is that the energy shift due to it changes sign under a parity transformation. However, since **d** is a vector, its expectation value in a state $|\psi\rangle$ must be proportional to $\langle\psi| J |\psi\rangle$. Thus, under time reversal, an invariant state must have vanishing EDM. In other words, a non-vanishing EDM signals both P and T symmetry-breaking.[2]

It is interesting to examine this argument further, since one feels that some molecules, such as water, must have EDM irrespective of whether **T** is a symmetry. This is correct: if a quantum system has degenerate ground states that transform into each other under parity, then time reversal need not be broken to give EDM.

Experimentally observed bounds on the electric dipole moment of the nucleon currently set stringent limits on the violation of time reversal symmetry in the strong interactions, and their modern theory: quantum chromodynamics. Then, using the CPT invariance of a relativistic quantum field theory, this puts strong bounds on strong CP violation.

Experimental bounds on the electron electric dipole moment also place limits on theories of particle physics and their parameters.[3] [4]

15.6.4 Kramers' theorem

Main article: Kramers' degeneracy theorem

For T, which is an anti-unitary Z_2 symmetry generator

$$T^2 = UKUK = U U^* = U (U^T)^{-1} = \Phi,$$

where Φ is a diagonal matrix of phases. As a result, $U = \Phi U^T$ and $U^T = U\Phi$, showing that

$$U = \Phi U \Phi.$$

This means that the entries in Φ are ± 1, as a result of which one may have either $T^2 = \pm 1$. This is specific to the anti-unitarity of T. For a unitary operator, such as the parity, any phase is allowed.

Next, take a Hamiltonian invariant under T. Let $|a\rangle$ and $T|a\rangle$ be two quantum states of the same energy. Now, if $T^2 = -1$, then one finds that the states are orthogonal: a result called **Kramers' theorem**. This implies that if $T^2 = -1$, then there is a twofold degeneracy in the state. This result in non-relativistic quantum mechanics presages the spin statistics theorem of quantum field theory.

Quantum states that give unitary representations of time reversal, i.e., have **$T^2 = 1$**, are characterized by a multiplicative quantum number, sometimes called the **T-parity**.

Time reversal transformation for fermions in quantum field theories can be represented by an 8-component spinor in which the above-mentioned **T-parity** can be a complex number with unit radius. The CPT invariance is not a theorem but a **better to have** property in these class of theories.

15.6.5   Time reversal of the known dynamical laws

Particle physics codified the basic laws of dynamics into the standard model. This is formulated as a quantum field theory that has CPT symmetry, i.e., the laws are invariant under simultaneous operation of time reversal, parity and charge conjugation. However, time reversal itself is seen not to be a symmetry (this is usually called CP violation). There are two possible origins of this asymmetry, one through the mixing of different flavours of quarks in their weak decays, the second through a direct CP violation in strong interactions. The first is seen in experiments, the second is strongly constrained by the non-observation of the EDM of a neutron.

It is important to stress that this time reversal violation is unrelated to the second law of thermodynamics, because due to the conservation of the CPT symmetry, the effect of time reversal is to rename particles as antiparticles and *vice versa*. Thus the second law of thermodynamics is thought to originate in the initial conditions in the universe.

15.6.6   Time reversal of noninvasive measurements

Strong measurements (both classical and quantum) are certainly disturbing, causing asymmetry due to second law of thermodynamics. However, noninvasive measurements should not disturb the evolution so they are expected to be time-symmetric. Surprisingly, it is true only in classical physics but not quantum, even in a thermodynamically invariant equilibrium state. [1] This type of asymmetry is independent of CPT symmetry but has not yet been confirmed experimentally due to extreme conditions of the checking proposal.

15.7   See also

- The second law of thermodynamics, Maxwell's demon and the arrow of time (also Loschmidt's paradox).

- Microscopic reversibility

- Detailed balance

- Applications to reversible computing and quantum computing, including limits to computing.

- The standard model of particle physics, CP violation, the CKM matrix and the strong CP problem

- Neutrino masses and CPT invariance.

- Wheeler–Feynman absorber theory

- Teleonomy

15.8   References

[1] Bednorz, Adam; Franke, Kurt; Belzig, Wolfgang (February 2013). "Noninvasiveness and time symmetry of weak measurements". *New Journal of Physics* **15**: 023043. Bibcode:2013NJPh...15b3043B. doi:10.1088/1367-2630/15/2/023043.

[2] Khriplovich, Iosip B.; Lamoreaux, Steve K. (2012). *CP violation without strangeness : electric dipole moments of particles, atoms, and molecules.* [S.l.]: Springer. ISBN 978-3-642-64577-8.

[3] Ibrahim, Tarik; Itani, Ahmad; Nath, Pran (12 Aug 2014). "Electron EDM as a Sensitive Probe of PeV Scale Physics". *arXiv*: 1406.0083.

[4] Kim, Jihn E.; Carosi, Gianpaolo (4 March 2010). "Axions and the strong CP problem". *Reviews of Modern Physics* **82**: 557. arXiv:0807.3125. Bibcode:2010RvMP...82..557K. doi:10.1103/RevModPhys.82.557.

- Maxwell's demon: entropy, information, computing, edited by H.S.Leff and A.F. Rex (IOP publishing, 1990) [ISBN 0-7503-0057-4]

- Maxwell's demon, 2: entropy, classical and quantum information, edited by H.S.Leff and A.F. Rex (IOP publishing, 2003) [ISBN 0-7503-0759-5]

- The emperor's new mind: concerning computers, minds, and the laws of physics, by Roger Penrose (Oxford university press, 2002) [ISBN 0-19-286198-0]

- Sozzi, M.S. (2008). *Discrete symmetries and CP violation*. Oxford University Press. ISBN 978-0-19-929666-8.

- Birss, R. R. (1964). *Symmetry and Magnetism*. John Wiley & Sons, Inc., New York.

- Multiferroic materials with time-reversal breaking optical properties

- CP violation, by I.I. Bigi and A.I. Sanda (Cambridge University Press, 2000) [ISBN 0-521-44349-0]

- Particle Data Group on CP violation

    - the Babar experiment in SLAC
    - the BELLE experiment in KEK
    - the KTeV experiment in Fermilab
    - the CPLEAR experiment in CERN

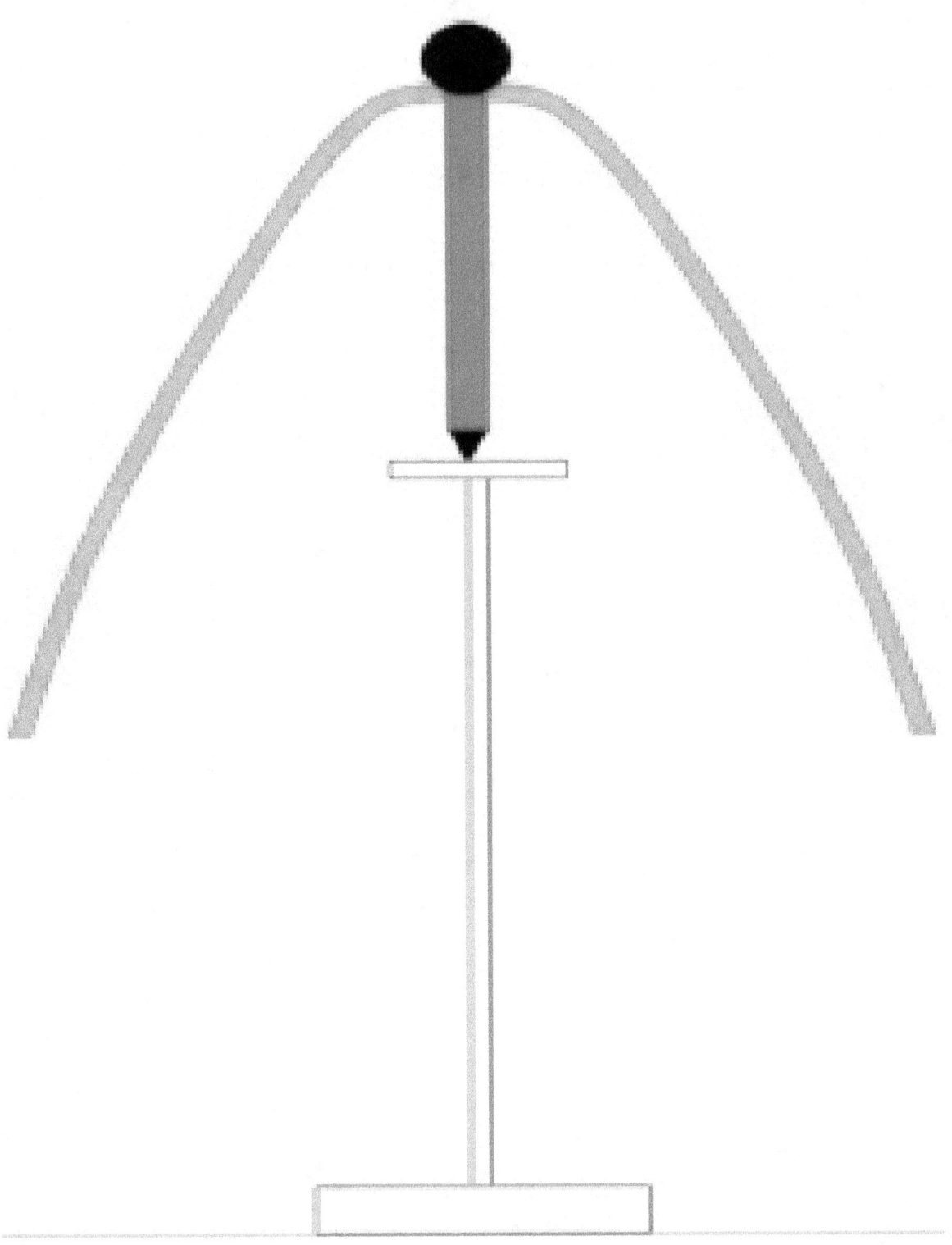

A toy called the teeter-totter illustrates the two aspects of time reversal invariance. When set into motion atop a pedestal, the figure oscillates for a very long time. The toy is engineered to minimize friction and illustrate the reversibility of Newton's laws of motion. However, the mechanically stable state of the toy is when the figure falls down from the pedestal into one of arbitrarily many positions. This is an illustration of the law of increase of entropy through Boltzmann's identification of the logarithm of the number of states with the entropy.

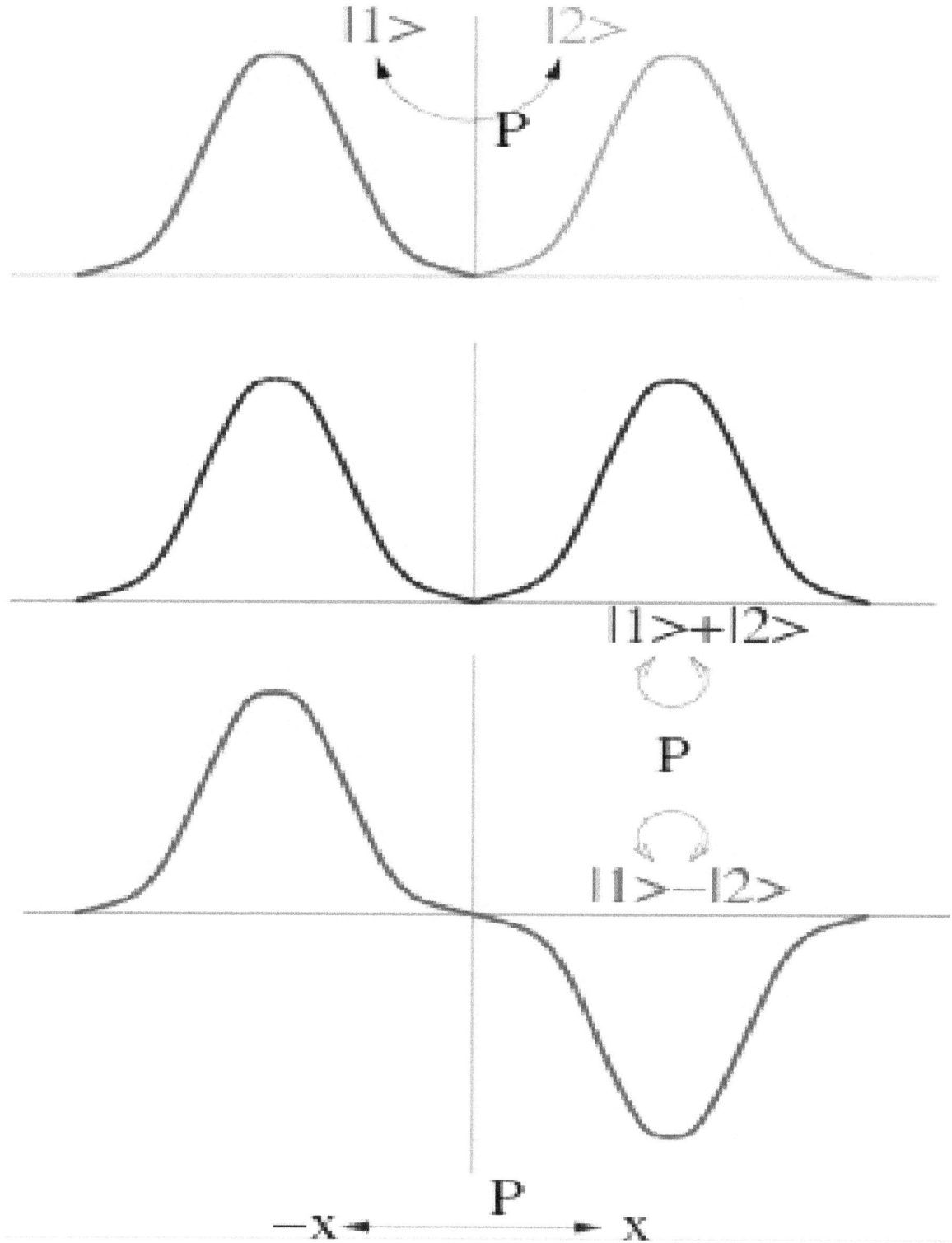

Two-dimensional representations of parity are given by a pair of quantum states that go into each other under parity. However, this representation can always be reduced to linear combinations of states, each of which is either even or odd under parity. One says that all irreducible representations of parity are one -dimensional. Kramers' theorem states that time reversal need not have this property because it is represented by an anti-unitary operator.

Chapter 16

Parity (physics)

In quantum mechanics, a **parity transformation** (also called **parity inversion**) is the flip in the sign of *one* spatial coordinate. In three dimensions, it is also often described by the simultaneous flip in the sign of all three spatial coordinates (a point reflection):

$$\mathbf{P} : \begin{pmatrix} x \\ y \\ z \end{pmatrix} \mapsto \begin{pmatrix} -x \\ -y \\ -z \end{pmatrix}.$$

It can also be thought of as a test for chirality of a physical phenomenon, in that a parity inversion transforms a phenomenon into its mirror image. A parity transformation on something achiral, on the other hand, can be viewed as an identity transformation. All fundamental interactions of elementary particles, with the exception of the weak interaction, are symmetric under parity. The weak interaction is chiral and thus provides a means for probing chirality in physics. In interactions that are symmetric under parity, such as electromagnetism in atomic and molecular physics, parity serves as a powerful controlling principle underlying quantum transitions.

A matrix representation of **P** (in any number of dimensions) has determinant equal to −1, and hence is distinct from a rotation, which has a determinant equal to 1. In a two-dimensional plane, a simultaneous flip of all coordinates in sign is *not* a parity transformation; it is the same as a 180°-rotation.

16.1   Simple symmetry relations

Under rotations, classical geometrical objects can be classified into scalars, vectors, and tensors of higher rank. In classical physics, physical configurations need to transform under representations of every symmetry group.

Quantum theory predicts that states in a Hilbert space do not need to transform under representations of the group of rotations, but only under projective representations. The word *projective* refers to the fact that if one projects out the phase of each state, where we recall that the overall phase of a quantum state is not an observable, then a projective representation reduces to an ordinary representation. All representations are also projective representations, but the converse is not true, therefore the projective representation condition on quantum states is weaker than the representation condition on classical states.

The projective representations of any group are isomorphic to the ordinary representations of a central extension of the group. For example, projective representations of the 3-dimensional rotation group, which is the special orthogonal group SO(3), are ordinary representations of the special unitary group SU(2) (see Representation theory of SU(2)). Projective representations of the rotation group that are not representations are called spinors, and so quantum states may transform not only as tensors but also as spinors.

If one adds to this a classification by parity, these can be extended, for example, into notions of

- *scalars* ($P = 1$) and *pseudoscalars* ($P = -1$) which are rotationally invariant.

- *vectors* ($P = -1$) and *axial vectors* (also called *pseudovectors*) ($P = 1$) which both transform as vectors under rotation.

One can define **reflections** such as

$$V_x : \begin{pmatrix} x \\ y \\ z \end{pmatrix} \mapsto \begin{pmatrix} -x \\ y \\ z \end{pmatrix},$$

which also have negative determinant and form a valid parity transformation. Then, combining them with rotations (or successively performing x-, y-, and z-reflections) one can recover the particular parity transformation defined earlier. The first parity transformation given does not work in an even number of dimensions, though, because it results in a positive determinant. In odd number of dimensions only the latter example of a parity transformation (or any reflection of an odd number of coordinates) can be used.

Parity forms the abelian group Z_2 due to the relation $\mathbf{P}^2 = 1$. All Abelian groups have only one-dimensional irreducible representations. For Z_2, there are two irreducible representations: one is even under parity ($\mathbf{P}\varphi = \varphi$), the other is odd ($\mathbf{P}\varphi = -\varphi$). These are useful in quantum mechanics. However, as is elaborated below, in quantum mechanics states need not transform under actual representations of parity but only under projective representations and so in principle a parity transformation may rotate a state by any phase.

16.2 Classical mechanics

Newton's equation of motion $\mathbf{F} = m\mathbf{a}$ (if the mass is constant) equates two vectors, and hence is invariant under parity. The law of gravity also involves only vectors and is also, therefore, invariant under parity.

However, angular momentum $\mathbf{L}$ is an axial vector,

$$\mathbf{L} = \mathbf{r} \times \mathbf{p},$$
$$P(\mathbf{L}) = (-\mathbf{r}) \times (-\mathbf{p}) = \mathbf{L}.$$

In classical electrodynamics, the charge density ρ is a scalar, the electric field, $\mathbf{E}$, and current $\mathbf{j}$ are vectors, but the magnetic field, $\mathbf{H}$ is an axial vector. However, Maxwell's equations are invariant under parity because the curl of an axial vector is a vector.

16.3 Effect of spatial inversion on some variables of classical physics

16.3.1 Even

Classical variables, predominantly scalar quantities, which do not change upon spatial inversion include:

t, the time when an event occurs

m, the mass of a particle

E, the energy of the particle

P, power (rate of work done)

ρ, the electric charge density

V, the electric potential (voltage)

ρ, energy density of the electromagnetic field

L , the angular momentum of a particle (both orbital and spin) (axial vector)

B , the magnetic field (axial vector)

H , the auxiliary magnetic field

M , the magnetization

T_{ij} Maxwell stress tensor.

All masses, charges, coupling constants, and other physical constants, except those associated with the weak force

16.3.2   Odd

Classical variables, predominantly vector quantities, which have their sign flipped by spatial inversion include:

h , the helicity

Φ , the magnetic flux

x , the position of a particle in three-space

v , the velocity of a particle

a , the acceleration of the particle

p , the linear momentum of a particle

F , the force exerted on a particle

J , the electric current density

E , the electric field

D , the electric displacement field

P , the electric polarization

A , the electromagnetic vector potential

S , Poynting vector.

16.4   Quantum mechanics

16.4.1   Possible eigenvalues

In quantum mechanics, spacetime transformations act on quantum states. The parity transformation, **P**, is a unitary operator, in general acting on a state ψ as follows: $\mathbf{P}\psi(r) = e^{i\varphi/2}\psi(-r)$.

One must then have $\mathbf{P}^2\psi(r) = e^{i\varphi}\psi(r)$, since an overall phase is unobservable. The operator $\mathbf{P}^2$, which reverses the parity of a state twice, leaves the spacetime invariant, and so is an internal symmetry which rotates its eigenstates by phases $e^{i\varphi}$. If $\mathbf{P}^2$ is an element e^{iQ} of a continuous U(1) symmetry group of phase rotations, then $e^{-iQ/2}$ is part of this U(1) and so is also a symmetry. In particular, we can define $\mathbf{P'} = \mathbf{P}e^{-iQ/2}$, which is also a symmetry, and so we can choose to call $\mathbf{P'}$ our parity operator, instead of **P**. Note that $\mathbf{P'}^2 = 1$ and so $\mathbf{P'}$ has eigenvalues ± 1. However, when no such symmetry group exists, it may be that all parity transformations have some eigenvalues which are phases other than ± 1.

For electronic wavefunctions, even states are usually indicated by a subscript g for *gerade* (German: even) and odd states by a subscript u for *ungerade* (German: odd). For example, the lowest energy level of the hydrogen molecule ion (H_2^+) is labelled $1\sigma_g$ and the next-lowest $1\sigma_u$.[1]

16.4.2 Consequences of parity symmetry

When parity generates the Abelian group $\mathbb{Z}_2$, one can always take linear combinations of quantum states such that they are either even or odd under parity (see the figure). Thus the parity of such states is ± 1. The parity of a multiparticle state is the product of the parities of each state: in other words parity is a multiplicative quantum number

In quantum mechanics, Hamiltonians are invariant (symmetric) under a parity transformation if **P** commutes with the Hamiltonian. In non-relativistic quantum mechanics, this happens for any potential which is scalar, i.e., $V = V(r)$, hence the potential is spherically symmetric. The following facts can be easily proven:

- If $|A\rangle$ and $|B\rangle$ have the same parity, then $\langle A|\,\mathbf{X}\,|B\rangle = 0$ where **X** is the position operator.
- For a state $|L, L_z\rangle$ of orbital angular momentum **L** with z-axis projection L_z, $\mathbf{P}|L, L_z\rangle = (-1)^L|L, L_z\rangle$.
- If $[\mathbf{H}, \mathbf{P}] = 0$, then atomic dipole transitions only occur between states of opposite parity.[2]
- If $[\mathbf{H}, \mathbf{P}] = 0$, then a non-degenerate eigenstate of **H** is also an eigenstate of the parity operator; i.e., a non-degenerate eigenfunction of **H** is either invariant to **P** or is changed in sign by **P**.

Some of the non-degenerate eigenfunctions of **H** are unaffected (invariant) by parity **P** and the others will be merely reversed in sign when the Hamiltonian operator and the parity operator commute:

$$\mathbf{P}\,\Psi = c\,\Psi,$$

where c is a constant, the eigenvalue of **P**,

$$\mathbf{P}^2\Psi = c\mathbf{P}\,\Psi.$$

16.5 Quantum field theory

The intrinsic parity assignments in this section are true for relativistic quantum mechanics as well as quantum field theory.

If we can show that the vacuum state is invariant under parity ($\mathbf{P}|0\rangle = |0\rangle$), the Hamiltonian is parity invariant ($[\mathbf{H}, \mathbf{P}] = 0$) and the quantization conditions remain unchanged under parity, then it follows that every state has good parity, and this parity is conserved in any reaction.

To show that quantum electrodynamics is invariant under parity, we have to prove that the action is invariant and the quantization is also invariant. For simplicity we will assume that canonical quantization is used; the vacuum state is then invariant under parity by construction. The invariance of the action follows from the classical invariance of Maxwell's equations. The invariance of the canonical quantization procedure can be worked out, and turns out to depend on the transformation of the annihilation operator:

$$\mathbf{P}a(\mathbf{p}, \pm)\mathbf{P}^+ = -a(-\mathbf{p}, \pm)$$

where **p** denotes the momentum of a photon and $\pm$ refers to its polarization state. This is equivalent to the statement that the photon has odd intrinsic parity. Similarly all vector bosons can be shown to have odd intrinsic parity, and all axial-vectors to have even intrinsic parity.

There is a straightforward extension of these arguments to scalar field theories which shows that scalars have even parity, since

$$\mathbf{P}a(\mathbf{p})\mathbf{P}^+ = a(-\mathbf{p}).$$

This is true even for a complex scalar field. (*Details of spinors are dealt with in the article on the* Dirac equation, *where it is shown that fermions and antifermions have opposite intrinsic parity.*)

With fermions, there is a slight complication because there is more than one spin group.

16.6 Parity in the standard model

16.6.1 Fixing the global symmetries

See also: $(-1)^F$

In the Standard Model of fundamental interactions there are precisely three global internal U(1) symmetry groups available, with charges equal to the baryon number B, the lepton number L and the electric charge Q. The product of the parity operator with any combination of these rotations is another parity operator. It is conventional to choose one specific combination of these rotations to define a standard parity operator, and other parity operators are related to the standard one by internal rotations. One way to fix a standard parity operator is to assign the parities of three particles with linearly independent charges B, L and Q. In general one assigns the parity of the most common massive particles, the proton, the neutron and the electron, to be +1.

Steven Weinberg has shown that if $\mathbf{P}^2 = (-1)^F$, where F is the fermion number operator, then, since the fermion number is the sum of the lepton number plus the baryon number, $F = B + L$, for all particles in the Standard Model and since lepton number and baryon number are charges Q of continuous symmetries e^{iQ}, it is possible to redefine the parity operator so that $\mathbf{P}^2 = 1$. However, if there exist Majorana neutrinos, which experimentalists today believe is quite possible, their fermion number is equal to one because they are neutrinos while their baryon and lepton numbers are zero because they are Majorana, and so $(-1)^F$ would not be embedded in a continuous symmetry group. Thus Majorana neutrinos would have parity $\pm i$.

16.6.2 Parity of the pion

In 1954, a paper by William Chinowsky and Jack Steinberger demonstrated that the pion has negative parity.[3] They studied the decay of an "atom" made from a deuteron (2
1H+) and a negatively charged pion ($\pi-$) in a state with zero orbital angular momentum $L = 0$ into two neutrons (n).

Neutrons are fermions and so obey Fermi–Dirac statistics, which implies that the final state is antisymmetric. Using the fact that the deuteron has spin one and the pion spin zero together with the antisymmetry of the final state they concluded that the two neutrons must have orbital angular momentum $L = 1$. The total parity is the product of the intrinsic parities of the particles and the extrinsic parity of the spherical harmonic function $(-1)^L$. Since the orbital momentum changes from zero to one in this process, if the process is to conserve the total parity then the products of the intrinsic parities of the initial and final particles must have opposite sign. A deuteron nucleus is made from a proton and a neutron, and so using the aforementioned convention that protons and neutrons have intrinsic parities equal to +1 they argued that the parity of the pion is equal to minus the product of the parities of the two neutrons divided by that of the proton and neutron in the deuteron, $(-1)(1)^2/(1)^2$, which is equal to minus one. Thus they concluded that the pion is a pseudoscalar particle.

16.6.3 Parity violation

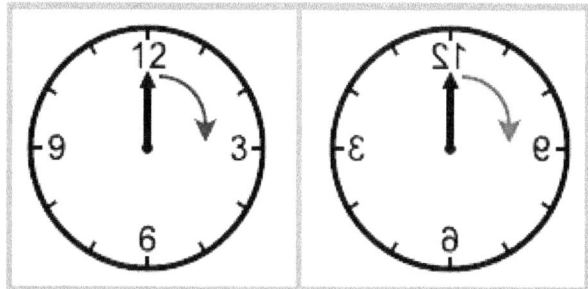

Top: P-symmetry: A clock built like its mirrored image will behave like the mirrored image of the original clock.
Bottom: P-asymmetry: A clock built like its mirrored image will *not* behave like the mirrored image of the original clock.

Although parity is conserved in electromagnetism, strong interactions and gravity, it turns out to be violated in weak interactions. The Standard Model incorporates **parity violation** by expressing the weak interaction as a chiral gauge interaction. Only the left-handed components of particles and right-handed components of antiparticles participate in weak interactions in the Standard Model. This implies that parity is not a symmetry of our universe, unless a hidden mirror sector exists in which parity is violated in the opposite way.

By the mid-20th Century, it had been suggested by several scientists that parity might not be conserved (in different contexts), but without solid evidence these suggestions were not considered important. Then, in 1956, a careful review and analysis by theoretical physicists Tsung Dao Lee and Chen Ning Yang[4] went further, showing that while parity conservation had been verified in decays by the strong or electromagnetic interactions, it was untested in the weak interaction. They proposed several possible direct experimental tests. They were mostly ignored, but Lee was able to convince his Columbia colleague Chien-Shiung Wu to try it. She needed special cryogenic facilities and expertise, so the experiment was done at the National Bureau of Standards.

In 1957 C. S. Wu, E. Ambler, R. W. Hayward, D. D. Hoppes, and R. P. Hudson found a clear violation of parity conservation in the beta decay of cobalt-60.[5] As the experiment was winding down, with double-checking in progress, Wu informed Lee and Yang of their positive results, and saying the results need further examination, she asked them not to publicize the results first. However, Lee revealed the results to his Columbia colleagues on 4 January 1957 at a "Friday Lunch" gathering of the Physics Department of Columbia. Three of them, R. L. Garwin, Leon Lederman, and R. Weinrich modified an existing cyclotron experiment, and they immediately verified the parity violation.[6] They delayed publication of their results until after Wu's group was ready, and the two papers appeared back to back in the same physics journal.

After the fact, it was noted that an obscure 1928 experiment had in effect reported parity violation in weak decays, but since the appropriate concepts had not yet been developed, those results had no impact.[7] The discovery of parity violation immediately explained the outstanding τ–θ puzzle in the physics of kaons.

In 2010, it was reported that physicists working with the Relativistic Heavy Ion Collider (RHIC) had created a short-lived parity symmetry-breaking bubble in quark-gluon plasmas. An experiment conducted by several physicists including Yale's Jack Sandweiss as part of the STAR collaboration, suggested that parity may also be violated in the strong interaction.[8]

16.6.4 Intrinsic parity of hadrons

To every particle one can assign an **intrinsic parity** as long as nature preserves parity. Although weak interactions do not, one can still assign a parity to any hadron by examining the strong interaction reaction that produces it, or through decays not involving the weak interaction, such as rho meson decay to pions.

16.7 See also

- Electroweak theory

- Standard Model

- Mirror matter

16.8 References

General

- Perkins, Donald H. (2000). *Introduction to High Energy Physics*. ISBN 9780521621960.

- Sozzi, M. S. (2008). *Discrete symmetries and CP violation*. Oxford University Press. ISBN 978-0-19-929666-8.

- Bigi, I. I.; Sanda, A. I. (2000). *CP Violation*. Cambridge Monographs on Particle Physics, Nuclear Physics and Cosmology. Cambridge University Press. ISBN 0-521-44349-0.

- Weinberg, S. (1995). *The Quantum Theory of Fields*. Cambridge University Press. ISBN 0-521-67053-5.

Specific

[1] Levine, I.N. *Quantum Chemistry* (Prentice-Hall, 4th edn. 1991), p.355

[2] Bransden, B. H.; Joachain, C. J. (2003). *Physics of Atoms and Molecules* (2nd ed.). Prentice Hall. p. 204. ISBN 978-0-582-35692-4.

[3] Chinowsky, W.; Steinberger, J. (1954). "Absorption of Negative Pions in Deuterium: Parity of the Pion". *Physical Review* **95** (6): 1561–1564. Bibcode:1954PhRv...95.1561C. doi:10.1103/PhysRev.95.1561.

[4] Lee, T. D.; Yang, C. N. (1956). "Question of Parity Conservation in Weak Interactions". *Physical Review* **104** (1): 254–258. Bibcode:1956PhRv..104..254L. doi:10.1103/PhysRev.104.254.

[5] Wu, C. S.; Ambler, E; Hayward, R. W.; Hoppes, D. D.; Hudson, R. P. (1957). "Experimental Test of Parity Conservation in Beta Decay". *Physical Review* **105** (4): 1413–1415. Bibcode:1957PhRv..105.1413W. doi:10.1103/PhysRev.105.1413.

[6] Garwin, R. L.; Lederman, L. M.; Weinrich, M. (1957). "Observations of the Failure of Conservation of Parity and Charge Conjugation in Meson Decays: The Magnetic Moment of the Free Muon". *Physical Review* **105** (4): 1415–1417. Bibcode:1955G.doi:10.1103/PhysRev.105.1415.

[7] Roy, A. (2005). "Discovery of parity violation". *Resonance* **10** (12): 164–175. doi:10.1007/BF02835140.

[8] Muzzin, S. T. (19 March 2010). "For One Tiny Instant, Physicists May Have Broken a Law of Nature". *PhysOrg*. Retrieved 2011-08-05.

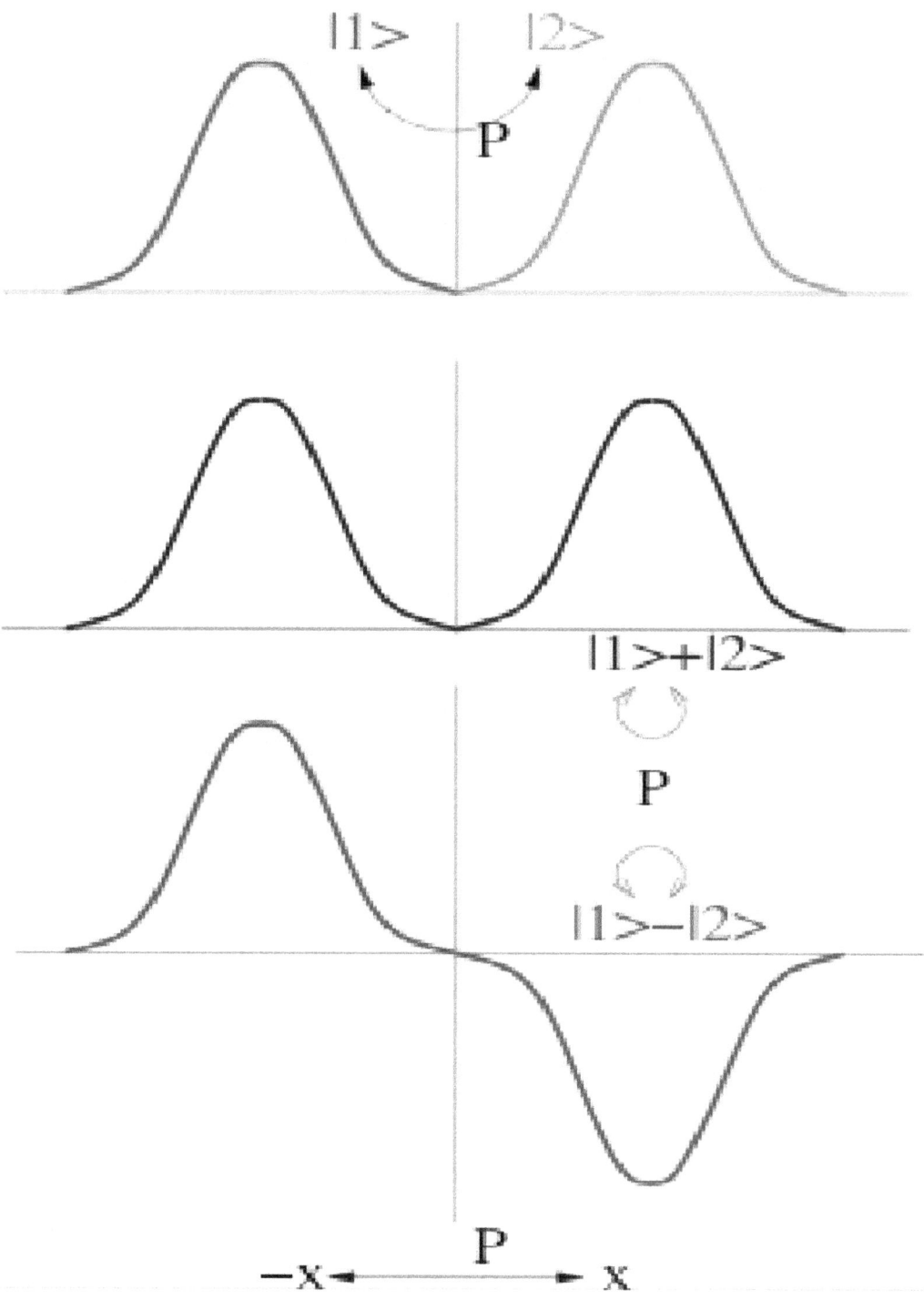

Two-dimensional representations of parity are given by a pair of quantum states that go into each other under parity. However, this representation can always be reduced to linear combinations of states, each of which is either even or odd under parity. One says that all irreducible representations of parity are one-dimensional. Kramers' theorem states that time reversal need not have this property because it is represented by an anti-unitary operator.

Chapter 17

Glide reflection

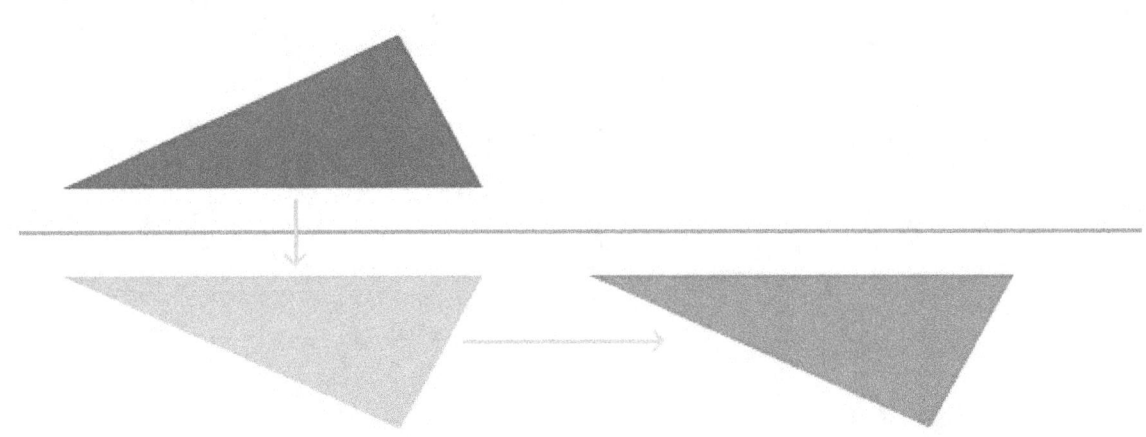

Example of a glide reflection: A composite of a reflection across a line and a translation parallel to the line of reflection

A glide reflection will map a set of left and right footprints into each other

In 2-dimensional geometry, a **glide reflection** (or **transflection**) is a type of opposite isometry of the Euclidean plane: the combination of a reflection in a line and a translation along that line.

A single glide is represented as frieze group p11g. A glide reflection can be seen as a limiting rotoreflection, where the rotation becomes a translation. It can also be given a Schoenflies notation as $S_{2\infty}$, Coxeter notation as $[\infty^+,2^+]$, and orbifold notation as $\infty\times$.

17.1 Description

The combination of a reflection in a line and a translation in a perpendicular direction is a reflection in a parallel line. However, a glide reflection cannot be reduced like that. Thus the effect of a reflection combined with *any* translation is a glide reflection, with as special case just a reflection. These are the two kinds of indirect isometries in 2D.

For example, there is an isometry consisting of the reflection on the x-axis, followed by translation of one unit parallel to it. In coordinates, it takes

$$(x, y) \rightarrow (x + 1, -y).$$

It fixes a system of parallel lines.

The isometry group generated by just a glide reflection is an infinite cyclic group.[1]

Combining two equal glide reflections gives a pure translation with a translation vector that is twice that of the glide reflection, so the even powers of the glide reflection form a translation group.

In the case of **glide reflection symmetry**, the symmetry group of an object contains a glide reflection, and hence the group generated by it. If that is all it contains, this type is frieze group p11g.

Example pattern with this symmetry group:

Frieze group nr. 6 (glide-reflections, translations and rotations) is generated by a glide reflection and a rotation about a point on the line of reflection. It is isomorphic to a semi-direct product of **Z** and C_2.

Example pattern with this symmetry group:

A typical example of glide reflection in everyday life would be the track of footprints left in the sand by a person walking on a beach.

For any symmetry group containing some glide reflection symmetry, the translation vector of any glide reflection is one half of an element of the translation group. If the translation vector of a glide reflection is itself an element of the translation group, then the corresponding glide reflection symmetry reduces to a combination of reflection symmetry and translational symmetry.

Glide reflection symmetry with respect to two parallel lines with the same translation implies that there is also translational symmetry in the direction perpendicular to these lines, with a translation distance which is twice the distance between glide reflection lines. This corresponds to wallpaper group pg; with additional symmetry it occurs also in pmg, pgg and p4g.

If there are also true reflection lines in the same direction then they are evenly spaced between the glide reflection lines. A glide reflection line parallel to a true reflection line already implies this situation. This corresponds to wallpaper group cm. The translational symmetry is given by oblique translation vectors from one point on a true reflection line to two points on the next, supporting a rhombus with the true reflection line as one of the diagonals. With additional symmetry it occurs also in cmm, p3m1, p31m, p4m and p6m.

In 3D the glide reflection is called a **glide plane**. It is a reflection in a plane combined with a translation parallel to the plane.

17.2   Wallpaper groups

In the Euclidean plane 3 of 17 wallpaper groups require glide reflection generators. p2gg has orthogonal glide reflections and 2-fold rotations. cm has parallel mirrors and glides, and pg has parallel glides. (Glide reflections are shown below as dashed lines)

17.3   Glide reflection in nature and games

Glide symmetry can be observed in nature among certain fossils of the Ediacara biota; the machaeridians; and certain palaeoscolecid worms.[2]

Glide reflection is common in Conway's Game of Life.

17.4   See also

- Screw axis, glide plane for the corresponding 3D symmetry operations

17.5   References

[1] Martin, George E. (1982), *Transformation Geometry: An Introduction to Symmetry*, Undergraduate Texts in Mathematics, Springer, p. 64, ISBN 9780387906362.

[2] Waggoner, B. M. (1996). "Phylogenetic Hypotheses of the Relationships of Arthropods to Precambrian and Cambrian Problematic Fossil Taxa". *Systematic Biology* **45** (2): 190–222. doi:10.2307/2413615. JSTOR 2413615.

17.6   External links

- Glide Reflection at cut-the-knot

Chapter 18

Wallpaper group

A **wallpaper group** (or **plane symmetry group** or **plane crystallographic group**) is a mathematical classification of a two-dimensional repetitive pattern, based on the symmetries in the pattern. Such patterns occur frequently in architecture and decorative art. There are 17 possible distinct groups.

Wallpaper groups are two-dimensional symmetry groups, intermediate in complexity between the simpler frieze groups and the three-dimensional crystallographic groups (also called space groups).

18.1 Introduction

Wallpaper groups categorize patterns by their symmetries. Subtle differences may place similar patterns in different groups, while patterns that are very different in style, color, scale or orientation may belong to the same group.

Consider the following examples:

- Example **A**: Cloth, Tahiti
- Example **B**: Ornamental painting, Nineveh, Assyria
- Example **C**: Painted porcelain, China

Examples **A** and **B** have the same wallpaper group; it is called **p4mm** in the IUC notation and *442 in the orbifold notation. Example **C** has a different wallpaper group, called **p4mg** or 4*2 . The fact that **A** and **B** have the same wallpaper group means that they have the same symmetries, regardless of details of the designs, whereas **C** has a different set of symmetries despite any superficial similarities.

A complete list of all seventeen possible wallpaper groups can be found below.

18.1.1 Symmetries of patterns

A symmetry of a pattern is, loosely speaking, a way of transforming the pattern so that the pattern looks exactly the same after the transformation. For example, translational symmetry is present when the pattern can be translated (shifted) some finite distance and appear unchanged. Think of shifting a set of vertical stripes horizontally by one stripe. The pattern is unchanged. Strictly speaking, a true symmetry only exists in patterns that repeat exactly and continue indefinitely. A set of only, say, five stripes does not have translational symmetry — when shifted, the stripe on one end "disappears" and a new stripe is "added" at the other end. In practice, however, classification is applied to finite patterns, and small imperfections may be ignored.

Sometimes two categorizations are meaningful, one based on shapes alone and one also including colors. When colors are ignored there may be more symmetry. In black and white there are also 17 wallpaper groups; e.g., a colored tiling is

111

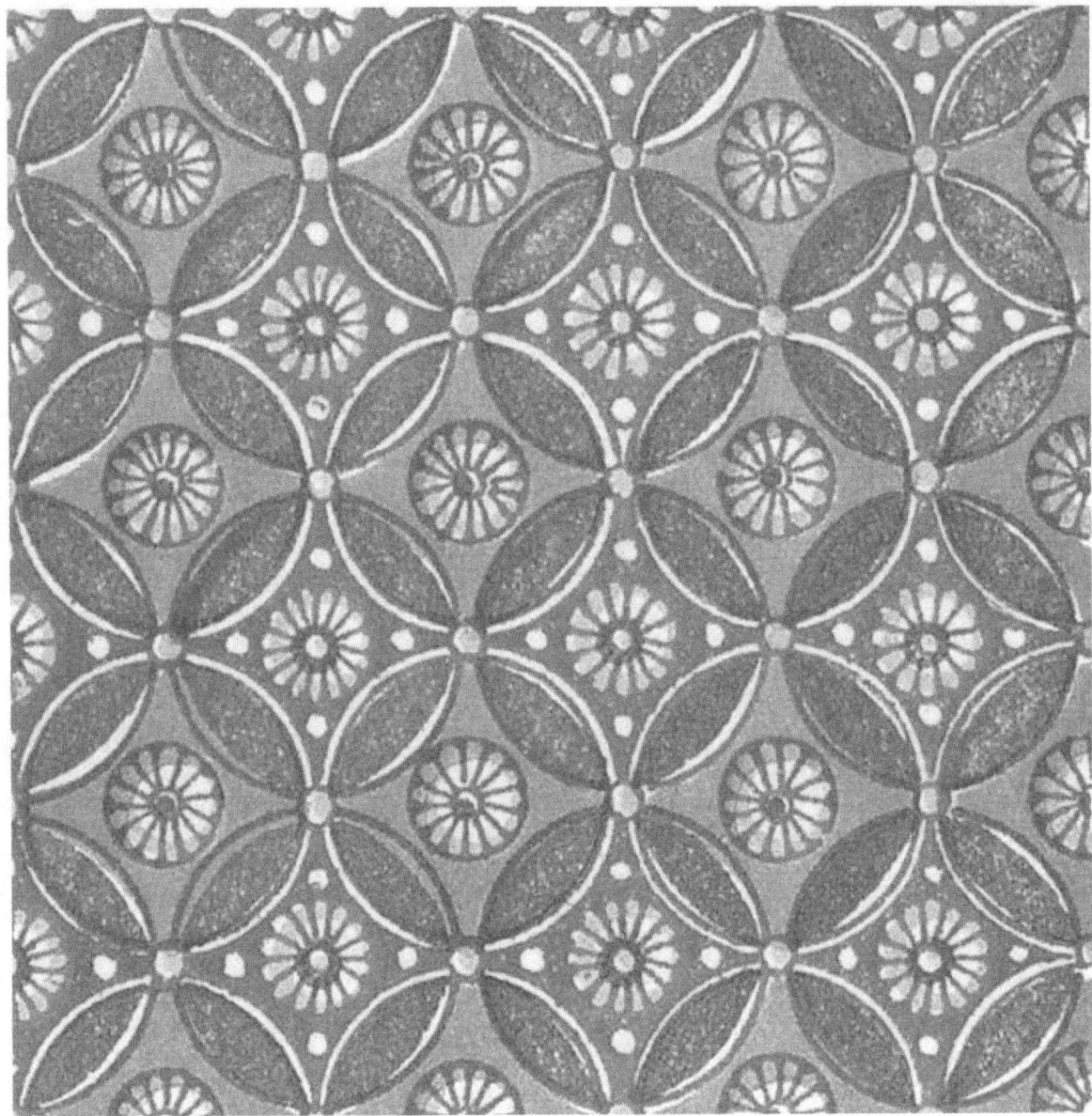

*Example of an Egyptian design with wallpaper group **p4mm***

equivalent with one in black and white with the colors coded radially in a circularly symmetric "bar code" in the centre of mass of each tile.

The types of transformations that are relevant here are called Euclidean plane isometries. For example:

- If we *shift* example **B** one unit to the right, so that each square covers the square that was originally adjacent to it, then the resulting pattern is *exactly the same* as the pattern we started with. This type of symmetry is called a **translation**. Examples **A** and **C** are similar, except that the smallest possible shifts are in diagonal directions.

- If we *turn* example **B** clockwise by 90°, around the centre of one of the squares, again we obtain exactly the same pattern. This is called a **rotation**. Examples **A** and **C** also have 90° rotations, although it requires a little more ingenuity to find the correct centre of rotation for **C**.

- We can also *flip* example **B** across a horizontal axis that runs across the middle of the image. This is called a **reflection**. Example **B** also has reflections across a vertical axis, and across two diagonal axes. The same can be said for **A**.

However, example **C** is *different*. It only has reflections in horizontal and vertical directions, *not* across diagonal axes. If we flip across a diagonal line, we do *not* get the same pattern back; what we *do* get is the original pattern shifted across by a certain distance. This is part of the reason that the wallpaper group of **A** and **B** is different from the wallpaper group of **C**.

18.2   History

A proof that there were only 17 possible patterns was first carried out by Evgraf Fedorov in 1891[1] and then derived independently by George Pólya in 1924.[2] The proof that the list of wallpaper groups was complete only came after the much harder case of space groups had been done.

18.3   Formal definition and discussion

Mathematically, a wallpaper group or plane crystallographic group is a type of topologically discrete group of isometries of the Euclidean plane that contains two linearly independent translations.

Two such isometry groups are of the same type (of the same wallpaper group) if they are the same up to an affine transformation of the plane. Thus e.g. a translation of the plane (hence a translation of the mirrors and centres of rotation) does not affect the wallpaper group. The same applies for a change of angle between translation vectors, provided that it does not add or remove any symmetry (this is only the case if there are no mirrors and no glide reflections, and rotational symmetry is at most of order 2).

Unlike in the three-dimensional case, we can equivalently restrict the affine transformations to those that preserve orientation.

It follows from the Bieberbach theorem that all wallpaper groups are different even as abstract groups (as opposed to e.g. frieze groups, of which two are isomorphic with **Z**).

2D patterns with double translational symmetry can be categorized according to their symmetry group type.

18.3.1   Isometries of the Euclidean plane

Isometries of the Euclidean plane fall into four categories (see the article Euclidean plane isometry for more information).

- **Translations**, denoted by Tv, where v is a vector in $\mathbf{R}^2$. This has the effect of shifting the plane applying displacement vector v.

- **Rotations**, denoted by Rc,θ, where c is a point in the plane (the centre of rotation), and θ is the angle of rotation.

- **Reflections**, or **mirror isometries**, denoted by FL, where L is a line in $\mathbf{R}^2$. (F is for "flip"). This has the effect of reflecting the plane in the line L, called the **reflection axis** or the associated **mirror**.

- **Glide reflections**, denoted by GL,d, where L is a line in $\mathbf{R}^2$ and d is a distance. This is a combination of a reflection in the line L and a translation along L by a distance d.

18.3.2   The independent translations condition

The condition on linearly independent translations means that there exist linearly independent vectors v and w (in $\mathbf{R}^2$) such that the group contains both Tv and Tw.

The purpose of this condition is to distinguish wallpaper groups from frieze groups, which possess a translation but not two linearly independent ones, and from two-dimensional discrete point groups, which have no translations at all. In other words, wallpaper groups represent patterns that repeat themselves in *two* distinct directions, in contrast to frieze groups, which only repeat along a single axis.

(It is possible to generalise this situation. We could for example study discrete groups of isometries of $\mathbf{R}^n$ with m linearly independent translations, where m is any integer in the range $0 \leq m \leq n$.)

18.3.3 The discreteness condition

The discreteness condition means that there is some positive real number ε, such that for every translation Tv in the group, the vector v has length *at least* ε (except of course in the case that v is the zero vector).

The purpose of this condition is to ensure that the group has a compact fundamental domain, or in other words, a "cell" of nonzero, finite area, which is repeated through the plane. Without this condition, we might have for example a group containing the translation Tx for every rational number x, which would not correspond to any reasonable wallpaper pattern.

One important and nontrivial consequence of the discreteness condition in combination with the independent translations condition is that the group can only contain rotations of order 2, 3, 4, or 6; that is, every rotation in the group must be a rotation by 180°, 120°, 90°, or 60°. This fact is known as the crystallographic restriction theorem, and can be generalised to higher-dimensional cases.

18.3.4 Notations for wallpaper groups

Crystallographic notation

Crystallography has 230 space groups to distinguish, far more than the 17 wallpaper groups, but many of the symmetries in the groups are the same. Thus we can use a similar notation for both kinds of groups, that of Carl Hermann and Charles-Victor Mauguin. An example of a full wallpaper name in Hermann-Mauguin style (also called IUC notation) is **p31m**, with four letters or digits; more usual is a shortened name like **c2mm** or **pg**.

For wallpaper groups the full notation begins with either **p** or **c**, for a *primitive cell* or a *face-centred cell*; these are explained below. This is followed by a digit, n, indicating the highest order of rotational symmetry: 1-fold (none), 2-fold, 3-fold, 4-fold, or 6-fold. The next two symbols indicate symmetries relative to one translation axis of the pattern, referred to as the "main" one; if there is a mirror perpendicular to a translation axis we choose that axis as the main one (or if there are two, one of them). The symbols are either **m**, **g**, or **1**, for mirror, glide reflection, or none. The axis of the mirror or glide reflection is perpendicular to the main axis for the first letter, and either parallel or tilted 180°/n (when $n > 2$) for the second letter. Many groups include other symmetries implied by the given ones. The short notation drops digits or an **m** that can be deduced, so long as that leaves no confusion with another group.

A primitive cell is a minimal region repeated by lattice translations. All but two wallpaper symmetry groups are described with respect to primitive cell axes, a coordinate basis using the translation vectors of the lattice. In the remaining two cases symmetry description is with respect to centred cells that are larger than the primitive cell, and hence have internal repetition; the directions of their sides is different from those of the translation vectors spanning a primitive cell. Hermann-Mauguin notation for crystal space groups uses additional cell types.

Examples

- **p2** (**p211**): Primitive cell, 2-fold rotation symmetry, no mirrors or glide reflections.

- **p4mg** (**p4mm**): Primitive cell, 4-fold rotation, glide reflection perpendicular to main axis, mirror axis at 45°.

- **c2mm** (**c2mm**): Centred cell, 2-fold rotation, mirror axes both perpendicular and parallel to main axis.

- **p31m** (**p31m**): Primitive cell, 3-fold rotation, mirror axis at 60°.

Here are all the names that differ in short and full notation.

The remaining names are **p1**, **p3**, **p3m1**, **p31m**, **p4**, and **p6**.

Orbifold notation

Orbifold notation for wallpaper groups, advocated by John Horton Conway (Conway, 1992) (Conway 2008), is based not on crystallography, but on topology. We fold the infinite periodic tiling of the plane into its essence, an orbifold, then describe that with a few symbols.

- A digit, *n*, indicates a centre of *n*-fold rotation corresponding to a cone point on the orbifold. By the crystallographic restriction theorem, *n* must be 2, 3, 4, or 6.

- An asterisk, *, indicates a mirror symmetry corresponding to a boundary of the orbifold. It interacts with the digits as follows:

  1. Digits before * denote centres of pure rotation (cyclic).
  2. Digits after * denote centres of rotation with mirrors through them, corresponding to "corners" on the boundary of the orbifold (dihedral).

- A cross, ×, occurs when a glide reflection is present and indicates a crosscap on the orbifold. Pure mirrors combine with lattice translation to produce glides, but those are already accounted for so we do not notate them.

- The "no symmetry" symbol, **o**, stands alone, and indicates we have only lattice translations with no other symmetry. The orbifold with this symbol is a torus; in general the symbol **o** denotes a handle on the orbifold.

Consider the group denoted in crystallographic notation by **c2mm**; in Conway's notation, this will be **2*22**. The **2** before the * says we have a 2-fold rotation centre with no mirror through it. The * itself says we have a mirror. The first **2** after the * says we have a 2-fold rotation centre on a mirror. The final **2** says we have an independent second 2-fold rotation centre on a mirror, one that is not a duplicate of the first one under symmetries.

The group denoted by **p2gg** will be **22×**. We have two pure 2-fold rotation centres, and a glide reflection axis. Contrast this with **p2mg**, Conway **22***, where crystallographic notation mentions a glide, but one that is implicit in the other symmetries of the orbifold.

Coxeter's bracket notation is also included, based on reflectional Coxeter groups, and modified with plus superscripts accounting for rotations, improper rotations and translations.

18.3.5 Why there are exactly seventeen groups

An orbifold can be viewed as a polygon with face, edges, and vertices, which can be unfolded to form a possibly infinite set of polygons which tile either the sphere, the plane or the hyperbolic plane. When it tiles the plane it will give a wallpaper group and when it tiles the sphere or hyperbolic plane it gives either a spherical symmetry group or Hyperbolic symmetry group. The type of space the polygons tile can be found by calculating the Euler characteristic, $\chi = V - E + F$, where V is the number of corners (vertices), E is the number of edges and F is the number of faces. If the Euler characteristic is positive then the orbifold has an elliptic (spherical) structure; if it is zero then it has a parabolic structure, i.e. a wallpaper group; and if it is negative it will have a hyperbolic structure. When the full set of possible orbifolds is enumerated it is found that only 17 have Euler characteristic 0.

When an orbifold replicates by symmetry to fill the plane, its features create a structure of vertices, edges, and polygon faces, which must be consistent with the Euler characteristic. Reversing the process, we can assign numbers to the features of the orbifold, but fractions, rather than whole numbers. Because the orbifold itself is a quotient of the full surface by the symmetry group, the orbifold Euler characteristic is a quotient of the surface Euler characteristic by the order of the symmetry group.

The orbifold Euler characteristic is 2 minus the sum of the feature values, assigned as follows:

- A digit *n* before a * counts as $(n-1)/n$.

- A digit *n* after a * counts as $(n-1)/2n$.

- Both * and × count as 1.

- The "no symmetry" ° counts as 2.

For a wallpaper group, the sum for the characteristic must be zero; thus the feature sum must be 2.

Examples

- 632: $5/6 + 2/3 + 1/2 = 2$

- 3*3: $2/3 + 1 + 1/3 = 2$

- 4*2: $3/4 + 1 + 1/4 = 2$

- 22×: $1/2 + 1/2 + 1 = 2$

Now enumeration of all wallpaper groups becomes a matter of arithmetic, of listing all feature strings with values summing to 2.

Feature strings with other sums are not nonsense; they imply non-planar tilings, not discussed here. (When the orbifold Euler characteristic is negative, the tiling is hyperbolic; when positive, spherical or *bad*).

18.4   Guide to recognizing wallpaper groups

To work out which wallpaper group corresponds to a given design, one may use the following table.[3]

See also this overview with diagrams.

18.5   The seventeen groups

Each of the groups in this section has two cell structure diagrams, which are to be interpreted as follows (it is the shape that is significant, not the colour):

On the right-hand side diagrams, different equivalence classes of symmetry elements are colored (and rotated) differently.

The **brown or yellow area** indicates a fundamental domain, i.e. the smallest part of the pattern that is repeated.

The diagrams on the right show the cell of the lattice corresponding to the smallest translations; those on the left sometimes show a larger area.

18.5.1   Group p1 (o)

- Orbifold notation: °

- Coxeter notation (rectangular): $[\infty^+, 2, \infty^+]$ or $[\infty]^+ \times [\infty]^+$

- Lattice: oblique

- Point group: C_1

*Example and diagram for **p1***

- The group **p1** contains only translations; there are no rotations, reflections, or glide reflections.

Examples of group p1

- Computer generated

- Mediæval wall diapering

The two translations (cell sides) can each have different lengths, and can form any angle.

18.5.2 Group p2 (2222)

*Example and diagram for **p2***

- Orbifold notation: 2222

- Coxeter notation (rectangular): $[\infty,2,\infty]^{+}$

- Lattice: oblique

- Point group: C_2

- The group **p2** contains four rotation centres of order two (180°), but no reflections or glide reflections.

Examples of group *p2*

- Computer generated
- Cloth, Sandwich Islands (Hawaii)
- Mat on which Egyptian king stood
- Egyptian mat (detail)
- Ceiling of Egyptian tomb
- Wire fence, U.S.

18.5.3   Group pm (**)

*Example and diagram for **pm***

- Orbifold notation: **
- Coxeter notation: [∞,2,∞⁺] or [∞⁺,2,∞]
- Lattice: rectangular
- Point group: D_1
- The group **pm** has no rotations. It has reflection axes, they are all parallel.

Examples of group *pm*

(The first three have a vertical symmetry axis, and the last two each have a different diagonal one.)

- Computer generated
- Dress of a figure in a tomb at Biban el Moluk, Egypt
- Egyptian tomb, Thebes
- Ceiling of a tomb at Gourna, Egypt. Reflection axis is diagonal.
- Indian metalwork at the Great Exhibition in 1851. This is almost **pm** (ignoring short diagonal lines between ovals motifs, which make it **p1**).

18.5.4 Group pg (××)

Example and diagram for **pg**

- Orbifold notation: ××

- Coxeter notation: $[(\infty,2)^+,\infty^+]$ or $[\infty^+,(2,\infty)^+]$

- Lattice: rectangular

- Point group: D_1

- The group **pg** contains glide reflections only, and their axes are all parallel. There are no rotations or reflections.

Examples of group *pg*

- Computer generated

- Mat with herringbone pattern on which Egyptian king stood

- Egyptian mat (detail)

- Pavement with herringbone pattern in Salzburg. Glide reflection axis runs northeast-southwest.

- One of the colorings of the snub square tiling; the glide reflection lines are in the direction upper left / lower right; ignoring colors there is much more symmetry than just **pg**, then it is **p4mg** (see there for this image with equally colored triangles)[1]

1. ^ It helps to consider the squares as the background, then we see a simple patterns of rows of rhombuses.

Without the details inside the zigzag bands the mat is **p2mg**; with the details but without the distinction between brown and black it is **p2gg**.

Ignoring the wavy borders of the tiles, the pavement is **p2gg**.

18.5.5 Group cm (*×)

- Orbifold notation: *×

- Coxeter notation: $[\infty^+,2^+,\infty]$ or $[\infty,2^+,\infty^+]$

Example and diagram for **cm**

- Lattice: rhombic

- Point group: D_1

- The group **cm** contains no rotations. It has reflection axes, all parallel. There is at least one glide reflection whose axis is *not* a reflection axis; it is halfway between two adjacent parallel reflection axes.

- This group applies for symmetrically staggered rows (i.e. there is a shift per row of half the translation distance inside the rows) of identical objects, which have a symmetry axis perpendicular to the rows.

Examples of group *cm*

- Computer generated

- Dress of Amun, from Abu Simbel, Egypt

- Dado from Biban el Moluk, Egypt

- Bronze vessel in Nimroud, Assyria

- Spandrils of arches, the Alhambra, Spain

- Soffitt of arch, the Alhambra, Spain

- Persian tapestry

- Indian metalwork at the Great Exhibition in 1851

- Dress of a figure in a tomb at Biban el Moluk, Egypt

18.5.6   Group p2mm (*2222)

- Orbifold notation: *2222

- Coxeter notation (rectangular): [∞,2,∞] or [∞]×[∞]

- Coxeter notation (square): [4,1$^+$,4] or [1$^+$,4,4,1$^+$]

- Lattice: rectangular

- Point group: D_2

*Example and diagram for **p2mm***

- The group **p2mm** has reflections in two perpendicular directions, and four rotation centres of order two (180°) located at the intersections of the reflection axes.

Examples of group *p2mm*

- 2D image of lattice fence. U.S. (in 3D there is additional symmetry)
- Mummy case stored in The Louvre
- Mummy case stored in The Louvre. Would be type **p4mm** except for the mismatched coloring.

18.5.7   Group p2mg (22*)

*Example and diagram for **p2mg***

- Orbifold notation: 22*
- Coxeter notation: $[(\infty,2)^+,\infty]$ or $[\infty,(2,\infty)^+]$
- Lattice: rectangular
- Point group: D_2

- The group **p2mg** has two rotation centres of order two (180°), and reflections in only one direction. It has glide reflections whose axes are perpendicular to the reflection axes. The centres of rotation all lie on glide reflection axes.

Examples of group *p2mg*

- Computer generated

- Cloth, Sandwich Islands (Hawaii)

- Ceiling of Egyptian tomb

- Floor tiling in Prague, the Czech Republic

- Bowl from Kerma

- Pentagon packing

18.5.8   Group p2gg (22×)

Example and diagram for **p2gg**

- Orbifold notation: 22×

- Coxeter notation (rectangular): $[(\infty,2)^+,(\infty,2)^+)]$

- Coxeter notation (square): $[4^+,4^+]$

- Lattice: rectangular

- Point group: D_2

- The group **p2gg** contains two rotation centres of order two (180°), and glide reflections in two perpendicular directions. The centres of rotation are not located on the glide reflection axes. There are no reflections.

Examples of group *p2gg*

- Computer generated

- Bronze vessel in Nimroud, Assyria

- Pavement in Budapest, Hungary. Glide reflection axes are diagonal.

18.5.9   Group c2mm (2*22)

Example and diagram for c2mm

- Orbifold notation: 2*22

- Coxeter notation (rhombic): $[\infty, 2^+, \infty]$

- Coxeter notation (square): $[(4,4,2^+)]$

- Lattice: rhombic

- Point group: D_2

- The group **c2mm** has reflections in two perpendicular directions, and a rotation of order two (180°) whose centre is *not* on a reflection axis. It also has two rotations whose centres *are* on a reflection axis.

- This group is frequently seen in everyday life, since the most common arrangement of bricks in a brick building (running bond) utilises this group (see example below).

The rotational symmetry of order 2 with centres of rotation at the centres of the sides of the rhombus is a consequence of the other properties.

The pattern corresponds to each of the following:

- symmetrically staggered rows of identical doubly symmetric objects

- a checkerboard pattern of two alternating rectangular tiles, of which each, by itself, is doubly symmetric

- a checkerboard pattern of alternatingly a 2-fold rotationally symmetric rectangular tile and its mirror image

Examples of group *c2mm*

- Computer generated

- one of the 8 semi-regular tessellations

- Suburban brick wall using running bond arrangement, U.S.

- Ceiling of Egyptian tomb. Ignoring colors, this would be **p4mg**.

- Egyptian

- Persian tapestry

- Egyptian tomb

- Turkish dish

- A compact packing of two sizes of circle.

- Another compact packing of two sizes of circle.

- Another compact packing of two sizes of circle.

18.5.10   Group p4 (442)

Example and diagram for p4

- Orbifold notation: 442

- Coxeter notation: $[4,4]^+$

- Lattice: square

- Point group: C_4

- The group **p4** has two rotation centres of order four (90°), and one rotation centre of order two (180°). It has no reflections or glide reflections.

Examples of group *p4*

A **p4** pattern can be looked upon as a repetition in rows and columns of equal square tiles with 4-fold rotational symmetry. Also it can be looked upon as a checkerboard pattern of two such tiles, a factor $\sqrt{2}$ smaller and rotated 45°.

- Computer generated

- Ceiling of Egyptian tomb; ignoring colors this is **p4**, otherwise **p2**

- Ceiling of Egyptian tomb

- Overlaid patterns

- Frieze, the Alhambra, Spain. Requires close inspection to see why there are no reflections.

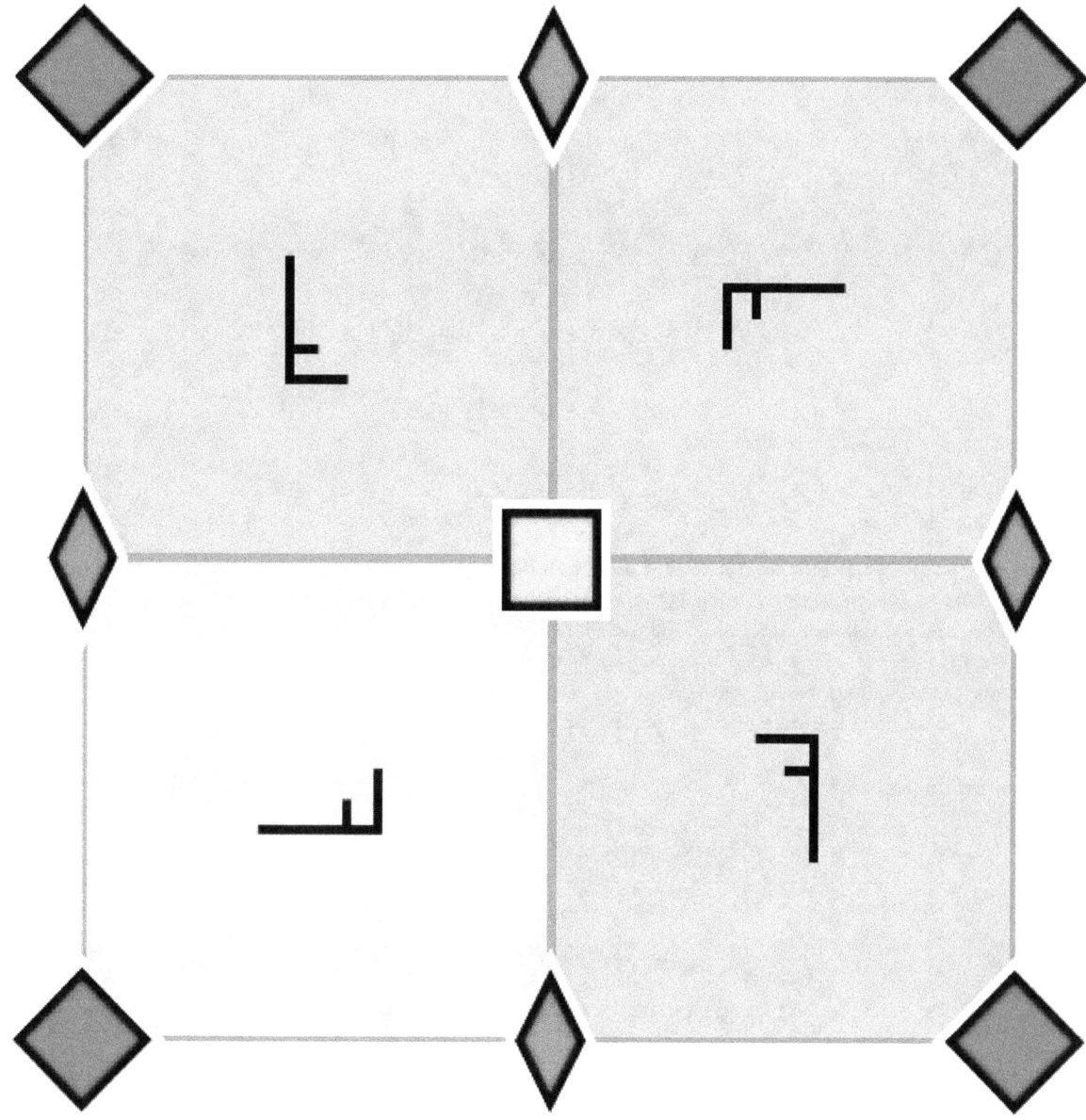

Cell structure for p4

- Viennese cane

- Renaissance earthenware

- Pythagorean tiling

- Generated from a photograph

18.5.11   Group p4mm (*442)

- Orbifold notation: *442

- Coxeter notation: [4,4]

- Lattice: square

Example and diagram for **p4mm**

- Point group: D_4

- The group **p4mm** has two rotation centres of order four (90°), and reflections in four distinct directions (horizontal, vertical, and diagonals). It has additional glide reflections whose axes are not reflection axes; rotations of order two (180°) are centred at the intersection of the glide reflection axes. All rotation centres lie on reflection axes.

This corresponds to a straightforward grid of rows and columns of equal squares with the four reflection axes. Also it corresponds to a checkerboard pattern of two of such squares.

Examples of group *p4mm*

Examples displayed with the smallest translations horizontal and vertical (like in the diagram):

- Computer generated
- one of the 3 regular tessellations
- Demiregular tiling with triangles; ignoring colors, this is **p4mm**, otherwise **c2mm**
- one of the 8 semi-regular tessellations (ignoring color also, with smaller translations)
- Ornamental painting, Nineveh, Assyria
- Storm drain, U.S.
- Egyptian mummy case
- Persian glazed tile
- Compact packing of two sizes of circle.

Examples displayed with the smallest translations diagonal:

- checkerboard
- Cloth, Otaheite (Tahiti)
- Egyptian tomb
- Cathedral of Bourges
- Dish from Turkey, Ottoman period

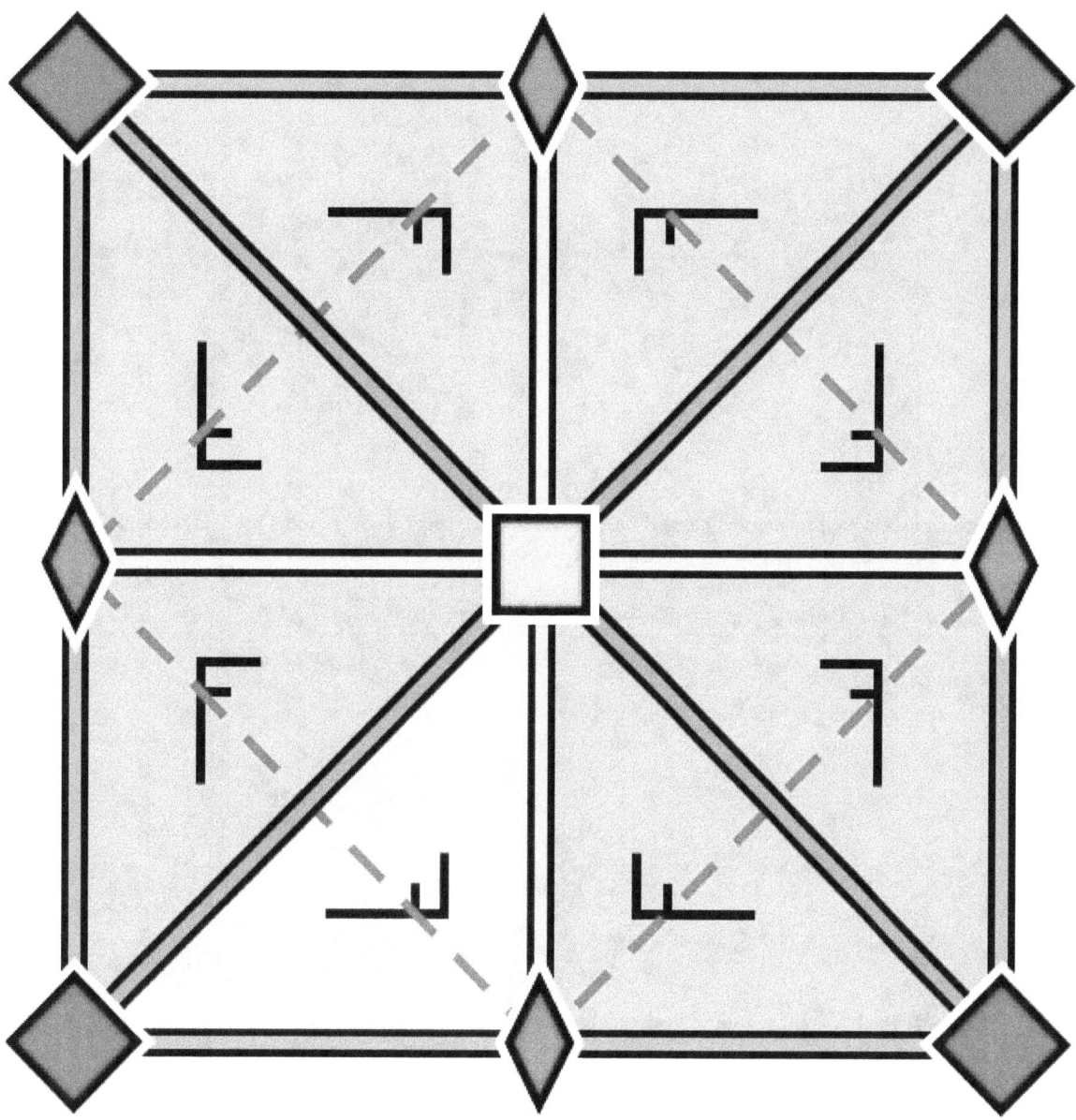

Cell structure for **p4mm**

18.5.12   Group p4mg (4*2)

- Orbifold notation: 4*2

- Coxeter notation: [4$^+$,4]

- Lattice: square

- Point group: D$_4$

- The group **p4mg** has two centres of rotation of order four (90°), which are each other's mirror image, but it has reflections in only two directions, which are perpendicular. There are rotations of order two (180°) whose centres are located at the intersections of reflection axes. It has glide reflections axes parallel to the reflection axes, in between them, and also at an angle of 45° with these.

*Example and diagram for **p4mg***

A **p4mg** pattern can be looked upon as a checkerboard pattern of copies of a square tile with 4-fold rotational symmetry, and its mirror image. Alternatively it can be looked upon (by shifting half a tile) as a checkerboard pattern of copies of a horizontally and vertically symmetric tile and its 90° rotated version. Note that neither applies for a plain checkerboard pattern of black and white tiles, this is group **p4mm** (with diagonal translation cells).

Examples of group *p4mg*

- Bathroom linoleum, U.S.

- Painted porcelain, China

  - Fly screen, U.S.

  - Painting, China

    - one of the colorings of the snub square tiling (see also at **pg**)

18.5.13   Group p3 (333)

- Orbifold notation: 333

- Coxeter notation: $[(3,3,3)]^+$ or $[3^{[3]}]^+$

- Lattice: hexagonal

- Point group: C_3

- The group **p3** has three different rotation centres of order three (120°), but no reflections or glide reflections.

Imagine a tessellation of the plane with equilateral triangles of equal size, with the sides corresponding to the smallest translations. Then half of the triangles are in one orientation, and the other half upside down. This wallpaper group corresponds to the case that all triangles of the same orientation are equal, while both types have rotational symmetry of order three, but the two are not equal, not each other's mirror image, and not both symmetric (if the two are equal we have p6, if they are each other's mirror image we have p31m, if they are both symmetric we have p3m1; if two of the three apply then the third also, and we have p6mm). For a given image, three of these tessellations are possible, each with rotation centres as vertices, i.e. for any tessellation two shifts are possible. In terms of the image: the vertices can be the red, the blue or the green triangles.

Equivalently, imagine a tessellation of the plane with regular hexagons, with sides equal to the smallest translation distance divided by √3. Then this wallpaper group corresponds to the case that all hexagons are equal (and in the same orientation)

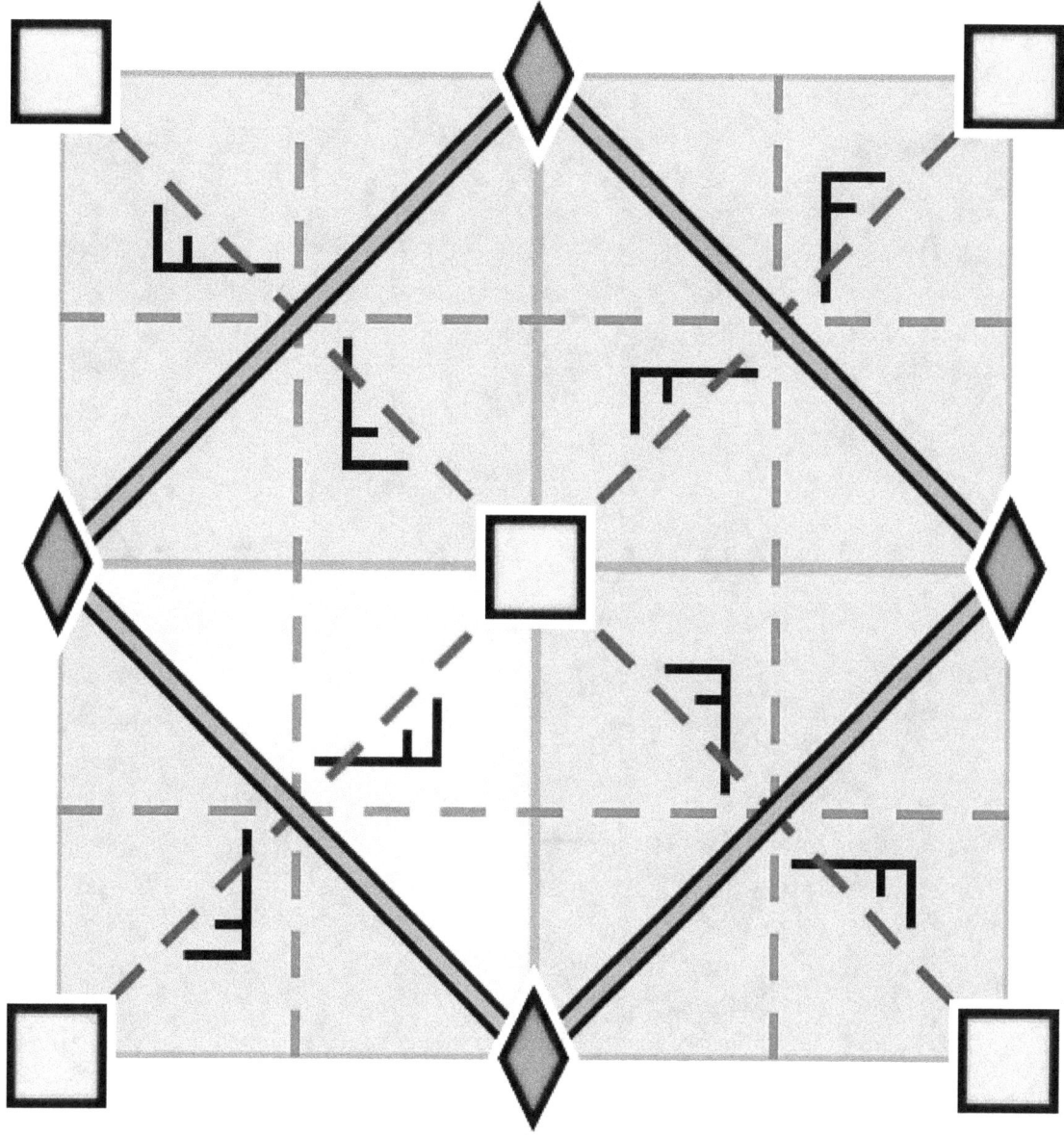

Cell structure for **p4mg**

and have rotational symmetry of order three, while they have no mirror image symmetry (if they have rotational symmetry of order six we have p6, if they are symmetric with respect to the main diagonals we have p31m, if they are symmetric with respect to lines perpendicular to the sides we have p3m1; if two of the three apply then the third also, and we have p6mm). For a given image, three of these tessellations are possible, each with one third of the rotation centres as centres of the hexagons. In terms of the image: the centres of the hexagons can be the red, the blue or the green triangles.

Examples of group *p3*

- Computer generated
- one of the 8 semi-regular tessellations (ignoring the colors: *p6*); the translation vectors are rotated a little to the right compared with the directions in the underlying hexagonal lattice of the image
- Street pavement in Zakopane, Poland

Example and diagram for p3

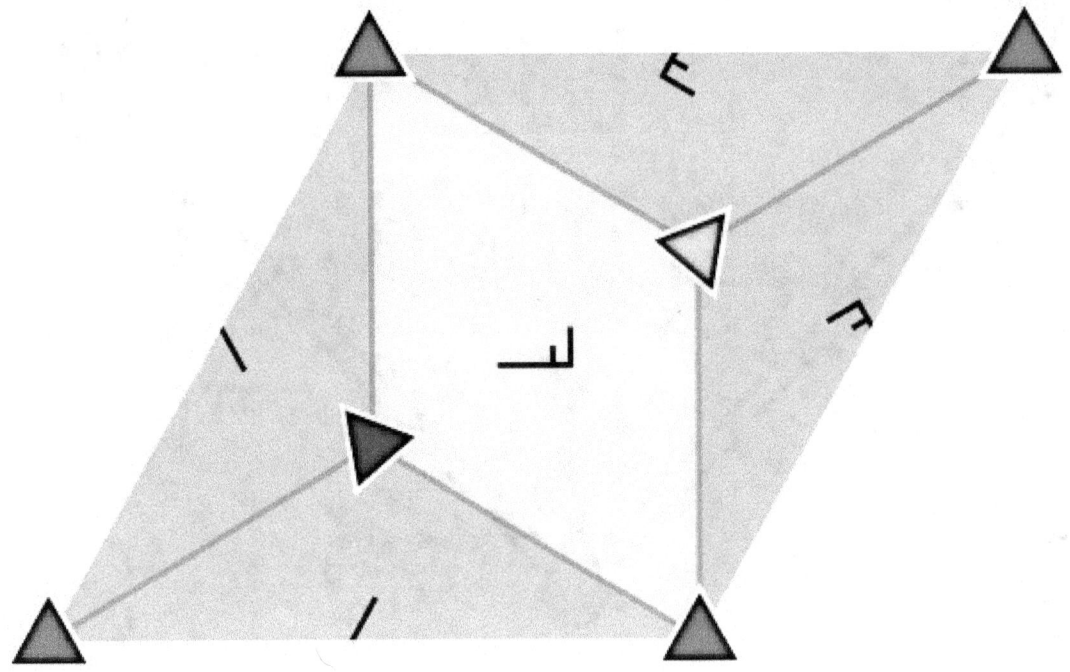

Cell structure for p3

- Wall tiling in the Alhambra, Spain (and the whole wall); ignoring all colors this is **p3** (ignoring only star colors it is **p1**)

18.5.14   Group p3m1 (*333)

- Orbifold notation: *333

- Coxeter notation: [(3,3,3)] or $[3^{[3]}]$

- Lattice: hexagonal

- Point group: D_3

*Example and diagram for **p3m1***

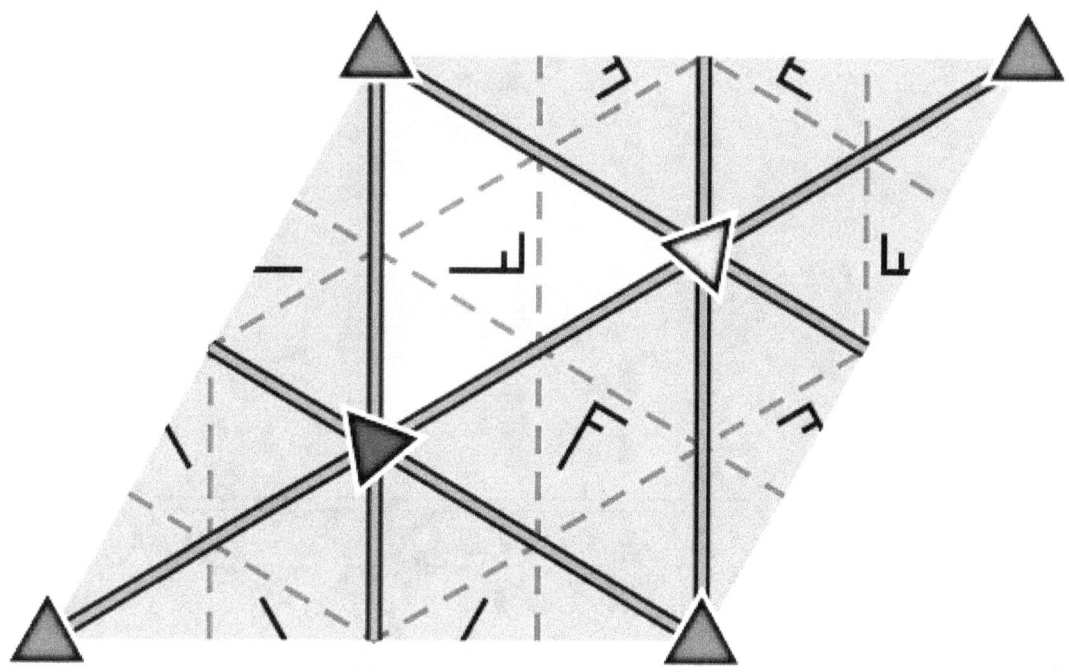

*Cell structure for **p3m1***

- The group **p3m1** has three different rotation centres of order three (120°). It has reflections in the three sides of an equilateral triangle. The centre of every rotation lies on a reflection axis. There are additional glide reflections in three distinct directions, whose axes are located halfway between adjacent parallel reflection axes.

Like for **p3**, imagine a tessellation of the plane with equilateral triangles of equal size, with the sides corresponding to the smallest translations. Then half of the triangles are in one orientation, and the other half upside down. This wallpaper group corresponds to the case that all triangles of the same orientation are equal, while both types have rotational symmetry of order three, and both are symmetric, but the two are not equal, and not each other's mirror image. For a given image, three of these tessellations are possible, each with rotation centres as vertices. In terms of the image: the vertices can be the red, the dark blue or the green triangles.

Examples of group *p3m1*

- one of the 3 regular tessellations (ignoring colors: p6mm)

- another regular tessellation (ignoring colors: p6mm)

- one of the 8 semi-regular tessellations (ignoring colors: p6mm)

- Persian glazed tile (ignoring colors: p6mm)

- Persian ornament

- Painting. China (see detailed image)

18.5.15   Group p31m (3*3)

Example and diagram for **p31m**

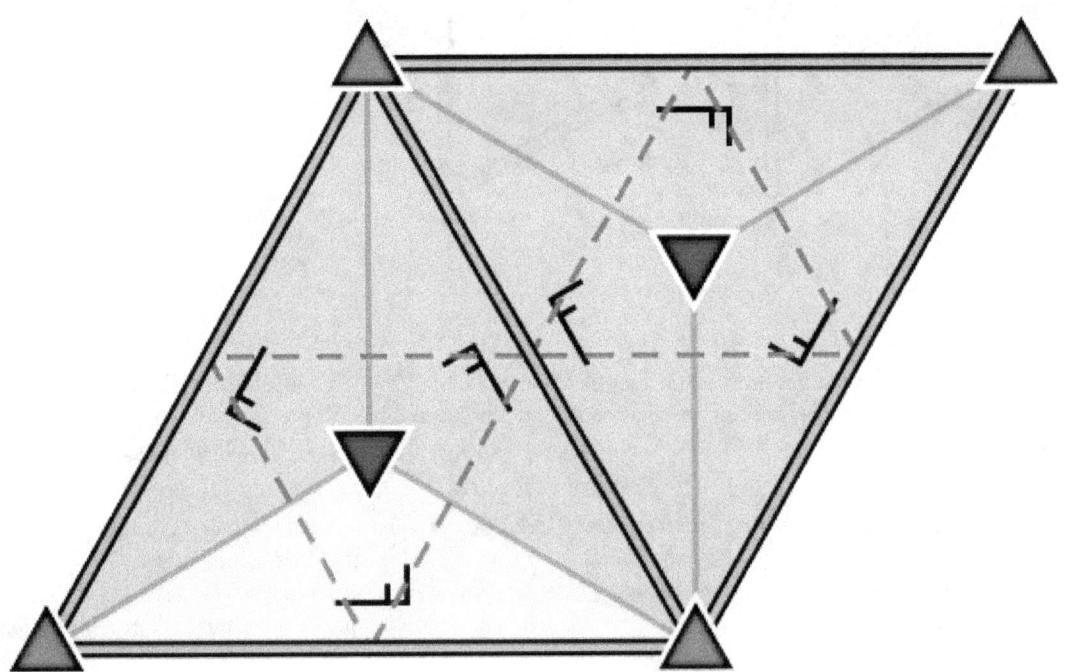

Cell structure for **p31m**

- Orbifold notation: 3*3

- Coxeter notation: [6,3$^+$]

- Lattice: hexagonal

- Point group: D$_3$

- The group **p31m** has three different rotation centres of order three (120°), of which two are each other's mirror image. It has reflections in three distinct directions. It has at least one rotation whose centre does *not* lie on a reflection axis. There are additional glide reflections in three distinct directions, whose axes are located halfway between adjacent parallel reflection axes.

Like for **p3** and **p3m1**, imagine a tessellation of the plane with equilateral triangles of equal size, with the sides corresponding to the smallest translations. Then half of the triangles are in one orientation, and the other half upside down. This wallpaper group corresponds to the case that all triangles of the same orientation are equal, while both types have rotational symmetry of order three and are each other's mirror image, but not symmetric themselves, and not equal. For a given image, only one such tessellation is possible. In terms of the image: the vertices can *not* be dark blue triangles.

Examples of group *p31m*

- Persian glazed tile

- Painted porcelain, China

- Painting, China

- Compact packing of two sizes of circle.

18.5.16   Group p6 (632)

*Example and diagram for **p6***

- Orbifold notation: 632

- Coxeter notation: [6,3]$^+$

- Lattice: hexagonal

- Point group: C$_6$

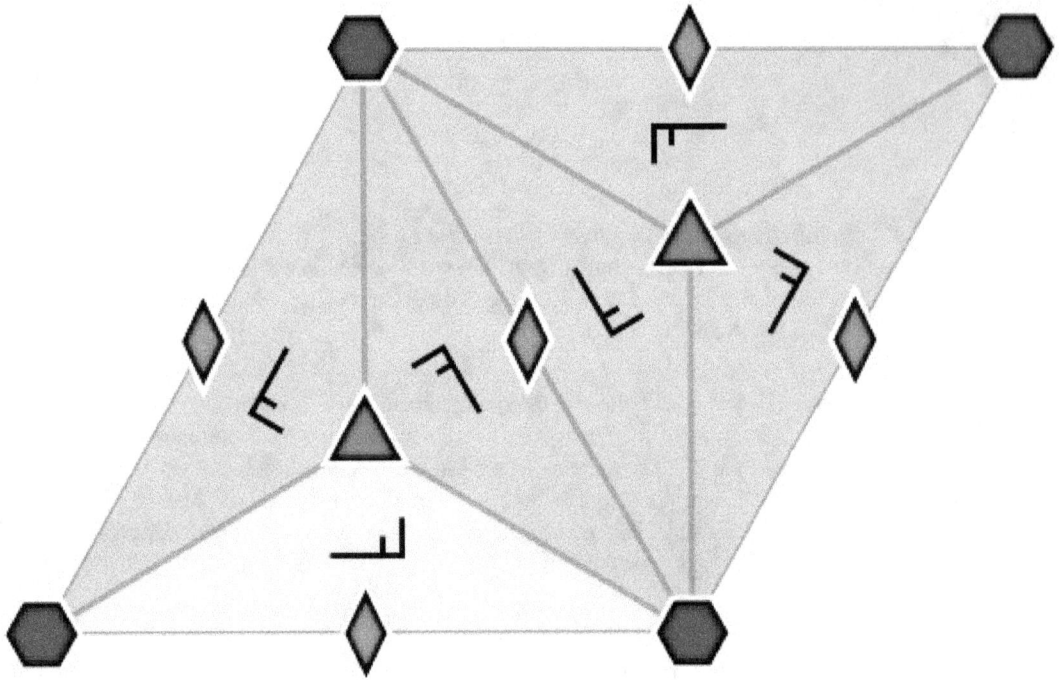

*Cell structure for **p6***

- The group **p6** has one rotation centre of order six, which only differ by a rotation of 60°; it has also two rotation centres of order three, which only differ by a rotation of 120° and three of order two (or, equivalently, 180°). It has no reflections or glide reflections.

A pattern with this symmetry can be looked upon as a tessellation of the plane with equal triangular tiles with C_3 symmetry, or equivalently, a tessellation of the plane with equal hexagonal tiles with C_6 symmetry (with the edges of the tiles not necessarily part of the pattern).

Examples of group *p6*

- Computer generated

- Regular polygons

- Wall panelling, the Alhambra, Spain

- Persian ornament

18.5.17   Group p6mm (*632)

- Orbifold notation: *632

- Coxeter notation: [6,3]

- Lattice: hexagonal

- Point group: D_6

- The group **p6mm** has one rotation centre of order six (60°); it has also two rotation centres of order three, which only differ by a rotation of 60° (or, equivalently, 180°), and three of order two, which only differ by a rotation of 60°. It has also reflections in six distinct directions. There are additional glide reflections in six distinct directions, whose axes are located halfway between adjacent parallel reflection axes.

Example and diagram for **p6mm**

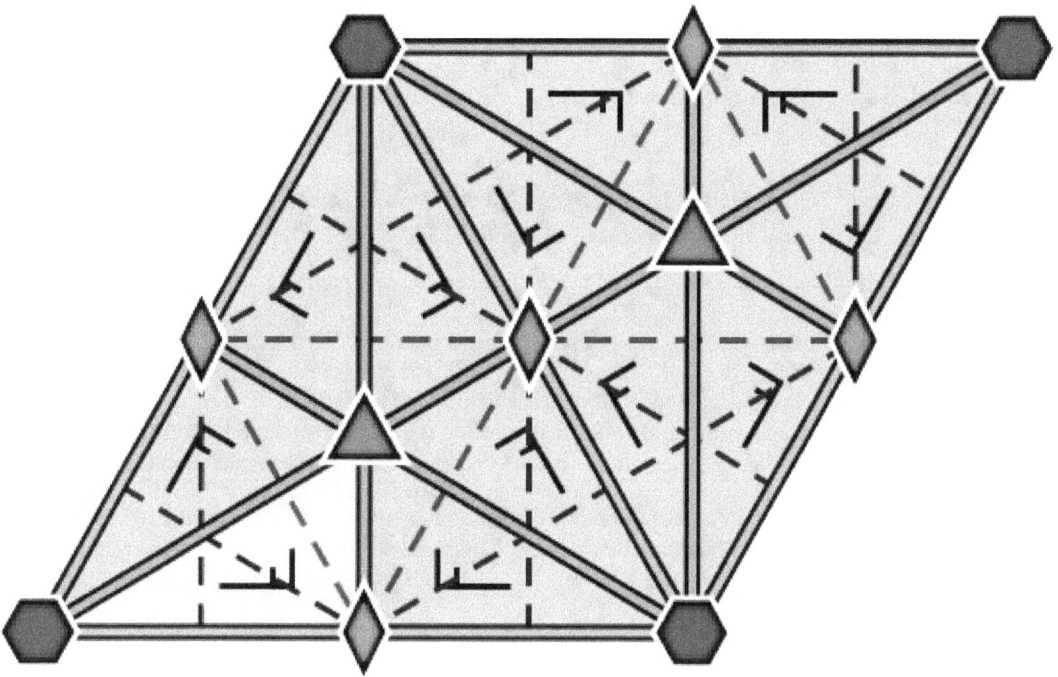

Cell structure for **p6mm**

A pattern with this symmetry can be looked upon as a tessellation of the plane with equal triangular tiles with D_3 symmetry, or equivalently, a tessellation of the plane with equal hexagonal tiles with D_6 symmetry (with the edges of the tiles not necessarily part of the pattern). Thus the simplest examples are a triangular lattice with or without connecting lines, and a hexagonal tiling with one color for outlining the hexagons and one for the background.

Examples of group *p6mm*

- Computer generated
- one of the 8 semi-regular tessellations
  - another semi-regular tessellation
  - another semi-regular tessellation

- Persian glazed tile

- King's dress, Khorsabad, Assyria; this is almost **p6mm** (ignoring inner parts of flowers, which make it **c2mm**)

- Bronze vessel in Nimroud, Assyria

- Byzantine marble pavement, Rome

- Painted porcelain, China

- Painted porcelain, China

- Compact packing of two sizes of circle.

- Another compact packing of two sizes of circle.

18.6   Lattice types

There are five lattice types or Bravais lattices, corresponding to the five possible wallpaper groups of the lattice itself. The wallpaper group of a pattern with this lattice of translational symmetry cannot have more, but may have less symmetry than the lattice itself.

- In the 5 cases of rotational symmetry of order 3 or 6, the unit cell consists of two equilateral triangles (hexagonal lattice, itself **p6mm**). They form a rhombus with angles 60° and 120°.

- In the 3 cases of rotational symmetry of order 4, the cell is a square (square lattice, itself **p4mm**).

- In the 5 cases of reflection or glide reflection, but not both, the cell is a rectangle (rectangular lattice, itself **p2mm**). Special cases: square.

- In the 2 cases of reflection combined with glide reflection, the cell is a rhombus (rhombic lattice, itself **c2mm**). It may also be interpreted as a centered rectangular lattice. Special cases: square, hexagonal unit cell.

- In the case of only rotational symmetry of order 2, and the case of no other symmetry than translational, the cell is in general a parallelogram (parallelogrammatic or oblique lattice, itself **p2**). Special cases: rectangle, square, rhombus, hexagonal unit cell.

18.7   Symmetry groups

The actual symmetry group should be distinguished from the wallpaper group. Wallpaper groups are collections of symmetry groups. There are 17 of these collections, but for each collection there are infinitely many symmetry groups, in the sense of actual groups of isometries. These depend, apart from the wallpaper group, on a number of parameters for the translation vectors, the orientation and position of the reflection axes and rotation centers.

The numbers of degrees of freedom are:

- 6 for **p2**

- 5 for **p2mm**, **p2mg**, **p2gg**, and **c2mm**

- 4 for the rest.

However, within each wallpaper group, all symmetry groups are algebraically isomorphic.

Some symmetry group isomorphisms:

- **p1**: $\mathbf{Z}^2$

- **pm**: $\mathbf{Z} \times D\infty$

- **p2mm**: $D\infty \times D\infty$.

18.8 Dependence of wallpaper groups on transformations

- The wallpaper group of a pattern is invariant under isometries and uniform scaling (similarity transformations).

- Translational symmetry is preserved under arbitrary bijective affine transformations.

- Rotational symmetry of order two ditto; this means also that 4- and 6-fold rotation centres at least keep 2-fold rotational symmetry.

- Reflection in a line and glide reflection are preserved on expansion/contraction along, or perpendicular to, the axis of reflection and glide reflection. It changes **p6mm**, **p4mg**, and **p3m1** into **c2mm**, **p3m1** into **cm**, and **p4mm**, depending on direction of expansion/contraction, into **p2mm** or **c2mm**. A pattern of symmetrically staggered rows of points is special in that it can convert by expansion/contraction from **p6mm** to **p4mm**.

Note that when a transformation decreases symmetry, a transformation of the same kind (the inverse) obviously for some patterns increases the symmetry. Such a special property of a pattern (e.g. expansion in one direction produces a pattern with 4-fold symmetry) is not counted as a form of extra symmetry.

Change of colors does not affect the wallpaper group if any two points that have the same color before the change, also have the same color after the change, and any two points that have different colors before the change, also have different colors after the change.

If the former applies, but not the latter, such as when converting a color image to one in black and white, then symmetries are preserved, but they may increase, so that the wallpaper group can change.

18.9 Web demo and software

Several software graphic tools will let you create 2D patterns using wallpaper symmetry groups. Usually you can edit the original tile and its copies in the entire pattern are updated automatically.

- MadPattern, a free set of Adobe Illustrator templates that support the 17 wallpaper groups

- Tess, a nagware tessellation program for multiple platforms, supports all wallpaper, frieze, and rosette groups, as well as Heesch tilings.

- Kali, online graphical symmetry editor applet.

- Kali, free downloadable Kali for Windows and Mac Classic.

- Inkscape, a free vector graphics editor, supports all 17 groups plus arbitrary scales, shifts, rotates, and color changes per row or per column, optionally randomized to a given degree. (See)

- SymmetryWorks is a commercial plugin for Adobe Illustrator, supports all 17 groups.

- Arabeske is a free standalone tool, supports a subset of wallpaper groups.

18.10   See also

- List of planar symmetry groups (summary of this page)

- Aperiodic tiling

- Crystallography

- Layer group

- Mathematics and art

- M. C. Escher

- Point group

- Symmetry groups in one dimension

- Symmetry groups in one dimension

- Tessellation

18.11   Notes

[1] E. Fedorov (1891) "Simmetrija na ploskosti" [Symmetry in the plane]. *Zapiski Imperatorskogo Sant-Petersburgskogo Mineralogicheskogo Obshchestva* [Proceedings of the Imperial St. Petersburg Mineralogical Society], series 2, vol. 28, pages 245-291 (in Russian).

[2] George Pólya (1924) "Über die Analogie der Kristallsymmetrie in der Ebene." *Zeitschrift für Kristallographie*, vol. 60, pages 278–282.

[3] Radaelli, Paulo G. *Symmetry in Crystallography*. Oxford University Press.

18.12   References

- *The Grammar of Ornament* (1856), by Owen Jones. Many of the images in this article are from this book; it contains many more.

- J. H. Conway (1992). "The Orbifold Notation for Surface Groups". In: M. W. Liebeck and J. Saxl (eds.), *Groups, Combinatorics and Geometry*, Proceedings of the L.M.S. Durham Symposium, July 5–15, Durham, UK, 1990; London Math. Soc. Lecture Notes Series **165**. Cambridge University Press, Cambridge. pp. 438–447

- J. H. Conway; H. Burgiel, C. Goodman-Strauss (2008): "The Symmetries of Things". Worcester MA: A.K. Peters. ISBN 1-56881-220-5.

- Grünbaum, Branko; Shephard, G. C. (1987): *Tilings and Patterns*. New York: Freeman. ISBN 0-7167-1193-1.

- Pattern Design, Lewis F. Day

18.13 External links

- The 17 plane symmetry groups by David E. Joyce

- Introduction to wallpaper patterns by Chaim Goodman-Strauss and Heidi Burgiel

- Description by Silvio Levy

- Example tiling for each group, with dynamic demos of properties

- Overview with example tiling for each group

- Escher Web Sketch, a java applet with interactive tools for drawing in all 17 plane symmetry groups

- Burak, a Java applet for drawing symmetry groups.

- A JavaScript app for drawing wallpaper patterns

- Beobachtungen zum geometrischen Motiv der Pelta

- Seventeen Kinds of Wallpaper Patterns the 17 symmetries found in traditional Japanese patterns.

Chapter 19

C-symmetry

In physics, **C-symmetry** means the symmetry of physical laws under a charge-conjugation transformation. Electromagnetism, gravity and the strong interaction all obey C-symmetry, but weak interactions violate C-symmetry.

19.1   Charge reversal in electromagnetism

The laws of electromagnetism (both classical and quantum) are invariant under this transformation: if each charge q were to be replaced with a charge $-q$, and thus the directions of the electric and magnetic fields were reversed, the dynamics would preserve the same form. In the language of quantum field theory, charge conjugation transforms:[1]

1. $\psi \rightarrow -i(\bar{\psi}\gamma^0\gamma^2)^T$
2. $\bar{\psi} \rightarrow -i(\gamma^0\gamma^2\psi)^T$
3. $A^\mu \rightarrow -A^\mu$

Notice that these transformations do not alter the chirality of particles. A left-handed neutrino would be taken by charge conjugation into a left-handed antineutrino, which does not interact in the Standard Model. This property is what is meant by the "maximal violation" of C-symmetry in the weak interaction.

(Some postulated extensions of the Standard Model, like left-right models, restore this C-symmetry.)

19.2   Combination of charge and parity reversal

It was believed for some time that C-symmetry could be combined with the parity-inversion transformation (see P-symmetry) to preserve a combined CP-symmetry. However, violations of this symmetry have been identified in the weak interactions (particularly in the kaons and B mesons). In the Standard Model, this CP violation is due to a single phase in the CKM matrix. If CP is combined with time reversal (T-symmetry), the resulting CPT-symmetry can be shown using only the Wightman axioms to be universally obeyed.

19.3   Charge definition

To give an example, take two real scalar fields, φ and χ. Suppose both fields have even C-parity (even C-parity refers to even symmetry under charge conjugation ex. $C\psi(q) = C\psi(-q)$, as opposed to odd C-parity which refers to antisymmetry under charge conjugation ex. $C\psi(q) = -C\psi(-q)$). Now reformulate things so that $\psi \overset{\text{def}}{=} \frac{\phi+i\chi}{\sqrt{2}}$. Now, φ and χ have even C-parities because the imaginary number i has an odd C-parity (C is antiunitary).

In other models, it is possible for both φ and χ to have odd C-parities.

19.4 See also

- C parity
- anti-particle
- antimatter

19.5 References

[1] Peskin, M.E. and Schroeder, D.V. (1997). *An Introduction to Quantum Field Theory*. Addison Wesley. ISBN 0-201-50397-2.

- Sozzi, M.S. (2008). *Discrete symmetries and CP violation*. Oxford University Press. ISBN 978-0-19-929666-8.

Chapter 20

CPT symmetry

"CPT theorem" redirects here. For the album by Greydon Square, see The C.P.T. Theorem.

CPT symmetry is a fundamental symmetry of physical laws under the simultaneous transformations of charge conjugation (C), parity transformation (P), and time reversal (T). CPT is the only combination of C, P and T that's observed to be an exact symmetry of nature at the fundamental level.[1] The **CPT theorem** says that CPT symmetry holds for all physical phenomena, or more precisely, that any Lorentz invariant local quantum field theory with a Hermitian Hamiltonian must have CPT symmetry.

20.1 History

Efforts during the late 1950s revealed the violation of P-symmetry by phenomena that involve the weak force, and there were well-known violations of C-symmetry as well. For a short time, the CP-symmetry was believed to be preserved by all physical phenomena, but that was later found to be false too, which implied, by **CPT invariance**, violations of T-symmetry as well.

The CPT theorem appeared for the first time, implicitly, in the work of Julian Schwinger in 1951 to prove the connection between spin and statistics.[2] In 1954, Gerhart Lüders and Wolfgang Pauli derived more explicit proofs,[3][4] so this theorem is sometimes known as the Lüders–Pauli theorem. At about the same time, and independently, this theorem was also proved by John Stewart Bell.[5] These proofs are based on the principle of Lorentz invariance and the principle of locality in the interaction of quantum fields. Subsequently Res Jost gave a more general proof in the framework of axiomatic quantum field theory.

20.2 Derivation of the CPT theorem

Consider a Lorentz boost in a fixed direction z. This can be interpreted as a rotation of the time axis into the z axis, with an imaginary rotation parameter. If this rotation parameter were real, it would be possible for a 180° rotation to reverse the direction of time and of z. Reversing the direction of one axis is a reflection of space in any number of dimensions. If space has 3 dimensions, it is equivalent to reflecting all the coordinates, because an additional rotation of 180° in the x-y plane could be included.

This defines a CPT transformation if we adopt the Feynman-Stueckelberg interpretation of antiparticles as the corresponding particles traveling backwards in time. This interpretation requires a slight analytic continuation, which is well-defined only under the following assumptions:

1. The theory is Lorentz invariant;

2. The vacuum is Lorentz invariant;

3. The energy is bounded below.

When the above hold, quantum theory can be extended to a Euclidean theory, defined by translating all the operators to imaginary time using the Hamiltonian. The commutation relations of the Hamiltonian, and the Lorentz generators, guarantee that Lorentz invariance implies rotational invariance, so that any state can be rotated by 180 degrees.

Since a sequence of two CPT reflections is equivalent to a 360-degree rotation, fermions change by a sign under two CPT reflections, while bosons do not. This fact can be used to prove the spin-statistics theorem.

20.3 Consequences and implications

A consequence of this derivation is that a violation of CPT automatically indicates a Lorentz violation.

The implication of CPT symmetry is that a "mirror-image" of our universe — with all objects having their positions reflected by an arbitrary plane (corresponding to a parity inversion), all momenta reversed (corresponding to a time inversion) and with all matter replaced by antimatter (corresponding to a charge inversion)— would evolve under exactly our physical laws. The CPT transformation turns our universe into its "mirror image" and vice versa. CPT symmetry is recognized to be a fundamental property of physical laws.

In order to preserve this symmetry, every violation of the combined symmetry of two of its components (such as CP) must have a corresponding violation in the third component (such as T); in fact, mathematically, these are the same thing. Thus violations in T symmetry are often referred to as CP violations.

The CPT theorem can be generalized to take into account pin groups.

In 2002 Oscar Greenberg proved that CPT violation implies the breaking of Lorentz symmetry.[6] This implies that any study of CPT violation includes also Lorentz violation. However, Chaichian *et al* later disputed the validity of Greenberg's result.[7] The overwhelming majority of experimental searches for Lorentz violation have yielded negative results. A detailed tabulation of these results is given by Kostelecky and Russell.[8]

20.4 See also

- Poincaré symmetry and Quantum field theory

- Parity (physics), Charge conjugation and T-symmetry

- CP violation and kaon

- Gravitational interaction of antimatter#CPT theorem

20.5 References

[1] Kostelecký, V. A. (1998). "The Status of CPT". arXiv:hep-ph/9810365 [hep-ph].

[2] Schwinger, Julian (1951). *The Theory of Quantized Fields I. Physical Review* **82** (6). p. 914. doi:10.1103/PhysRev.82.914.

[3] Lüders, G. (1954). "On the Equivalence of Invariance under Time Reversal and under Particle-Antiparticle Conjugation for Relativistic Field Theories". *Kongelige Danske Videnskabernes Selskab, Matematisk-Fysiske Meddelelser* **28** (5): 1–17.

[4] Pauli, W.; Rosenfelf, L.; Weisskopf, V., eds. (1955). *Niels Bohr and the Development of Physics*. McGraw-Hill. LCCN 56040984.

[5] Bell, J. S. (1954). (Thesis). Birmingham University. Missing or empty |title= (help)

[6] Greenberg, O. W. (2002). "CPT Violation Implies Violation of Lorentz Invariance". *Physical Review Letters* **89** (23): 231602. arXiv:hep-ph/0201258. Bibcode:2002PhRvL..89w1602G. doi:10.1103/PhysRevLett.89.231602.

[7] Chaichian, M.; Dolgov, A. D.; Novikov, V. A.; Tureanu, A. (2011). "CPT Violation Does Not Lead to Violation of Lorentz Invariance and Vice Versa".*Physics Letters B***699**(3): 177–180. arXiv:1103.0168. Bibcode:2011PhLB..699..177C.do.03.026.

[8] Kostelecký, V. A.; Russell, N. (2011). "Data tables for Lorentz and *CPT* violation". *Reviews of Modern Physics* **83** (1): 11–31. arXiv:0801.0287. Bibcode:2011RvMP...83...11K. doi:10.1103/RevModPhys.83.11.

20.6   Sources

- Sozzi, M.S. (2008). *Discrete symmetries and CP violation*. Oxford University Press. ISBN 978-0-19-929666-8.

- Griffiths, David J. (1987). *Introduction to Elementary Particles*. Wiley, John & Sons, Inc. ISBN 0-471-60386-4.

- R. F. Streater and A. S. Wightman (1964). *PCT, spin and statistics, and all that*. Benjamin/Cummings. ISBN 0-691-07062-8.

20.7   External links

- Background information on Lorentz and CPT violation by Alan Kostelecký at Theoretical Physics Indiana University

- Data Tables for Lorentz and CPT Violation at the arXiv

- The Pin Groups in Physics: C, P, and T at the arXiv

- Charge, Parity, and Time Reversal (CPT) Symmetry at LBL

- CPT Invariance Tests in Neutral Kaon Decay at LBL

- Space--Time Symmetry, CPT and Mirror Fermions at the arXiv
  8-component theory for fermions in which *T-parity* can be a complex number with unit radius. The CPT invariance is not a theorem but a *better to have* property in these class of theories.

Chapter 21

CP violation

In particle physics, **CP violation** (CP standing for **charge parity**) is a violation of the postulated **CP-symmetry** (or **charge conjugation parity symmetry**): the combination of C-symmetry (charge conjugation symmetry) and P-symmetry (parity symmetry). CP-symmetry states that the laws of physics should be the same if a particle is interchanged with its antiparticle (C symmetry), and when its spatial coordinates are inverted ("mirror" or P symmetry). The discovery of CP violation in 1964 in the decays of neutral kaons resulted in the Nobel Prize in Physics in 1980 for its discoverers James Cronin and Val Fitch.

It plays an important role both in the attempts of cosmology to explain the dominance of matter over antimatter in the present Universe, and in the study of weak interactions in particle physics.

21.1 CP-symmetry

CP-symmetry, often called just *CP*, is the product of two symmetries: C for charge conjugation, which transforms a particle into its antiparticle, and P for parity, which creates the mirror image of a physical system. The strong interaction and electromagnetic interaction seem to be invariant under the combined CP transformation operation, but this symmetry is slightly violated during certain types of weak decay. Historically, CP-symmetry was proposed to restore order after the discovery of parity violation in the 1950s.

The idea behind parity symmetry is that the equations of particle physics are invariant under mirror inversion. This leads to the prediction that the mirror image of a reaction (such as a chemical reaction or radioactive decay) occurs at the same rate as the original reaction. Parity symmetry appears to be valid for all reactions involving electromagnetism and strong interactions. Until 1956, parity conservation was believed to be one of the fundamental geometric conservation laws (along with conservation of energy and conservation of momentum). However, in 1956 a careful critical review of the existing experimental data by theoretical physicists Tsung-Dao Lee and Chen Ning Yang revealed that while parity conservation had been verified in decays by the strong or electromagnetic interactions, it was untested in the weak interaction. They proposed several possible direct experimental tests. The first test based on beta decay of cobalt-60 nuclei was carried out in 1956 by a group led by Chien-Shiung Wu, and demonstrated conclusively that weak interactions violate the P symmetry or, as the analogy goes, some reactions did not occur as often as their mirror image.

Overall, the symmetry of a quantum mechanical system can be restored if another symmetry S can be found such that the combined symmetry PS remains unbroken. This rather subtle point about the structure of Hilbert space was realized shortly after the discovery of P violation, and it was proposed that charge conjugation was the desired symmetry to restore order.

Simply speaking, charge conjugation is a symmetry between particles and antiparticles, and so CP-symmetry was proposed in 1957 by Lev Landau as the true symmetry between matter and antimatter. In other words, a process in which all particles are exchanged with their antiparticles was assumed to be equivalent to the mirror image of the original process.

21.1.1   CP violation in the Standard Model

"Direct" CP violation is allowed in the Standard Model if a complex phase appears in the CKM matrix describing quark mixing, or the PMNS matrix describing neutrino mixing. A necessary condition for the appearance of the complex phase is the presence of at least three generations of quarks (if fewer generations are present, the complex phase parameter can be absorbed into redefinitions of the quark fields).

The reason why such a complex phase causes CP violation is not immediately obvious, but can be seen as follows. Consider any given particles (or sets of particles) a and b, and their antiparticles $\bar{a}$ and $\bar{b}$. Now consider the processes $a \rightarrow b$ and the corresponding antiparticle process $\bar{a} \rightarrow \bar{b}$, and denote their amplitudes M and $\bar{M}$ respectively. Before CP violation, these terms must be the *same* complex number. We can separate the magnitude and phase by writing $M = |M|e^{i\theta}$. If a phase term is introduced from (e.g.) the CKM matrix, denote it $e^{i\phi}$. Note that $\bar{M}$ contains the conjugate matrix to M, so it picks up a phase term $e^{-i\phi}$. Now we have:

$$M = |M|e^{i\theta}e^{i\phi}$$

$$\bar{M} = |M|e^{i\theta}e^{-i\phi}$$

Physically measurable reaction rates are proportional to $|M|^2$, thus so far nothing is different. However, consider that there are *two different routes* (e.g. intermediate states) for $a \rightarrow b$. Now we have:

$$M = |M_1|e^{i\theta_1}e^{i\phi_1} + |M_2|e^{i\theta_2}e^{i\phi_2}$$

$$\bar{M} = |M_1|e^{i\theta_1}e^{-i\phi_1} + |M_2|e^{i\theta_2}e^{-i\phi_2}$$

Some further calculation gives:

$$|M|^2 - |\bar{M}|^2 = 4|M_1||M_2|\sin(\theta_1 - \theta_2)\sin(\phi_1 - \phi_2)$$

Thus, we see that a complex phase gives rise to processes that proceed at different rates for particles and antiparticles, and CP is violated.

21.2   Experimental status

21.2.1   Indirect CP violation

In 1964, James Cronin, Val Fitch and coworkers provided clear evidence (which was first announced at the 12th ICHEP conference in Dubna) that CP-symmetry could be broken. This work[1] won them the 1980 Nobel Prize. This discovery showed that weak interactions violate not only the charge-conjugation symmetry C between particles and antiparticles and the P or parity, but also their combination. The discovery shocked particle physics and opened the door to questions still at the core of particle physics and of cosmology today. The lack of an exact CP-symmetry, but also the fact that it is so nearly a symmetry, created a great puzzle.

Only a weaker version of the symmetry could be preserved by physical phenomena, which was CPT symmetry. Besides C and P, there is a third operation, time reversal (T), which corresponds to reversal of motion. Invariance under time reversal implies that whenever a motion is allowed by the laws of physics, the reversed motion is also an allowed one. The combination of CPT is thought to constitute an exact symmetry of all types of fundamental interactions. Because of the CPT symmetry, a violation of the CP-symmetry is equivalent to a violation of the T symmetry. CP violation implied nonconservation of T, provided that the long-held CPT theorem was valid. In this theorem, regarded as one of the basic principles of quantum field theory, charge conjugation, parity, and time reversal are applied together.

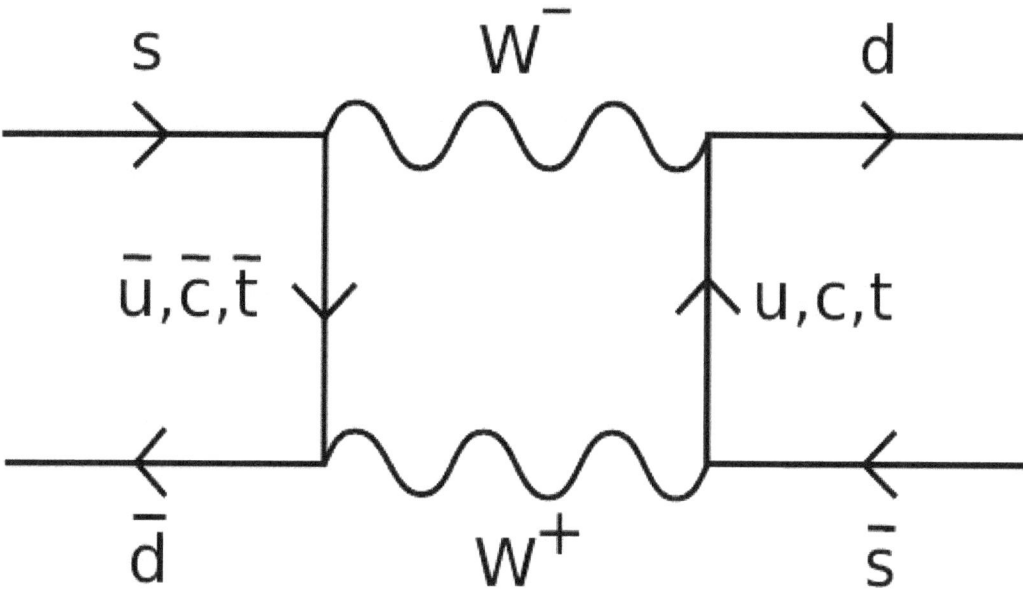

Kaon oscillation box diagram

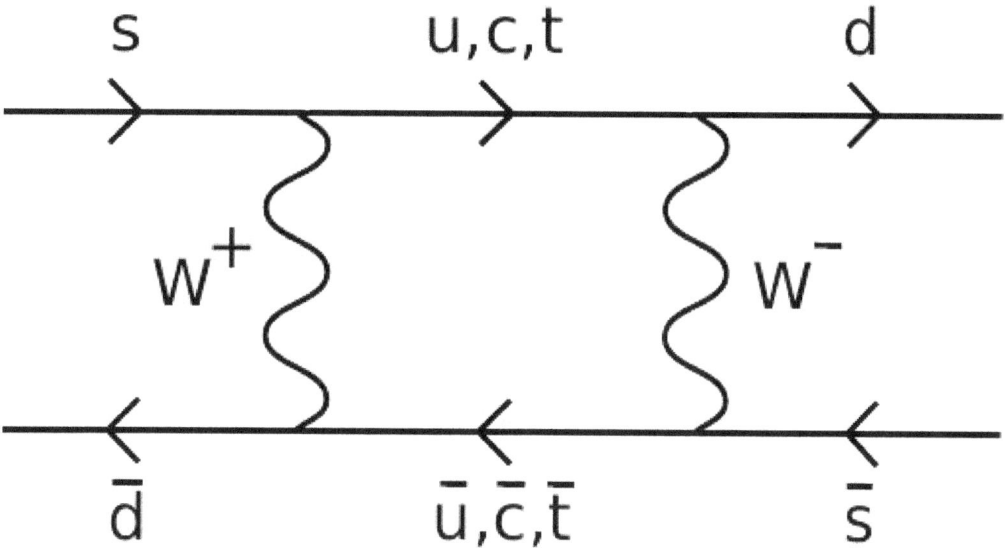

The two box diagrams above are the Feynman diagrams providing the leading contributions to the amplitude of K0-K0 oscillation

21.2.2 Direct CP violation

The kind of CP violation discovered in 1964 was linked to the fact that neutral kaons can transform into their antiparticles (in which each quark is replaced with the other's antiquark) and vice versa, but such transformation does not occur with exactly the same probability in both directions; this is called *indirect* CP violation. Despite many searches, no other manifestation of CP violation was discovered until the 1990s, when the NA31 experiment at CERN suggested evidence for CP violation in the decay process of the very same neutral kaons (*direct* CP violation). The observation was somewhat controversial, and final proof for it came in 1999 from the KTeV experiment at Fermilab[2] and the NA48 experiment at CERN.[3]

In 2001, a new generation of experiments, including the BaBar Experiment at the Stanford Linear Accelerator Center (SLAC)[4] and the Belle Experiment at the High Energy Accelerator Research Organisation (KEK)[5] in Japan, observed direct CP violation in a different system, namely in decays of the B mesons.[6] By now a large number of CP violation processes in B meson decays have been discovered. Before these "B-factory" experiments, there was a logical possibility that all CP violation was confined to kaon physics. However, this raised the question of why it's *not* extended to the strong force, and furthermore, why this is not predicted in the unextended Standard Model, despite the model being undeniably accurate with "normal" phenomena.

In 2011, a first indication of CP violation in decays of neutral D mesons was reported by the LHCb experiment at CERN.[7]

21.3   Strong CP problem

There is no experimentally known violation of the CP-symmetry in quantum chromodynamics. As there is no known reason for it to be conserved in QCD specifically, this is a "fine tuning" problem known as the strong CP problem.

QCD does not violate the CP-symmetry as easily as the electroweak theory; unlike the electroweak theory in which the gauge fields couple to chiral currents constructed from the fermionic fields, the gluons couple to vector currents. Experiments do not indicate any CP violation in the QCD sector. For example, a generic CP violation in the strongly interacting sector would create the electric dipole moment of the neutron which would be comparable to 10^{-18} e·m while the experimental upper bound is roughly one trillionth that size.

This is a problem because at the end, there are natural terms in the QCD Lagrangian that are able to break the CP-symmetry.

$$\mathcal{L} = -\frac{1}{4}F_{\mu\nu}F^{\mu\nu} - \frac{n_f g^2 \theta}{32\pi^2}F_{\mu\nu}\check{F}^{\mu\nu} + \bar{\psi}(i\gamma^\mu D_\mu - me^{i\theta'\gamma_5})\psi$$

For a nonzero choice of the θ angle and the chiral quark mass phase θ' one expects the CP-symmetry to be violated. One usually assumes that the chiral quark mass phase can be converted to a contribution to the total effective $\bar{\theta}$ angle, but it remains to be explained why this angle is extremely small instead of being of order one; the particular value of the θ angle that must be very close to zero (in this case) is an example of a fine-tuning problem in physics, and is typically solved by physics beyond the Standard Model.

There are several proposed solutions to solve the strong CP problem. The most well-known is Peccei–Quinn theory, involving new scalar particles called axions. A newer, more radical approach not requiring the axion is a theory involving two time dimensions first proposed in 1998 by Bars, Deliduman, and Andreev.[8]

21.3.1   Little CP problem

The little CP problem is a term coined by Lisa Randall. It refers to an issue related to the enhanced new physics contributions to the electric dipole moment (EDM) of the neutron in flavor anarchic models.[9]

21.4   CP violation and the matter–antimatter imbalance

Main article: Baryogenesis

The universe is made chiefly of matter, rather than consisting of equal parts of matter and antimatter as might be expected. It can be demonstrated that, to create an imbalance in matter and antimatter from an initial condition of balance, the

Sakharov conditions must be satisfied, one of which is the existence of CP violation during the extreme conditions of the first seconds after the Big Bang. Explanations which do not involve CP violation are less plausible, since they rely on the assumption that the matter–antimatter imbalance was present at the beginning, or on other admittedly exotic assumptions.

The Big Bang should have produced equal amounts of matter and antimatter if CP-symmetry was preserved; as such, there should have been total cancellation of both—protons should have cancelled with antiprotons, electrons with positrons, neutrons with antineutrons, and so on. This would have resulted in a sea of radiation in the universe with no matter. Since this is not the case, after the Big Bang, physical laws must have acted differently for matter and antimatter, i.e. violating CP-symmetry.

The Standard Model contains at least three sources of CP violation. The first of these, involving the Cabibbo–Kobayashi–Maskawa matrix in the quark sector, has been observed experimentally and can only account for a small portion of the CP violation required to explain the matter-antimatter asymmetry. The strong interaction should also violate CP, in principle, but the failure to observe the electric dipole moment of the neutron in experiments suggests that any CP violation in the strong sector is also too small to account for the necessary CP violation in the early universe. The third source of CP violation is the Pontecorvo–Maki–Nakagawa–Sakata matrix in the lepton sector. Current neutrino experiments are not yet sensitive enough to allow experimental observation of CP violation in the lepton sector, but the NOvA experiment currently under construction could observe some small fraction of possible CP violating phases and proposed neutrino experiments Hyper-Kamiokande and LBNE will be sensitive to a relatively large fraction of CP violating phases. Further into the future, a neutrino factory could be sensitive to nearly all possible CP violating phases. If neutrinos are Majorana fermions, the PMNS matrix could have two independent CP violating phases leading to a fourth source of CP violation within the Standard Model. The experimental evidence for Majorana neutrinos would be the observation of neutrinoless double-beta decay. As of September 2013, the best limits come from the GERDA experiment. CP violation in the lepton sector generates a matter-antimatter asymmetry through a process called leptogenesis. This could become the preferred explanation in the Standard Model for the matter-antimatter asymmetry of the universe once CP violation is experimentally confirmed in the lepton sector.

If CP violation in the lepton sector is experimentally determined to be too small to account for matter-antimatter asymmetry, some new physics beyond the Standard Model would be required to explain additional sources of CP violation. Fortunately, it is generally the case that adding new particles and/or interactions to the Standard Model introduces new sources of CP violation since CP is not a symmetry of nature.

21.5 See also

- B-factory
- LHCb
- BTeV experiment
- Cabibbo–Kobayashi–Maskawa matrix
- Penguin diagram
- Neutral particle oscillation

21.6 References

[1] "Evidence for the 2π Decay of the K0
2 Meson System". *Physical Review Letters* **13**: 138. 1964. Bibcode:1964PhRvL..13..138C. doi:10.1103/PhysRevLett.13.138.

[2] "Observation of Direct CP Violation in KS,L→ππ Decays". *Physical Review Letters* **83**: 22. 1999. arXiv:hep-ex/9905060. Bibcode:1999PhRvL..83...22A. doi:10.1103/PhysRevLett.83.22.

[3] NA48 Collaboration, V. Fanti, A. Lai, D. Marras, L. Musa et al. (1999). "A new measurement of direct CP violation in two pion decays of the neutral kaon". *Physics Letters B* **465** (1–4): 335–348. arXiv:hep-ex/9909022. Bibcode:1999PhLB..465..335F. doi:10.1016/S0370-2693(99)01030-8.

[4] "Measurement of CP-Violating Asymmetries in B^0 Decays to CP Eigenstates". *Physical Review Letters* **86**: 2515. 2001. arXiv:hep-ex/0102030. Bibcode:2001PhRvL..86.2515A. doi:10.1103/PhysRevLett.86.2515.

[5] "Observation of Large CP Violation in the Neutral B Meson System". *Physical Review Letters* **87**. 2001. arXiv:hep-ex/0107061. Bibcode:2001PhRvL..87i1802A. doi:10.1103/PhysRevLett.87.091802.

[6] Rodgers, Peter (August 2001). "Where did all the antimatter go?". *Physics World*. p. 11.

[7] Carbone, A. (2012). "A search for time-integrated CP violation in $D^0 \rightarrow h^- h^+$ decays". arXiv:1210.8257.

[8] I. Bars; C. Deliduman; O. Andreev (1998). "Gauged Duality, Conformal Symmetry, and Spacetime with Two Times". *Physical Review D* **58** (6): 066004. arXiv:hep-th/9803188. Bibcode:1998PhRvD..58f6004B. doi:10.1103/PhysRevD.58.066004.

[9] Kadosh, Avihay; Pallante, Elisabetta (2011). "CP violation and FCNC in a warped A_4 flavor model". *Journal of High Energy Physics* **2011** (6). arXiv:1101.5420. Bibcode:2011JHEP...06..121K. doi:10.1007/JHEP06(2011)121.

21.7   Further reading

- Sozzi, M.S. (2008). *Discrete symmetries and CP violation*. Oxford University Press. ISBN 978-0-19-929666-8.

- G. C. Branco, L. Lavoura and J. P. Silva (1999). *CP violation*. Clarendon Press. ISBN 0-19-850399-7.

- I. Bigi and A. Sanda (1999). *CP violation*. Cambridge University Press. ISBN 0-521-44349-0.

- Michael Beyer, ed. (2002). *CP Violation in Particle, Nuclear and Astrophysics*. Springer. ISBN 3-540-43705-3. *(A collection of essays introducing the subject, with an emphasis on experimental results.)*

- L. Wolfenstein (1989). *CP violation*. North–Holland Publishing. ISBN 0-444-88081-X. *(A compilation of reprints of numerous important papers on the topic, including papers by T.D. Lee, Cronin, Fitch, Kobayashi and Maskawa, and many others.)*

- David J. Griffiths (1987). *Introduction to Elementary Particles*. John Wiley & Sons. ISBN 0-471-60386-4.

- Bigi, I. (1997). "CP Violation — An Essential Mystery in Nature's Grand Design". *Surveys of High Energy Physics* **12**: 269–336. arXiv:hep-ph/9712475. Bibcode:1997hep.ph...12475B. doi:10.1080/01422419808228861.

- Mark Trodden (1998). "Electroweak Baryogenesis". *Reviews of Modern Physics* **71** (5): 1463. arXiv:hep-ph/9803479. Bibcode:1999RvMP...71.1463T. doi:10.1103/RevModPhys.71.1463.

- Davide Castelvecchi. "What is direct CP-violation?". SLAC. Retrieved 2009-07-01.

21.8   External links

- Cern Courier article

Chapter 22

Supersymmetry

"SUSY" redirects here. For other uses, see Susy (disambiguation).
For the episode of the American TV series *Angel*, see Supersymmetry (Angel).

Supersymmetry (**SUSY**), a theory of particle physics, is a proposed type of spacetime symmetry that relates two basic classes of elementary particles: bosons, which have an integer-valued spin, and fermions, which have a half-integer spin.[1] Each particle from one group is associated with a particle from the other, known as its superpartner, the spin of which differs by a half-integer. In a theory with perfectly "unbroken" supersymmetry, each pair of superpartners would share the same mass and internal quantum numbers besides spin. For example, there would be a "selectron" (superpartner electron), a bosonic version of the electron with the same mass as the electron, that would be easy to find in a laboratory. Thus, since no superpartners have been observed, if supersymmetry exists it must be a spontaneously broken symmetry so that superpartners may differ in mass.[2][3] Spontaneously-broken supersymmetry could solve many mysterious problems in particle physics including the hierarchy problem. The simplest realization of spontaneously-broken supersymmetry, the so-called Minimal Supersymmetric Standard Model, is one of the best studied candidates for physics beyond the Standard Model.

There is only indirect evidence and motivation for the existence of supersymmetry. Direct confirmation would entail production of superpartners in collider experiments, such as the Large Hadron Collider (LHC). The first run of the LHC found no evidence for supersymmetry (all results were consistent with the Standard Model), and thus set limits on superpartner masses in supersymmetric theories. Whilst many remain enthusiastic about supersymmetry,[4] this first run at the LHC led some physicists to explore other ideas.[5] In any case, in 2015 the LHC resumed its search for supersymmetry and other new physics in its second run.

22.1 Motivations

There are numerous phenomenological motivations for supersymmetry close to the electroweak scale, as well as technical motivations for supersymmetry at any scale.

22.1.1 The hierarchy problem

Supersymmetry close to the electroweak scale ameliorates the hierarchy problem that afflicts the Standard Model. In the Standard Model, the electroweak scale receives enormous Planck-scale quantum corrections. The observed hierarchy between the electroweak scale and the Planck scale must be achieved with extraordinary fine tuning. In a supersymmetric theory, on the other hand, Planck-scale quantum corrections cancel between partners and superpartners (owing to a minus sign associated with fermionic loops). The hierarchy between the electroweak scale and the Planck scale is achieved in a natural manner, without miraculous fine-tuning.

22.1.2   Gauge coupling unification

The idea that the gauge symmetry groups unify at high-energy is called Grand unification theory. In the Standard Model, however, the weak, strong and electromagnetic couplings fail to unify at high energy. In a supersymmetry theory, the running of the gauge couplings are modified, and precise high-energy unification of the gauge couplings is achieved. The modified running also provides a natural mechanism for radiative electroweak symmetry breaking.

22.1.3   Dark matter

TeV-scale supersymmetry (augmented with a discrete symmetry) typically provides a candidate dark matter particle at a mass scale consistent with thermal relic abundance calculations.[6][7]

22.1.4   Other technical motivations

Supersymmetry is also motivated by solutions to several theoretical problems, for generally providing many desirable mathematical properties, and for ensuring sensible behavior at high energies. Supersymmetric quantum field theory is often much easier to analyze, as many more problems become exactly solvable. When supersymmetry is imposed as a *local* symmetry, Einstein's theory of general relativity is included automatically, and the result is said to be a theory of supergravity. It is also a necessary feature of the most popular candidate for a theory of everything, superstring theory.

Another theoretically appealing property of supersymmetry is that it offers the only "loophole" to the Coleman–Mandula theorem, which prohibits spacetime and internal symmetries from being combined in any nontrivial way, for quantum field theories like the Standard Model with very general assumptions. The Haag-Lopuszanski-Sohnius theorem demonstrates that supersymmetry is the only way spacetime and internal symmetries can be combined consistently.[8]

22.2   History

A supersymmetry relating mesons and baryons was first proposed, in the context of hadronic physics, by Hironari Miyazawa during 1966. This supersymmetry did not involve spacetime, that is, it concerned internal symmetry, and was broken badly. Miyazawa's work was largely ignored at the time.[9][10][11][12]

J. L. Gervais and B. Sakita (during 1971),[13] Yu. A. Golfand and E. P. Likhtman (also during 1971), and D.V. Volkov and V.P. Akulov (1972),[14] independently rediscovered supersymmetry in the context of quantum field theory, a radically new type of symmetry of spacetime and fundamental fields, which establishes a relationship between elementary particles of different quantum nature, bosons and fermions, and unifies spacetime and internal symmetries of microscopic phenomena. Supersymmetry with a consistent Lie-algebraic graded structure on which the Gervais–Sakita rediscovery was based directly first arose during 1971[15] in the context of an early version of string theory by Pierre Ramond, John H. Schwarz and André Neveu.

Finally, Julius Wess and Bruno Zumino (during 1974)[16] identified the characteristic renormalization features of four-dimensional supersymmetric field theories, which identified them as remarkable QFTs, and they and Abdus Salam and their fellow researchers introduced early particle physics applications. The mathematical structure of supersymmetry (Graded Lie superalgebras) has subsequently been applied successfully to other topics of physics, ranging from nuclear physics,[17][18] critical phenomena,[19] quantum mechanics to statistical physics. It remains a vital part of many proposed theories of physics.

The first realistic supersymmetric version of the Standard Model was proposed during 1977 by Pierre Fayet and is known as the Minimal Supersymmetric Standard Model or MSSM for short. It was proposed to solve, amongst other things, the hierarchy problem.

22.3   Applications

22.3.1   Extension of possible symmetry groups

One reason that physicists explored supersymmetry is because it offers an extension to the more familiar symmetries of quantum field theory. These symmetries are grouped into the Poincaré group and internal symmetries and the Coleman–Mandula theorem showed that under certain assumptions, the symmetries of the S-matrix must be a direct product of the Poincaré group with a compact internal symmetry group or if there is not any mass gap, the conformal group with a compact internal symmetry group. During 1971 Golfand and Likhtman were the first to show that the Poincaré algebra can be extended through introduction of four anticommuting spinor generators (in four dimensions), which later became known as supercharges. During 1975 the Haag-Lopuszanski-Sohnius theorem analyzed all possible superalgebras in the general form, including those with an extended number of the supergenerators and central charges. This extended super-Poincaré algebra paved the way for obtaining a very large and important class of supersymmetric field theories.

The supersymmetry algebra

Main article: Supersymmetry algebra

Traditional symmetries of physics are generated by objects that transform by the tensor representations of the Poincaré group and internal symmetries. Supersymmetries, however, are generated by objects that transform by the spinor representations. According to the spin-statistics theorem, bosonic fields commute while fermionic fields anticommute. Combining the two kinds of fields into a single algebra requires the introduction of a $\mathbf{Z}_2$-grading under which the bosons are the even elements and the fermions are the odd elements. Such an algebra is called a Lie superalgebra.

The simplest supersymmetric extension of the Poincaré algebra is the Super-Poincaré algebra. Expressed in terms of two Weyl spinors, has the following anti-commutation relation:

$$\{Q_\alpha, \bar{Q}\beta\} = 2(\sigma^\mu)_{\alpha\beta}P_\mu$$

and all other anti-commutation relations between the Qs and commutation relations between the Qs and Ps vanish. In the above expression $P_\mu = -i\partial_\mu$ are the generators of translation and σ^μ are the Pauli matrices.

There are representations of a Lie superalgebra that are analogous to representations of a Lie algebra. Each Lie algebra has an associated Lie group and a Lie superalgebra can sometimes be extended into representations of a Lie supergroup.

22.3.2   The Supersymmetric Standard Model

Main article: Minimal Supersymmetric Standard Model

Incorporating supersymmetry into the Standard Model requires doubling the number of particles since there is no way that any of the particles in the Standard Model can be superpartners of each other. With the addition of new particles, there are many possible new interactions. The simplest possible supersymmetric model consistent with the Standard Model is the Minimal Supersymmetric Standard Model (MSSM) which can include the necessary additional new particles that are able to be superpartners of those in the Standard Model.

One of the main motivations for SUSY comes from the quadratically divergent contributions to the Higgs mass squared. The quantum mechanical interactions of the Higgs boson causes a large renormalization of the Higgs mass and unless there is an accidental cancellation, the natural size of the Higgs mass is the greatest scale possible. This problem is known as the hierarchy problem. Supersymmetry reduces the size of the quantum corrections by having automatic cancellations between fermionic and bosonic Higgs interactions. If supersymmetry is restored at the weak scale, then the Higgs mass is related to supersymmetry breaking which can be induced from small non-perturbative effects explaining the vastly different scales in the weak interactions and gravitational interactions.

In many supersymmetric Standard Models there is a heavy stable particle (such as neutralino) which could serve as a weakly interacting massive particle (WIMP) dark matter candidate. The existence of a supersymmetric dark matter candidate is related closely to R-parity.

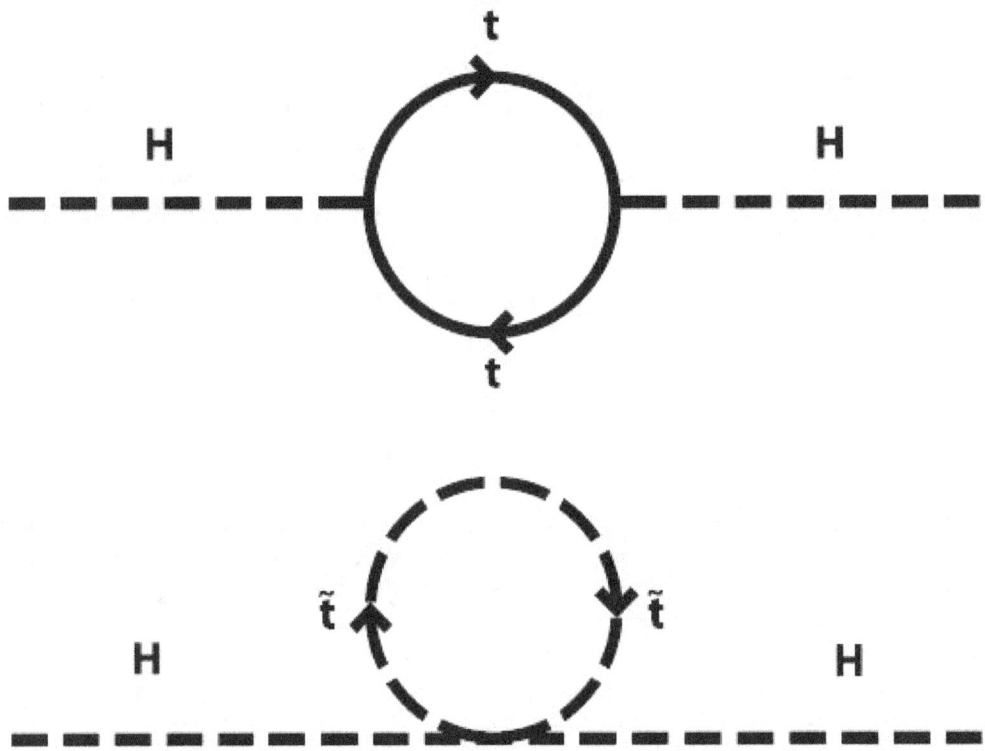

Cancellation of the Higgs boson quadratic mass renormalization between fermionic top quark loop and scalar stop squark tadpole Feynman diagrams in a supersymmetric extension of the Standard Model

The standard paradigm for incorporating supersymmetry into a realistic theory is to have the underlying dynamics of the theory be supersymmetric, but the ground state of the theory does not respect the symmetry and supersymmetry is broken spontaneously. The supersymmetry break can not be done permanently by the particles of the MSSM as they currently appear. This means that there is a new sector of the theory that is responsible for the breaking. The only constraint on this new sector is that it must break supersymmetry permanently and must give superparticles TeV scale masses. There are many models that can do this and most of their details do not matter. In order to parameterize the relevant features of supersymmetry breaking, arbitrary soft SUSY breaking terms are added to the theory which temporarily break SUSY explicitly but could never arise from a complete theory of supersymmetry breaking.

Gauge-coupling unification

Main article: Minimal Supersymmetric Standard Model § Gauge-coupling unification

One piece of evidence for supersymmetry existing is gauge coupling unification. The renormalization group evolution of the three gauge coupling constants of the Standard Model is somewhat sensitive to the present particle content of the theory. These coupling constants do not quite meet together at a common energy scale if we run the renormalization group using the Standard Model.[20] With the addition of minimal SUSY joint convergence of the coupling constants is projected at approximately 10^{16} GeV.[20]

22.3.3 Supersymmetric quantum mechanics

Main article: Supersymmetric quantum mechanics

Supersymmetric quantum mechanics adds the SUSY superalgebra to quantum mechanics as opposed to quantum field theory. Supersymmetric quantum mechanics often becomes relevant when studying the dynamics of supersymmetric solitons, and due to the simplified nature of having fields which are only functions of time (rather than space-time), a great deal of progress has been made in this subject and it is now studied in its own right.

SUSY quantum mechanics involves pairs of Hamiltonians which share a particular mathematical relationship, which are called *partner Hamiltonians*. (The potential energy terms which occur in the Hamiltonians are then known as *partner potentials*.) An introductory theorem shows that for every eigenstate of one Hamiltonian, its partner Hamiltonian has a corresponding eigenstate with the same energy. This fact can be exploited to deduce many properties of the eigenstate spectrum. It is analogous to the original description of SUSY, which referred to bosons and fermions. We can imagine a "bosonic Hamiltonian", whose eigenstates are the various bosons of our theory. The SUSY partner of this Hamiltonian would be "fermionic", and its eigenstates would be the theory's fermions. Each boson would have a fermionic partner of equal energy.

22.3.4 Supersymmetry: Applications to condensed matter physics

SUSY concepts have provided useful extensions to the WKB approximation. Additionally, SUSY has been applied to disorder averaged systems both quantum and non-quantum (through statistical mechanics), the Fokker-Planck equation being an example of a non-quantum theory. The 'supersymmetry' in all these systems arises from the fact that one is modelling one particle and as such the 'statistics' don't matter. The use of the supersymmetry method provides a mathematical rigorous alternative to the replica trick, but only in non-interacting systems, which attempts to address the so-called 'problem of the denominator' under disorder averaging. For more on the applications of supersymmetry in condensed matter physics see the book[21]

22.3.5 Supersymmetry in optics

Integrated optics was recently found[22] to provide a fertile ground on which certain ramifications of SUSY can be explored in readily-accessible laboratory settings. Making use of the analogous mathematical structure of the quantum-mechanical Schrödinger equation and the wave equation governing the evolution of light in one-dimensional settings, one may interpret the refractive index distribution of a structure as a potential landscape in which optical wave packets propagate. In this manner, a new class of functional optical structures with possible applications in phase matching, mode conversion[23] and space-division multiplexing becomes possible. SUSY transformations have been also proposed as a way to address inverse scattering problems in optics and as a one-dimensional transformation optics [24]

22.3.6 Mathematics

SUSY is also sometimes studied mathematically for its intrinsic properties. This is because it describes complex fields satisfying a property known as holomorphy, which allows holomorphic quantities to be exactly computed. This makes supersymmetric models useful "toy models" of more realistic theories. A prime example of this has been the demonstration of S-duality in four-dimensional gauge theories[25] that interchanges particles and monopoles.

The proof of the Atiyah-Singer index theorem is much simplified by the use of supersymmetric quantum mechanics.

22.4 General supersymmetry

Supersymmetry appears in many related contexts of theoretical physics. It is possible to have multiple supersymmetries and also have supersymmetric extra dimensions.

22.4.1 Extended supersymmetry

Main article: Extended supersymmetry

It is possible to have more than one kind of supersymmetry transformation. Theories with more than one supersymmetry transformation are known as extended supersymmetric theories. The more supersymmetry a theory has, the more constrained are the field content and interactions. Typically the number of copies of a supersymmetry is a power of 2, i.e. 1, 2, 4, 8. In four dimensions, a spinor has four degrees of freedom and thus the minimal number of supersymmetry generators is four in four dimensions and having eight copies of supersymmetry means that there are 32 supersymmetry generators.

The maximal number of supersymmetry generators possible is 32. Theories with more than 32 supersymmetry generators automatically have massless fields with spin greater than 2. It is not known how to make massless fields with spin greater than two interact, so the maximal number of supersymmetry generators considered is 32. This corresponds to an $N = 8$ supersymmetry theory. Theories with 32 supersymmetries automatically have a graviton.

For four dimensions there are the following theories, with the corresponding multiplets[26](CPT adds a copy, whenever they are not invariant under such symmetry)

- $N = 1$

Chiral multiplet: $(0, \frac{1}{2})$ Vector multiplet: $(\frac{1}{2}, 1)$ Gravitino multiplet: $(1, \frac{3}{2})$ Graviton multiplet: $(\frac{3}{2}, 2)$

- $N = 2$

hypermultiplet: $(-\frac{1}{2}, 0^2, \frac{1}{2})$ vector multiplet: $(0, \frac{1}{2}^2, 1)$ supergravity multiplet: $(1, \frac{3}{2}^2, 2)$

- $N = 4$

Vector multiplet: $(-1, -\frac{1}{2}^4, 0^6, \frac{1}{2}^4, 1)$ Supergravity multiplet: $(0, \frac{1}{2}^4, 1^6, \frac{3}{2}^4, 2)$

- $N = 8$

Supergravity multiplet: $(-2, -\frac{3}{2}^8, -1^{28}, -\frac{1}{2}^{56}, 0^{70}, \frac{1}{2}^{56}, 1^{28}, \frac{3}{2}^8, 2)$

22.4.2 Supersymmetry in alternate numbers of dimensions

It is possible to have supersymmetry in dimensions other than four. Because the properties of spinors change drastically between different dimensions, each dimension has its characteristic. In d dimensions, the size of spinors is approximately $2^{d/2}$ or $2^{(d-1)/2}$. Since the maximum number of supersymmetries is 32, the greatest number of dimensions in which a supersymmetric theory can exist is eleven.

22.5 Supersymmetry as a quantum group

Main article: Supersymmetry as a quantum group

Supersymmetry can be reinterpreted in the language of noncommutative geometry and quantum groups. In particular, it involves a mild form of noncommutativity, namely supercommutativity. See the main article for more details.

22.6 Supersymmetry in quantum gravity

Supersymmetry is part of a larger enterprise of theoretical physics to unify everything we know about the universe into a single consistent set of physical principles, known as the quest for a Theory of Everything (TOE). A significant part of this larger enterprise is the quest for a theory of quantum gravity, which would unify the classical theory of general relativity and the Standard Model, which explains the other three basic forces in physics (electromagnetism, the strong interaction, and the weak interaction), and provides a palette of fundamental particles upon which all four forces act. Two of the most active methods of forming a theory of quantum gravity are string theory and loop quantum gravity (LQG), although in theory, supersymmetry could be a component of other theories as well.

For string theory to be consistent, supersymmetry seems to be required at some level (although it may be a strongly broken symmetry). In particle theory, supersymmetry is recognized as a way to stabilize the hierarchy between the unification scale and the electroweak scale (or the Higgs boson mass), and can also provide a natural dark matter candidate. String theory also requires extra spatial dimensions which have to be compactified as in Kaluza–Klein theory.

Loop quantum gravity (LQG) predicts no additional spatial dimensions, nor anything else about particle physics. These theories can be formulated in three spatial dimensions and one dimension of time, although in some LQG theories dimensionality is an emergent property of the theory, rather than a fundamental assumption of the theory. Also, LQG is a theory of quantum gravity which does not require supersymmetry. Lee Smolin, one of the originators of LQG, has proposed that a loop quantum gravity theory incorporating either supersymmetry or extra dimensions, or both, be called "loop quantum gravity II".

If experimental evidence confirms supersymmetry in the form of supersymmetric particles such as the neutralino that is often believed to be the lightest superpartner, some people believe this would be a major boost to string theory. Since supersymmetry is a required component of string theory, any discovered supersymmetry would be consistent with string theory. If the Large Hadron Collider and other major particle physics experiments fail to detect supersymmetric partners or evidence of extra dimensions, many versions of string theory which had predicted certain low mass superpartners to existing particles may need to be significantly revised. The failure of experiments to discover either supersymmetric partners or extra spatial dimensions, as of 2013, has encouraged loop quantum gravity researchers.

22.7 Current status

Supersymmetric models are constrained by a variety of experiments, including measurements of low-energy observables – for example, the anomalous magnetic moment of the muon at Brookhaven; the WMAP dark matter density measurement and direct detection experiments – for example, XENON−100 and LUX; and by particle collider experiments, including B-physics, Higgs phenomenology and direct searches for superpartners (sparticles), at the Large Electron–Positron Collider, Tevatron and the LHC.

Historically, the tightest limits were from direct production at colliders. The first mass limits for squarks and gluinos were made at CERN by the UA1 experiment and the UA2 experiment at the Super Proton Synchrotron. LEP later set very strong limits,.[27] which in 2006 were extended by the D0 experiment at the Tevatron.[28][29] From 2003, WMAP's and Planck's dark matter density measurements have strongly constrained supersymmetry models, which, if they explain dark matter, have to be tuned to invoke a particular mechanism to sufficiently reduce the neutralino density.

Prior to the beginning of the LHC, in 2009 fits of available data to CMSSM and NUHM1 indicated that squarks and gluinos were most likely to have masses in the 500 to 800 GeV range, though values as high as 2.5 TeV were allowed with low probabilities. Neutralinos and sleptons were expected to be quite light, with the lightest neutralino and the lightest stau most likely to be found between 100 to 150 GeV.[30]

The first run of the LHC found no evidence for supersymmetry, and, as a result, surpassed existing experimental limits from the Large Electron–Positron Collider and Tevatron and partially excluded the aforementioned expected ranges.[31]

During 2011 and 2012, the LHC discovered a Higgs boson with a mass of about 125 GeV, and with couplings to fermions and bosons which are consistent with the Standard Model. The MSSM predicts that the mass of the lightest Higgs boson should not be much higher than the mass of the Z boson, and, in the absence of fine tuning (with the supersymmetry breaking scale on the order of 1 TeV), should not exceed 130 GeV. Furthermore, for values of the MSSM parameter *tan* β ≤ 3, it predicts a Higgs mass below 114 GeV over most of the parameter space.[32] This region of Higgs mass was excluded by LEP by 2000. The LHC result is somewhat problematic for the minimal supersymmetric model, as the value of 125 GeV is relatively large for the model and can only be achieved with large radiative loop corrections from top squarks, which many theorists consider to be "unnatural" (see naturalness and fine tuning).[33] On the other hand, the lightest Higgs boson in the MSSM is Standard Model-like, which is consistent with measurements of the Higgs boson couplings at the LHC.

In spite of the null searches and the heavy Higgs, a recent analysis of the constrained minimal supersymmetric Standard Model, the CMSSM, suggests that the model is still compatible with all present experimental constraints.[34][35] The preferred masses for squarks and gluinos is about 2 TeV. The resulting fine-tuning of the electroweak scale, however, is considered "unnatural" (see little hierarchy problem), and some theorists now favor extended supersymmetry models – for example, the NMSSM.

22.8   See also

- Supersymmetric gauge theory

- Wess–Zumino model

- Minimal Supersymmetric Standard Model

- Supersymmetry as a quantum group

- Quantum group

- Supercharge

- Superfield

- Supergeometry

- Supergravity

- Supergroup

- Superspace

22.9   References

[1] Haber, Howie. "SUPERSYMMETRY, PART I (THEORY)" (PDF). *Reviews, Tables and Plots*. Particle Data Group (PDG). Retrieved 8 July 2015.

[2] Martin, Stephen P. (1997). "A Supersymmetry Primer". arXiv:hep-ph/9709356.

[3] Dine, Michael (2007). *Supersymmetry and String Theory: Beyond the Standard Model*. p. 169.

[4] Ellis, John. "The Physics Landscape after the Higgs Discovery at the LHC". *arXiv*. Invited plenary talk at SILAFAE 2014. Retrieved 8 July 2015.

[5] Wolchover, Natalie (November 20, 2012). "Supersymmetry Fails Test, Forcing Physics to Seek New Ideas". *Quanta Magazine*.

[6] Jonathan Feng: Supersymmetric Dark Matter *(pdf)*, University of California, Irvine, 11 May 2007

[7] Torsten Bringmann: The WIMP "Miracle" *(pdf)* University of Hamburg

[8] R. Haag, J. T. Lopuszanski and M. Sohnius, "All Possible Generators Of Supersymmetries Of The S Matrix", Nucl. Phys. B 88 (1975) 257

[9] H.Miyazawa(1966). "Baryon Number Changing Currents".*Prog.Theor.Phys.***36**(6): 1266–1276. Bibcode:1966PThPh..36.M. doi:10.1143/PTP.36.1266.

[10] H.Miyazawa(1968). "Spinor Currents and Symmetries of Baryons and Mesons".*Phys.Rev.***170**(5): 1586–1590. BibcodeM. doi:10.1103/PhysRev.170.1586.

[11] Michio Kaku, *Quantum Field Theory*, ISBN 0-19-509158-2, pg 663.

[12] Peter Freund, *Introduction to Supersymmetry*, ISBN 0-521-35675-X, pages 26-27, 138.

[13] Gervais, J. -L.; Sakita, B. (1971). "Field theory interpretation of supergauges in dual models". *Nuclear Physics B* **34** (2): 632. Bibcode:1971NuPhB..34..632G. doi:10.1016/0550-3213(71)90351-8.

[14] D.V. Volkov, V.P. Akulov, Pisma Zh.Eksp.Teor.Fiz. 16 (1972) 621; Phys.Lett. B46 (1973) 109; V.P. Akulov, D.V. Volkov, Teor.Mat.Fiz. 18 (1974) 39

[15] Ramond, P. (1971). "Dual Theory for Free Fermions". *Physical Review D* **3** (10): 2415. Bibcode:1971PhRvD...3.2415R. doi:10.1103/PhysRevD.3.2415.

[16] Wess,J.;Zumino,B. (1974). "Supergauge transformations in four dimensions".*Nuclear Physics B***70**: 39. Bibcode:1974NuP. doi:10.1016/0550-3213(74)90355-1.

[17] http://users.physik.fu-berlin.de/~{}kleinert/kleinert/?p=supersym suggested here

[18] Iachello,F. (1980). "Dynamical Supersymmetries in Nuclei".*Physical Review Letters***44**(12): 772. Bibcode:1980PhRvL. doi:10.1103/PhysRevLett.44.772.

[19] Friedan, D.; Qiu, Z.; Shenker, S. (1984). "Conformal Invariance, Unitarity, and Critical Exponents in Two Dimensions". *Physical Review Letters* **52** (18): 1575. Bibcode:1984PhRvL..52.1575F. doi:10.1103/PhysRevLett.52.1575.

[20] Gordon L. Kane, *The Dawn of Physics Beyond the Standard Model*, Scientific American, June 2003, page 60 and *The frontiers of physics*, special edition, Vol 15, #3, page 8

[21] *Supersymmetry in Disorder and Chaos*, Konstantin Efetov, Cambridge university press, 1997.

[22] Miri, M.-A.; Heinrich, M.; El-Ganainy, R.; Christodoulides, D. N. (2013). "Superymmetric optical structures". *Physical Review Letters* (APS) **110** (23): 233902. arXiv:1304.6646. Bibcode:2013PhRvL.110w3902M. doi:10.1103/PhysRevLett.110.233902. PMID 25167493. Retrieved April 2014.

[23] Heinrich, M.; Miri, M.-A.; Stützer, S.; El-Ganainy, R.; Nolte, S.; Szameit, A.; Christodoulides, D. N. (2014). "Superymmetric mode converters". *Nature Communications* (NPG) **5**: 3698. arXiv:1401.5734. Bibcode:2014NatCo...5 .PMID24739256. Retrieved April2014.

[24] Miri, M.-A.; Heinrich, Matthias; Christodoulides, D. N. (2014). "SUSY-inspired one-dimensional transformation optics". *Optica* (OSA) **1** (2): 89. arXiv:1408.0832. doi:10.1364/OPTICA.1.000089. Retrieved August 2014.

[25] Krasnitz, Michael (2002). *Correlation functions in supersymmetric gauge theories from supergravity flactuafluctuations hHKtions* (PDF). Princeton University Department of Physics: Princeton University Department of Physics. p. 91.

[26] Polchinski,J. *String theory. Vol. 2: Superstring theory and beyond*. Appendix B

[27] LEPSUSYWG, ALEPH, DELPHI, L3 and OPAL experiments, charginos, large m0 LEPSUSYWG/01-03.1

[28] The D0-Collaboration (2009). "Search for associated production of charginos and neutralinos in the trilepton final state using 2.3 fb^{-1} of data". arXiv:0901.0646. Bibcode:2009PhLB..680...34D. doi:10.1016/j.physletb.2009.08.011.

[29] The D0 Collaboration (2006). "Search for squarks and gluinos in events with jets and missing transverse energy using 2.1 fb-1 of pp¯ collision data at s=1.96 TeV". arXiv:0712.3805. Bibcode:2008PhLB..660..449D. doi:10.1016/j.physletb.2008.01.042.

[30] O. Buchmueller; et al. (2009). "Likelihood Functions for Supersymmetric Observables in Frequentist Analyses of the CMSSM and NUHM1". *The European Physical Journal C* **64** (3): 391–415. arXiv:0907.5568. Bibcode:2009EPJC...64..391B. doi:10. 1140/epjc/s10052-009-1159-z.

[31] Roszkowski, Leszek; Sessolo, Enrico Maria; Williams, Andrew J. (11 August 2014). "What next for the CMSSM and the NUHM:improved prospects for superpartner and dark matter detection".*Journal of High Energy Physics***2014**(8)..

[32] Marcela Carena and Howard E. Haber; Haber (1970). "Higgs Boson Theory and Phenomenology". *Progress in Particle and Nuclear Physics* **50**: 63. arXiv:hep-ph/0208209v3. Bibcode:2003PrPNP..50...63C. doi:10.1016/S0146-6410(02)00177-1.

[33] Patrick Draper; et al. (December 2011). "Implications of a 125 GeV Higgs for the MSSM and Low-Scale SUSY Breaking". *Physical Review D* **85** (9): 095007. arXiv:1112.3068. Bibcode:2012PhRvD..85i5007D. doi:10.1103/PhysRevD.85.095007.

[34] Bechtle, Philip. "How alive is constrained SUSY really?". *arXiv*. Retrieved 8 July 2015.

[35] Jan de Vries, Kees. "SUSY fits with full LHC Run I data". *arXiv*. Retrieved 8 July 2015.

22.10   Further reading

- Supersymmetry and Supergravity page in String Theory Wiki lists more books and reviews.

22.10.1   Theoretical introductions, free and online

- S. Martin (2011). "A Supersymmetry Primer". arXiv:hep-ph/9709356.

- Joseph D. Lykken (1996). "Introduction to Supersymmetry". arXiv:hep-th/9612114.

- Manuel Drees (1996). "An Introduction to Supersymmetry". arXiv:hep-ph/9611409.

- Adel Bilal (2001). "Introduction to Supersymmetry". arXiv:hep-th/0101055.

- An Introduction to Global Supersymmetry by Philip Arygres, 2001

22.10.2   Monographs

- Weak Scale Supersymmetry by Howard Baer and Xerxes Tata, 2006.

- Cooper, F.; Khare, A.; Sukhatme, U. (1995). "Supersymmetry and quantum mechanics". *Physics Reports* **251** (5–6): 267. doi:10.1016/0370-1573(94)00080-M. (arXiv:hep-th/9405029).

- Junker, G. (1996). "Supersymmetric Methods in Quantum and Statistical Physics". doi:10.1007/978-3-642-61194-0. ISBN 978-3-540-61591-0..

- Gordon L. Kane.*Supersymmetry: Unveiling the Ultimate Laws of Nature* Basic Books, New York (2001). ISBN 0-7382-0489-7.

- Gordon L. Kane and Shifman, M., eds. *The Supersymmetric World: The Beginnings of the Theory*, World Scientific, Singapore (2000). ISBN 981-02-4522-X.

- Weinberg, Steven, *The Quantum Theory of Fields, Volume 3: Supersymmetry*, Cambridge University Press, Cambridge, (1999). ISBN 0-521-66000-9.

- Wess, Julius, and Jonathan Bagger, *Supersymmetry and Supergravity*, Princeton University Press, Princeton, (1992). ISBN 0-691-02530-4.

- "Concise Encyclopedia of Supersymmetry". 2003. doi:10.1007/1-4020-4522-0. ISBN 978-1-4020-1338-6.

22.10.3   On experiments

- Bennett GW; Muon (g–2) Collaboration; Bousquet; Brown; Bunce; Carey; Cushman; Danby; Debevec; Deile; Deng; Dhawan; Druzhinin; Duong; Farley; Fedotovich; Gray; Grigoriev; Grosse-Perdekamp; Grossmann; Hare; Hertzog; Huang; Hughes; Iwasaki; Jungmann; Kawall; Khazin; Krienen; Kronkvist; et al. (2004). "Measurement of the negative muon anomalous magnetic moment to 0.7 ppm". *Physical Review Letters* **92** (16): 161802. arXiv:hep-ex/0401008. Bibcode:2004PhRvL..92p1802B. doi:10.1103/PhysRevLett.92.161802. PMID 15169217.

- Brookhaven National Laboratory (Jan. 8, 2004). *New g−2 measurement deviates further from Standard Model.* Press Release.

- Fermi National Accelerator Laboratory (Sept 25, 2006). *Fermilab's CDF scientists have discovered the quick-change behavior of the B-sub-s meson.* Press Release.

22.11   External links

- Supersymmetry (physics) at *Encyclopædia Britannica*

- What do current LHC results (mid-August 2011) imply about supersymmetry? Matt Strassler

- ATLAS Experiment Supersymmetry search documents

- CMS Experiment Supersymmetry search documents

- "Particle wobble shakes up supersymmetry", *Cosmos* magazine, September 2006

- LHC results put supersymmetry theory 'on the spot' BBC news 27/8/2011

- SUSY running out of hiding places BBC news 12/11/2012

- Supersymmetry in optics? "Skulls in the Stars" blog 22/08/2013

Chapter 23

Symmetric group

Not to be confused with Symmetry group.

In abstract algebra, the **symmetric group** Sn on a finite set of n symbols is the group whose elements are all the permutation operations that can be performed on n distinct symbols, and whose group operation is the composition of such permutation operations, which are defined as bijective functions from the set of symbols to itself.[1] Since there are $n!$ (n factorial) possible permutation operations that can be performed on a tuple composed of n symbols, it follows that the order (the number of elements) of the symmetric group Sn is $n!$.

Although symmetric groups can be defined on infinite sets as well, this article discusses only the finite symmetric groups: their applications, their elements, their conjugacy classes, a finite presentation, their subgroups, their automorphism groups, and their representation theory. For the remainder of this article, "symmetric group" will mean a symmetric group on a finite set.

The symmetric group is important to diverse areas of mathematics such as Galois theory, invariant theory, the representation theory of Lie groups, and combinatorics. Cayley's theorem states that every group G is isomorphic to a subgroup of the symmetric group on G.

23.1 Definition and first properties

The **symmetric group** on a finite set X is the group whose elements are all bijective functions from X to X and whose group operation is that of function composition.[1] For finite sets, "permutations" and "bijective functions" refer to the same operation, namely rearrangement. The symmetric group of **degree** n is the symmetric group on the set $X = \{ 1, 2, ..., n \}$.

The symmetric group on a set X is denoted in various ways including SX, $\mathfrak{S}X$, ΣX, $X!$ and Sym(X).[1] If X is the set $\{ 1, 2, ..., n \}$, then the symmetric group on X is also denoted Sn,[1] $\mathfrak{S}n$, Σn, and Sym(n).

Symmetric groups on infinite sets behave quite differently from symmetric groups on finite sets, and are discussed in (Scott 1987, Ch. 11), (Dixon & Mortimer 1996, Ch. 8), and (Cameron 1999). This article concentrates on the finite symmetric groups.

The symmetric group on a set of n elements has order $n!$ [2] It is abelian if and only if $n \leq 2$. For $n = 0$ and $n = 1$ (the empty set and the singleton set) the symmetric group is trivial (note that this agrees with $0! = 1! = 1$), and in these cases the alternating group equals the symmetric group, rather than being an index two subgroup. The group Sn is solvable if and only if $n \leq 4$. This is an essential part of the proof of the Abel–Ruffini theorem that shows that for every $n > 4$ there are polynomials of degree n which are not solvable by radicals, i.e., the solutions cannot be expressed by performing a finite number of operations of addition, subtraction, multiplication, division and root extraction on the polynomial's coefficients.

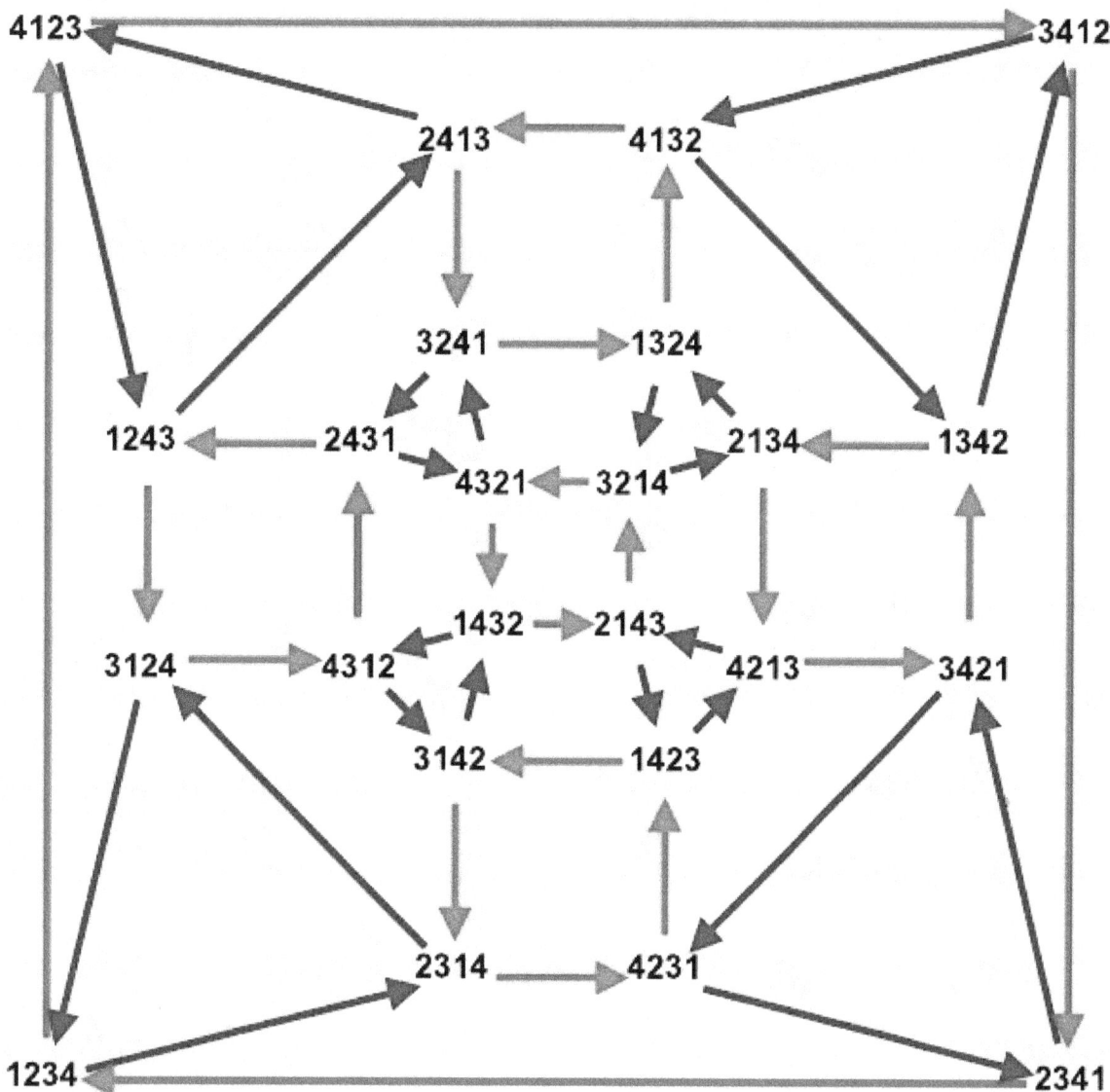

A Cayley graph of the symmetric group S_1

23.2 Applications

The symmetric group on a set of size n is the Galois group of the general polynomial of degree n and plays an important role in Galois theory. In invariant theory, the symmetric group acts on the variables of a multi-variate function, and the functions left invariant are the so-called symmetric functions. In the representation theory of Lie groups, the representation theory of the symmetric group plays a fundamental role through the ideas of Schur functors. In the theory of Coxeter groups, the symmetric group is the Coxeter group of type An and occurs as the Weyl group of the general linear group. In combinatorics, the symmetric groups, their elements (permutations), and their representations provide a rich source of problems involving Young tableaux, plactic monoids, and the Bruhat order. Subgroups of symmetric groups are called permutation groups and are widely studied because of their importance in understanding group actions, homogeneous spaces, and automorphism groups of graphs, such as the Higman–Sims group and the Higman–Sims graph.

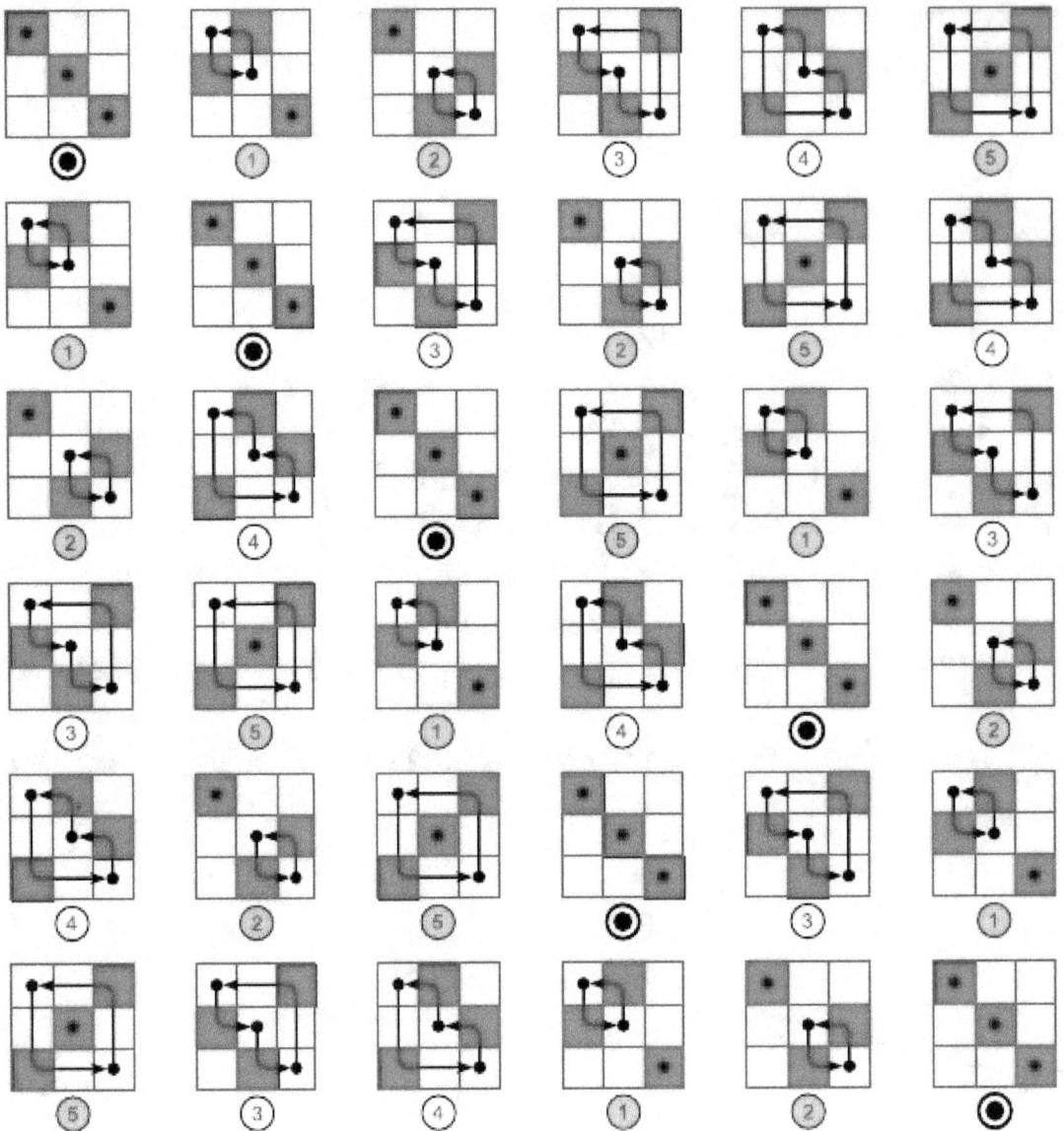

Cayley table of the symmetric group S_3
(multiplication table of permutation matrices)
These are the positions of the six matrices:
Only the unity matrices are arranged symmetrically to the main diagonal - thus the symmetric group is not abelian.

23.3   Elements

The elements of the symmetric group on a set X are the permutations of X.

23.3.1   Multiplication

The group operation in a symmetric group is function composition, denoted by the symbol ∘ or simply by juxtaposition of the permutations. The composition $f \circ g$ of permutations f and g, pronounced "f of g", maps any element x of X to $f(g(x))$. Concretely, let (see permutation for an explanation of notation):

$$f = (1\ 3)(4\ 5) = \begin{pmatrix} 1 & 2 & 3 & 4 & 5 \\ 3 & 2 & 1 & 5 & 4 \end{pmatrix}$$

$$g = (1\ 2\ 5)(3\ 4) = \begin{pmatrix} 1 & 2 & 3 & 4 & 5 \\ 2 & 5 & 4 & 3 & 1 \end{pmatrix}.$$

Applying f after g maps 1 first to 2 and then 2 to itself; 2 to 5 and then to 4; 3 to 4 and then to 5, and so on. So composing f and g gives

$$fg = f \circ g = (1\ 2\ 4)(3\ 5) = \begin{pmatrix} 1 & 2 & 3 & 4 & 5 \\ 2 & 4 & 5 & 1 & 3 \end{pmatrix}.$$

A cycle of length $L = k \cdot m$, taken to the k-th power, will decompose into k cycles of length m: For example ($k = 2$, $m = 3$),

$$(1\ 2\ 3\ 4\ 5\ 6)^2 = (1\ 3\ 5)(2\ 4\ 6).$$

23.3.2   Verification of group axioms

To check that the symmetric group on a set X is indeed a group, it is necessary to verify the group axioms of closure, associativity, identity, and inverses.[3] 1) The operation of function composition is closed in the set of permutations of the given set X, 2) function composition is always associative, 3) The trivial bijection that assigns each element of X to itself serves as an identity for the group, and 4) Every bijection has an inverse function that undoes its action, and thus each element of a symmetric group does have an inverse which is a permutation too.

23.3.3   Transpositions

Main article: Transposition (mathematics)

A **transposition** is a permutation which exchanges two elements and keeps all others fixed; for example (1 3) is a transposition. Every permutation can be written as a product of transpositions; for instance, the permutation g from above can be written as $g = (1\ 2)(2\ 5)(3\ 4)$. Since g can be written as a product of an odd number of transpositions, it is then called an odd permutation, whereas f is an even permutation.

The representation of a permutation as a product of transpositions is not unique; however, the number of transpositions needed to represent a given permutation is either always even or always odd. There are several short proofs of the invariance of this parity of a permutation.

The product of two even permutations is even, the product of two odd permutations is even, and all other products are odd. Thus we can define the **sign** of a permutation:

$$\operatorname{sgn} f = \begin{cases} +1, & \text{if } f \text{ is even} \\ -1, & \text{if } f \text{ odd is} \end{cases}.$$

With this definition,

$$\operatorname{sgn}: S_n \to \{+1, -1\}$$

is a group homomorphism ($\{+1, -1\}$ is a group under multiplication, where $+1$ is e, the neutral element). The kernel of this homomorphism, i.e. the set of all even permutations, is called the **alternating group** An. It is a normal subgroup of Sn, and for $n \geq 2$ it has $n!/2$ elements. The group Sn is the semidirect product of An and any subgroup generated by a single transposition.

Furthermore, every permutation can be written as a product of *adjacent transpositions*, that is, transpositions of the form (a $a+1$). For instance, the permutation g from above can also be written as $g = (4\ 5)(3\ 4)(4\ 5)(1\ 2)(2\ 3)(3\ 4)(4\ 5)$. The sorting algorithm Bubble sort is an application of this fact. The representation of a permutation as a product of adjacent transpositions is also not unique.

23.3.4   Cycles

A cycle of *length* k is a permutation f for which there exists an element x in $\{1,...,n\}$ such that $x, f(x), f^2(x), ..., f^k(x) = x$ are the only elements moved by f: it is required that $k \geq 2$ since with $k = 1$ the element x itself would not be moved either. The permutation h defined by

$$h = \begin{pmatrix} 1 & 2 & 3 & 4 & 5 \\ 4 & 2 & 1 & 3 & 5 \end{pmatrix}$$

is a cycle of length three, since $h(1) = 4$, $h(4) = 3$ and $h(3) = 1$, leaving 2 and 5 untouched. We denote such a cycle by (1 4 3), but it could equally well be written (4 3 1) or (3 1 4) by starting at a different point. The order of a cycle is equal to its length. Cycles of length two are transpositions. Two cycles are *disjoint* if they move disjoint subsets of elements. Disjoint cycles commute, e.g. in S_6 we have $(4\ 1\ 3)(2\ 5\ 6) = (2\ 5\ 6)(4\ 1\ 3)$. Every element of S$n$ can be written as a product of disjoint cycles; this representation is unique up to the order of the factors, and the freedom present in representing each individual cycle by choosing its starting point.

Cycles admits the following conjugation property with any permutation σ, this property is often used to obtain its Generators and relations.

$$\sigma \begin{pmatrix} a & b & c & ... \end{pmatrix} \sigma^{-1} = \begin{pmatrix} \sigma(a) & \sigma(b) & \sigma(c) & ... \end{pmatrix}$$

23.3.5   Special elements

Certain elements of the symmetric group of $\{1, 2, ..., n\}$ are of particular interest (these can be generalized to the symmetric group of any finite totally ordered set, but not to that of an unordered set).

The **order reversing permutation** is the one given by:

$$\begin{pmatrix} 1 & 2 & \cdots & n \\ n & n-1 & \cdots & 1 \end{pmatrix}.$$

This is the unique maximal element with respect to the Bruhat order and the longest element in the symmetric group with respect to generating set consisting of the adjacent transpositions (i $i+1$), $1 \leq i \leq n - 1$.

This is an involution, and consists of $\lfloor n/2 \rfloor$ (non-adjacent) transpositions

$$(1\ n)(2\ n - 1)\cdots, \text{ or } \sum_{k=1}^{n-1} k = \frac{n(n-1)}{2} \text{ transpositions: adjacent}$$

$$(n\ n - 1)(n - 1\ n - 2)\cdots(2\ 1)(n - 1\ n - 2)(n - 2\ n - 3)\cdots,$$

so it thus has sign:

$$\text{sgn}(\rho_n) = (-1)^{\lfloor n/2 \rfloor} = (-1)^{n(n-1)/2} = \begin{cases} +1 & n \equiv 0, 1 \pmod 4 \\ -1 & n \equiv 2, 3 \pmod 4 \end{cases}$$

which is 4-periodic in n.

In S_{2n}, the *perfect shuffle* is the permutation that splits the set into 2 piles and interleaves them. Its sign is also $(-1)^{\lfloor n/2 \rfloor}$.

Note that the reverse on n elements and perfect shuffle on $2n$ elements have the same sign; these are important to the classification of Clifford algebras, which are 8-periodic.

23.4 Conjugacy classes

The conjugacy classes of Sn correspond to the cycle structures of permutations; that is, two elements of Sn are conjugate in Sn if and only if they consist of the same number of disjoint cycles of the same lengths. For instance, in S_5, (1 2 3)(4 5) and (1 4 3)(2 5) are conjugate; (1 2 3)(4 5) and (1 2)(4 5) are not. A conjugating element of Sn can be constructed in "two line notation" by placing the "cycle notations" of the two conjugate permutations on top of one another. Continuing the previous example:

$$k = \begin{pmatrix} 1 & 2 & 3 & 4 & 5 \\ 1 & 4 & 3 & 2 & 5 \end{pmatrix}$$

which can be written as the product of cycles, namely: (2 4).

This permutation then relates (1 2 3)(4 5) and (1 4 3)(2 5) via conjugation, i.e.

$$(2\ 4) \circ (1\ 2\ 3)(4\ 5) \circ (2\ 4) = (1\ 4\ 3)(2\ 5).$$

It is clear that such a permutation is not unique.

23.5 Low degree groups

See also: Representation theory of the symmetric group § Special cases

The low-degree symmetric groups have simpler and exceptional structure, and often must be treated separately.

S_0 **and** S_1 The symmetric groups on the empty set and the singleton set are trivial, which corresponds to $0! = 1! = 1$. In this case the alternating group agrees with the symmetric group, rather than being an index 2 subgroup, and the sign map is trivial. In the case of S_0, its only member is the Empty function.

S_2 This group consists of exactly two elements: the identity and the permutation swapping the two points. It is a cyclic group and so abelian. In Galois theory, this corresponds to the fact that the quadratic formula gives a direct solution to the general quadratic polynomial after extracting only a single root. In invariant theory, the representation theory of the symmetric group on two points is quite simple and is seen as writing a function of two variables as a sum of its symmetric and anti-symmetric parts: Setting $fs(x,y) = f(x,y) + f(y,x)$, and $fa(x,y) = f(x,y) - f(y,x)$, one gets that $2 \cdot f = f_s + f_a$. This process is known as symmetrization.

S_3 S_3 is the first nonabelian symmetric group. This group is isomorphic to the dihedral group of order 6, the group of reflection and rotation symmetries of an equilateral triangle, since these symmetries permute the three vertices of the triangle. Cycles of length two correspond to reflections, and cycles of length three are rotations. In Galois theory, the sign map from S_3 to S_2 corresponds to the resolving quadratic for a cubic polynomial, as discovered by Gerolamo Cardano, while the A_3 kernel corresponds to the use of the discrete Fourier transform of order 3 in the solution, in the form of Lagrange resolvents.

S_4   The group S_4 is isomorphic to the group of proper rotations about opposite faces, opposite diagonals and opposite edges, 9, 8 and 6 permutations, of the cube.[4] Beyond the group A_4, S_4 has a Klein four-group V as a proper normal subgroup, namely the even transpositions {(1), (1 2)(3 4), (1 3)(2 4), (1 4)(2 3)}, with quotient S3. In Galois theory, this map corresponds to the resolving cubic to a quartic polynomial, which allows the quartic to be solved by radicals, as established by Lodovico Ferrari. The Klein group can be understood in terms of the Lagrange resolvents of the quartic. The map from S_4 to S_3 also yields a 2-dimensional irreducible representation, which is an irreducible representation of a symmetric group of degree n of dimension below $n - 1$, which only occurs for $n = 4$.

S_5 S_5 is the first non-solvable symmetric group. Along with the special linear group SL(2, 5) and the icosahedral group $A_5 \times S_2$, S_5 is one of the three non-solvable groups of order 120, up to isomorphism. S_5 is the Galois group of the general quintic equation, and the fact that S_5 is not a solvable group translates into the non-existence of a general formula to solve quintic polynomials by radicals. There is an exotic inclusion map $S_5 \to S_6$ as a transitive subgroup; the obvious inclusion map $Sn \to Sn_{+1}$ fixes a point and thus is not transitive. This yields the outer automorphism of S_6, discussed below, and corresponds to the resolvent sextic of a quintic.

S_6   Unlike all other symmetric groups, S_6, has an outer automorphism. Using the language of Galois theory, this can also be understood in terms of Lagrange resolvents. The resolvent of a quintic is of degree 6—this corresponds to an exotic inclusion map $S_5 \to S_6$ as a transitive subgroup (the obvious inclusion map $S_n \to S_{n+1}$ fixes a point and thus is not transitive) and, while this map does not make the general quintic solvable, it yields the exotic outer automorphism of S_6—see automorphisms of the symmetric and alternating groups for details.

Note that while A_6 and A_7 have an exceptional Schur multiplier (a triple cover) and that these extend to triple covers of S_6 and S_7, these do not correspond to exceptional Schur multipliers of the symmetric group.

23.5.1   Maps between symmetric groups

Other than the trivial map $Sn \to 1 \cong S_0 \cong S_1$ and the sign map $Sn \to S_2$, the most notable homomorphisms between symmetric groups, in order of relative dimension, are:

- $S_4 \to S_3$ corresponding to the exceptional normal subgroup $V < A_4 < S_4$;

- $S_6 \to S_6$ (or rather, a class of such maps up to inner automorphism) corresponding to the outer automorphism of S_6.

- $S_5 \to S_6$ as a transitive subgroup, yielding the outer automorphism of S_6 as discussed above.

There are also a host of other homomorphisms $Sm \to Sn$ where $n > m$.

23.6   Properties

Symmetric groups are Coxeter groups and reflection groups. They can be realized as a group of reflections with respect to hyperplanes $xi = xj$, $1 \le i < j \le n$. Braid groups Bn admit symmetric groups Sn as quotient groups.

Cayley's theorem states that every group G is isomorphic to a subgroup of the symmetric group on the elements of G, as a group acts on itself faithfully by (left or right) multiplication.

23.7 Relation with alternating group

For $n \geq 5$, the alternating group An is simple, and the induced quotient is the sign map: A$n \to$ S$n \to$ S$_2$ which is split by taking a transposition of two elements. Thus Sn is the semidirect product A$n \rtimes$ S$_2$, and has no other proper normal subgroups, as they would intersect A$_n$ in either the identity (and thus themselves be the identity or a 2-element group, which is not normal), or in An (and thus themselves be An or Sn).

Sn acts on its subgroup An by conjugation, and for $n \neq 6$, Sn is the full automorphism group of An: Aut(An) $\cong$ Sn. Conjugation by even elements are inner automorphisms of An while the outer automorphism of An of order 2 corresponds to conjugation by an odd element. For $n = 6$, there is an exceptional outer automorphism of An so Sn is not the full automorphism group of An.

Conversely, for $n \neq 6$, Sn has no outer automorphisms, and for $n \neq 2$ it has no center, so for $n \neq 2, 6$ it is a complete group, as discussed in automorphism group, below.

For $n \geq 5$, Sn is an almost simple group, as it lies between the simple group An and its group of automorphisms.

S_n can be embedded into A_{n+2} by appending the transposition $(n + 1, n + 2)$ to all odd permutations, while embedding into A_{n+1} is impossible for $n > 1$.

23.8 Generators and relations

The symmetric group on n-letters, Sn, may be described as follows. It has generators: $\sigma_1, \ldots, \sigma_{n-1}$ and relations:

- $\sigma_i^2 = 1$,

- $\sigma_i \sigma_j = \sigma_j \sigma_i$ if $j \neq i \pm 1$,

- $(\sigma_i \sigma_{i+1})^3 = 1$.

One thinks of σ_i as swapping the ith and $(i + 1)$th position.

Other popular generating sets include the set of transpositions that swap 1 and i for $2 \leq i \leq n$ and a set containing any n-cycle and a 2-cycle of adjacent elements in the n-cycle.

23.9 Subgroup structure

A subgroup of a symmetric group is called a permutation group.

23.9.1 Normal subgroups

The normal subgroups of the finite symmetric groups are well understood. If $n \leq 2$, Sn has at most 2 elements, and so has no nontrivial proper subgroups. The alternating group of degree n is always a normal subgroup, a proper one for $n \geq 2$ and nontrivial for $n \geq 3$; for $n \geq 3$ it is in fact the only non-identity proper normal subgroup of Sn, except when $n = 4$ where there is one additional such normal subgroup, which is isomorphic to the Klein four group.

The symmetric group on an infinite set does not have an associated alternating group: not all elements can be written as a (finite) product of transpositions. However it does contain a normal subgroup S of permutations that fix all but finitely many elements, and such permutations can be classified as either even or odd. The even elements of S form the alternating subgroup A of S, and since A is even a characteristic subgroup of S, it is also a normal subgroup of the full symmetric group of the infinite set. The groups A and S are the only non-identity proper normal subgroups of the symmetric group on a countably infinite set. For more details see (Scott 1987, Ch. 11.3) or (Dixon & Mortimer 1996, Ch. 8.1).

23.9.2   Maximal subgroups

The maximal subgroups of the finite symmetric groups fall into three classes: the intransitive, the imprimitive, and the primitive. The intransitive maximal subgroups are exactly those of the form Sym(k) × Sym($n - k$) for $1 \leq k < n/2$. The imprimitive maximal subgroups are exactly those of the form Sym(k) wr Sym(n/k) where $2 \leq k \leq n/2$ is a proper divisor of n and "wr" denotes the wreath product acting imprimitively. The primitive maximal subgroups are more difficult to identify, but with the assistance of the O'Nan–Scott theorem and the classification of finite simple groups, (Liebeck, Praeger & Saxl 1988) gave a fairly satisfactory description of the maximal subgroups of this type according to (Dixon & Mortimer 1996, p. 268).

23.9.3   Sylow subgroups

The Sylow subgroups of the symmetric groups are important examples of p-groups. They are more easily described in special cases first:

The Sylow p-subgroups of the symmetric group of degree p are just the cyclic subgroups generated by p-cycles. There are $(p - 1)!/(p - 1) = (p - 2)!$ such subgroups simply by counting generators. The normalizer therefore has order $p \cdot (p-1)$ and is known as a Frobenius group $F_{p(p - 1)}$ (especially for $p = 5$), and is the affine general linear group, AGL($1, p$).

The Sylow p-subgroups of the symmetric group of degree p^2 are the wreath product of two cyclic groups of order p. For instance, when $p = 3$, a Sylow 3-subgroup of Sym(9) is generated by $a = (1\ 4\ 7)(2\ 5\ 8)(3\ 6\ 9)$ and the elements $x = (1\ 2\ 3)$, $y = (4\ 5\ 6)$, $z = (7\ 8\ 9)$, and every element of the Sylow 3-subgroup has the form $a^i x^j y^k z^l$ for $0 \leq i,j,k,l \leq 2$.

The Sylow p-subgroups of the symmetric group of degree p^n are sometimes denoted W$p(n)$, and using this notation one has that W$p(n + 1)$ is the wreath product of W$p(n)$ and W$p(1)$.

In general, the Sylow p-subgroups of the symmetric group of degree n are a direct product of a_i copies of W$p(i)$, where $0 \leq a_i \leq p - 1$ and $n = a_0 + p \cdot a_1 + ... + p^k \cdot a_k$.

For instance, $W_2(1) = C_2$ and $W_2(2) = D_8$, the dihedral group of order 8, and so a Sylow 2-subgroup of the symmetric group of degree 7 is generated by { (1,3)(2,4), (1,2), (3,4), (5,6) } and is isomorphic to $D_8 \times C_2$.

These calculations are attributed to (Kaloujnine 1948) and described in more detail in (Rotman 1995, p. 176). Note however that (Kerber 1971, p. 26) attributes the result to an 1844 work of Cauchy, and mentions that it is even covered in textbook form in (Netto 1882, §39–40).

23.9.4   Transitive subgroups

A **transitive subgroup** of Sn is a subgroup whose action on {1, 2, ..., n} is transitive. For example, the Galois group of a (finite) Galois extension is a transitive subgroup of Sn, for some n.

23.10   Automorphism group

For more details on this topic, see Automorphisms of the symmetric and alternating groups.

For $n \neq 2, 6$, Sn is a complete group: its center and outer automorphism group are both trivial.

For $n = 2$, the automorphism group is trivial, but S$_2$ is not trivial: it is isomorphic to C_2, which is abelian, and hence the center is the whole group.

For $n = 6$, it has an outer automorphism of order 2: Out(S$_6$) = C_2, and the automorphism group is a semidirect product

Aut(S$_6$) = S$_6 \rtimes C_2$.

In fact, for any set X of cardinality other than 6, every automorphism of the symmetric group on X is inner, a result first due to (Schreier & Ulam 1937) according to (Dixon & Mortimer 1996, p. 259).

23.11 Homology

See also: Alternating group § Group homology

The group homology of Sn is quite regular and stabilizes: the first homology (concretely, the abelianization) is:

$$H_1(S_n, \mathbf{Z}) = \begin{cases} 0 & n < 2 \\ \mathbf{Z}/2 & n \geq 2. \end{cases}$$

The first homology group is the abelianization, and corresponds to the sign map S$n \to$ S$_2$ which is the abelianization for $n \geq 2$; for $n < 2$ the symmetric group is trivial. This homology is easily computed as follows: Sn is generated by involutions (2-cycles, which have order 2), so the only non-trivial maps S$n \to$ Cp are to S$_2$ and all involutions are conjugate, hence map to the same element in the abelianization (since conjugation is trivial in abelian groups). Thus the only possible maps S$n \to$ S$_2 \cong \{\pm 1\}$ send an involution to 1 (the trivial map) or to -1 (the sign map). One must also show that the sign map is well-defined, but assuming that, this gives the first homology of Sn.

The second homology (concretely, the Schur multiplier) is:

$$H_2(S_n, \mathbf{Z}) = \begin{cases} 0 & n < 4 \\ \mathbf{Z}/2 & n \geq 4. \end{cases}$$

This was computed in (Schur 1911), and corresponds to the double cover of the symmetric group, $2 \cdot$ Sn.

Note that the exceptional low-dimensional homology of the alternating group ($H_1(A_3) \cong H_1(A_4) \cong C_3$, corresponding to non-trivial abelianization, and $H_2(A_6) \cong H_2(A_7) \cong C_6$, due to the exceptional 3-fold cover) does not change the homology of the symmetric group; the alternating group phenomena do yield symmetric group phenomena – the map $A_4 \to C_3$ extends to $S_4 \to S_3$, and the triple covers of A_6 and A_7 extend to triple covers of S_6 and S_7 – but these are not *homological* – the map $S_4 \to S_3$ does not change the abelianization of S_4, and the triple covers do not correspond to homology either.

The homology "stabilizes" in the sense of stable homotopy theory: there is an inclusion map S$n \to$ S$n+1$, and for fixed k, the induced map on homology $Hk(Sn) \to Hk(Sn+1)$ is an isomorphism for sufficiently high n. This is analogous to the homology of families Lie groups stabilizing.

The homology of the infinite symmetric group is computed in (Nakaoka 1961), with the cohomology algebra forming a Hopf algebra.

23.12 Representation theory

Main article: Representation theory of the symmetric group

The representation theory of the symmetric group is a particular case of the representation theory of finite groups, for which a concrete and detailed theory can be obtained. This has a large area of potential applications, from symmetric function theory to problems of quantum mechanics for a number of identical particles.

The symmetric group Sn has order $n!$. Its conjugacy classes are labeled by partitions of n. Therefore according to the representation theory of a finite group, the number of inequivalent irreducible representations, over the complex numbers, is equal to the number of partitions of n. Unlike the general situation for finite groups, there is in fact a natural way to parametrize irreducible representation by the same set that parametrizes conjugacy classes, namely by partitions of n or equivalently Young diagrams of size n.

Each such irreducible representation can be realized over the integers (every permutation acting by a matrix with integer coefficients); it can be explicitly constructed by computing the Young symmetrizers acting on a space generated by the Young tableaux of shape given by the Young diagram.

Over other fields the situation can become much more complicated. If the field K has characteristic equal to zero or greater than n then by Maschke's theorem the group algebra KSn is semisimple. In these cases the irreducible representations defined over the integers give the complete set of irreducible representations (after reduction modulo the characteristic if necessary).

However, the irreducible representations of the symmetric group are not known in arbitrary characteristic. In this context it is more usual to use the language of modules rather than representations. The representation obtained from an irreducible representation defined over the integers by reducing modulo the characteristic will not in general be irreducible. The modules so constructed are called *Specht modules*, and every irreducible does arise inside some such module. There are now fewer irreducibles, and although they can be classified they are very poorly understood. For example, even their dimensions are not known in general.

The determination of the irreducible modules for the symmetric group over an arbitrary field is widely regarded as one of the most important open problems in representation theory.

23.13   See also

- History of group theory

- Symmetric inverse semigroup

- Signed symmetric group

- Generalized symmetric group

23.14   References

[1] Jacobson (2009), p. 31.

[2] Jacobson (2009), p. 32. Theorem 1.1.

[3] modern algebra Author A. R. Vasishtha, A. K. Vasishtha Publisher Krishna Prakashan Media

[4] Die Untergruppenverbände der Gruppen der ordnung weniger als 100, Habilitationsschrift, J. Neubuser, Universität Kiel, Germany, 1967.

- Cameron, Peter J. (1999), *Permutation Groups*, London Mathematical Society Student Texts **45**, Cambridge University Press, ISBN 978-0-521-65378-7

- Dixon, John D.; Mortimer, Brian (1996), *Permutation groups*, Graduate Texts in Mathematics **163**, Berlin, New York: Springer-Verlag, ISBN 978-0-387-94599-6, MR 1409812

- Jacobson, Nathan (2009), *Basic algebra* **1** (2nd ed.), Dover, ISBN 978-0-486-47189-1.

- Kaloujnine, Léo (1948), "La structure des p-groupes de Sylow des groupes symétriques finis", *Annales Scientifiques de l'École Normale Supérieure. Troisième Série* **65**: 239–276, ISSN 0012-9593, MR 0028834

- Kerber, Adalbert (1971), *Representations of permutation groups. I*, Lecture Notes in Mathematics, Vol. 240 **240**, Berlin, New York: Springer-Verlag, doi:10.1007/BFb0067943, ISBN 978-3-540-05693-5, MR 0325752

- Liebeck, M.W.; Praeger, C.E.; Saxl, J. (1988), "On the O'Nan-Scott theorem for finite primitive permutation groups", *J. Austral. Math. Soc.* **44** (3): 389–396, doi:10.1017/S144678870003216X

- Nakaoka, Minoru (March 1961), "Homology of the Infinite Symmetric Group", *The Annals of Mathematics*, 2 (Annals of Mathematics) **73** (2): 229–257, doi:10.2307/1970333, JSTOR 1970333

- Netto, E. (1882), *Substitutionentheorie und ihre Anwendungen auf die Algebra* (in German), Leipzig. Teubner, JFM 14.0090.01

- Scott, W.R. (1987), *Group Theory*, New York: Dover Publications, pp. 45–46, ISBN 978-0-486-65377-8

- Schur, Issai (1911), "Über die Darstellung der symmetrischen und der alternierenden Gruppe durch gebrochene lineare Substitutionen", *Journal für die reine und angewandte Mathematik* **139**: 155–250, doi:10.1515/crll.1911.139.155

- Schreier, J.; Ulam, Stanislaw (1936), "Über die Automorphismen der Permutationsgruppe der natürlichen Zahlenfolge" (PDF), *Fundam. Math.* (in German) **28**: 258–260, Zbl 0016.20301

23.15   External links

- Hazewinkel, Michiel, ed. (2001), "Symmetric group", *Encyclopedia of Mathematics*, Springer, ISBN 978-1-55608-010-4

- Weisstein, Eric W., "Symmetric group", *MathWorld*.

- Weisstein, Eric W., "Symmetric group graph", *MathWorld*.

- Marcus du Sautoy: Symmetry, reality's riddle (video of a talk)

- OEIS Entries dealing with the Symmetric Group

Chapter 24

Lie group

In mathematics, a **Lie group** /'liː/ is a group that is also a differentiable manifold, with the property that the group operations are compatible with the smooth structure. Lie groups are named after Sophus Lie, who laid the foundations of the theory of continuous transformation groups. The term *groupes de Lie* first appeared in French in 1893 in the thesis of Lie's student Arthur Tresse, page 3.[1]

Lie groups represent the best-developed theory of continuous symmetry of mathematical objects and structures, which makes them indispensable tools for many parts of contemporary mathematics, as well as for modern theoretical physics. They provide a natural framework for analysing the continuous symmetries of differential equations (differential Galois theory), in much the same way as permutation groups are used in Galois theory for analysing the discrete symmetries of algebraic equations. An extension of Galois theory to the case of continuous symmetry groups was one of Lie's principal motivations.

24.1   Overview

Lie groups are smooth[Note 1] differentiable manifolds and as such can be studied using differential calculus, in contrast with the case of more general topological groups. One of the key ideas in the theory of Lie groups is to replace the *global* object, the group, with its *local* or linearized version, which Lie himself called its "infinitesimal group" and which has since become known as its Lie algebra.

Lie groups play an enormous role in modern geometry, on several different levels. Felix Klein argued in his Erlangen program that one can consider various "geometries" by specifying an appropriate transformation group that leaves certain geometric properties invariant. Thus Euclidean geometry corresponds to the choice of the group E(3) of distance-preserving transformations of the Euclidean space $\mathbf{R}^3$, conformal geometry corresponds to enlarging the group to the conformal group, whereas in projective geometry one is interested in the properties invariant under the projective group. This idea later led to the notion of a G-structure, where G is a Lie group of "local" symmetries of a manifold. On a "global" level, whenever a Lie group acts on a geometric object, such as a Riemannian or a symplectic manifold, this action provides a measure of rigidity and yields a rich algebraic structure. The presence of continuous symmetries expressed via a Lie group action on a manifold places strong constraints on its geometry and facilitates analysis on the manifold. Linear actions of Lie groups are especially important, and are studied in representation theory.

In the 1940s–1950s, Ellis Kolchin, Armand Borel, and Claude Chevalley realised that many foundational results concerning Lie groups can be developed completely algebraically, giving rise to the theory of algebraic groups defined over an arbitrary field. This insight opened new possibilities in pure algebra, by providing a uniform construction for most finite simple groups, as well as in algebraic geometry. The theory of automorphic forms, an important branch of modern number theory, deals extensively with analogues of Lie groups over adele rings; p-adic Lie groups play an important role, via their connections with Galois representations in number theory.

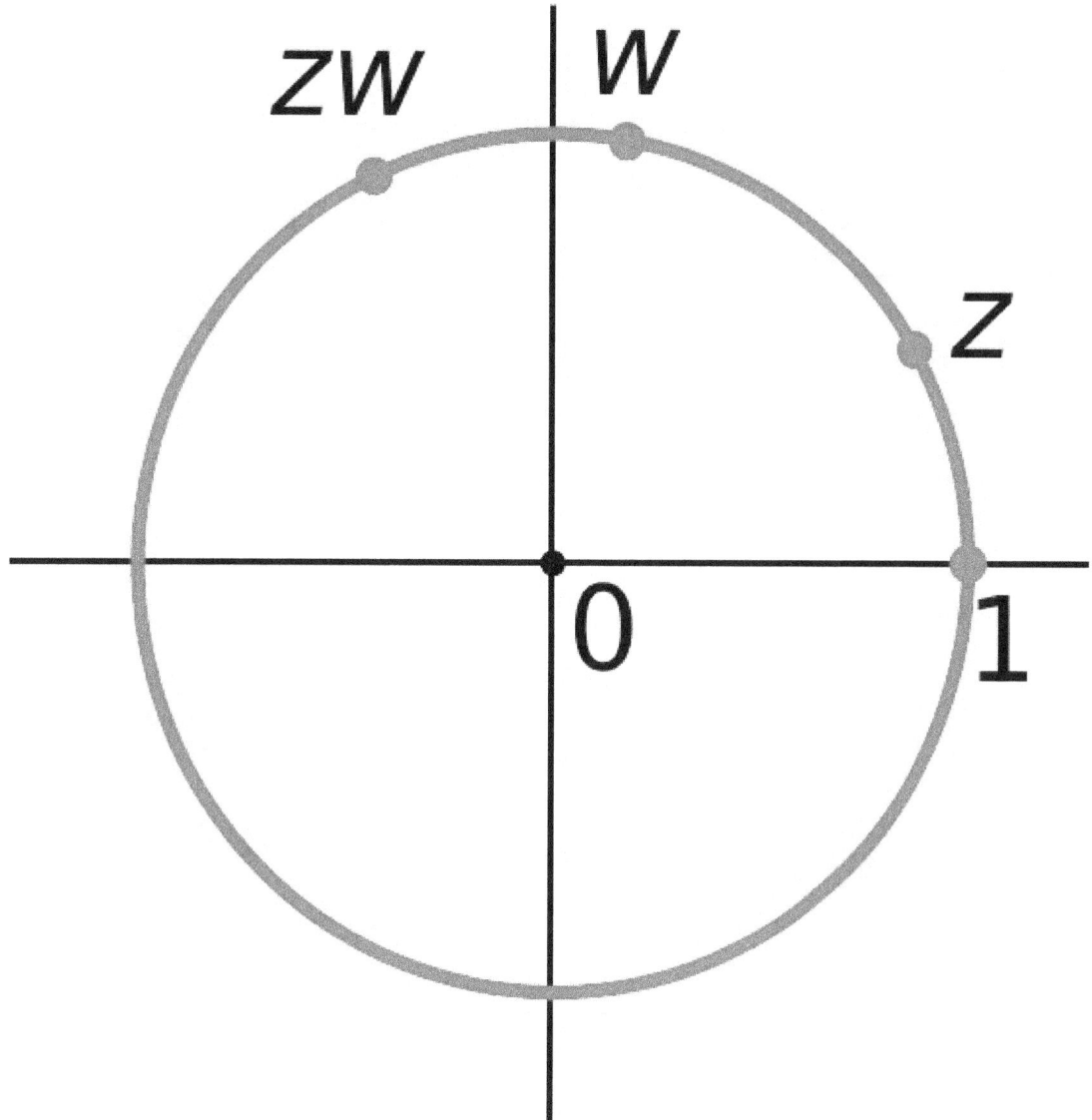

The circle of center 0 and radius 1 in the complex plane is a Lie group with complex multiplication.

24.2   Definitions and examples

A **real Lie group** is a group that is also a finite-dimensional real smooth manifold, in which the group operations of multiplication and inversion are smooth maps. Smoothness of the group multiplication

$$\mu : G \times G \to G \quad \mu(x, y) = xy$$

means that μ is a smooth mapping of the product manifold $G \times G$ into G. These two requirements can be combined to the single requirement that the mapping

$$(x, y) \mapsto x^{-1} y$$

be a smooth mapping of the product manifold into G.

24.2.1   First examples

- The 2×2 real invertible matrices form a group under multiplication, denoted by GL(2, **R**) or by GL2(**R**):

$$\mathrm{GL}(2, \mathbf{R}) = \left\{ A = \begin{pmatrix} a & b \\ c & d \end{pmatrix} : \det A = ad - bc \neq 0 \right\}.$$

This is a four-dimensional noncompact real Lie group. This group is disconnected; it has two connected components corresponding to the positive and negative values of the determinant.

- The rotation matrices form a subgroup of GL(2, **R**), denoted by SO(2, **R**). It is a Lie group in its own right: specifically, a one-dimensional compact connected Lie group which is diffeomorphic to the circle. Using the rotation angle φ as a parameter, this group can be parametrized as follows:

$$\mathrm{SO}(2, \mathbf{R}) = \left\{ \begin{pmatrix} \cos \varphi & -\sin \varphi \\ \sin \varphi & \cos \varphi \end{pmatrix} : \varphi \in \mathbf{R}/2\pi\mathbf{Z} \right\}.$$

Addition of the angles corresponds to multiplication of the elements of SO(2, **R**), and taking the opposite angle corresponds to inversion. Thus both multiplication and inversion are differentiable maps.

- The orthogonal group also forms an interesting example of a Lie group.

All of the previous examples of Lie groups fall within the class of classical groups.

24.2.2   Related concepts

A **complex Lie group** is defined in the same way using complex manifolds rather than real ones (example: SL(2, **C**)), and similarly, using an alternate metric completion of **Q**, one can define a **p-adic Lie group** over the p-adic numbers, a topological group in which each point has a p-adic neighborhood. Hilbert's fifth problem asked whether replacing differentiable manifolds with topological or analytic ones can yield new examples. The answer to this question turned out to be negative: in 1952, Gleason, Montgomery and Zippin showed that if G is a topological manifold with continuous group operations, then there exists exactly one analytic structure on G which turns it into a Lie group (see also Hilbert–Smith conjecture). If the underlying manifold is allowed to be infinite-dimensional (for example, a Hilbert manifold), then one arrives at the notion of an infinite-dimensional Lie group. It is possible to define analogues of many Lie groups over finite fields, and these give most of the examples of finite simple groups.

The language of category theory provides a concise definition for Lie groups: a Lie group is a group object in the category of smooth manifolds. This is important, because it allows generalization of the notion of a Lie group to Lie supergroups.

24.3   More examples of Lie groups

See also: Table of Lie groups and List of simple Lie groups

Lie groups occur in abundance throughout mathematics and physics. Matrix groups or algebraic groups are (roughly) groups of matrices (for example, orthogonal and symplectic groups), and these give most of the more common examples of Lie groups.

24.3.1   Examples with a specific number of dimensions

- The circle group S^1 consisting of angles mod 2π under addition or, alternatively, the complex numbers with absolute value 1 under multiplication. This is a one-dimensional compact connected abelian Lie group.

- The 3-sphere S^3 forms a Lie group by identification with the set of quaternions of unit norm, called versors. The only other spheres that admit the structure of a Lie group are the 0-sphere S^0 (real numbers with absolute value 1) and the circle S^1 (complex numbers with absolute value 1). For example, for even $n > 1$, S^n is not a Lie group because it does not admit a nonvanishing vector field and so *a fortiori* cannot be parallelizable as a differentiable manifold. Of the spheres only S^0, S^1, S^3, and S^7 are parallelizable. The last carries the structure of a Lie quasigroup (a nonassociative group), which can be identified with the set of unit octonions.

- The (3-dimensional) metaplectic group is a double cover of SL(2, **R**) playing an important role in the theory of modular forms. It is a connected Lie group that cannot be faithfully represented by matrices of finite size, i.e., a nonlinear group.

- The Heisenberg group is a connected nilpotent Lie group of dimension 3, playing a key role in quantum mechanics.

- The Lorentz group is a 6-dimensional Lie group of linear isometries of the Minkowski space.

- The Poincaré group is a 10-dimensional Lie group of affine isometries of the Minkowski space.

- The group U(1)×SU(2)×SU(3) is a Lie group of dimension 1+3+8=12 that is the gauge group of the Standard Model in particle physics. The dimensions of the factors correspond to the 1 photon + 3 vector bosons + 8 gluons of the standard model

- The exceptional Lie groups of types G_2, F_1, E_6, E_7, E_8 have dimensions 14, 52, 78, 133, and 248. Along with the A-B-C-D series of simple Lie groups, the exceptional groups complete the list of simple Lie groups. There is also a Lie group named $E_7\frac{1}{2}$ of dimension 190, but it is not a *simple* Lie group.

24.3.2   Examples with n dimensions

- Euclidean space $\mathbf{R}^n$ with ordinary vector addition as the group operation becomes an n-dimensional noncompact abelian Lie group.

- The Euclidean group E(n, **R**) is the Lie group of all Euclidean motions, i.e., isometric affine maps, of n-dimensional Euclidean space $\mathbf{R}^n$.

- The orthogonal group O(n, **R**), consisting of all $n \times n$ orthogonal matrices with real entries is an $n(n-1)/2$-dimensional Lie group. This group is disconnected, but it has a connected subgroup SO(n, **R**) of the same dimension consisting of orthogonal matrices of determinant 1, called the special orthogonal group (for $n = 3$, the rotation group SO(3)).

- The unitary group U(n) consisting of $n \times n$ unitary matrices (with complex entries) is a compact connected Lie group of dimension n^2. Unitary matrices of determinant 1 form a closed connected subgroup of dimension $n^2 - 1$ denoted SU(n), the special unitary group.

- Spin groups are double covers of the special orthogonal groups, used for studying fermions in quantum field theory (among other things).

- The group GL(n, **R**) of invertible matrices (under matrix multiplication) is a Lie group of dimension n^2, called the general linear group. It has a closed connected subgroup SL(n, **R**), the special linear group, consisting of matrices of determinant 1 which is also a Lie group.

- The symplectic group Sp($2n$, **R**) consists of all $2n \times 2n$ matrices preserving a *symplectic form* on $\mathbf{R}^{2n}$. It is a connected Lie group of dimension $2n^2 + n$.

- The group of invertible upper triangular n by n matrices is a solvable Lie group of dimension $n(n+1)/2$. (cf. Borel subgroup)

- The A-series, B-series, C-series and D-series, whose elements are denoted by An, Bn, Cn, and Dn, are infinite families of simple Lie groups.

24.3.3   Constructions

There are several standard ways to form new Lie groups from old ones:

- The product of two Lie groups is a Lie group.

- Any topologically closed subgroup of a Lie group is a Lie group. This is known as the Closed subgroup theorem or **Cartan's theorem**.

- The quotient of a Lie group by a closed normal subgroup is a Lie group.

- The universal cover of a connected Lie group is a Lie group. For example, the group **R** is the universal cover of the circle group **S**^{1}. In fact any covering of a differentiable manifold is also a differentiable manifold, but by specifying *universal* cover, one guarantees a group structure (compatible with its other structures).

24.3.4   Related notions

Some examples of groups that are *not* Lie groups (except in the trivial sense that any group can be viewed as a 0-dimensional Lie group, with the discrete topology), are:

- Infinite-dimensional groups, such as the additive group of an infinite-dimensional real vector space. These are not Lie groups as they are not *finite-dimensional* manifolds.

- Some totally disconnected groups, such as the Galois group of an infinite extension of fields, or the additive group of the p-adic numbers. These are not Lie groups because their underlying spaces are not real manifolds. (Some of these groups are "p-adic Lie groups".) In general, only topological groups having similar local properties to **R**^{n} for some positive integer n can be Lie groups (of course they must also have a differentiable structure).

24.4   Basic concepts

24.4.1   The Lie algebra associated with a Lie group

Main article: Lie group–Lie algebra correspondence

To every Lie group we can associate a Lie algebra whose underlying vector space is the tangent space of the Lie group at the identity element and which completely captures the local structure of the group. Informally we can think of elements of the Lie algebra as elements of the group that are "infinitesimally close" to the identity, and the Lie bracket of the Lie algebra is related to the commutator of two such infinitesimal elements. Before giving the abstract definition we give a few examples:

- The Lie algebra of the vector space **R**^{n} is just **R**^{n} with the Lie bracket given by
  $[A, B] = 0$.
  (In general the Lie bracket of a connected Lie group is always 0 if and only if the Lie group is abelian.)

- The Lie algebra of the general linear group GL(n, **R**) of invertible matrices is the vector space M(n, **R**) of square matrices with the Lie bracket given by
  $[A, B] = AB - BA$.
  If G is a closed subgroup of GL(n, **R**) then the Lie algebra of G can be thought of informally as the matrices m

of M(n, **R**) such that $1 + \varepsilon m$ is in G, where ε is an infinitesimal positive number with $\varepsilon^2 = 0$ (of course, no such real number ε exists). For example, the orthogonal group O(n, **R**) consists of matrices A with $AA^T = 1$, so the Lie algebra consists of the matrices m with $(1 + \varepsilon m)(1 + \varepsilon m)^T = 1$, which is equivalent to $m + m^T = 0$ because $\varepsilon^2 = 0$.

- Formally, when working over the reals, as here, this is accomplished by considering the limit as $\varepsilon \to 0$; but the "infinitesimal" language generalizes directly to Lie groups over general rings.

The concrete definition given above is easy to work with, but has some minor problems: to use it we first need to represent a Lie group as a group of matrices, but not all Lie groups can be represented in this way, and it is not obvious that the Lie algebra is independent of the representation we use. To get around these problems we give the general definition of the Lie algebra of a Lie group (in 4 steps):

1. Vector fields on any smooth manifold M can be thought of as derivations X of the ring of smooth functions on the manifold, and therefore form a Lie algebra under the Lie bracket $[X, Y] = XY - YX$, because the Lie bracket of any two derivations is a derivation.

2. If G is any group acting smoothly on the manifold M, then it acts on the vector fields, and the vector space of vector fields fixed by the group is closed under the Lie bracket and therefore also forms a Lie algebra.

3. We apply this construction to the case when the manifold M is the underlying space of a Lie group G, with G acting on $G = M$ by left translations $Lg(h) = gh$. This shows that the space of left invariant vector fields (vector fields satisfying $Lg*Xh = Xgh$ for every h in G, where $Lg*$ denotes the differential of Lg) on a Lie group is a Lie algebra under the Lie bracket of vector fields.

4. Any tangent vector at the identity of a Lie group can be extended to a left invariant vector field by left translating the tangent vector to other points of the manifold. Specifically, the left invariant extension of an element v of the tangent space at the identity is the vector field defined by $v^\wedge g = Lg*v$. This identifies the tangent space TeG at the identity with the space of left invariant vector fields, and therefore makes the tangent space at the identity into a Lie algebra, called the Lie algebra of G, usually denoted by a Fraktur $\mathfrak{g}$. Thus the Lie bracket on $\mathfrak{g}$ is given explicitly by $[v, w] = [v^\wedge, w^\wedge]e$.

This Lie algebra $\mathfrak{g}$ is finite-dimensional and it has the same dimension as the manifold G. The Lie algebra of G determines G up to "local isomorphism", where two Lie groups are called **locally isomorphic** if they look the same near the identity element. Problems about Lie groups are often solved by first solving the corresponding problem for the Lie algebras, and the result for groups then usually follows easily. For example, simple Lie groups are usually classified by first classifying the corresponding Lie algebras.

We could also define a Lie algebra structure on Te using right invariant vector fields instead of left invariant vector fields. This leads to the same Lie algebra, because the inverse map on G can be used to identify left invariant vector fields with right invariant vector fields, and acts as -1 on the tangent space Te.

The Lie algebra structure on Te can also be described as follows: the commutator operation

$$(x, y) \to xyx^{-1}y^{-1}$$

on $G \times G$ sends (e, e) to e, so its derivative yields a bilinear operation on TeG. This bilinear operation is actually the zero map, but the second derivative, under the proper identification of tangent spaces, yields an operation that satisfies the axioms of a Lie bracket, and it is equal to twice the one defined through left-invariant vector fields.

24.4.2 Homomorphisms and isomorphisms

If G and H are Lie groups, then a Lie group homomorphism $f : G \to H$ is a smooth group homomorphism. In the case of complex Lie groups, such a homomorphism is required to be a holomorphic map. However, these requirements are a bit stringent; over real or complex numbers, every continuous homomorphism between Lie groups turns out to be (real or complex) analytic.

The composition of two Lie homomorphisms is again a homomorphism, and the class of all Lie groups, together with these morphisms, forms a category. Moreover, every Lie group homomorphism induces a homomorphism between the corresponding Lie algebras. Let $\phi: G \to H$ be a Lie group homomorphism and let ϕ_* be its derivative at the identity. If we identify the Lie algebras of G and H with their tangent spaces at the identity elements then ϕ_* is a map between the corresponding Lie algebras:

$$\phi_*: \mathfrak{g} \to \mathfrak{h}$$

One can show that ϕ_* is actually a Lie algebra homomorphism (meaning that it is a linear map which preserves the Lie bracket). In the language of category theory, we then have a covariant functor from the category of Lie groups to the category of Lie algebras which sends a Lie group to its Lie algebra and a Lie group homomorphism to its derivative at the identity.

Two Lie groups are called *isomorphic* if there exists a bijective homomorphism between them whose inverse is also a Lie group homomorphism. Equivalently, it is a diffeomorphism which is also a group homomorphism.

Ado's theorem says every finite-dimensional Lie algebra is isomorphic to a matrix Lie algebra. For every finite-dimensional matrix Lie algebra, there is a linear group (matrix Lie group) with this algebra as its Lie algebra. So every abstract Lie algebra is the Lie algebra of some (linear) Lie group.

The *global structure* of a Lie group is not determined by its Lie algebra; for example, if Z is any discrete subgroup of the center of G then G and G/Z have the same Lie algebra (see the table of Lie groups for examples). A *connected* Lie group is simple, semisimple, solvable, nilpotent, or abelian if and only if its Lie algebra has the corresponding property.

If we require that the Lie group be simply connected, then the global structure is determined by its Lie algebra: for every finite-dimensional Lie algebra $\mathfrak{g}$ over $\mathbf{F}$ there is a simply connected Lie group G with $\mathfrak{g}$ as Lie algebra, unique up to isomorphism. Moreover every homomorphism between Lie algebras lifts to a unique homomorphism between the corresponding simply connected Lie groups.

24.4.3   The exponential map

Main article: Exponential map (Lie theory)

The exponential map from the Lie algebra $M(n, \mathbf{R})$ of the general linear group $GL(n, \mathbf{R})$ to $GL(n, \mathbf{R})$ is defined by the usual power series:

$$\exp(A) = 1 + A + \frac{A^2}{2!} + \frac{A^3}{3!} + \cdots$$

for matrices A. If G is any subgroup of $GL(n, \mathbf{R})$, then the exponential map takes the Lie algebra of G into G, so we have an exponential map for all matrix groups.

The definition above is easy to use, but it is not defined for Lie groups that are not matrix groups, and it is not clear that the exponential map of a Lie group does not depend on its representation as a matrix group. We can solve both problems using a more abstract definition of the exponential map that works for all Lie groups, as follows.

Every vector v in $\mathfrak{g}$ determines a linear map from $\mathbf{R}$ to $\mathfrak{g}$ taking 1 to v, which can be thought of as a Lie algebra homomorphism. Because $\mathbf{R}$ is the Lie algebra of the simply connected Lie group $\mathbf{R}$, this induces a Lie group homomorphism $c: \mathbf{R} \to G$ so that

$$c(s + t) = c(s)c(t)$$

for all s and t. The operation on the right hand side is the group multiplication in G. The formal similarity of this formula with the one valid for the exponential function justifies the definition

$$\exp(v) = c(1).$$

This is called the **exponential map**, and it maps the Lie algebra $\mathfrak{g}$ into the Lie group G. It provides a diffeomorphism between a neighborhood of 0 in $\mathfrak{g}$ and a neighborhood of e in G. This exponential map is a generalization of the exponential function for real numbers (because **R** is the Lie algebra of the Lie group of positive real numbers with multiplication), for complex numbers (because **C** is the Lie algebra of the Lie group of non-zero complex numbers with multiplication) and for matrices (because M(n, **R**) with the regular commutator is the Lie algebra of the Lie group GL(n, **R**) of all invertible matrices).

Because the exponential map is surjective on some neighbourhood N of e, it is common to call elements of the Lie algebra **infinitesimal generators** of the group G. The subgroup of G generated by N is the identity component of G.

The exponential map and the Lie algebra determine the *local group structure* of every connected Lie group, because of the Baker–Campbell–Hausdorff formula: there exists a neighborhood U of the zero element of $\mathfrak{g}$, such that for u, v in U we have

$$\exp(u)\,\exp(v) = \exp\left(u + v + \tfrac{1}{2}[u,v] + \tfrac{1}{12}[[u,v],v] - \tfrac{1}{12}[[u,v],u] - \cdots\right).$$

where the omitted terms are known and involve Lie brackets of four or more elements. In case u and v commute, this formula reduces to the familiar exponential law $\exp(u)\exp(v) = \exp(u + v)$.

The exponential map relates Lie group homomorphisms. That is, if $\phi : G \to H$ is a Lie group homomorphism and $\phi_* : \mathfrak{g} \to \mathfrak{h}$ the induced map on the corresponding Lie algebras, then for all $x \in \mathfrak{g}$ we have

$$\phi(\exp(x)) = \exp(\phi_*(x)).$$

In other words the following diagram commutes.[Note 2]

(In short, exp is a natural transformation from the functor Lie to the identity functor on the category of Lie groups.)

The exponential map from the Lie algebra to the Lie group is not always onto, even if the group is connected (though it does map onto the Lie group for connected groups that are either compact or nilpotent). For example, the exponential map of SL(2, **R**) is not surjective. Also, exponential map is not surjective nor injective for infinite-dimensional (see below) Lie groups modelled on C^∞ Fréchet space, even from arbitrary small neighborhood of 0 to corresponding neighborhood of 1.

See also: derivative of the exponential map and normal coordinates.

24.4.4 Lie subgroup

A **Lie subgroup** H of a Lie group G is a Lie group that is a subset of G and such that the inclusion map from H to G is an injective immersion and group homomorphism. According to Cartan's theorem, a closed subgroup of G admits a unique smooth structure which makes it an embedded Lie subgroup of G—i.e. a Lie subgroup such that the inclusion map is a smooth embedding.

Examples of non-closed subgroups are plentiful; for example take G to be a torus of dimension ≥ 2, and let H be a one-parameter subgroup of *irrational slope*, i.e. one that winds around in G. Then there is a Lie group homomorphism $\varphi : \mathbf{R} \to G$ with H as its image. The closure of H will be a sub-torus in G.

In terms of the exponential map of G, in general, only some of the Lie subalgebras of the Lie algebra g of G correspond to closed Lie subgroups H of G. There is no criterion solely based on the structure of g which determines which those are.

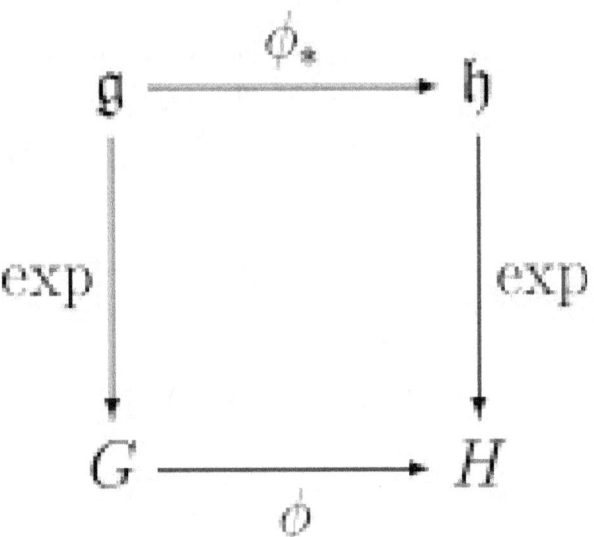

24.5   Early history

According to the most authoritative source on the early history of Lie groups (Hawkins, p. 1), Sophus Lie himself considered the winter of 1873–1874 as the birth date of his theory of continuous groups. Hawkins, however, suggests that it was "Lie's prodigious research activity during the four-year period from the fall of 1869 to the fall of 1873" that led to the theory's creation (*ibid*). Some of Lie's early ideas were developed in close collaboration with Felix Klein. Lie met with Klein every day from October 1869 through 1872: in Berlin from the end of October 1869 to the end of February 1870, and in Paris, Göttingen and Erlangen in the subsequent two years (*ibid*, p. 2). Lie stated that all of the principal results were obtained by 1884. But during the 1870s all his papers (except the very first note) were published in Norwegian journals, which impeded recognition of the work throughout the rest of Europe (*ibid*, p. 76). In 1884 a young German mathematician, Friedrich Engel, came to work with Lie on a systematic treatise to expose his theory of continuous groups. From this effort resulted the three-volume *Theorie der Transformationsgruppen*, published in 1888, 1890, and 1893.

Lie's ideas did not stand in isolation from the rest of mathematics. In fact, his interest in the geometry of differential equations was first motivated by the work of Carl Gustav Jacobi, on the theory of partial differential equations of first order and on the equations of classical mechanics. Much of Jacobi's work was published posthumously in the 1860s, generating enormous interest in France and Germany (Hawkins, p. 43). Lie's *idée fixe* was to develop a theory of symmetries of differential equations that would accomplish for them what Évariste Galois had done for algebraic equations: namely, to classify them in terms of group theory. Lie and other mathematicians showed that the most important equations for special functions and orthogonal polynomials tend to arise from group theoretical symmetries. In Lie's early work, the idea was to construct a theory of *continuous groups*, to complement the theory of discrete groups that had developed in the theory of modular forms, in the hands of Felix Klein and Henri Poincaré. The initial application that Lie had in mind was to the theory of differential equations. On the model of Galois theory and polynomial equations, the driving conception was of a theory capable of unifying, by the study of symmetry, the whole area of ordinary differential equations. However, the hope that Lie Theory would unify the entire field of ordinary differential equations was not fulfilled. Symmetry methods for ODEs continue to be studied, but do not dominate the subject. There is a differential Galois theory, but it was developed by others, such as Picard and Vessiot, and it provides a theory of quadratures, the indefinite integrals required to express solutions.

Additional impetus to consider continuous groups came from ideas of Bernhard Riemann, on the foundations of geometry, and their further development in the hands of Klein. Thus three major themes in 19th century mathematics were combined by Lie in creating his new theory: the idea of symmetry, as exemplified by Galois through the algebraic notion of a group; geometric theory and the explicit solutions of differential equations of mechanics, worked out by Poisson and Jacobi; and the new understanding of geometry that emerged in the works of Plücker, Möbius, Grassmann and others, and culminated in Riemann's revolutionary vision of the subject.

Although today Sophus Lie is rightfully recognized as the creator of the theory of continuous groups, a major stride in the development of their structure theory, which was to have a profound influence on subsequent development of mathematics, was made by Wilhelm Killing, who in 1888 published the first paper in a series entitled *Die Zusammensetzung der stetigen endlichen Transformationsgruppen* (*The composition of continuous finite transformation groups*) (Hawkins, p. 100). The work of Killing, later refined and generalized by Élie Cartan, led to classification of semisimple Lie algebras, Cartan's theory of symmetric spaces, and Hermann Weyl's description of representations of compact and semisimple Lie groups using highest weights.

In 1900 David Hilbert challenged Lie theorists with his Fifth Problem presented at the International Congress of Mathematicians in Paris.

Weyl brought the early period of the development of the theory of Lie groups to fruition, for not only did he classify irreducible representations of semisimple Lie groups and connect the theory of groups with quantum mechanics, but he also put Lie's theory itself on firmer footing by clearly enunciating the distinction between Lie's *infinitesimal groups* (i.e., Lie algebras) and the Lie groups proper, and began investigations of topology of Lie groups.[2] The theory of Lie groups was systematically reworked in modern mathematical language in a monograph by Claude Chevalley.

24.6 The concept of a Lie group, and possibilities of classification

Lie groups may be thought of as smoothly varying families of symmetries. Examples of symmetries include rotation about an axis. What must be understood is the nature of 'small' transformations, e.g., rotations through tiny angles, that link nearby transformations. The mathematical object capturing this structure is called a Lie algebra (Lie himself called them "infinitesimal groups"). It can be defined because Lie groups are manifolds, so have tangent spaces at each point.

The Lie algebra of any compact Lie group (very roughly: one for which the symmetries form a bounded set) can be decomposed as a direct sum of an abelian Lie algebra and some number of simple ones. The structure of an abelian Lie algebra is mathematically uninteresting (since the Lie bracket is identically zero); the interest is in the simple summands. Hence the question arises: what are the simple Lie algebras of compact groups? It turns out that they mostly fall into four infinite families, the "classical Lie algebras" A_n, B_n, C_n and D_n, which have simple descriptions in terms of symmetries of Euclidean space. But there are also just five "exceptional Lie algebras" that do not fall into any of these families. E_8 is the largest of these.

Lie groups are classified according to their algebraic properties (simple, semisimple, solvable, nilpotent, abelian), their connectedness (connected or simply connected) and their compactness.

- Compact Lie groups are all known: they are finite central quotients of a product of copies of the circle group S^1 and simple compact Lie groups (which correspond to connected Dynkin diagrams).

- Any simply connected solvable Lie group is isomorphic to a closed subgroup of the group of invertible upper triangular matrices of some rank, and any finite-dimensional irreducible representation of such a group is 1-dimensional. Solvable groups are too messy to classify except in a few small dimensions.

- Any simply connected nilpotent Lie group is isomorphic to a closed subgroup of the group of invertible upper triangular matrices with 1's on the diagonal of some rank, and any finite-dimensional irreducible representation of such a group is 1-dimensional. Like solvable groups, nilpotent groups are too messy to classify except in a few small dimensions.

- Simple Lie groups are sometimes defined to be those that are simple as abstract groups, and sometimes defined to be connected Lie groups with a simple Lie algebra. For example, SL(2, **R**) is simple according to the second definition but not according to the first. They have all been classified (for either definition).

- Semisimple Lie groups are Lie groups whose Lie algebra is a product of simple Lie algebras.[3] They are central extensions of products of simple Lie groups.

The identity component of any Lie group is an open normal subgroup, and the quotient group is a discrete group. The universal cover of any connected Lie group is a simply connected Lie group, and conversely any connected Lie group is a quotient of a simply connected Lie group by a discrete normal subgroup of the center. Any Lie group G can be decomposed into discrete, simple, and abelian groups in a canonical way as follows. Write

G_{con} for the connected component of the identity

G_{sol} for the largest connected normal solvable subgroup

G_{nil} for the largest connected normal nilpotent subgroup

so that we have a sequence of normal subgroups

$1 \subseteq G_{nil} \subseteq G_{sol} \subseteq G_{con} \subseteq G.$

Then

G/G_{con} is discrete

G_{con}/G_{sol} is a central extension of a product of simple connected Lie groups.

G_{sol}/G_{nil} is abelian. A connected abelian Lie group is isomorphic to a product of copies of **R** and the circle group S^1.

$G_{nil}/1$ is nilpotent, and therefore its ascending central series has all quotients abelian.

This can be used to reduce some problems about Lie groups (such as finding their unitary representations) to the same problems for connected simple groups and nilpotent and solvable subgroups of smaller dimension.

- The diffeomorphism group of a Lie group acts transitively on the Lie group

- Every Lie group is parallelizable, and hence an orientable manifold (there is a bundle isomorphism between its tangent bundle and the product of itself with the tangent space at the identity)

24.7   Infinite-dimensional Lie groups

Lie groups are often defined to be finite-dimensional, but there are many groups that resemble Lie groups, except for being infinite-dimensional. The simplest way to define infinite-dimensional Lie groups is to model them on Banach spaces, and in this case much of the basic theory is similar to that of finite-dimensional Lie groups. However this is inadequate for many applications, because many natural examples of infinite-dimensional Lie groups are not Banach manifolds. Instead one needs to define Lie groups modeled on more general locally convex topological vector spaces. In this case the relation between the Lie algebra and the Lie group becomes rather subtle, and several results about finite-dimensional Lie groups no longer hold.

The literature is not entirely uniform in its terminology as to exactly which properties of infinite-dimensional groups qualify the group for the prefix *Lie* in *Lie group*. On the Lie algebra side of affairs, things are simpler since the qualifying criteria for the prefix *Lie* in *Lie algebra* are purely algebraic. For example, an infinite-dimensional Lie algebra may or may not have a corresponding Lie group. That is, there may be a group corresponding to the Lie algebra, but it might not be nice enough to be called a Lie group, or the connection between the group and the Lie algebra might not be nice enough (e.g failure of the exponential map to be onto a neighborhood of the identity). It is the "nice enough" that is not universally defined.

Some of the examples that have been studied include:

- The group of diffeomorphisms of a manifold. Quite a lot is known about the group of diffeomorphisms of the circle. Its Lie algebra is (more or less) the Witt algebra, which has a central extension called the Virasoro algebra, used in string theory and conformal field theory. Diffeomorphism groups of compact manifolds of larger dimension are regular Fréchet Lie groups; very little about their structure is known.

The diffeomorphism group of spacetime sometimes appears in attempts to quantize gravity.

- The group of smooth maps from a manifold to a finite-dimensional Lie group is an example of a gauge group (with operation of pointwise multiplication), and is used in quantum field theory and Donaldson theory. If the manifold is a circle these are called loop groups, and have central extensions whose Lie algebras are (more or less) Kac–Moody algebras.

- There are infinite-dimensional analogues of general linear groups, orthogonal groups, and so on. One important aspect is that these may have *simpler* topological properties: see for example Kuiper's theorem. In M-Theory theory, for example, a 10 dimensional SU(N) gauge theory becomes an 11 dimensional theory when N becomes infinite.

- A specific example is that $SU(\infty)$ is equal to the group of area preserving diffeomorphisms of a torus.

24.8 See also

- Lie subgroup

- E_8

- Adjoint representation of a Lie group

- Adjoint endomorphism

- Haar measure

- Homogeneous space

- List of Lie group topics

- List of simple Lie groups

- Moufang polygon

- Riemannian manifold

- Representations of Lie groups

- Table of Lie groups

- Lie algebra

- Symmetry in quantum mechanics

- Lie group action

24.9 Notes

24.9.1 Explanatory notes

[1] having derivatives of all orders

[2] http://www.math.sunysb.edu/~{}vkiritch/MAT552/ProblemSet1.pdf

24.9.2 Citations

[1] Arthur Tresse (1893). "Sur les invariants différentiels des groupes continus de transformations". *Acta Mathematica* **18**: 1–88. doi:10.1007/bf02418270.

[2] Borel (2001).

[3] Helgason, Sigurdur (1978). *Differential Geometry, Lie Groups, and Symmetric Spaces*. New York: Academic Press. p. 131. ISBN 0-12-338460-5.

24.10 References

- Adams, John Frank (1969), *Lectures on Lie Groups*, Chicago Lectures in Mathematics, Chicago: Univ. of Chicago Press, ISBN 0-226-00527-5, MR 0252560.

- Borel, Armand (2001), *Essays in the history of Lie groups and algebraic groups*, History of Mathematics **21**, Providence, R.I.: American Mathematical Society, ISBN 978-0-8218-0288-5, MR 1847105

- Bourbaki, Nicolas, *Elements of mathematics: Lie groups and Lie algebras*. Chapters 1–3 ISBN 3-540-64242-0, Chapters 4–6 ISBN 3-540-42650-7, Chapters 7–9 ISBN 3-540-43405-4

- Chevalley, Claude (1946), *Theory of Lie groups*, Princeton: Princeton University Press, ISBN 0-691-04990-4.

- P. M. Cohn (1957) *Lie Groups*, Cambridge Tracts in Mathematical Physics.

- J. L. Coolidge (1940) *A History of Geometrical Methods*, pp 304–17, Oxford University Press (Dover Publications 2003).

- Fulton, William; Harris, Joe (1991), *Representation theory. A first course*, Graduate Texts in Mathematics, Readings in Mathematics **129**, New York: Springer-Verlag, ISBN 978-0-387-97495-8, MR 1153249, ISBN 978-0-387-97527-6

- Robert Gilmore (2008) *Lie groups, physics, and geometry: an introduction for physicists, engineers and chemists*, Cambridge University Press ISBN 9780521884006 .

- Hall, Brian C. (2003), *Lie Groups, Lie Algebras, and Representations: An Elementary Introduction*, Springer, ISBN 0-387-40122-9.

- F. Reese Harvey (1990) *Spinors and calibrations*, Academic Press, ISBN 0-12-329650-1 .

- Hawkins, Thomas (2000), *Emergence of the theory of Lie groups*, Sources and Studies in the History of Mathematics and Physical Sciences, Berlin, New York: Springer-Verlag, ISBN 978-0-387-98963-1, MR 1771134 Borel's review

- Helgason, Sigurdur (2001), *Differential geometry, Lie groups, and symmetric spaces*, Graduate Studies in Mathematics **34**, Providence, R.I.: American Mathematical Society, ISBN 978-0-8218-2848-9, MR 1834454

- Knapp, Anthony W. (2002), *Lie Groups Beyond an Introduction*, Progress in Mathematics **140** (2nd ed.), Boston: Birkhäuser, ISBN 0-8176-4259-5.

- Nijenhuis, Albert (1959). "Review: *Lie groups*, by P. M. Cohn". *Bulletin of the American Mathematical Society* **65** (6): 338–341. doi:10.1090/s0002-9904-1959-10358-x.

- Rossmann, Wulf (2001), *Lie Groups: An Introduction Through Linear Groups*, Oxford Graduate Texts in Mathematics, Oxford University Press, ISBN 978-0-19-859683-7. The 2003 reprint corrects several typographical mistakes.

- Sattinger, David H.; Weaver, O. L. (1986). *Lie groups and algebras with applications to physics, geometry, and mechanics*. Springer-Verlag. ISBN 3-540-96240-9. MR 0835009.

- Serre, Jean-Pierre (1965), *Lie Algebras and Lie Groups: 1964 Lectures given at Harvard University*, Lecture notes in mathematics **1500**, Springer, ISBN 3-540-55008-9.

- Stillwell, John (2008). *Naive Lie Theory*. Springer. ISBN 0-387-98289-2.

- Heldermann Verlag Journal of Lie Theory

- Warner, Frank W. (1983), *Foundations of differentiable manifolds and Lie groups*, Graduate Texts in Mathematics **94**, New York Berlin Heidelberg: Springer-Verlag, ISBN 978-0-387-90894-6, MR 0722297

- Steeb, Willi-Hans (2007), *Continuous Symmetries, Lie algebras, Differential Equations and Computer Algebra: second edition*, World Scientific Publishing, ISBN 981-270-809-X, MR 2382250.

- Lie Groups, Representation Theory and Symmetric Spaces Wolfgang Ziller, Vorlesung 2010

Chapter 25

Orthogonal group

"Rotation group" redirects here. For other uses, see Rotation group (disambiguation).

In mathematics, the **orthogonal group** in dimension n, denoted $O(n)$, is the group of distance-preserving transformations of a Euclidean space of dimension n that preserve a fixed point, where the group operation is given by composing transformations. Equivalently, it is the group of $n{\times}n$ orthogonal matrices, where the group operation is given by matrix multiplication, and an orthogonal matrix is a real matrix whose inverse equals its transpose.

The determinant of an orthogonal matrix being either 1 or −1, an important subgroup of $O(n)$ is the **special orthogonal group**, denoted $SO(n)$, of the orthogonal matrices of determinant 1. This group is also called the **rotation group**, because, in dimensions 2 and 3, its elements are the usual rotations around a point (in dimension 2) or a line (in dimension 3). In low dimension, these groups have been widely studied, see SO(2), SO(3) and SO(4).

The term "orthogonal group" may also refer to a generalization of the above case: the group of invertible linear operators that preserve a non-degenerate symmetric bilinear form or quadratic form[1] on a vector space over a field. In particular, when the bilinear form is the scalar product on the vector space F^n of dimension n over a field F, with quadratic form the sum of squares, then the corresponding orthogonal group, denoted $O(n, F)$, is the set of $n \times n$ orthogonal matrices with entries from F, with the group operation of matrix multiplication. This is a subgroup of the general linear group $GL(n, F)$ given by

$$O(n, F) = \{Q \in GL(n, F) \mid Q^{\mathrm{T}}Q = QQ^{\mathrm{T}} = I\}$$

where Q^{T} is the transpose of Q and I is the identity matrix.

This article mainly discusses the orthogonal groups of quadratic forms that may be expressed over some bases as the dot product; over the reals, they are the positive definite quadratic forms. Over the reals, for any non-degenerate quadratic form, there is a basis, on which the matrix of the form is a diagonal matrix such that the diagonal entries are either 1 or −1. Thus the orthogonal group depends only on the numbers of 1 and of −1, and is denoted $O(p, q)$, where p is the number of ones and q the number of negative ones. For details, see indefinite orthogonal group.

The derived subgroup $\Omega(n, F)$ of $O(n, F)$ is an often studied object because, when F is a finite field, $\Omega(n, F)$ is often a central extension of a finite simple group.

Both $O(n, F)$ and $SO(n, F)$ are algebraic groups, because the condition that a matrix be orthogonal, i.e. have its own transpose as inverse, can be expressed as a set of polynomial equations in the entries of the matrix. The Cartan–Dieudonné theorem describes the structure of the orthogonal group for a non-singular form.

25.1 Name

The determinant of any orthogonal matrix is either 1 or −1. The orthogonal *n*-by-*n* matrices with determinant 1 form a normal subgroup of O(*n*, *F*) known as the **special orthogonal group** SO(*n*, *F*), consisting of all proper rotations. (More precisely, SO(*n*, *F*) is the kernel of the Dickson invariant, discussed below.). By analogy with GL–SL (general linear group, special linear group), the orthogonal group is sometimes called the ***general* orthogonal group** and denoted GO, though this term is also sometimes used for *indefinite* orthogonal groups O(*p*, *q*). The term **rotation group** can be used to describe either the special or general orthogonal group.

25.2 In even and odd dimension

The structure of the orthogonal group differs in certain respects between even and odd dimensions; for example, over ordered fields (such as **R**) the −*1* element is orientation-preserving in even dimensions, but orientation-reversing in odd dimensions. When this distinction is to be emphasized, the groups may be denoted O(2*k*) and O(2*k* + 1), reserving *n* for the dimension of the space (*n* = 2*k* or *n* = 2*k* + 1). The letters *p* or *r* are also used, indicating the rank of the corresponding Lie algebra; in odd dimension the corresponding Lie algebra is $\mathfrak{so}(2r + 1)$, while in even dimension the Lie algebra is $\mathfrak{so}(2r)$.

25.2.1 Difference between O(*n*) and SO(*n*) in even dimensions

In two dimensions, O(2) is the group of all rotations about the origin and all reflections along a line through the origin. SO(2) is the group of all rotations about the origin.

These groups are closely related: SO(2) is a subgroup of O(2), since any two rotations gives a rotation.

More generally, in any number of dimensions an even number of reflections gives a rotation and a rotation followed by reflection, or vice versa, produces a reflection. Therefore, the rotations define a subgroup of O(2), but the reflections do not define a subgroup.

A "reflection through the origin" may be generated as a combination of one reflection along each of the axes. The 'reflection through the origin' is not a reflection in the usual sense in even dimensions, but rather a rotation. In two dimensions it is the only nontrivial rotation that when applied twice gives the identity. It is its own inverse in any number of dimensions. In 4D it is isoclinic, and if that classification were generalised it would be isoclinic in every even number of dimensions.

25.3 Over the real number field

Over the field **R** of real numbers, the orthogonal group O(*n*, **R**) and the special orthogonal group SO(*n*, **R**) are often simply denoted by O(*n*) and SO(*n*) if no confusion is possible. They form real compact Lie groups of dimension *n*(*n* − 1)/2. O(*n*, **R**) has two connected components, with SO(*n*, **R**) being the identity component, i.e., the connected component containing the identity matrix.

25.3.1 Geometric interpretation

The real orthogonal and real special orthogonal groups have the following geometric interpretations:

O(*n*, **R**) is a subgroup of the Euclidean group *E*(*n*), the group of isometries of **R**^{n}; it contains those that leave the origin fixed – O(*n*, **R**) = *E*(*n*) ∩ GL(*n*, **R**). It is the symmetry group of the sphere (*n* = 3) or (*n* − 1)-sphere and all objects with spherical symmetry, if the origin is chosen at the center.

SO(*n*, **R**) is a subgroup of *E*$^+$(*n*), which consists of *direct* isometries, i.e., isometries preserving orientation; it contains those that leave the origin fixed – SO(*n*, **R**) = *E*$^+$(*n*) ∩ GL(*n*, **R**) = *E*(*n*) ∩ GL$^+$(*n*, **R**). It is the rotation group of the sphere and all objects with spherical symmetry, if the origin is chosen at the center.

$\{\pm I\}$ is a normal subgroup and even a characteristic subgroup of $O(n, \mathbf{R})$, and, if n is even, also of $SO(n, \mathbf{R})$. If n is odd, $O(n, \mathbf{R})$ is the internal direct product of $SO(n, \mathbf{R})$ and $\{\pm I\}$. For every positive integer k the cyclic group Ck of k-fold rotations is a normal subgroup of $O(2, \mathbf{R})$ and $SO(2, \mathbf{R})$.

Relative to suitable orthogonal bases, the isometries are of the form:

$$
\begin{bmatrix}
R_1 & & & & & \\
& \ddots & & & & 0 \\
& & R_k & & & \\
& & & \pm 1 & & \\
& 0 & & & \ddots & \\
& & & & & \pm 1
\end{bmatrix}
$$

where the matrices $R_1, ..., Rk$ are 2-by-2 rotation matrices in orthogonal planes of rotation. As a special case, known as Euler's rotation theorem, any (non-identity) element of $SO(3, \mathbf{R})$ is rotation about a uniquely defined axis.

The orthogonal group is generated by reflections (two reflections give a rotation), as in a Coxeter group,[note 1] and elements have length at most n (require at most n reflections to generate; this follows from the above classification, noting that a rotation is generated by 2 reflections, and is true more generally for indefinite orthogonal groups, by the Cartan–Dieudonné theorem). A longest element (element needing the most reflections) is reflection through the origin (the map $v \mapsto -v$), though so are other maximal combinations of rotations (and a reflection, in odd dimension).

The symmetry group of a circle is $O(2, \mathbf{R})$. The orientation preserving subgroup $SO(2, \mathbf{R})$ is isomorphic (as a *real* Lie group) to the circle group, also known as $U(1)$. This isomorphism sends the complex number $\exp(\varphi\, i) = \cos\varphi + i \sin\varphi$ of absolute value 1 to the special orthogonal matrix

$$
\begin{bmatrix}
\cos(\phi) & -\sin(\phi) \\
\sin(\phi) & \cos(\phi)
\end{bmatrix}.
$$

The group $SO(3, \mathbf{R})$, understood as the set of rotations of 3-dimensional space, is of major importance in the sciences and engineering, and there are numerous charts on $SO(3)$.

25.3.2 Maximal tori and Weyl groups

A maximal torus T for $SO(2n)$, of rank n, is given by the block-diagonal matrices

$$
\begin{bmatrix}
R_1 & & 0 \\
& \ddots & \\
0 & & R_n
\end{bmatrix}.
$$

where the Rj are 2-by-2 rotation matrices. The image $T \times \{1\}$ of the same torus under the block-diagonal inclusion

$$
SO(2n) \cong SO(2n) \times \{1\} < SO(2n+1)
$$

is a maximal torus for $SO(2n+1)$. The Weyl group of $SO(2n+1)$ is the semidirect product $\{\pm 1\}^n \rtimes S_n$ of a normal elementary abelian 2-subgroup and a symmetric group, where the nontrivial element of each $\{\pm 1\}$ factor of $\{\pm 1\}^n$ acts on the corresponding circle factor of $T \times \{1\}$ by inversion, and the symmetric group Sn acts on both $\{\pm 1\}^n$ and $T \times \{1\}$ by permuting factors. The elements of the Weyl group are represented by matrices in $O(2n) \times \{\pm 1\}$. The Sn factor is represented by block permutation matrices with 2-by-2 blocks, and a final 1 on the diagonal. The $\{\pm 1\}^n$ component is represented by block-diagonal matrices with 2-by-2 blocks either

$$\begin{bmatrix} 1 & 0 \\ 0 & 1 \end{bmatrix} \quad \text{or} \quad \begin{bmatrix} 0 & 1 \\ 1 & 0 \end{bmatrix}.$$

with the last component ± 1 chosen to make the determinant 1.

The Weyl group of SO(2n) is the subgroup $H_{n-1} \rtimes S_n < \{\pm 1\}^n \rtimes S_n$ of that of SO(2n + 1), where $Hn-1 < \{\pm 1\}^n$ is the kernel of the product homomorphism $\{\pm 1\}^n \to \{\pm 1\}$ given by $(\epsilon_1, \ldots, \epsilon_n) \mapsto \epsilon_1 \cdots \epsilon_n$; that is $Hn-1 < \{\pm 1\}^n$ is the subgroup with an even number of minus signs. The Weyl group of SO(2n) is represented in SO(2n) by the preimages under the standard injection SO(2n) $\to$ SO(2n+ 1) of the representatives for the Weyl group of SO(2n + 1). Those matrices with an odd number of $\begin{bmatrix} 0 & 1 \\ 1 & 0 \end{bmatrix}$ blocks have no remaining final -1 coordinate to make their determinants positive, and hence cannot be represented in SO(2n).

25.3.3 Low-dimensional topology

The low-dimensional (real) orthogonal groups are familiar spaces:

- O(1) = S^0, a two-point discrete space

- SO(1) = {1}

- SO(2) is S^1

- SO(3) is $\mathbf{RP}^3$

- SO(4) is doubly covered by SU(2) × SU(2) = $S^3 \times S^3$.

25.3.4 Homotopy groups

In terms of algebraic topology, for $n > 2$ the fundamental group of SO(n, $\mathbf{R}$) is cyclic of order 2, and the spin group Spin(n) is its universal cover. For $n = 2$ the fundamental group is infinite cyclic and the universal cover corresponds to the real line (the group Spin(2) is the unique connected 2-fold cover).

Generally, the homotopy groups $\pi k(O)$ of the real orthogonal group are related to homotopy groups of spheres, and thus are in general hard to compute. However, one can compute the homotopy groups of the stable orthogonal group (aka the infinite orthogonal group), defined as the direct limit of the sequence of inclusions:

$$O(0) \subset O(1) \subset O(2) \subset \cdots \subset O = \bigcup_{k=0}^{\infty} O(k)$$

Since the inclusions are all closed, hence cofibrations, this can also be interpreted as a union. On the other hand S^n is a homogeneous space for O(n + 1), and one has the following fiber bundle:

$$O(n) \to O(n + 1) \to S^n.$$

which can be understood as "The orthogonal group O(n + 1) acts transitively on the unit sphere S^n, and the stabilizer of a point (thought of as a unit vector) is the orthogonal group of the perpendicular complement, which is an orthogonal group one dimension lower. Thus the natural inclusion O(n) $\to$ O(n + 1) is $(n - 1)$-connected, so the homotopy groups stabilize, and $\pi k(O(n+1)) = \pi k(O(n))$ for $n > k + 1$: thus the homotopy groups of the stable space equal the lower homotopy groups of the unstable spaces.

From Bott periodicity we obtain $\Omega^8 O \cong O$, therefore the homotopy groups of O are 8-fold periodic, meaning $\pi k_{+8}(O) = \pi k(O)$, and one needs only to list the lower 8 homotopy groups:

$\pi_0(O) = \mathbf{Z}/2$

$\pi_1(O) = \mathbf{Z}/2$

$\pi_2(O) = 0$

$\pi_3(O) = \mathbf{Z}$

$\pi_4(O) = 0$

$\pi_5(O) = 0$

$\pi_6(O) = 0$

$\pi_7(O) = \mathbf{Z}$

Relation to KO-theory

Via the clutching construction, homotopy groups of the stable space O are identified with stable vector bundles on spheres (up to isomorphism), with a dimension shift of 1: $\pi k(O) = \pi k_{+1}(BO)$. Setting $KO = BO \times \mathbf{Z} = \Omega^{-1}O \times \mathbf{Z}$ (to make π_0 fit into the periodicity), one obtains:

$\pi_0(KO) = \mathbf{Z}$

$\pi_1(KO) = \mathbf{Z}/2$

$\pi_2(KO) = \mathbf{Z}/2$

$\pi_3(KO) = 0$

$\pi_4(KO) = \mathbf{Z}$

$\pi_5(KO) = 0$

$\pi_6(KO) = 0$

$\pi_7(KO) = 0$

Computation and interpretation of homotopy groups

Low-dimensional groups   The first few homotopy groups can be calculated by using the concrete descriptions of low-dimensional groups.

- $\pi_0(O) = \pi_0(O(1)) = \mathbf{Z}/2$, from orientation-preserving/reversing (this class survives to $O(2)$ and hence stably)

- $\pi_1(O) = \pi_1(SO(3)) = \mathbf{Z}/2$, which is spin comes from $SO(3) = \mathbf{RP}^3 = S^3/(\mathbf{Z}/2)$.

- $\pi_2(O) = \pi_2(SO(3)) = 0$, which surjects onto $\pi_2(SO(4))$; this latter thus vanishes.

Lie groups   From general facts about Lie groups, $\pi_2(G)$ always vanishes, and $\pi_3(G)$ is free (free abelian).

Vector bundles   From the vector bundle point of view, $\pi_0(KO)$ is vector bundles over S^0, which is two points. Thus over each point, the bundle is trivial, and the non-triviality of the bundle is the difference between the dimensions of the vector spaces over the two points, so $\pi_0(KO) = \mathbf{Z}$ is dimension.

Loop spaces   Using concrete descriptions of the loop spaces in Bott periodicity, one can interpret higher homotopy of O as lower homotopy of simple to analyze spaces. Using π_0, O and O/U have two components, $KO = BO \times \mathbf{Z}$ and $KSp = BSp \times \mathbf{Z}$ have countably many components, and the rest are connected.

Interpretation of homotopy groups

In a nutshell:[2]

- $\pi_0(KO) = \mathbf{Z}$ is about dimension

- $\pi_1(KO) = \mathbf{Z}/2$ is about orientation

- $\pi_2(KO) = \mathbf{Z}/2$ is about spin

- $\pi_4(KO) = \mathbf{Z}$ is about topological quantum field theory.

Let R be any of the four division algebras $\mathbf{R}$, $\mathbf{C}$, $\mathbf{H}$, $\mathbf{O}$, and let LR be the tautological line bundle over the projective line $R\mathrm{P}^1$, and $[LR]$ its class in K-theory. Noting that $\mathbf{R}\mathrm{P}^1 = S^1$, $\mathbf{C}\mathrm{P}^1 = S^2$, $\mathbf{H}\mathrm{P}^1 = S^4$, $\mathbf{O}\mathrm{P}^1 = S^8$, these yield vector bundles over the corresponding spheres, and

- $\pi_1(KO)$ is generated by $[L\mathbf{R}]$

- $\pi_2(KO)$ is generated by $[L\mathbf{C}]$

- $\pi_4(KO)$ is generated by $[L\mathbf{H}]$

- $\pi_8(KO)$ is generated by $[L\mathbf{O}]$

From the point of view of symplectic geometry, $\pi_0(KO) \cong \pi_8(KO) = \mathbf{Z}$ can be interpreted as the Maslov index, thinking of it as the fundamental group $\pi_1(U/O)$ of the stable Lagrangian Grassmannian as $U/O \cong \Omega^7(KO)$, so $\pi_1(U/O) = \pi_{1+7}(KO)$.

25.4 Over the complex number field

Over the field $\mathbf{C}$ of complex numbers, $O(n, \mathbf{C})$ and $SO(n, \mathbf{C})$ are complex Lie groups of dimension $n(n - 1)/2$ over $\mathbf{C}$ (it means the dimension over $\mathbf{R}$ is twice that). $O(n, \mathbf{C})$ has two connected components, and $SO(n, \mathbf{C})$ is the connected component containing the identity matrix. For $n \geq 2$ these groups are noncompact.

Just as in the real case $SO(n, \mathbf{C})$ is not simply connected. For $n > 2$ the fundamental group of $SO(n, \mathbf{C})$ is cyclic of order 2 whereas the fundamental group of $SO(2, \mathbf{C})$ is infinite cyclic.

25.5 Over finite fields

Orthogonal groups can also be defined over finite fields $\mathbf{F}q$, where is a power of a prime p.

Over finite fields of characteristic not equal to 2, orthogonal groups come in two types in even dimension: $O^+(2n, q)$ and $O^-(2n, q)$; and one type in odd dimension: $O(2n + 1, q)$.[3]

If V is the vector space on which the orthogonal group G acts, it can be written as a direct orthogonal sum as follows:

$$V = L_1 \oplus L_2 \oplus \cdots \oplus L_m \oplus W,$$

where Li are hyperbolic lines and W contains no singular vectors. If W is the zero subspace, then G is of plus type. If W is one-dimensional then G has odd dimension. If W has dimension 2, G is of minus type.

In the special case where $n = 1$, $O'(2, q)$ is a dihedral group of order $2(q - \epsilon)$.

We have the following formulas for the order of $O(n, q)$, when the characteristic is not two:

$$|O(2n + 1, q)| = 2q^n \prod_{i=0}^{n-1} (q^{2n} - q^{2i}).$$

If −1 is a square in $\mathbf{F}q$

$$|O(2n, q)| = 2(q^n - 1) \prod_{i=1}^{n-1} (q^{2n} - q^{2i}).$$

If −1 is a non-square in $\mathbf{F}q$

$$|O(2n, q)| = 2(q^n + (-1)^{n+1}) \prod_{i=1}^{n-1} (q^{2n} - q^{2i}).$$

25.6   The Dickson invariant

For orthogonal groups, the **Dickson invariant** is a homomorphism from the orthogonal group to the quotient group $\mathbf{Z}/2\mathbf{Z}$ (integers modulo 2), taking the value 0 in case the element is the product of an even number of reflections, and the value of 1 otherwise.[4]

Algebraically, the Dickson invariant can be defined as $D(f) = \mathrm{rank}(I - f)$ modulo 2, where I is the identity (Taylor 1992, Theorem 11.43). Over fields that are not of characteristic 2 it is equivalent to the determinant: the determinant is −1 to the power of the Dickson invariant. Over fields of characteristic 2, the determinant is always 1, so the Dickson invariant gives more information than the determinant.

The special orthogonal group is the kernel of the Dickson invariant[4] and usually has index 2 in $O(n, F)$.[5] When the characteristic of F is not 2, the Dickson Invariant is 0 whenever the determinant is 1. Thus when the characteristic is not 2, $SO(n, F)$ is commonly defined to be the elements of $O(n, F)$ with determinant 1. Each element in $O(n, F)$ has determinant ±1. Thus in characteristic 2, the determinant is always 1.

The Dickson invariant can also be defined for Clifford groups and Pin groups in a similar way (in all dimensions).

25.7   Orthogonal groups of characteristic 2

Over fields of characteristic 2 orthogonal groups often exhibit special behaviors, some of which are listed in this section. (Formerly these groups were known as the **hypoabelian groups** but this term is no longer used.)

- Any orthogonal group over any field is generated by reflections, except for a unique example where the vector space is 4-dimensional over the field with 2 elements and the Witt index is 2.[6] Note that a reflection in characteristic two has a slightly different definition. In characteristic two, the reflection orthogonal to a vector $\mathbf{u}$ takes a vector $\mathbf{v}$ to $\mathbf{v} + B(\mathbf{v}, \mathbf{u})/Q(\mathbf{u}) \cdot \mathbf{u}$ where B is the bilinear form and Q is the quadratic form associated to the orthogonal geometry. Compare this to the Householder reflection of odd characteristic or characteristic zero, which takes $\mathbf{v}$ to $\mathbf{v} - 2 \cdot B(\mathbf{v}, \mathbf{u})/Q(\mathbf{u}) \cdot \mathbf{u}$.

- The center of the orthogonal group usually has order 1 in characteristic 2, rather than 2, since $I = -I$.

- In odd dimensions $2n + 1$ in characteristic 2, orthogonal groups over perfect fields are the same as symplectic groups in dimension $2n$. In fact the symmetric form is alternating in characteristic 2, and as the dimension is odd it must have a kernel of dimension 1, and the quotient by this kernel is a symplectic space of dimension $2n$, acted upon by the orthogonal group.

- In even dimensions in characteristic 2 the orthogonal group is a subgroup of the symplectic group, because the symmetric bilinear form of the quadratic form is also an alternating form.

25.8   The spinor norm

The **spinor norm** is a homomorphism from an orthogonal group over a field F to the quotient group F^*/F^{*2} (the multiplicative group of the field F up to square elements), that takes reflection in a vector of norm n to the image of n in F^*/F^{*2}.[7]

For the usual orthogonal group over the reals it is trivial, but it is often non-trivial over other fields, or for the orthogonal group of a quadratic form over the reals that is not positive definite.

25.9   Galois cohomology and orthogonal groups

In the theory of Galois cohomology of algebraic groups, some further points of view are introduced. They have explanatory value, in particular in relation with the theory of quadratic forms; but were for the most part *post hoc*, as far as the discovery of the phenomena is concerned. The first point is that quadratic forms over a field can be identified as a Galois H^1, or twisted forms (torsors) of an orthogonal group. As an algebraic group, an orthogonal group is in general neither connected nor simply-connected; the latter point brings in the spin phenomena, while the former is related to the discriminant.

The 'spin' name of the spinor norm can be explained by a connection to the spin group (more accurately a pin group). This may now be explained quickly by Galois cohomology (which however postdates the introduction of the term by more direct use of Clifford algebras). The spin covering of the orthogonal group provides a short exact sequence of algebraic groups.

$$1 \rightarrow \mu_2 \rightarrow \text{Pin}_V \rightarrow O_V \rightarrow 1$$

Here μ_2 is the algebraic group of square roots of 1; over a field of characteristic not 2 it is roughly the same as a two-element group with trivial Galois action. The connecting homomorphism from $H^0(OV)$, which is simply the group $OV(F)$ of F-valued points, to $H^1(\mu_2)$ is essentially the spinor norm, because $H^1(\mu_2)$ is isomorphic to the multiplicative group of the field modulo squares.

There is also the connecting homomorphism from H^1 of the orthogonal group, to the H^2 of the kernel of the spin covering. The cohomology is non-abelian, so that this is as far as we can go, at least with the conventional definitions.

25.10   Lie algebra

The Lie algebra corresponding to Lie groups $O(n, F)$ and $SO(n, F)$ consists of the skew-symmetric $n \times n$ matrices, with the Lie bracket $[,]$ given by the commutator. One Lie algebra corresponds to both groups. It is often denoted by $\mathfrak{o}(n, F)$ or $\mathfrak{so}(n, F)$, and called the **orthogonal Lie algebra** or **special orthogonal Lie algebra**. Over real numbers, these Lie algebras for different n are the compact real forms of two of the four families of semisimple Lie algebras: in odd dimension Bk, where $n = 2k + 1$, while in even dimension Dr, where $n = 2r$.

More intrinsically, given a vector space with an inner product, the special orthogonal Lie algebra is given by the bivectors on the space, which are sums of simple bivectors (2-blades) $\mathbf{v} \wedge \mathbf{w}$. The correspondence is given by the map $\mathbf{v} \wedge \mathbf{w} \mapsto \mathbf{v}^* \otimes \mathbf{w} - \mathbf{w}^* \otimes \mathbf{v}$, where $\mathbf{v}^*$ is the covector dual to the vector $\mathbf{v}$; in coordinates these are exactly the elementary skew-symmetric matrices.

Over real numbers, this characterization is used in interpreting the curl of a vector field (naturally a 2-vector) as an infinitesimal rotation or "curl", hence the name. Generalizing the inner product with a nondegenerate form yields the indefinite orthogonal Lie algebras $\mathfrak{so}(p, q)$.

The representation theory of the orthogonal Lie algebras includes both representations corresponding to linear representations of the orthogonal groups, and representations corresponding to projective representations of the orthogonal groups (linear representations of spin groups), the so-called spin representation, which are important in physics.

25.11   Related groups

The orthogonal groups and special orthogonal groups have a number of important subgroups, supergroups, quotient groups, and covering groups. These are listed below.

The inclusions $O(n) \subset U(n) \subset Sp(n) = USp(2n)$ and $USp(n) \subset U(n) \subset O(2n)$ are part of a sequence of 8 inclusions used in a geometric proof of the Bott periodicity theorem, and the corresponding quotient spaces are symmetric spaces of independent interest – for example, $U(n)/O(n)$ is the Lagrangian Grassmannian.

25.11.1   Lie subgroups

In physics, particularly in the areas of Kaluza–Klein compactification, it is important to find out the subgroups of the orthogonal group. The main ones are:

$O(n) \supset O(n - 1)$ – preserve an axis

$O(2n) \supset U(n) \supset SU(n)$ – $U(n)$ are those that preserve a compatible complex structure *or* a compatible symplectic structure – see 2-out-of-3 property; $SU(n)$ also preserves a complex orientation.

$O(2n) \supset USp(n)$

$O(7) \supset G_2$

25.11.2   Lie supergroups

The orthogonal group $O(n)$ is also an important subgroup of various Lie groups:

$U(n) \supset SU(n) \supset O(n)$

$USp(2n) \supset O(n)$

$G_2 \supset O(3)$

$F_4 \supset O(9)$

$E_6 \supset O(10)$

$E_7 \supset O(12)$

$E_8 \supset O(16)$

Conformal group

Main article: Conformal group

Being isometries, real orthogonal transforms preserve angles, and are thus conformal maps, though not all conformal linear transforms are orthogonal. In classical terms this is the difference between congruence and similarity, as exemplified by SSS (Side-Side-Side) congruence of triangles and AAA (Angle-Angle-Angle) similarity of triangles. The group of conformal linear maps of $\mathbf{R}^n$ is denoted $CO(n)$ for the **conformal orthogonal group**, and consists of the product of the orthogonal group with the group of dilations. If n is odd, these two subgroups do not intersect, and they are a direct product: $CO(2k + 1) = O(2k + 1) \times \mathbf{R}^{\cdot}$, where $\mathbf{R}^{\cdot} = \mathbf{R} \setminus \{0\}$ is the real multiplicative group, while if n is even, these subgroups intersect in ± 1, so this is not a direct product, but it is a direct product with the subgroup of dilation by a positive scalar: $CO(2k) = O(2k) \times \mathbf{R}^{+}$.

Similarly one can define $CSO(n)$; note that this is always: $CSO(n) = CO(n) \cap GL^{+}(n) = SO(n) \times \mathbf{R}^{+}$.

25.11.3   Discrete subgroups

As the orthogonal group is compact, discrete subgroups are equivalent to finite subgroups.[note 2] These subgroups are known as point group and can be realized as the symmetry groups of polytopes. A very important class of examples are the finite Coxeter groups, which include the symmetry groups of regular polytopes.

Dimension 3 is particularly studied – see point groups in three dimensions, polyhedral groups, and list of spherical symmetry groups. In 2 dimensions, the finite groups are either cyclic or dihedral – see point groups in two dimensions.

Other finite subgroups include:

- Permutation matrices (the Coxeter group An)

- Signed permutation matrices (the Coxeter group Bn); also equals the intersection of the orthogonal group with the integer matrices.[note 3]

25.11.4   Covering and quotient groups

The orthogonal group is neither simply connected nor centerless, and thus has both a covering group and a quotient group, respectively:

- Two covering Pin groups, $\mathrm{Pin}_{+}(n) \to O(n)$ and $\mathrm{Pin}_{-}(n) \to O(n)$.

- The quotient projective orthogonal group, $O(n) \to PO(n)$.

These are all 2-to-1 covers.

For the special orthogonal group, the corresponding groups are:

- Spin group, $\mathrm{Spin}(n) \to SO(n)$.

- Projective special orthogonal group, $SO(n) \to PSO(n)$.

Spin is a 2-to-1 cover, while in even dimension, $PSO(2k)$ is a 2-to-1 cover, and in odd dimension $PSO(2k + 1)$ is a 1-to-1 cover, i.e., isomorphic to $SO(2k + 1)$. These groups, $\mathrm{Spin}(n)$, $SO(n)$, and $PSO(n)$ are Lie group forms of the compact special orthogonal Lie algebra, $\mathfrak{so}(n, \mathbb{R})$ – Spin is the simply connected form, while PSO is the centerless form, and SO is in general neither.[note 4]

In dimension 3 and above these are the covers and quotients, while dimension 2 and below are somewhat degenerate; see specific articles for details.

25.12   Principal homogeneous space: Stiefel manifold

Main article: Stiefel manifold

The principal homogeneous space for the orthogonal group $O(n)$ is the Stiefel manifold $V_n(\mathbf{R}^n)$ of orthonormal bases (orthonormal n-frames).

In other words, the space of orthonormal bases is like the orthogonal group, but without a choice of base point: given an orthogonal space, there is no natural choice of orthonormal basis, but once one is given one, there is a one-to-one correspondence between bases and the orthogonal group. Concretely, a linear map is determined by where it sends a basis: just as an invertible map can take any basis to any other basis, an orthogonal map can take any *orthogonal* basis to any other *orthogonal* basis.

The other Stiefel manifolds $Vk(\mathbf{R}^n)$ for $k < n$ of *incomplete* orthonormal bases (orthonormal k-frames) are still homogeneous spaces for the orthogonal group, but not *principal* homogeneous spaces: any k-frame can be taken to any other k-frame by an orthogonal map, but this map is not uniquely determined.

25.13   See also

25.13.1   Specific transforms

- Coordinate rotations and reflections
- Reflection through the origin

25.13.2   Specific groups

- rotation group, SO(3, **R**)
- SO(8)

25.13.3   Related groups

- indefinite orthogonal group
- unitary group
- symplectic group

25.13.4   Lists of groups

- list of finite simple groups
- list of simple Lie groups

25.14   Notes

[1] The analogy is stronger: Weyl groups, a class of (representations of) Coxeter groups, can be considered as simple algebraic groups over the field with one element, and there are a number of analogies between algebraic groups and vector spaces on the one hand, and Weyl groups and sets on the other.

[2] Infinite subsets of a compact space have an accumulation point and are not discrete.

[3] $O(n) \cap GL(n, \mathbf{Z})$ equals the signed permutation matrices because an integer vector of norm 1 must have a single non-zero entry, which must be ± 1 (if it has two non-zero entries or a larger entry, the norm will be larger than 1), and in an orthogonal matrix these entries must be in different coordinates, which is exactly the signed permutation matrices.

[4] In odd dimension, $SO(2k + 1) \cong PSO(2k + 1)$ is centerless (but not simply connected), while in even dimension $SO(2k)$ is neither centerless nor simply connected.

25.15 References

[1] For base fields of characteristic not 2, it is equivalent to use symmetric bilinear forms or quadratic forms. But in characteristic 2 these notions differ.

[2] John Baez "This Week's Finds in Mathematical Physics" week 105

[3] Wilson, Robert A. (2009), *The finite simple groups*. Graduate Texts in Mathematics **251**. London: Springer. pp. 69–75. ISBN 978-1-84800-987-5. Zbl 1203.20012.

[4] Knus, Max-Albert (1991), *Quadratic and Hermitian forms over rings*, Grundlehren der Mathematischen Wissenschaften **294**, Berlin etc.: Springer-Verlag. p. 224, ISBN 3-540-52117-8, Zbl 0756.11008

[5] (Taylor 1992, page 160)

[6] (Grove 2002, Theorem 6.6 and 14.16)

[7] Cassels 1978, p. 178

- Cassels, J.W.S. (1978), *Rational Quadratic Forms*, London Mathematical Society Monographs **13**, Academic Press, ISBN 0-12-163260-1, Zbl 0395.10029

- Grove, Larry C. (2002), *Classical groups and geometric algebra*, Graduate Studies in Mathematics **39**, Providence, R.I.: American Mathematical Society, ISBN 978-0-8218-2019-3, MR 1859189

- Taylor, Donald E. (1992), *The Geometry of the Classical Groups*, Sigma Series in Pure Mathematics **9**, Berlin: Heldermann Verlag, ISBN 3-88538-009-9, MR 1189139, Zbl 0767.20001

25.16 External links

- Hazewinkel, Michiel, ed. (2001), "Orthogonal group", *Encyclopedia of Mathematics*, Springer, ISBN 978-1-55608-010-4

- John Baez "This Week's Finds in Mathematical Physics" week 105

- John Baez on Octonions

- (Italian) n-dimensional Special Orthogonal Group parametrization

Chapter 26

Lorentz group

In physics and mathematics, the **Lorentz group** is the group of all Lorentz transformations of Minkowski spacetime, the classical setting for all (nongravitational) physical phenomena. The Lorentz group is named for the Dutch physicist Hendrik Lorentz.

Under the Lorentz transformations, these laws and equations are invariant:

- The kinematical laws of special relativity

- Maxwell's field equations in the theory of electromagnetism

- The Dirac equation in the theory of the electron

Therefore, the Lorentz group expresses the fundamental symmetry of many known fundamental laws of nature.

26.1 Basic properties

The Lorentz group is a subgroup of the Poincaré group—the group of all isometries of Minkowski spacetime. Lorentz transformations are, precisely, isometries that leave the origin fixed. Thus, the Lorentz group is an isotropy subgroup of the isometry group of Minkowski spacetime. For this reason, the Lorentz group is sometimes called the **homogeneous Lorentz group** while the Poincaré group is sometimes called the *inhomogeneous Lorentz group*. Lorentz transformations are examples of linear transformations; general isometries of Minkowski spacetime are affine transformations. Mathematically, the Lorentz group may be described as the generalized orthogonal group O(1,3), the matrix Lie group that preserves the quadratic form

$$(t, x, y, z) \mapsto t^2 - x^2 - y^2 - z^2$$

on $\mathbf{R}^4$. This quadratic form is, when put on matrix form (see classical orthogonal group), interpreted in physics as the metric tensor of Minkowski spacetime.

The Lorentz group is a six-dimensional noncompact non-abelian real Lie group that is not connected. All four of its connected components are not simply connected. The identity component (i.e., the component containing the identity element) of the Lorentz group is itself a group, and is often called the **restricted Lorentz group**, and is denoted $SO^+(1,3)$. The restricted Lorentz group consists of those Lorentz transformations that preserve the orientation of space and direction of time. The restricted Lorentz group has often been presented through a facility of biquaternion algebra.

The restricted Lorentz group arises in other ways in pure mathematics. For example, it arises as the point symmetry group of a certain ordinary differential equation. This fact also has physical significance.

26.1.1   Connected components

Because it is a Lie group, the Lorentz group O(1,3) is both a group and a smooth manifold. As a manifold, it has four connected components. Intuitively, this means that it consists of four topologically separated pieces.

Each of the four connected components can be categorized by which of these two properties its elements have:

- The element reverses the direction of time, or more precisely, transforms a future-pointing timelike vector into a past-pointing one.

- The element reverses the orientation of a vierbein (tetrad).

Lorentz transformations that preserve the direction of time are called **orthochronous**. The subgroup of orthochronous transformations is often denoted $O^+(1,3)$. Those that preserve orientation are called **proper**, and as linear transformations they have determinant +1. (The improper Lorentz transformations have determinant −1.) The subgroup of proper Lorentz transformations is denoted SO(1,3).

The subgroup of all Lorentz transformations preserving both orientation and direction of time is called the **proper, orthochronous Lorentz group** or **restricted Lorentz group**, and is denoted by $SO^+(1,3)$. (Note that some authors refer to SO(1,3) or even O(1,3) when they actually mean $SO^+(1,3)$.)

The set of the four connected components can be given a group structure as the quotient group $O(1,3)/SO^+(1,3)$, which is isomorphic to the Klein four-group. Every element in O(1,3) can be written as the semidirect product of a proper, orthochronous transformation and an element of the discrete group

$$\{1, P, T, PT\}$$

where P and T are the space inversion and time reversal operators:

$$P = \text{diag}(1, -1, -1, -1)$$
$$T = \text{diag}(-1, 1, 1, 1).$$

Thus an arbitrary Lorentz transformation can be specified as a proper, orthochronous Lorentz transformation along with a further two bits of information, which pick out one of the four connected components. This pattern is typical of finite-dimensional Lie groups.

26.2   Restricted Lorentz group

The restricted Lorentz group is the identity component of the Lorentz group, which means that it consists of all Lorentz transformations that can be connected to the identity by a continuous curve lying in the group. The restricted Lorentz group is a connected normal subgroup of the full Lorentz group with the same dimension, in this case with dimension six.

The restricted Lorentz group is generated by ordinary spatial rotations and Lorentz boosts (which can be thought of as hyperbolic rotations in a plane that includes a time-like direction). Since every proper, orthochronous Lorentz transformation can be written as a product of a rotation (specified by 3 real parameters) and a boost (also specified by 3 real parameters), it takes 6 real parameters to specify an arbitrary proper orthochronous Lorentz transformation. This is one way to understand why the restricted Lorentz group is six-dimensional. (See also the Lie algebra of the Lorentz group.)

The set of all rotations forms a Lie subgroup isomorphic to the ordinary rotation group SO(3). The set of all boosts, however, does *not* form a subgroup, since composing two boosts does not, in general, result in another boost. (Rather, a pair of non-colinear boosts is equivalent to a boost and a rotation, and this relates to Thomas rotation.) A boost in some direction, or a rotation about some axis, generates a one-parameter subgroup.

26.2.1   Surfaces of transitivity

If a group G acts on a space V, then a surface $S \subset V$ is a **surface of transitivity** if S is invariant under G, i.e., $gs \in S \ \forall g \in G$, $\forall s \in S$, and for any two points s_1, $s_2 \in S$ there is a $g \in G$ such that $gs_1 = s_2$. By definition of the Lorentz group, it preserves the quadratic form

$$Q(x) = x_0^2 - x_1^2 - x_2^2 - x_3^2.$$

The surfaces of transitivity of the orthochronous Lorentz group $O^+(1, 3)$, $Q(x) = $ const. of spacetime are the following:[1]

- $Q(x) > 0$, $x_0 > 0$ is the upper branch of a hyperboloid of two sheets.

- $Q(x) > 0$, $x_0 < 0$ is the lower branch of this hyperboloid.

- $Q(x) = 0$, $x_0 > 0$ is the upper branch of the light cone.

- $Q(x) = 0$, $x_0 < 0$ is the lower branch of the light cone.

- $Q(x) < 0$ is a hyperboloid of one sheet.

- The origin $x_0 = x_1 = x_2 = x_3 = 0$.

These surfaces are 3-dimensional, so the images are not faithful, but they are faithful for the corresponding facts about $O^+(1, 2)$. For the full Lorentz group, the surfaces of transitivity are only four since the transformation T takes an upper branch of a hyperboloid (cone) to a lower one and vice versa.

These observations constitute a good starting point for finding all infinite-dimensional unitary representations of the Lorentz group, in fact, of the Poincaré group, using the method of induced representations.[2] One begins with a "standard vector", one for each surface of transitivity, and then ask which subgroup preserves these vectors. These subgroups are called little groups by physicists. The problem is then essentially reduced to the easier problem of finding representations of the little groups. For example, a standard vector in one of the hyperbolas of two sheets could be suitably chosen as $(m, 0, 0, 0)$. For each $m \neq 0$, the vector pierces exactly one sheet. In this case the little group is SO(3), the rotation group, all of whose representations are known. The precise infinite-dimensional unitary representation under which a particle transform is part of its classification. Not all representations can correspond to physical particles (as far as is known). Standard vectors on the one-sheeted hyperbolas would correspond to tachyons. Particles on the light cone are photons, and more hypothetically, gravitons. The "particle" corresponding to the origin is the vacuum.

26.2.2   Relation to the Möbius group

See also: Algebra of physical space

The restricted Lorentz group $SO^+(1, 3)$ is isomorphic to the projective special linear group PSL(2,**C**), which is in turn isomorphic to the Möbius group, the symmetry group of conformal geometry on the Riemann sphere. (This observation was utilized by Roger Penrose as the starting point of twistor theory.)

This may be shown by constructing a surjective homomorphism of Lie groups from SL(2,**C**) to $SO^+(1,3)$, which we will call the **spinor map**. This proceeds as follows:

We can define an action of SL(2,**C**) on Minkowski spacetime by writing a point of spacetime as a two-by-two Hermitian matrix in the form

$$X = \begin{bmatrix} t + z & x - iy \\ x + iy & t - z \end{bmatrix}.$$

This presentation has the pleasant feature that

$\det X = t^2 - x^2 - y^2 - z^2.$

Therefore, we have identified the space of Hermitian matrices (which is four-dimensional, as a *real* vector space) with Minkowski spacetime in such a way that the determinant of a Hermitian matrix is the squared length of the corresponding vector in Minkowski spacetime. SL(2,C) acts on the space of Hermitian matrices via

$$X \mapsto PXP^*$$

where P^* is the Hermitian transpose of P, and this action preserves the determinant. Therefore, SL(2,C) acts on Minkowski spacetime by (linear) isometries, and so is homomorphic to a subgroup of the Lorentz group (by the definition of the Lorentz group.)

This completes the proof that there is a homomorphism from SL(2,C) to SO$^+$(1,3). The kernel of the spinor map is the two element subgroup $\pm I$, and it happens that the map is surjective. By the first isomorphism theorem, the quotient group PSL(2,C) is isomorphic to SO$^+$(1,3).

In optics, this construction is known as the Poincaré sphere.

Appearance of the night sky

This isomorphism has the consequence that Möbius transformations of the Riemann sphere represent the way that Lorentz transformations change the appearance of the night sky, as seen by an observer who is maneuvering at relativistic velocities relative to the "fixed stars".

Suppose the "fixed stars" live in Minkowski spacetime and are modeled by points on the celestial sphere. Then a given point on the celestial sphere can be associated with $\xi = u + iv$, a complex number that corresponds to the point on the Riemann sphere, and can be identified with a null vector (a light-like vector) in Minkowski space

$$\begin{bmatrix} u^2 + v^2 + 1 \\ 2u \\ -2v \\ u^2 + v^2 - 1 \end{bmatrix}$$

or the Hermitian matrix

$$N = 2 \begin{bmatrix} u^2 + v^2 & u + iv \\ u - iv & 1 \end{bmatrix}.$$

The set of real scalar multiples of this null vector, called a *null line* through the origin, represents a *line of sight* from an observer at a particular place and time (an arbitrary event we can identify with the origin of Minkowski spacetime) to various distant objects, such as stars. Then the points of the celestial sphere (equivalently, lines of sight) are identified with certain Hermitian matrices.

26.2.3 Conjugacy classes

Because the restricted Lorentz group SO$^+$(1, 3) is isomorphic to the Möbius group PSL(2,C), its conjugacy classes also fall into five classes:

- **Elliptic** transformations
- **Hyperbolic** transformations

- **Loxodromic** transformations

- **Parabolic** transformations

- The trivial **identity** transformation

In the article on Möbius transformations, it is explained how this classification arises by considering the fixed points of Möbius transformations in their action on the Riemann sphere, which corresponds here to null eigenspaces of restricted Lorentz transformations in their action on Minkowski spacetime.

An example of each type is given in the subsections below, along with the effect of the one-parameter subgroup it generates (e.g., on the appearance of the night sky).

The Möbius transformations are the conformal transformations of the Riemann sphere (or celestial sphere). Then conjugating with an arbitrary element of SL(2,**C**) obtains the following examples of arbitrary elliptic, hyperbolic, loxodromic, and parabolic (restricted) Lorentz transformations, respectively. The effect on the **flow lines** of the corresponding one-parameter subgroups is to transform the pattern seen in the examples by some conformal transformation. For example, an elliptic Lorentz transformation can have any two distinct fixed points on the celestial sphere, but points still flow along circular arcs from one fixed point toward the other. The other cases are similar.

Elliptic

An elliptic element of SL(2,**C**) is

$$P_1 = \begin{bmatrix} \exp(i\theta/2) & 0 \\ 0 & \exp(-i\theta/2) \end{bmatrix}$$

and has fixed points $\xi = 0, \infty$. Writing the action as $X \mapsto P_1 X P_1{}^*$ and collecting terms, the spinor map converts this to the (restricted) Lorentz transformation

$$Q_1 = \begin{bmatrix} 1 & 0 & 0 & 0 \\ 0 & \cos(\theta) & -\sin(\theta) & 0 \\ 0 & \sin(\theta) & \cos(\theta) & 0 \\ 0 & 0 & 0 & 1 \end{bmatrix} = \exp\left(\theta \begin{bmatrix} 0 & 0 & 0 & 0 \\ 0 & 0 & -1 & 0 \\ 0 & 1 & 0 & 0 \\ 0 & 0 & 0 & 0 \end{bmatrix} \right) .$$

This transformation then represents a rotation about the z axis, $\exp(i\theta J_z)$. The one-parameter subgroup it generates is obtained by taking θ to be a real variable, the rotation angle, instead of a constant.

The corresponding continuous transformations of the celestial sphere (except for the identity) all share the same two fixed points, the North and South poles. The transformations move all other points around latitude circles so that this group yields a continuous counterclockwise rotation about the z axis as θ increases. The *angle doubling* evident in the spinor map is a characteristic feature of *spinorial double coverings*.

Hyperbolic

A hyperbolic element of SL(2,**C**) is

$$P_2 = \begin{bmatrix} \exp(\beta/2) & 0 \\ 0 & \exp(-\beta/2) \end{bmatrix}$$

and has fixed points $\xi = 0, \infty$. Under stereographic projection from the Riemann sphere to the Euclidean plane, the effect of this Möbius transformation is a dilation from the origin.

The spinor map converts this to the Lorentz transformation

$$Q_2 = \begin{bmatrix} \cosh(\beta) & 0 & 0 & \sinh(\beta) \\ 0 & 1 & 0 & 0 \\ 0 & 0 & 1 & 0 \\ \sinh(\beta) & 0 & 0 & \cosh(\beta) \end{bmatrix} = \exp\left(\beta \begin{bmatrix} 0 & 0 & 0 & 1 \\ 0 & 0 & 0 & 0 \\ 0 & 0 & 0 & 0 \\ 1 & 0 & 0 & 0 \end{bmatrix}\right).$$

This transformation represents a boost along the z axis with rapidity β. The one-parameter subgroup it generates is obtained by taking β to be a real variable, instead of a constant. The corresponding continuous transformations of the celestial sphere (except for the identity) all share the same fixed points (the North and South poles), and they move all other points along longitudes away from the South pole and toward the North pole.

Loxodromic

A loxodromic element of SL(2,C) is

$$P_3 = P_2 P_1 = P_1 P_2 = \begin{bmatrix} \exp\left((\beta + i\theta)/2\right) & 0 \\ 0 & \exp\left(-(\beta + i\theta)/2\right) \end{bmatrix}$$

and has fixed points $\xi = 0, \infty$. The spinor map converts this to the Lorentz transformation

$$Q_3 = Q_2 Q_1 = Q_1 Q_2.$$

The one-parameter subgroup this generates is obtained by replacing $\beta + i\theta$ with any real multiple of this complex constant. (If β, θ vary independently, then a *two-dimensional* abelian subgroup is obtained, consisting of simultaneous rotations about the z axis and boosts along the z-axis; in contrast, the *one-dimensional* subgroup discussed here consists of those elements of this two-dimensional subgroup such that the **rapidity** of the boost and **angle** of the rotation have a *fixed ratio*.)

The corresponding continuous transformations of the celestial sphere (excepting the identity) all share the same two fixed points (the North and South poles). They move all other points away from the South pole and toward the North pole (or vice versa), along a family of curves called **loxodromes**. Each loxodrome spirals infinitely often around each pole.

Parabolic

A parabolic element of SL(2,C) is

$$P_4 = \begin{bmatrix} 1 & \alpha \\ 0 & 1 \end{bmatrix}$$

and has the single fixed point $\xi = \infty$ on the Riemann sphere. Under stereographic projection, it appears as an ordinary translation along the real axis.

The spinor map converts this to the matrix (representing a Lorentz transformation)

$$Q_4 = \begin{bmatrix} 1 + |\alpha|^2/2 & \operatorname{Re}(\alpha) & \operatorname{Im}(\alpha) & -|\alpha|^2/2 \\ \operatorname{Re}(\alpha) & 1 & 0 & -\operatorname{Re}(\alpha) \\ -\operatorname{Im}(\alpha) & 0 & 1 & \operatorname{Im}(\alpha) \\ |\alpha|^2/2 & \operatorname{Re}(\alpha) & \operatorname{Im}(\alpha) & 1 - |\alpha|^2/2 \end{bmatrix}$$

$$= \exp \begin{bmatrix} 0 & \operatorname{Re}(\alpha) & \operatorname{Im}(\alpha) & 0 \\ \operatorname{Re}(\alpha) & 0 & 0 & -\operatorname{Re}(\alpha) \\ -\operatorname{Im}(\alpha) & 0 & 0 & \operatorname{Im}(\alpha) \\ 0 & \operatorname{Re}(\alpha) & \operatorname{Im}(\alpha) & 0 \end{bmatrix}.$$

This generates a two-parameter abelian subgroup, which is obtained by considering α a complex variable rather than a constant. The corresponding continuous transformations of the celestial sphere (except for the identity transformation) move points along a family of circles that are all tangent at the North pole to a certain great circle. All points other than the North pole itself move along these circles.

Parabolic Lorentz transformations are often called **null rotations**, since they preserve null vectors, just as rotations preserve timelike vectors and boosts preserve spacelike vectors. Since these are likely to be the least familiar of the four types of nonidentity Lorentz transformations (elliptic, hyperbolic, loxodromic, parabolic), it is illustrated here how to determine the effect of an example of a parabolic Lorentz transformation on Minkowski spacetime.

The matrix given above yields the transformation

$$
\begin{bmatrix} t \\ x \\ y \\ z \end{bmatrix} \rightarrow \begin{bmatrix} t \\ x \\ y \\ z \end{bmatrix} + \mathrm{Re}(\alpha) \begin{bmatrix} x \\ t-z \\ 0 \\ x \end{bmatrix} + \mathrm{Im}(\alpha) \begin{bmatrix} y \\ 0 \\ z-t \\ y \end{bmatrix} + \frac{|\alpha|^2}{2} \begin{bmatrix} t-z \\ 0 \\ 0 \\ t-z \end{bmatrix}.
$$

Now, without loss of generality, pick $Im(\alpha)=0$. Differentiating this transformation with respect to the now real group parameter α and evaluating at $\alpha=0$ produces the corresponding vector field (first order linear partial differential operator),

$$
x\,(\partial_t + \partial_z) + (t-z)\,\partial_x.
$$

Apply this to a function $f(t,x,y,z)$, and demand that it stays invariant, i.e., it is annihilated by this transformation. The solution of the resulting first order linear partial differential equation can be expressed in the form

$$
f(t,x,y,z) = F(y,\ t-z,\ t^2 - x^2 - z^2),
$$

where F is an *arbitrary* smooth function. The arguments of F give three *rational invariants* describing how points (events) move under this parabolic transformation, as they themselves do not move,

$$
y = c_1, \quad t - z = c_2, \quad t^2 - x^2 - z^2 = c_3.
$$

Choosing real values for the constants on the right hand sides yields three conditions, and thus specifies a curve in Minkowski spacetime. This curve is an orbit of the transformation.

The form of the rational invariants shows that these flowlines (orbits) have a simple description: suppressing the inessential coordinate y, each orbit is the intersection of a *null plane*, $t = z+c_2$, with a *hyperboloid*, $t^2-x^2-z^2 = c_3$. The case $c_3 = 0$ has the hyperboloid degenerate to a light cone with the orbits becoming parabolas lying in corresponding null planes.

A particular null line lying on the light cone is left *invariant*; this corresponds to the unique (double) fixed point on the Riemann sphere mentioned above. The other null lines through the origin are "swung around the cone" by the transformation. Following the motion of one such null line as α increases corresponds to following the motion of a point along one of the circular flow lines on the celestial sphere, as described above.

A choice $Re(\alpha)=0$ instead, produces similar orbits, now with the roles of x and y interchanged.

Parabolic transformations lead to the gauge symmetry of massless particles (like photons) with helicity $|h| \geq 1$. In the above explicit example, a massless particle moving in the z direction, so with 4-momentum $P=(p,0,0,p)$, is not affected at all by the x-boost and y-rotation combination $Kx-Jy$ displayed above, in the "little group" of its motion. This is evident from the explicit transformation law discussed: like any light-like vector, P itself is now invariant, i.e., all traces or effects of α have disappeared. $c_1 = c_2 = c_3 = 0$, in the special case discussed. (The other similar generator, $Ky+Jx$ as well as it and Jz comprise altogether the little group of the lightlike vector, isomorphic to E(2).)

26.3 Lie algebra

As with any Lie group, the best way to study many aspects of the Lorentz group is via its Lie algebra. The Lorentz group is a subgroup of the diffeomorphism group of $\mathbf{R}^4$ and therefore its Lie algebra can be identified with vector fields on $\mathbf{R}^4$. In particular, the vectors that generate isometries on a space are its Killing vectors, which provides a convenient alternative to the left-invariant vector field for calculating the Lie algebra. We can write down a set of six generators:

- vector fields on $\mathbf{R}^4$ generating three rotations $i\,\boldsymbol{J}$,

$$-y\partial_x + x\partial_y \equiv iJ_z\,,\qquad -z\partial_y + y\partial_z \equiv iJ_x\,,\qquad -x\partial_z + z\partial_x \equiv J_y\,;$$

- vector fields on $\mathbf{R}^4$ generating three boosts $i\,\boldsymbol{K}$,

$$x\partial_t + t\partial_x \equiv iK_x\,,\qquad y\partial_t + t\partial_y \equiv iK_y\,,\qquad z\partial_t + t\partial_z \equiv iK_z.$$

It may be helpful to briefly recall here how to obtain a one-parameter group from a vector field, written in the form of a first order linear partial differential operator such as

$$-y\partial_x + x\partial_y.$$

The corresponding initial value problem is

$$\frac{\partial x}{\partial \lambda} = -y,\quad \frac{\partial y}{\partial \lambda} = x,\ x(0) = x_0,\ y(0) = y_0.$$

The solution can be written

$$x(\lambda) = x_0\cos(\lambda) - y_0\sin(\lambda),\ y(\lambda) = x_0\sin(\lambda) + y_0\cos(\lambda)$$

or

$$\begin{bmatrix} t \\ x \\ y \\ z \end{bmatrix} = \begin{bmatrix} 1 & 0 & 0 & 0 \\ 0 & \cos(\lambda) & -\sin(\lambda) & 0 \\ 0 & \sin(\lambda) & \cos(\lambda) & 0 \\ 0 & 0 & 0 & 1 \end{bmatrix} \begin{bmatrix} t_0 \\ x_0 \\ y_0 \\ z_0 \end{bmatrix}$$

where we easily recognize the one-parameter matrix group of rotations $\exp(i\,\lambda\,Jz)$ about the z axis. Differentiating with respect to the group parameter λ and setting it $\lambda=0$ in that result, we recover the standard matrix,

$$iJ_z = \begin{bmatrix} 0 & 0 & 0 & 0 \\ 0 & 0 & -1 & 0 \\ 0 & 1 & 0 & 0 \\ 0 & 0 & 0 & 0 \end{bmatrix}.$$

which corresponds to the vector field we started with. This illustrates how to pass between matrix and vector field representations of elements of the Lie algebra.

Reversing the procedure in the previous section, we see that the Möbius transformations that correspond to our six generators arise from exponentiating respectively $\beta/2$ (for the three boosts) or $i\theta/2$ (for the three rotations) times the three Pauli matrices

$$\sigma_1 = \begin{bmatrix} 0 & 1 \\ 1 & 0 \end{bmatrix}, \quad \sigma_2 = \begin{bmatrix} 0 & -i \\ i & 0 \end{bmatrix}, \quad \sigma_3 = \begin{bmatrix} 1 & 0 \\ 0 & -1 \end{bmatrix}.$$

For our purposes, another generating set is more convenient. The following table lists the six generators, in which

- The first column gives a generator of the flow under the Möbius action (after stereographic projection from the Riemann sphere) as a *real* vector field on the Euclidean plane.

- The second column gives the corresponding one-parameter subgroup of Möbius transformations.

- The third column gives the corresponding one-parameter subgroup of Lorentz transformations (the image under our homomorphism of preceding one-parameter subgroup).

- The fourth column gives the corresponding generator of the flow under the Lorentz action as a real vector field on Minkowski spacetime.

Notice that the generators consist of

- Two parabolics (null rotations)

- One hyperbolic (boost in the ∂z direction)

- Three elliptics (rotations about the x, y, z axes, respectively)

Let's verify one line in this table. Start with

$$\sigma_2 = \begin{bmatrix} 0 & i \\ -i & 0 \end{bmatrix}.$$

Exponentiate:

$$\exp\left(\frac{i\theta}{2}\sigma_2\right) = \begin{bmatrix} \cos(\theta/2) & -\sin(\theta/2) \\ \sin(\theta/2) & \cos(\theta/2) \end{bmatrix}.$$

This element of SL(2,C) represents the one-parameter subgroup of (elliptic) Möbius transformations:

$$\xi \mapsto \frac{\cos(\theta/2)\,\xi - \sin(\theta/2)}{\sin(\theta/2)\,\xi + \cos(\theta/2)}.$$

Next,

$$\frac{d\xi}{d\theta}\Big|_{\theta=0} = -\frac{1 + \xi^2}{2}.$$

The corresponding vector field on $\mathbf{C}$ (thought of as the image of S^2 under stereographic projection) is

$$-\frac{1 + \xi^2}{2}\,\partial_\xi.$$

Writing $\xi = u + iv$, this becomes the vector field on $\mathbf{R}^2$

$$-\frac{1 + u^2 - v^2}{2}\,\partial_u - uv\,\partial_v.$$

Returning to our element of SL(2,C), writing out the action $X \mapsto PXP^*$ and collecting terms, we find that the image under the spinor map is the element of SO$^+$(1,3)

$$\begin{bmatrix} 1 & 0 & 0 & 0 \\ 0 & \cos(\theta) & 0 & \sin(\theta) \\ 0 & 0 & 1 & 0 \\ 0 & -\sin(\theta) & 0 & \cos(\theta) \end{bmatrix}.$$

Differentiating with respect to θ at $\theta=0$, yields the corresponding vector field on $\mathbf{R}^4$,

$$z\partial_x - x\partial_z.$$

This is evidently the generator of counterclockwise rotation about the y axis.

26.4   Subgroups of the Lorentz group

The subalgebras of the Lie algebra of the Lorentz group can be enumerated, up to conjugacy, from which we can list the closed subgroups of the restricted Lorentz group, up to conjugacy. (See the book by Hall cited below for the details.) We can readily express the result in terms of the generating set given in the table above.

The one-dimensional subalgebras of course correspond to the four conjugacy classes of elements of the Lorentz group:

- X_1 generates a one-parameter subalgebra of parabolics SO(0,1).

- X_3 generates a one-parameter subalgebra of boosts SO(1,1).

- X_4 generates a one-parameter of rotations SO(2).

- $X_3 + aX_4$ (for any $a \neq 0$) generates a one-parameter subalgebra of loxodromic transformations.

(Strictly speaking the last corresponds to infinitely many classes, since distinct a give different classes.)   The two-dimensional subalgebras are:

- X_1, X_2 generate an abelian subalgebra consisting entirely of parabolics.

- X_1, X_3 generate a nonabelian subalgebra isomorphic to the Lie algebra of the affine group A(1).

- X_3, X_4 generate an abelian subalgebra consisting of boosts, rotations, and loxodromics all sharing the same pair of fixed points.

The three-dimensional subalgebras are:

- X_1, X_2, X_3 generate a **Bianchi V** subalgebra, isomorphic to the Lie algebra of Hom(2), the group of *euclidean homotheties*.

- X_1, X_2, X_4 generate a **Bianchi VII_0** subalgebra, isomorphic to the Lie algebra of E(2), the euclidean group.

- $X_2, X_2, X_3 + aX_4$, where $a \neq 0$, generate a **Bianchi VII_a** subalgebra.

- X_1, X_3, X_5 generate a **Bianchi VIII** subalgebra, isomorphic to the Lie algebra of SL(2,**R**), the group of isometries of the hyperbolic plane,

- X_4, X_5, X_6 generate a **Bianchi IX** subalgebra, isomorphic to the Lie algebra of SO(3), the rotation group.

(Here, the Bianchi types refer to the classification of three-dimensional Lie algebras by the Italian mathematician Luigi Bianchi.) The four-dimensional subalgebras are all conjugate to

- X_1, X_2, X_3, X_4 generate a subalgebra isomorphic to the Lie algebra of Sim(2), the group of Euclidean similitudes.

The subalgebras form a lattice (see the figure), and each subalgebra generates by exponentiation a closed subgroup of the restricted Lie group. From these, all subgroups of the Lorentz group can be constructed, up to conjugation, by multiplying by one of the elements of the Klein four-group.

As with any connected Lie group, the coset spaces of the closed subgroups of the restricted Lorentz group, or homogeneous spaces, have considerable mathematical interest. A few, brief descriptions:

- The group Sim(2) is the stabilizer of a *null line*, i.e., of a point on the Riemann sphere—so the homogeneous space $SO^+(1,3)/Sim(2)$ is the Kleinian geometry that represents conformal geometry on the sphere $S.^2$

- The (identity component of the) Euclidean group SE(2) is the stabilizer of a null vector, so the homogeneous space $SO^+(1,3)/SE(2)$ is the momentum space of a massless particle; geometrically, this Kleinian geometry represents the *degenerate* geometry of the light cone in Minkowski spacetime.

- The rotation group SO(3) is the stabilizer of a timelike vector, so the homogeneous space $SO^+(1,3)/SO(3)$ is the momentum space of a massive particle; geometrically, this space is none other than three-dimensional hyperbolic space H^3.

26.5   Covering groups

In a previous section, we constructed a homomorphism SL(2, **C**) → $SO^+(1, 3)$, which we called the spinor map. Since SL(2,**C**) is simply connected, it is the covering group of the restricted Lorentz group $SO^+(1, 3)$. By restriction we obtain a homomorphism SU(2) → SO(3). Here, the special unitary group SU(2), which is isomorphic to the group of unit norm quaternions, is also simply connected, so it is the covering group of the rotation group SO(3). Each of these covering maps are twofold covers in the sense that precisely two elements of the covering group map to each element of the quotient. One often says that the restricted Lorentz group and the rotation group are **doubly connected**. This means that the fundamental group of the each group is isomorphic to the two-element cyclic group Z_2.

(In applications to quantum mechanics, the special linear group SL(2, **C**) is sometimes called the Lorentz group.)

Twofold coverings are characteristic of spin groups. Indeed, in addition to the double coverings

$$Spin^+(1, 3) = SL(2, \mathbf{C}) \rightarrow SO^+(1, 3)$$
$$Spin(3) = SU(2) \rightarrow SO(3)$$

we have the double coverings

$$Pin(1, 3) \rightarrow O(1, 3)$$
$$Spin(1, 3) \rightarrow SO(1, 3)$$
$$Spin^+(1, 2) = SU(1, 1) \rightarrow SO(1, 2)$$

These spinorial double coverings are all closely related to Clifford algebras.

26.6 Topology

The left and right groups in the double covering

$$SU(2) \to SO(3)$$

are deformation retracts of the left and right groups, respectively, in the double covering

$$SL(2,\mathbf{C}) \to SO^+(1,3).$$

But the homogeneous space $SO^+(1,3)/SO(3)$ is homeomorphic to hyperbolic 3-space H^3, so we have exhibited the restricted Lorentz group as a principal fiber bundle with fibers $SO(3)$ and base H^3. Since the latter is homeomorphic to $\mathbf{R}^3$, while $SO(3)$ is homeomorphic to three-dimensional real projective space $\mathbf{RP}^3$, we see that the restricted Lorentz group is *locally* homeomorphic to the product of $\mathbf{RP}^3$ with $\mathbf{R}^3$. Since the base space is contractible, this can be extended to a global homeomorphism.

26.7 Generalization to higher dimensions

The concept of the Lorentz group has a natural generalization to spacetime of any number of dimensions. Mathematically, the Lorentz group of $n+1$-dimensional Minkowski space is the group $O(n,1)$ (or $O(1,n)$) of linear transformations of $\mathbf{R}^{n+1}$ that preserves the quadratic form

$$(x_1, x_2, \ldots, x_n, x_{n+1}) \mapsto x_1^2 + x_2^2 + \cdots + x_n^2 - x_{n+1}^2.$$

Many of the properties of the Lorentz group in four dimensions (where $n = 3$) generalize straightforwardly to arbitrary n. For instance, the Lorentz group $O(n,1)$ has four connected components, and it acts by conformal transformations on the celestial $(n-1)$-sphere in $n+1$-dimensional Minkowski space. The identity component $SO^+(n,1)$ is an $SO(n)$-bundle over hyperbolic n-space H^n.

The low-dimensional cases $n = 1$ and $n = 2$ are often useful as "toy models" for the physical case $n = 3$, while higher-dimensional Lorentz groups are used in physical theories such as string theory that posit the existence of hidden dimensions. The Lorentz group $O(n,1)$ is also the isometry group of n-dimensional de Sitter space dSn, which may be realized as the homogeneous space $O(n,1)/O(n-1,1)$. In particular $O(4,1)$ is the isometry group of the de Sitter universe dS_1, a cosmological model.

26.8 Notes

[1] Gelfand, Minlos & Shapiro 1963

[2] Wigner 1939

26.9 See also

26.10 References

- Artin, Emil (1957). *Geometric Algebra*. New York: Wiley. ISBN 0-471-60839-4. *See Chapter III* for the orthogonal groups $O(p,q)$.

- Carmeli, Moshe (1977). *Group Theory and General Relativity, Representations of the Lorentz Group and Their Applications to the Gravitational Field*. McGraw-Hill, New York. ISBN 0-07-009986-3. A canonical reference; *see chapters 1–6* for representations of the Lorentz group.

- Frankel, Theodore (2004). *The Geometry of Physics (2nd Ed.)*. Cambridge: Cambridge University Press. ISBN 0-521-53927-7. An excellent resource for Lie theory, fiber bundles, spinorial coverings, and many other topics.

- Fulton, William; Harris, Joe (1991), *Representation theory. A first course*, Graduate Texts in Mathematics, Readings in Mathematics **129**, New York: Springer-Verlag, ISBN 978-0-387-97495-8, MR 1153249, ISBN 978-0-387-97527-6 *See Lecture 11* for the irreducible representations of SL(2,**C**).

- Gelfand, I.M.; Minlos, R.A.; Shapiro, Z.Ya. (1963), *Representations of the Rotation and Lorentz Groups and their Applications*, New York: Pergamon Press

- Hall, G. S. (2004). *Symmetries and Curvature Structure in General Relativity*. Singapore: World Scientific. ISBN 981-02-1051-5. *See Chapter 6* for the subalgebras of the Lie algebra of the Lorentz group.

- Hatcher, Allen (2002). *Algebraic topology*. Cambridge: Cambridge University Press. ISBN 0-521-79540-0. *See also* the "online version". Retrieved July 3, 2005. *See Section 1.3* for a beautifully illustrated discussion of covering spaces. *See Section 3D* for the topology of rotation groups.

- Naber, Gregory (1992). *The Geometry of Minkowski Spacetime*. New York: Springer-Verlag. ISBN 0486432351. (Dover reprint edition.) An excellent reference on Minkowski spacetime and the Lorentz group.

- Needham, Tristan (1997). *Visual Complex Analysis*. Oxford: Oxford University Press. ISBN 0-19-853446-9. *See Chapter 3* for a superbly illustrated discussion of Möbius transformations.

- Wigner, E. P. (1939), "On unitary representations of the inhomogeneous Lorentz group", *Annals of Mathematics* **40** (1): 149–204, Bibcode:1939AnMat..40..922E, doi:10.2307/1968551, MR 1503456.

Hendrik Antoon Lorentz (1853–1928), after whom the Lorentz group is named.

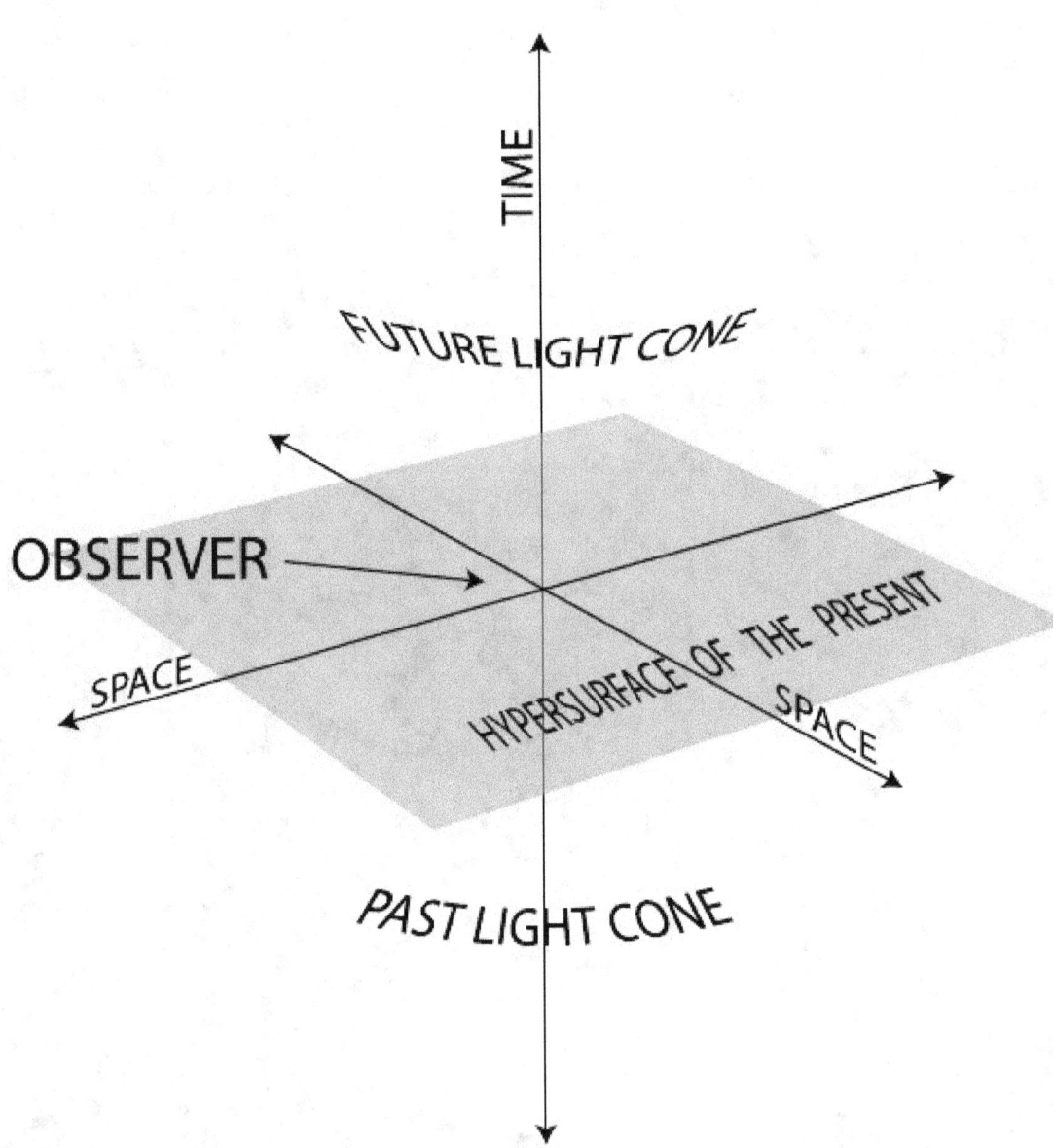

Light cone in 2D space plus a time dimension.

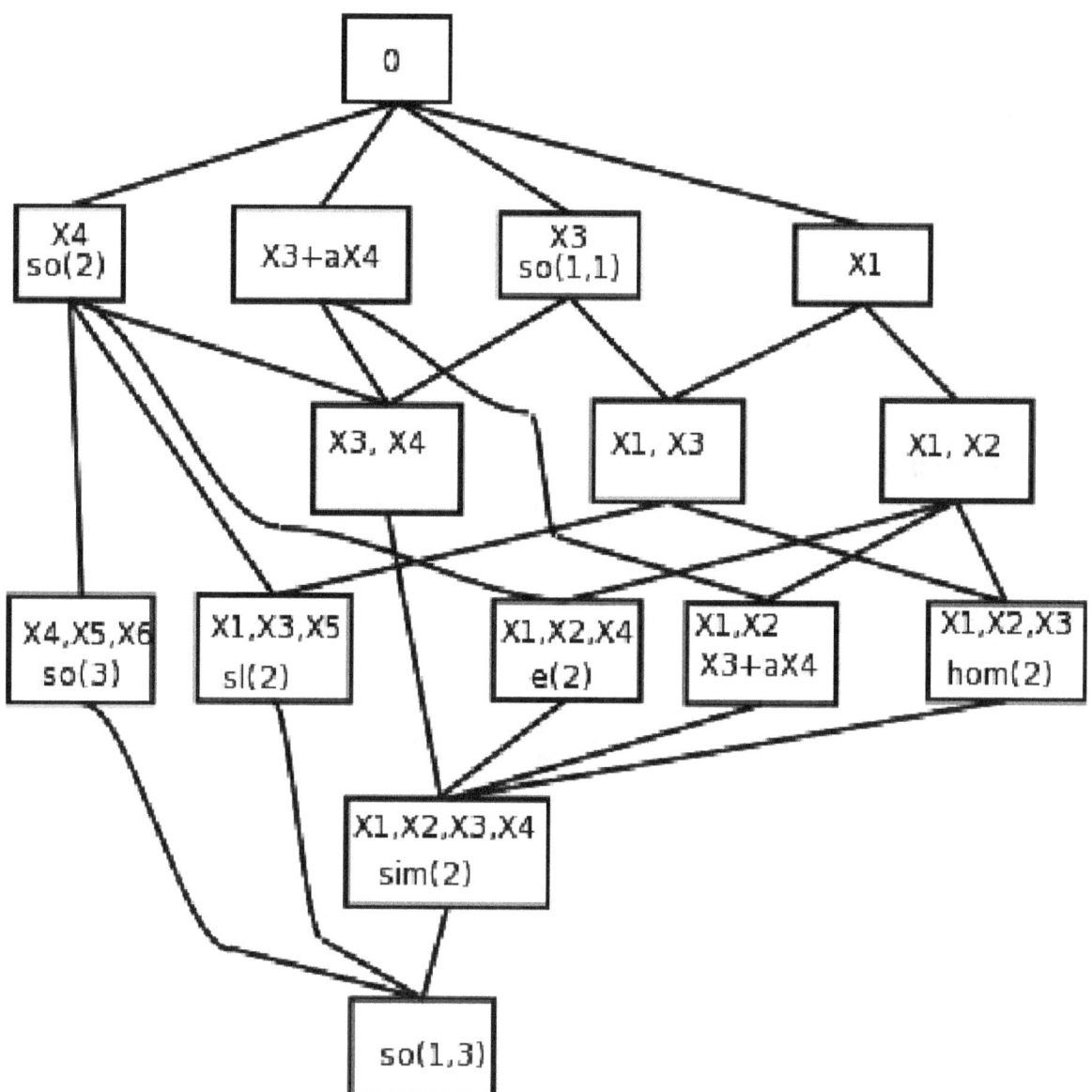

The lattice of subalgebras of the Lie algebra SO(1,3), up to conjugacy.

Chapter 27

Poincaré group

For the Poincaré group (fundamental group) of a topological space, see Fundamental group.

The **Poincaré group**, named after Henri Poincaré,[1] is the group of Minkowski spacetime isometries.[2][3] It is a ten-generator non-abelian Lie group of fundamental importance in physics.

27.1 Overview

A Minkowski spacetime isometry has the property that the interval between events is left invariant. For example, if everything was postponed by two hours including two events and the path you took to go from one to the other, then the time interval between the events recorded by a stop-watch you carried with you would be the same. Or if everything was shifted five miles to the west, or turned 60 degrees to the right, you would also see no change in the interval. It turns out that the proper length of an object is also unaffected by such a shift. A time or space reversal (a reflection) is also an isometry of this group.

In Minkowski space (i.e. ignoring the effects of gravity), there are ten degrees of freedom of the isometries, which may be thought of as translation through time or space (four degrees, one per dimension); reflection through a plane (three degrees, the freedom in orientation of this plane); or a "boost" in any of the three spatial directions (three degrees). Composition of transformations is the operator of the Poincaré group, with proper rotations being produced as the composition of an even number of reflections.

In classical physics, the Galilean group is a comparable ten-parameter group that acts on absolute time and space. Instead of boosts, it features shear mappings to relate co-moving frames of reference.

27.2 Details

The Poincaré group is the group of Minkowski spacetime isometries. It is a ten-dimensional noncompact Lie group. The abelian group of translations is a normal subgroup, while the Lorentz group is also a subgroup, the stabilizer of the origin. The Poincaré group itself is the minimal subgroup of the affine group which includes all translations and Lorentz transformations. More precisely, it is a semidirect product of the translations and the Lorentz group,

$$\mathbf{R}^{1,3} \rtimes SO(1,3) \, .$$

Another way of putting this is that the Poincaré group is a group extension of the Lorentz group by a vector representation of it; it is sometimes dubbed, informally, as the "*inhomogeneous Lorentz group*". In turn, it can also be obtained as a group contraction of the de Sitter group SO(4,1) ~ Sp(2,2), as the de Sitter radius goes to infinity.

Its positive energy unitary irreducible representations are indexed by mass (nonnegative number) and spin (integer or half integer) and are associated with particles in quantum mechanics (see Wigner's classification).

In accordance with the Erlangen program, the geometry of Minkowski space is defined by the Poincaré group: Minkowski space is considered as a homogeneous space for the group.

The **Poincaré algebra** is the Lie algebra of the Poincaré group. It is a Lie algebra extension of the Lie algebra of the Lorentz group. More specifically, the proper ($\det \Lambda = 1$), orthochronous ($\Lambda^0{}_0 \geq 1$) part of the Lorentz subgroup (its identity component), $SO^+(1, 3)$, is connected to the identity and is thus provided by the exponentiation $\exp(ia_\mu P^\mu) \exp(i\omega_{\mu\nu} M^{\mu\nu}/2)$ of this Lie algebra. In component form, the Poincaré algebra is given by the commutation relations:[4][5]

where P is the generator of translations, M is the generator of Lorentz transformations, and η is the $(+,-,-,-)$ Minkowski metric (see Sign convention).

The bottom commutation relation is the ("homogeneous") Lorentz group, consisting of rotations, $Ji = -\epsilon imn M^{mn}/2$, and boosts, $Ki = Mi0$. In this notation, the entire Poincaré algebra is expressible in noncovariant (but more practical) language as

$$[J_m, P_n] = i\epsilon_{mnk} P_k ,$$

$$[J_i, P_0] = 0 ,$$

$$[K_i, P_k] = i\eta_{ik} P_0 ,$$

$$[K_i, P_0] = -iP_i ,$$

$$[J_m, J_n] = i\epsilon_{mnk} J_k ,$$

$$[J_m, K_n] = i\epsilon_{mnk} K_k ,$$

$$[K_m, K_n] = -i\epsilon_{mnk} J_k .$$

where the bottom line commutator of two boosts is often referred to as a "Wigner rotation". Note the important simplification $[Jm+i \, Km , \, Jn-i \, Kn] = 0$, which permits reduction of the Lorentz subalgebra to **su(2)**⊕**su(2)** and efficient treatment of its associated representations.

The Casimir invariants of this algebra are $P_\mu P^\mu$ and $W_\mu W^\mu$ where W_μ is the Pauli–Lubanski pseudovector; they serve as labels for the representations of the group.

The Poincaré group is the full symmetry group of any relativistic field theory. As a result, all elementary particles fall in representations of this group. These are usually specified by the *four-momentum* squared of each particle (i.e. its mass squared) and the intrinsic quantum numbers J^{PC}, where J is the spin quantum number, P is the parity and C is the charge-conjugation quantum number. In practice, charge conjugation and parity are violated by many quantum field theories; where this occurs, P and C are forfeited. Since CPT symmetry is invariant in quantum field theory, a time-reversal quantum number may be constructed from those given.

As a topological space, the group has four connected components: the component of the identity; the time reversed component; the spatial inversion component; and the component which is both time-reversed and spatially inverted.

27.3 Poincaré symmetry

Poincaré symmetry is the full symmetry of special relativity. It includes:

- *translations* (displacements) in time and space (*P*), forming the abelian Lie group of translations on space-time;

- *rotations* in space, forming the non-Abelian Lie group of three-dimensional rotations (*J*);

- *boosts*, transformations connecting two uniformly moving bodies (**K**).

The last two symmetries, **J** and **K**, together make the Lorentz group (see also Lorentz invariance); the semi-direct product of the translations group and the Lorentz group then produce the Poincaré group. Objects which are invariant under this group are then said to possess **Poincaré invariance** or **relativistic invariance**.

27.4   See also

- Euclidean group

- Representation theory of the Poincaré group

- Wigner's classification

- Symmetry in quantum mechanics

- Center of mass (relativistic)

- Pauli–Lubanski pseudovector

- Particle physics and representation theory

27.5   Notes

[1] Poincaré,Henri, "Sur la dynamique de l'électron",*Rendiconti del Circolo matematico di Palermo***21**: 129–176,doi:10.1007 (Wikisource translation: On the Dynamics of the Electron).

[2] Minkowski, Hermann, "Die Grundgleichungen für die elektromagnetischen Vorgänge in bewegten Körpern", *Nachrichten von der Gesellschaft der Wissenschaften zu Göttingen, Mathematisch-Physikalische Klasse*: 53–111 (Wikisource translation: The Fundamental Equations for Electromagnetic Processes in Moving Bodies).

[3] Minkowski, Hermann, "Raum und Zeit", *Physikalische Zeitschrift* **10**: 75–88

[4] N.N. Bogolubov (1989). *General Principles of Quantum Field Theory* (2nd ed.). Springer. p. 272. ISBN 0-7923-0540-X.

[5] T. Ohlsson (2011). *Relativistic Quantum Physics: From Advanced Quantum Mechanics to Introductory Quantum Field Theory*. Cambridge University Press. p. 10. ISBN 1-13950-4320.

27.6   References

- Wu-Ki Tung (1985). *Group Theory in Physics*. World Scientific Publishing. ISBN 9971-966-57-3.

- Weinberg, Steven (1995). *The Quantum Theory of Fields***1**. Cambridge: Cambridge University press. 978-0-521-55001-7.

- L.H. Ryder (1996). *Quantum Field Theory* (2nd ed.). Cambridge University Press. p. 62. ISBN 0-52147-8146.

Chapter 28

Standard Model (mathematical formulation)

For a less mathematical description, see Standard Model.

This article describes the mathematics of the **Standard Model** of particle physics, a gauge quantum field theory con-

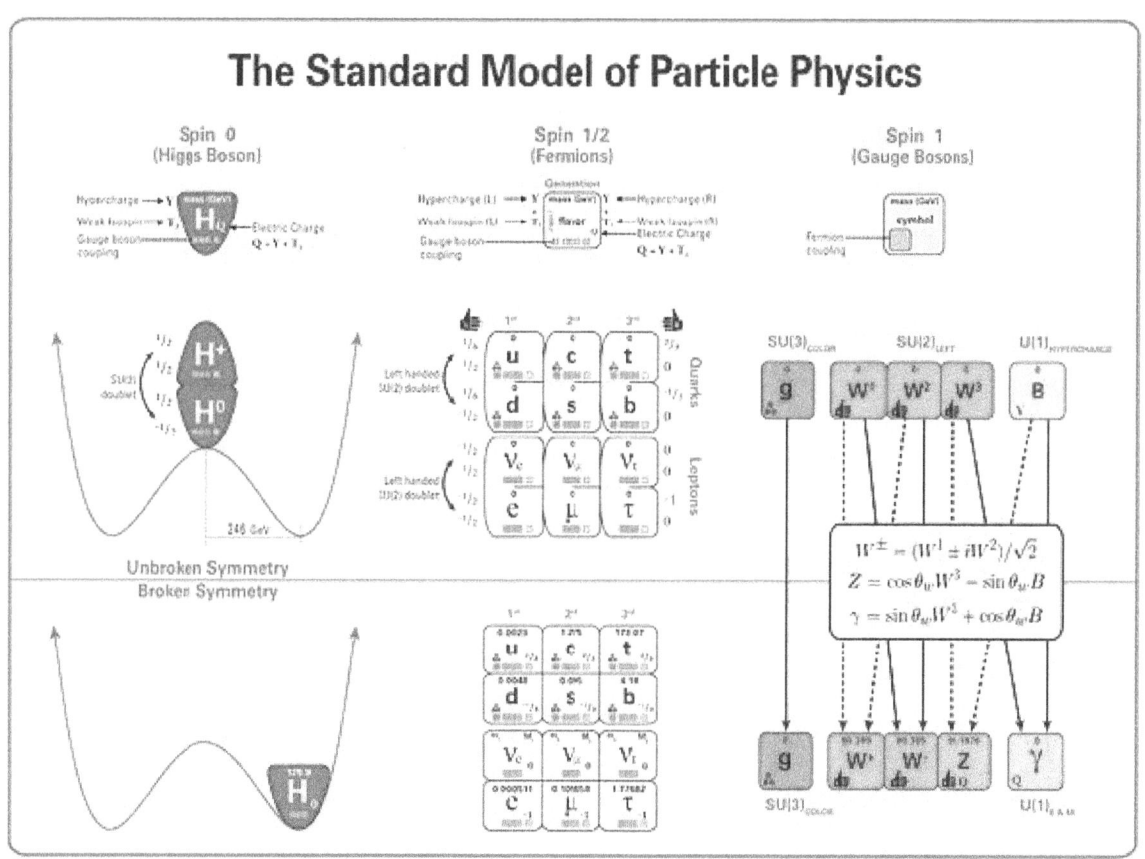

Standard Model of Particle Physics. The diagram shows the elementary particles of the Standard Model (the Higgs boson, the three generations of quarks and leptons, and the gauge bosons), including their names, masses, spins, charges, chiralities, and interactions with the strong, weak and electromagnetic forces. It also depicts the crucial role of the Higgs boson in electroweak symmetry breaking, and shows how the properties of the various particles differ in the (high-energy) symmetric phase (top) and the (low-energy) broken-symmetry phase (bottom).

taining the internal symmetries of the unitary product group SU(3) × SU(2) × U(1). The theory is commonly viewed as containing the fundamental set of particles – the leptons, quarks, gauge bosons and the Higgs particle.

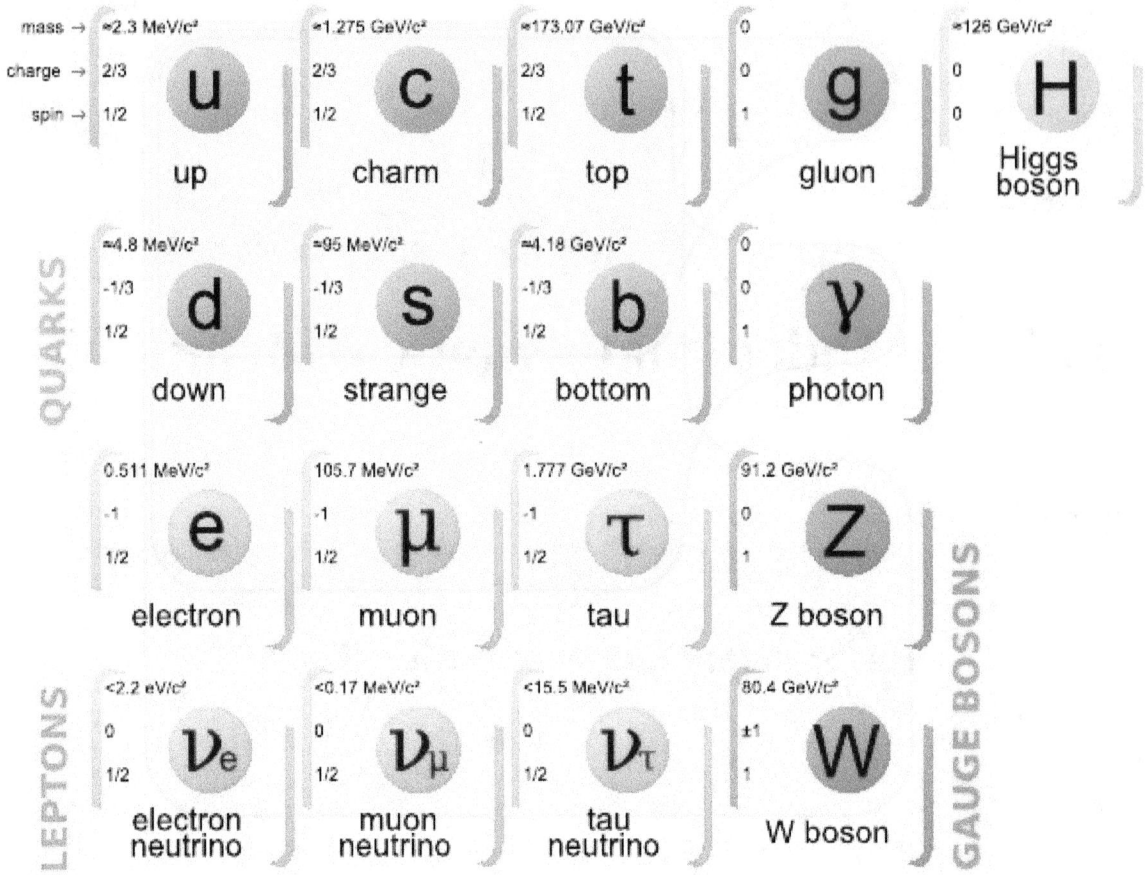

The Standard Model of Particle Physics: More Schematic Depiction

The Standard Model is renormalizable and mathematically self-consistent,[1] however despite having huge and continued successes in providing experimental predictions it does leave some unexplained phenomena. In particular, although the physics of special relativity is incorporated, general relativity is not, and the Standard Model will fail at energies or distances where the graviton is expected to emerge. Therefore in a modern field theory context, it is seen as an effective field theory.

This article requires some background in physics and mathematics, but is designed as both an introduction and a reference.

28.1   Quantum field theory

The standard model is a quantum field theory, meaning its fundamental objects are *quantum fields* which are defined at all points in spacetime. These fields are

- the fermion field, ψ, which accounts for "matter particles";
- the electroweak boson fields W_1, W_2, W_3, and B;
- the gluon field, G_a; and
- the Higgs field, φ.

That these are *quantum* rather than *classical* fields has the mathematical consequence that they are operator-valued. In particular, values of the fields generally do not commute. As operators, they act upon the quantum state (ket vector).

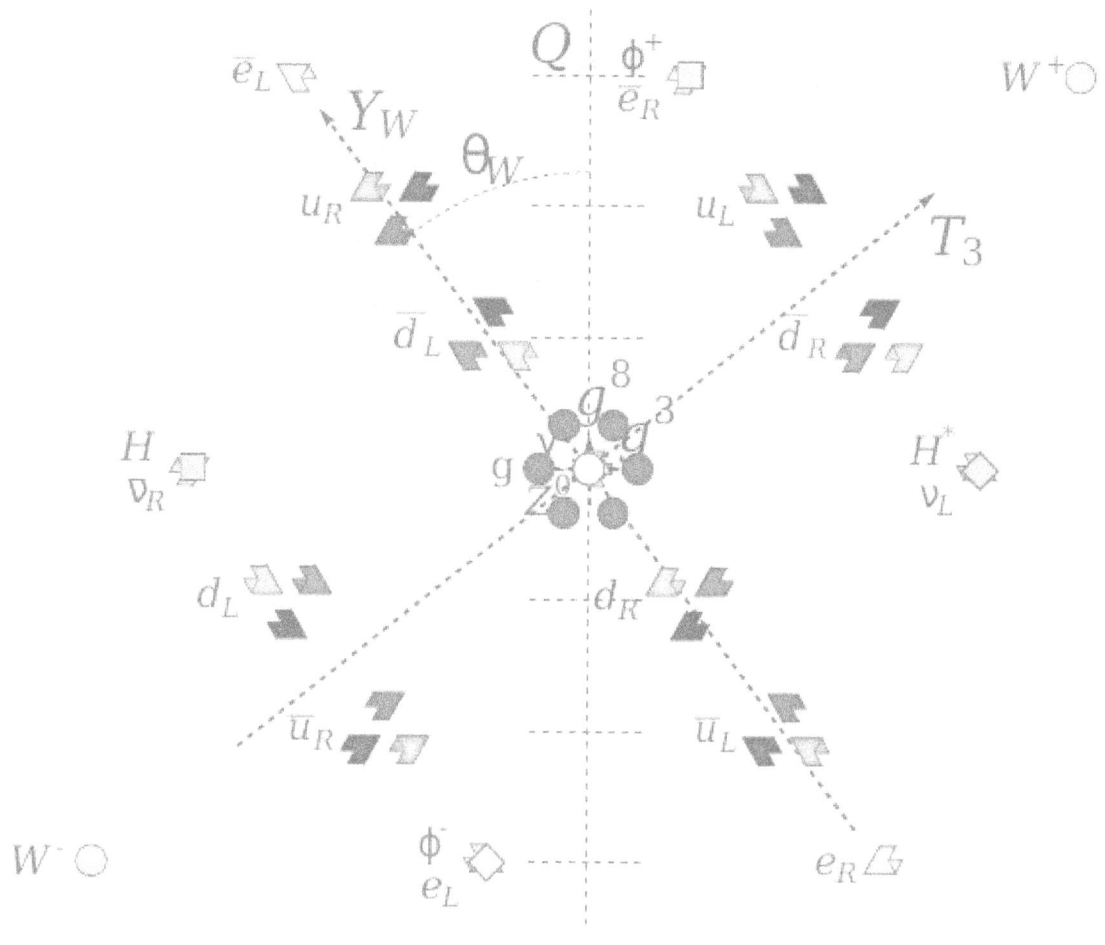

The pattern of weak isospin T_3, *weak hypercharge YW, and color charge of all known elementary particles, rotated by the weak mixing angle to show electric charge Q, roughly along the vertical. The neutral Higgs field (gray square) breaks the electroweak symmetry and interacts with other particles to give them mass.*

The dynamics of the quantum state and the fundamental fields are determined by the Lagrangian density $\mathcal{L}$ (usually for short just called the Lagrangian). This plays a role similar to that of the Schrödinger equation in non-relativistic quantum mechanics, but a Lagrangian is not an equation – rather, it is a polynomial function of the fields and their derivatives, and used with the principle of least action. While it would be possible to derive a system of differential equations governing the fields from the Langrangian, it is more common to use other techniques to compute with quantum field theories.

The standard model is furthermore a gauge theory, which means there are degrees of freedom in the mathematical formalism which do not correspond to changes in the physical state. The gauge group of the standard model is SU(3) × SU(2) × U(1), where U(1) acts on B and φ, SU(2) acts on W and φ, and SU(3) acts on G. The fermion field ψ also transforms under these symmetries, although all of them leave some parts of it unchanged.

28.1.1   The role of the quantum fields

In classical mechanics, the state of a system can usually be captured by a small set of variables, and the dynamics of the system is thus determined by the time evolution of these variables. In classical field theory, the *field* is part of the state of the system, so in order to describe it completely one effectively introduces separate variables for every point in spacetime (even though there are many restrictions on how the values of the field "variables" may vary from point to point, for example in the form of field equations involving partial derivatives of the fields).

In quantum mechanics, the classical variables are turned into operators, but these do not capture the state of the system, which is instead encoded into a wavefunction ψ or more abstract ket vector. If ψ is an eigenstate with respect to an operator P, then $P\psi = \lambda\psi$ for the corresponding eigenvalue λ, and hence letting an operator P act on ψ is analogous to multiplying ψ by the value of the classical variable to which P corresponds. By extension, a classical formula where all variables have been replaced by the corresponding operators will behave like an operator which, when it acts upon the state of the system, multiplies it by the analogue of the quantity that the classical formula would compute. The formula as such does however not contain any information about the state of the system; it would evaluate to the same operator regardless of what state the system is in.

Quantum fields relate to quantum mechanics as classical fields do to classical mechanics, i.e., there is a separate operator for every point in spacetime, and these operators do not carry any information about the state of the system; they are merely used to exhibit some aspect of the state, at the point to which they belong. In particular, the quantum fields are *not* wavefunctions, even though the equations which govern their time evolution may be deceptively similar to those of the corresponding wavefunction in a semiclassical formulation. There is no variation in strength of the fields between different points in spacetime; the variation that happens is rather one of phase factors.

28.1.2   Vectors, scalars, and spinors

Mathematically it may look as though all of the fields are vector-valued (in addition to being operator-valued), since they all have several components, can be multiplied by matrices, etc., but physicists assign a more specific physical meaning to the word: a **vector** is something which transforms like a four-vector under Lorentz transformations, and a **scalar** is something which is invariant under Lorentz transformations. The B, W_j, and G_a fields are all vectors in this sense, so the corresponding particles are said to be vector bosons. The Higgs field φ is a scalar.

The fermion field ψ does transform under Lorentz transformations, but not like a vector should; rotations will only turn it by half the angle a proper vector should. Therefore these constitute a third kind of quantity, which is known as a spinor.

It is common to make use of abstract index notation for the vector fields, in which case the vector fields all come with a Lorentzian index μ, like so: B^μ, W_j^μ, and G_a^μ. If abstract index notation is used also for spinors then these will carry a spinorial index and the Dirac gamma will carry one Lorentzian and two spinorian indices, but it is more common to regard spinors as column matrices and the Dirac gamma $\gamma\mu$ as a matrix which additionally carries a Lorentzian index. The Feynman slash notation can be used to turn a vector field into a linear operator on spinors, like so: $\slashed{B} = \gamma^\mu B_\mu$; this may involve raising and lowering indices.

28.2   Alternative presentations of the fields

As is common in quantum theory, there is more than one way to look at things. At first the basic fields given above may not seem to correspond well with the "fundamental particles" in the chart above, but there are several alternative presentations which, in particular contexts, may be more appropriate than those that are given above.

28.2.1   Fermions

Rather than having one fermion field ψ, it can be split up into separate components for each type of particle. This mirrors the historical evolution of quantum field theory, since the electron component ψ_e (describing the electron and its antiparticle the positron) is then the original ψ field of quantum electrodynamics, which was later accompanied by $\psi\mu$

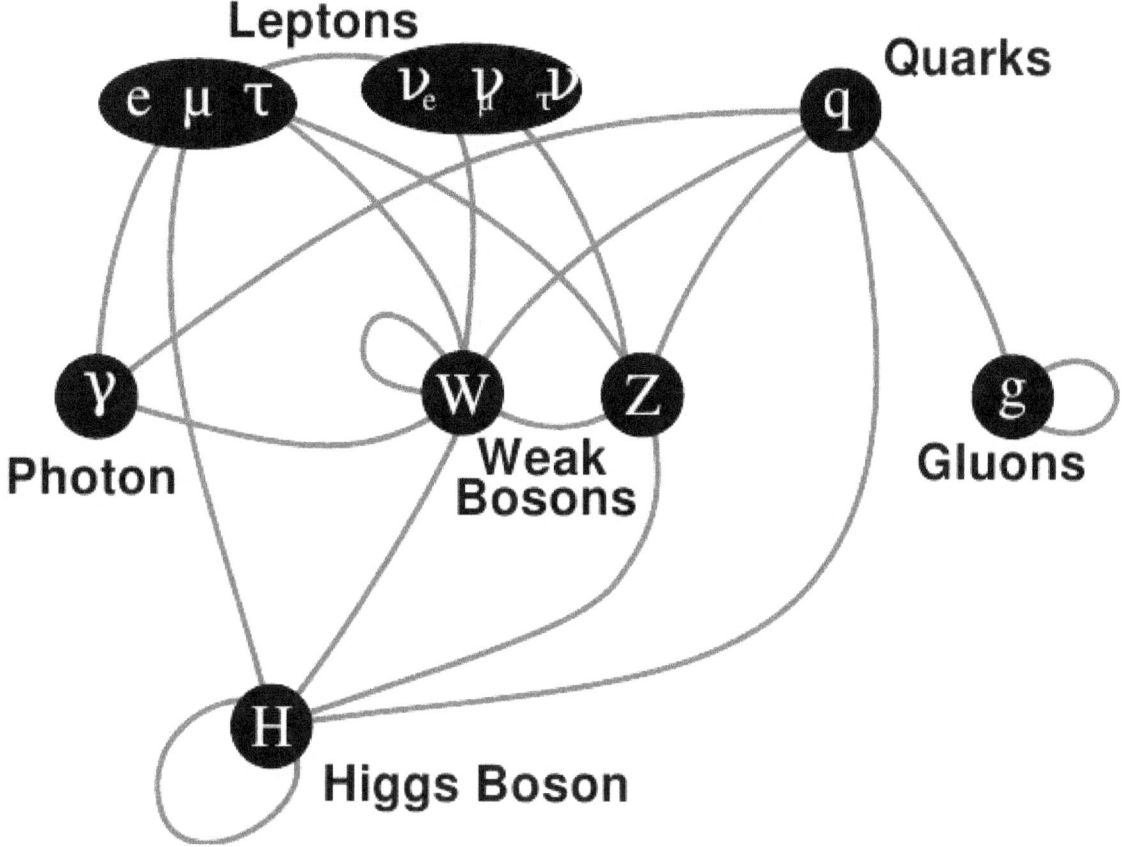

Connections denoting which particles interact with each other.

and ψτ fields for the muon and tauon respectively (and their antiparticles). Electroweak theory added ψ_{ν_e}, ψ_{ν_μ}, and ψ_{ν_τ} for the corresponding neutrinos, and the quarks add still further components. In order to be four-spinors like the electron and other lepton components, there must be one quark component for every combination of flavour and colour, bringing the total to 24 (3 for charged leptons, 3 for neutrinos, and $2 \cdot 3 \cdot 3 = 18$ for quarks).

An important definition is the barred fermion field $\bar{\psi}$ is defined to be $\psi^\dagger \gamma^0$, where † denotes the Hermitian adjoint and γ^0 is the zeroth gamma matrix. If ψ is thought of as an $n \times 1$ matrix then $\bar{\psi}$ should be thought of as a $1 \times n$ matrix.

A chiral theory

An independent decomposition of ψ is that into chirality components:

"Left" chirality: $\psi^L = \frac{1}{2}(1 - \gamma_5)\psi$

"Right" chirality: $\psi^R = \frac{1}{2}(1 + \gamma_5)\psi$

where γ_5 is the fifth gamma matrix. This is very important in the Standard Model because *left and right chirality components are treated differently by the gauge interactions.*

In particular, under weak isospin SU(2) transformations the left-handed particles are weak-isospin doublets, whereas the right-handed are singlets – i.e. the weak isospin of ψR is zero. Put more simply, the weak interaction could rotate e.g. a left-handed electron into a left-handed neutrino (with emission of a W⁻), but could not do so with the same right-handed particles. As an aside, the right-handed neutrino originally did not exist in the standard model – but the discovery of neutrino oscillation implies that neutrinos must have mass, and since chirality can change during the propagation of a massive particle, right-handed neutrinos must exist in reality. This does not however change the (experimentally-proven) chiral nature of the weak interaction.

Furthermore, U(1) acts differently on ψ_e^L than on ψ_e^R (because they have different weak hypercharges).

Mass and interaction eigenstates

A distinction can thus be made between, for example, the mass and interaction eigenstates of the neutrino. The former is the state which propagates in free space, whereas the latter is the *different* state that participates in interactions. Which is the "fundamental" particle? For the neutrino, it is conventional to define the "flavour" (ν
e, ν
μ, or ν
τ) by the interaction eigenstate, whereas for the quarks we define the flavour (up, down, etc.) by the mass state. We can switch between these states using the CKM matrix for the quarks, or the PMNS matrix for the neutrinos (the charged leptons on the other hand are eigenstates of both mass and flavour).

As an aside, if a complex phase term exists within either of these matrices, it will give rise to direct CP violation, which could explain the dominance of matter over antimatter in our current universe. This has been proven for the CKM matrix, and is expected for the PMNS matrix.

Positive and negative energies

Finally, the quantum fields are sometimes decomposed into "positive" and "negative" energy parts: $\psi = \psi^+ + \psi^-$. This is not so common when a quantum field theory has been set up, but often features prominently in the process of quantizing a field theory.

28.2.2   Bosons

Due to the Higgs mechanism, the electroweak boson fields W_1, W_2, W_3 , and B "mix" to create the states which are physically observable. To retain gauge invariance, the underlying fields must be massless, but the observable states can *gain masses* in the process. These states are:

The massive neutral boson:

$$Z = \cos\theta_W W_3 - \sin\theta_W B$$

The massless neutral boson:

$$A = \sin\theta_W W_3 + \cos\theta_W B$$

The massive charged W bosons:

$$W^\pm = \frac{1}{\sqrt{2}} (W_1 \mp iW_2)$$

where θW is the Weinberg angle.

The A field is the photon, which corresponds classically to the well-known electromagnetic four-potential – i.e. the electric and magnetic fields. The Z field actually contributes in every process the photon does, but due to its large mass, the contribution is usually negligible.

28.3   Perturbative QFT and the interaction picture

Much of the qualitative descriptions of the standard model in terms of "particles" and "forces" comes from the perturbative quantum field theory view of the model. In this, the Langrangian is decomposed as $\mathcal{L} = \mathcal{L}_0 + \mathcal{L}_1$ into separate *free field* and *interaction* Langrangians. The free fields care for particles in isolation, whereas processes involving several particles arise through interactions. The idea is that the state vector should only change when particles interact, meaning a free particle is one whose quantum state is constant. This corresponds to the interaction picture in quantum mechanics.

In the more common Schrödinger picture, even the states of free particles change over time: typically the phase changes at a rate which depends on their energy. In the alternative Heisenberg picture, state vectors are kept constant, at the price of having the operators (in particular the observables) be time-dependent. The interaction picture constitutes an intermediate between the two, where some time dependence is placed in the operators (the quantum fields) and some in the state vector. In QFT, the former is called the free field part of the model, and the latter is called the interaction part. The free field model can be solved exactly, and then the solutions to the full model can be expressed as perturbations of the free field solutions, for example using the Dyson series.

It should be observed that the decomposition into free fields and interactions is in principle arbitrary. For example renormalization in QED modifies the mass of the free field electron to match that of a physical electron (with an electromagnetic field), and will in doing so add a term to the free field Lagrangian which must be cancelled by a counterterm in the interaction Lagrangian, that then shows up as a two-line vertex in the Feynman diagrams. This is also how the Higgs field is thought to give particles mass: the part of the interaction term which corresponds to the (nonzero) vacuum expectation value of the Higgs field is moved from the interaction to the free field Lagrangian, where it looks just like a mass term having nothing to do with Higgs.

28.3.1   Free fields

Under the usual free/interaction decomposition, which is suitable for low energies, the free fields obey the following equations:

- The fermion field ψ satisfies the Dirac equation; $(i\hbar\partial\!\!\!/ - m_f c)\psi_f = 0$ for each type f of fermion.

- The photon field A satisfies the wave equation $\partial_\mu \partial^\mu A^\nu = 0$.

- The Higgs field φ satisfies the Klein–Gordon equation.

- The weak interaction fields Z, $W^\pm$ also satisfy the Klein–Gordon equation.

These equations can be solved exactly. One usually does so by considering first solutions that are periodic with some period L along each spatial axis; later taking the limit: $L \to \infty$ will lift this periodicity restriction.

In the periodic case, the solution for a field F (any of the above) can be expressed as a Fourier series of the form

$$F(x) = \beta \sum_{\mathbf{p}} \sum_r E_{\mathbf{p}}^{-\frac{1}{2}} \left(a_r(\mathbf{p}) u_r(\mathbf{p}) e^{-\frac{ipx}{\hbar}} + b_r^\dagger(\mathbf{p}) v_r(\mathbf{p}) e^{\frac{ipx}{\hbar}} \right)$$

where:

- β is a normalization factor; for the fermion field ψ_f it is $\sqrt{m_f c^2 / V}$, where $V = L^3$ is the volume of the fundamental cell considered; for the photon field A^μ it is $hc/\sqrt{2V}$.

- The sum over $\mathbf{p}$ is over all momenta consistent with the period L, i.e., over all vectors $\frac{2\pi\hbar}{L}(n_1, n_2, n_3)$ where n_1, n_2, n_3 are integers.

- The sum over r covers other degrees of freedom specific for the field, such as polarization or spin; it usually comes out as a sum from 1 to 2 or from 1 to 3.

- $E_{\mathbf{p}}$ is the relativistic energy for a momentum $\mathbf{p}$ quantum of the field, $= \sqrt{m^2 c^4 + c^2 \mathbf{p}^2}$ when the rest mass is m.

- $ar(\mathbf{p})$ and $b_r^{\dagger}(\mathbf{p})$ are annihilation and creation respectively operators for "a-particles" and "b-particles" respectively of momentum $\mathbf{p}$; "b-particles" are the antiparticles of "a-particles". Different fields have different "a-" and "b-particles". For some fields, a and b are the same.

- $ur(\mathbf{p})$ and $vr(\mathbf{p})$ are non-operators which carry the vector or spinor aspects of the field (where relevant).

- $p = (E_{\mathbf{p}}/c, \mathbf{p})$ is the four-momentum for a quanta with momentum $\mathbf{p}$. $px = p_{\mu} x^{\mu}$ denotes an inner product of four-vectors.

In the limit $L \to \infty$, the sum would turn into an integral with help from the V hidden inside β. The numeric value of β also depends on the normalization chosen for $u_r(\mathbf{p})$ and $v_r(\mathbf{p})$.

Technically, $a_r^{\dagger}(\mathbf{p})$ is the Hermitian adjoint of the operator $ar(\mathbf{p})$ in the inner product space of ket vectors. The identification of $a_r^{\dagger}(\mathbf{p})$ and $ar(\mathbf{p})$ as creation and annihilation operators comes from comparing conserved quantities for a state before and after one of these have acted upon it. $a_r^{\dagger}(\mathbf{p})$ can for example be seen to add one particle, because it will add 1 to the eigenvalue of the a-particle number operator, and the momentum of that particle ought to be $\mathbf{p}$ since the eigenvalue of the vector-valued momentum operator increases by that much. For these derivations, one starts out with expressions for the operators in terms of the quantum fields. That the operators with † are creation operators and the one without annihilation operators is a convention, imposed by the sign of the commutation relations postulated for them.

An important step in preparation for calculating in perturbative quantum field theory is to separate the "operator" factors a and b above from their corresponding vector or spinor factors u and v. The vertices of Feynman graphs come from the way that u and v from different factors in the interaction Lagrangian fit together, whereas the edges come from the way that the as and bs must be moved around in order to put terms in the Dyson series on normal form.

28.3.2   Interaction terms and the path integral approach

The Lagrangian can also be derived without using creation and annihilation operators (the "canonical" formalism), by using a "path integral" approach, pioneered by Feynman building on the earlier work of Dirac. See e.g. Path integral formulation on Wikipedia or A. Zee's QFT in a nutshell. This is one possible way that the Feynman diagrams, which are pictorial representations of interaction terms, can be derived relatively easily. A quick derivation is indeed presented at the article on Feynman diagrams.

28.4   Lagrangian formalism

We can now give some more detail about the aforementioned free and interaction terms appearing in the Standard Model Lagrangian density. Any such term must be both gauge and reference-frame invariant, otherwise the laws of physics would depend on an arbitrary choice or the frame of an observer. Therefore the global Poincaré symmetry, consisting of translational symmetry, rotational symmetry and the inertial reference frame invariance central to the theory of special relativity must apply. The local SU(3) × SU(2) × U(1) gauge symmetry is the internal symmetry. The three factors of the gauge symmetry together give rise to the three fundamental interactions, after some appropriate relations have been defined, as we shall see.

A complete formulation of the Standard Model Lagrangian with all the terms written together can be found e.g. here.

28.4.1   Kinetic terms

A free particle can be represented by a mass term, and a *kinetic* term which relates to the "motion" of the fields.

Standard Model Interactions
(Forces Mediated by Gauge Bosons)

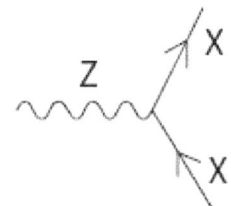

X is any fermion in
the Standard Model.

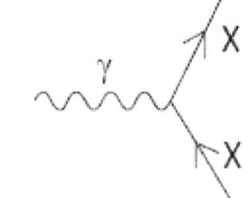

X is electrically charged.

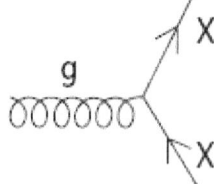

X is any quark.

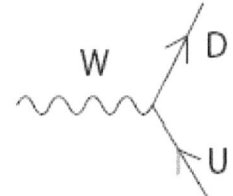

U is a up-type quark;
D is a down-type quark.

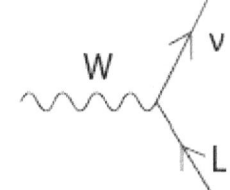

L is a lepton and ν is the
corresponding neutrino.

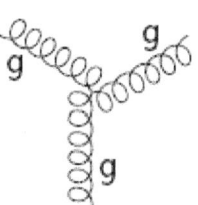

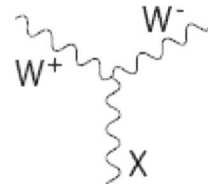

X is a photon or Z-boson.

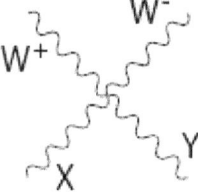

X and Y are any two
electroweak bosons such
that charge is conserved.

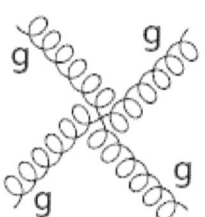

*The above interactions show some basic interaction vertices – Feynman diagrams in the standard model are built from these vertices.
Higgs boson interactions are however not shown, and neutrino oscillations are commonly added. The charge of the W bosons are dictated
by the fermions they interact with.*

Fermion fields

The kinetic term for a Dirac fermion is

$$i\bar{\psi}\gamma^\mu\partial_\mu\psi$$

where the notations are carried from earlier in the article. ψ can represent any, or all, Dirac fermions in the standard
model. Generally, as below, this term is included within the couplings (creating an overall "dynamical" term).

Gauge fields

For the spin-1 fields, first define the field strength tensor

$$F_{\mu\nu}^a = \partial_\mu A_\nu^a - \partial_\nu A_\mu^a + g f^{abc} A_\mu^b A_\nu^c$$

for a given gauge field (here we use A), with gauge coupling constant g. The quantity f^{abc} is the structure constant of the particular gauge group, defined by the commutator

$$[t_a, t_b] = i f^{abc} t_c,$$

where t_i are the generators of the group. In an Abelian (commutative) group (such as the U(1) we use here), since the generators t_a all commute with each other, the structure constants vanish. Of course, this is not the case in general – the standard model includes the non-Abelian SU(2) and SU(3) groups (such groups lead to what is called a Yang–Mills gauge theory).

We need to introduce three gauge fields corresponding to each of the subgroups SU(3) × SU(2) × U(1).

- The gluon field tensor will be denoted by $G_{\mu\nu}^a$, where the index a labels elements of the **8** representation of colour SU(3). The strong coupling constant is conventionally labelled g_s (or simply g where there is no ambiguity). *The observations leading to the discovery of this part of the Standard Model are discussed in the article in quantum chromodynamics.*

- The notation $W_{\mu\nu}^a$ will be used for the gauge field tensor of SU(2) where a runs over the 3 generators of this group. The coupling can be denoted g_w or again simply g. The gauge field will be denoted by W_μ^a.

- The gauge field tensor for the U(1) of weak hypercharge will be denoted by Bµv, the coupling by g′, and the gauge field by Bµ.

The kinetic term can now be written simply as

$$\mathcal{L}_{\mathrm{kin}} = -\frac{1}{4} B_{\mu\nu} B^{\mu\nu} - \frac{1}{2} \mathrm{tr} W_{\mu\nu} W^{\mu\nu} - \frac{1}{2} \mathrm{tr} G_{\mu\nu} G^{\mu\nu}$$

where the traces are over the SU(2) and SU(3) indices hidden in W and G respectively. The two-index objects are the field strengths derived from W and G the vector fields. There are also two extra hidden parameters: the theta angles for SU(2) and SU(3).

28.4.2   Coupling terms

The next step is to "couple" the gauge fields to the fermions, allowing for interactions.

Electroweak sector

Main article: Electroweak interaction

The electroweak sector interacts with the symmetry group U(1) × SU(2)L, where the subscript L indicates coupling only to left-handed fermions.

$$\mathcal{L}_{\mathrm{EW}} = \sum_\psi \bar{\psi} \gamma^\mu \left(i \partial_\mu - g' \frac{1}{2} Y_{\mathrm{W}} B_\mu - g \frac{1}{2} \tau \mathbf{W}_\mu \right) \psi$$

Where Bμ is the U(1) gauge field; YW is the weak hypercharge (the generator of the U(1) group); $\mathbf{W}\mu$ is the three-component SU(2) gauge field; and the components of $\boldsymbol{\tau}$ are the Pauli matrices (infinitesimal generators of the SU(2) group) whose eigenvalues give the weak isospin. Note that we have to redefine a new U(1) symmetry of *weak hypercharge*, different from QED, in order to achieve the unification with the weak force. The electric charge Q, third component of weak isospin T_3 (also called T_z, I_3 or I_z) and weak hypercharge YW are related by

$$Q = T_3 + \tfrac{1}{2}Y_W,$$

or by the alternate convention $Q = T_3 + YW$. The first convention (used in this article) is equivalent to the earlier Gell-Mann–Nishijima formula. We can then define the conserved current for weak isospin as

$$\mathbf{j}_\mu = \frac{1}{2}\bar{\psi}_L \gamma_\mu \boldsymbol{\tau} \psi_L$$

and for weak hypercharge as

$$j_\mu^Y = 2(j_\mu^{em} - j_\mu^3)$$

where j_μ^{em} is the electric current and j_μ^3 the third weak isospin current. As explained above, *these currents mix* to create the physically observed bosons, which also leads to testable relations between the coupling constants.

To explain in a simpler way, we can see the effect of the electroweak interaction by picking out terms from the Lagrangian. We see that the SU(2) symmetry acts on each (left-handed) fermion doublet contained in ψ, for example

$$-\frac{g}{2}(\bar{\nu}_e\ \bar{e})\tau^+\gamma_\mu (W^-)^\mu \begin{pmatrix} \nu_e \\ e \end{pmatrix} = -\frac{g}{2}\bar{\nu}_e\gamma_\mu (W^-)^\mu e$$

where the particles are understood to be left-handed, and where

$$\tau^\pm \equiv \frac{1}{2}(\tau^1 \pm i\tau^2) = \begin{pmatrix} 0 & 1 \\ 0 & 0 \end{pmatrix}$$

This is an interaction corresponding to a "rotation in weak isospin space" or in other words, a *transformation between eL and veL via emission of a* W^- *boson*. The U(1) symmetry, on the other hand, is similar to electromagnetism, but acts on all "*weak hypercharged*" fermions (both left and right handed) via the neutral Z^0, as well as the *charged* fermions via the photon.

Quantum chromodynamics sector

Main article: Quantum chromodynamics

The quantum chromodynamics (QCD) sector defines the interactions between quarks and gluons, with SU(3) symmetry, generated by T_a. Since leptons do not interact with gluons, they are not affected by this sector. The Dirac Lagrangian of the quarks coupled to the gluon fields is given by

$$\mathcal{L}_{QCD} = i\overline{U}\left(\partial_\mu - ig_s G_\mu^a T^a\right)\gamma^\mu U + i\overline{D}\left(\partial_\mu - ig_s G_\mu^a T^a\right)\gamma^\mu D.$$

where D and U are the Dirac spinors associated with up- and down-type quarks, and other notations are continued from the previous section.

28.4.3 Mass terms and the Higgs mechanism

Mass terms

The mass term arising from the Dirac Lagrangian (for any fermion ψ) is $-m\bar{\psi}\psi$ which is *not* invariant under the electroweak symmetry. This can be seen by writing ψ in terms of left and right handed components (skipping the actual calculation):

$$-m\bar{\psi}\psi = -m(\bar{\psi}_L\psi_R + \bar{\psi}_R\psi_L)$$

i.e. contribution from $\bar{\psi}_L\psi_L$ and $\bar{\psi}_R\psi_R$ terms do not appear. We see that the mass-generating interaction is achieved by constant flipping of particle chirality. The spin-half particles have no right/left chirality pair with the same SU(2) representations and equal and opposite weak hypercharges, so assuming these gauge charges are conserved in the vacuum, none of the spin-half particles could ever swap chirality, and must remain massless. Additionally, we know experimentally that the W and Z bosons are massive, but a boson mass term contains the combination e.g. $A^\mu A\mu$, which clearly depends on the choice of gauge. Therefore, none of the standard model fermions *or* bosons can "begin" with mass, but must acquire it by some other mechanism.

The Higgs mechanism

Main article: Higgs mechanism

The solution to both these problems comes from the Higgs mechanism, which involves scalar fields (the number of which depend on the exact form of Higgs mechanism) which (to give the briefest possible description) are "absorbed" by the massive bosons as degrees of freedom, and which couple to the fermions via Yukawa coupling to create what looks like mass terms.

In the Standard Model, the Higgs field is a complex scalar of the group SU(2)L:

$$\phi = \frac{1}{\sqrt{2}}\begin{pmatrix} \phi^+ \\ \phi^0 \end{pmatrix}.$$

where the superscripts + and 0 indicate the electric charge (Q) of the components. The weak hypercharge (YW) of both components is 1.

The Higgs part of the Lagrangian is

$$\mathcal{L}_H = \left[\left(\partial_\mu - igW_\mu^a t^a - ig'Y_\phi B_\mu\right)\phi\right]^2 + \mu^2\phi^\dagger\phi - \lambda(\phi^\dagger\phi)^2,$$

where $\lambda > 0$ and $\mu^2 > 0$, so that the mechanism of spontaneous symmetry breaking can be used. There is a parameter here, at first hidden within the shape of the potential, that is very important. In a unitarity gauge one can set $\varphi^+ = 0$ and make φ^0 real. Then $\langle\phi^0\rangle = v$ is the non-vanishing vacuum expectation value of the Higgs field. v has units of mass, and it is the only parameter in the Standard Model which is not dimensionless. It is also much smaller than the Planck scale: it is approximately equal to the Higgs mass, and sets the scale for the mass of everything else. This is the only real fine-tuning to a small nonzero value in the Standard Model, and it is called the Hierarchy problem. Quadratic terms in $W\mu$ and $B\mu$ arise, which give masses to the W and Z bosons:

$$M_W = \tfrac{1}{2}v|g|$$
$$M_Z = \tfrac{1}{2}v\sqrt{g^2 + g'^2}$$

The Yukawa interaction terms are

$$\mathcal{L}_{YU} = \overline{U}_L G_u U_R \phi^0 - \overline{D}_L G_u U_R \phi^- + \overline{U}_L G_d D_R \phi^+ + \overline{D}_L G_d D_R \phi^0 + hc$$

where $G_{u,d}$ are 3×3 matrices of Yukawa couplings, with the ij term giving the coupling of the generations i and j.

Neutrino masses

As previously mentioned, evidence shows neutrinos must have mass. But within the standard model, the right-handed neutrino does not exist, so even with a Yukawa coupling neutrinos remain massless. An obvious solution[2] is to simply *add a right-handed neutrino* vR resulting in a **Dirac mass** term as usual. This field however must be a sterile neutrino, since being right-handed it experimentally belongs to an isospin singlet ($T_3 = 0$) and also has charge $Q = 0$, implying $YW = 0$ (see above) i.e. it does not even participate in the weak interaction. Current experimental status is that evidence for observation of sterile neutrinos is not convincing.[3]

Another possibility to consider is that the neutrino satisfies the **Majorana equation**, which at first seems possible due to its zero electric charge. In this case the mass term is

$$-\frac{m}{2}\left(\overline{\nu}^C \nu + \overline{\nu}\nu^C\right)$$

where C denotes a charge conjugated (i.e. anti-) particle, and the terms are consistently all left (or all right) chirality (note that a left-chirality projection of an antiparticle is a right-handed field; care must be taken here due to different notations sometimes used). Here we are essentially flipping between LH neutrinos and RH anti-neutrinos (it is furthermore possible but *not* necessary that neutrinos are their own antiparticle, so these particles are the same). However for the left-chirality neutrinos, this term changes weak hypercharge by 2 units - not possible with the standard Higgs interation, requiring the Higgs field to be extended to include an extra triplet with weak hypercharge 2[4] - whereas for right-chirality neutrinos, no Higgs extensions are necessary. For both left and right chirality cases, Majorana terms violate lepton number, but possibly at a level beyond the current sensitivity of experiments to detect such violations.

It is possible to include **both** Dirac and Majorana mass terms in the same theory, which (in contrast to the Dirac-mass-only approach) can provide a "natural" explanation for the smallness of the observed neutrino masses, by linking the RH neutrinos to yet-unknown physics around the GUT scale[5] (see seesaw mechanism).

Since in any case new fields must be postulated to explain the experimental results, neutrinos are an obvious gateway to searching physics beyond the Standard Model.

28.5 Detailed Information

This section provides more detail on some aspects, and some reference material.

28.5.1 Field content in detail

The Standard Model has the following fields. These describe one *generation* of leptons and quarks, and there are three generations, so there are three copies of each field. By CPT symmetry, there is a set of right-handed fermions with the opposite quantum numbers. The column "**representation**" indicates under which representations of the gauge groups that each field transforms, in the order (SU(3), SU(2), U(1)). Symbols used are common but not universal; superscript C denotes an antiparticle; and for the U(1) group, the value of the weak hypercharge is listed. Note that there are twice as many left-handed lepton field components as left-handed antilepton field components in each generation, but an equal number of left-handed quark and antiquark fields.

28.5.2 Fermion content

This table is based in part on data gathered by the Particle Data Group.[6]

[1] These are not ordinary abelian charges, which can be added together, but are labels of group representations of Lie groups.

[2] Mass is really a coupling between a left-handed fermion and a right-handed fermion. For example, the mass of an electron is really a coupling between a left-handed electron and a right-handed electron, which is the antiparticle of a left-handed positron. Also neutrinos show large mixings in their mass coupling, so it's not accurate to talk about neutrino masses in the flavor basis or to suggest a left-handed electron antineutrino.

[3] The Standard Model assumes that neutrinos are massless. However, several contemporary experiments prove that neutrinos oscillate between their flavour states, which could not happen if all were massless. It is straightforward to extend the model to fit these data but there are many possibilities, so the mass eigenstates are still open. See neutrino mass.

[4] W.-M. Yao *et al.* (Particle Data Group) (2006). "Review of Particle Physics: Neutrino mass, mixing, and flavor change" (PDF). *Journal of Physics G* **33**: 1. arXiv:astro-ph/0601168. Bibcode:2006JPhG...33....1Y. doi:10.1088/0954-3899/33/1/001.

[5] The masses of baryons and hadrons and various cross-sections are the experimentally measured quantities. Since quarks can't be isolated because of QCD confinement, the quantity here is supposed to be the mass of the quark at the renormalization scale of the QCD scale.

28.5.3 Free parameters

Upon writing the most general Lagrangian without neutrinos, one finds that the dynamics depend on 19 parameters, whose numerical values are established by experiment. With neutrinos 7 more parameters are needed, 3 masses and 4 PMNS matrix parameters, for a total of 26 parameters.[7] The neutrino parameter values are still uncertain. The 19 certain parameters are summarized here (note: with the Higgs mass is at 125 GeV, the Higgs self-coupling strength $\lambda \sim 1/8$).

28.5.4 Additional symmetries of the Standard Model

From the theoretical point of view, the Standard Model exhibits four additional global symmetries, not postulated at the outset of its construction, collectively denoted **accidental symmetries**, which are continuous U(1) global symmetries. The transformations leaving the Lagrangian invariant are:

$$\psi_q(x) \rightarrow e^{i\alpha/3}\psi_q$$

$$E_L \rightarrow e^{i\beta}E_L \text{ and } (e_R)^c \rightarrow e^{i\beta}(e_R)^c$$

$$M_L \rightarrow e^{i\beta}M_L \text{ and } (\mu_R)^c \rightarrow e^{i\beta}(\mu_R)^c$$

$$T_L \rightarrow e^{i\beta}T_L \text{ and } (\tau_R)^c \rightarrow e^{i\beta}(\tau_R)^c$$

The first transformation rule is shorthand meaning that all quark fields for all generations must be rotated by an identical phase simultaneously. The fields ML, TL and $(\mu_R)^c$, $(\tau_R)^c$ are the 2nd (muon) and 3rd (tau) generation analogs of EL and $(e_R)^c$ fields.

By Noether's theorem, each symmetry above has an associated conservation law: the conservation of baryon number, electron number, muon number, and tau number. Each quark is assigned a baryon number of $\frac{1}{3}$, while each antiquark is assigned a baryon number of $-\frac{1}{3}$. Conservation of baryon number implies that the number of quarks minus the number of antiquarks is a constant. Within experimental limits, no violation of this conservation law has been found.

Similarly, each electron and its associated neutrino is assigned an electron number of +1, while the anti-electron and the associated anti-neutrino carry a −1 electron number. Similarly, the muons and their neutrinos are assigned a muon number of +1 and the tau leptons are assigned a tau lepton number of +1. The Standard Model predicts that each of these three numbers should be conserved separately in a manner similar to the way baryon number is conserved. These numbers are collectively known as lepton family numbers (LF).

In addition to the accidental (but exact) symmetries described above, the Standard Model exhibits several **approximate symmetries**. These are the "SU(2) custodial symmetry" and the "SU(2) or SU(3) quark flavor symmetry."

28.5.5 The U(1) symmetry

For the leptons, the gauge group can be written $SU(2)_l \times U(1)L \times U(1)R$. The two U(1) factors can be combined into $U(1)Y \times U(1)_l$ where l is the lepton number. Gauging of the lepton number is ruled out by experiment, leaving only the possible gauge group $SU(2)L \times U(1)Y$. A similar argument in the quark sector also gives the same result for the electroweak theory.

28.5.6 The charged and neutral current couplings and Fermi theory

The charged currents $j^{\pm} = j^1 \pm ij^2$ are

$$j_\mu^+ = \overline{U}_{iL}\gamma_\mu D_{iL} + \overline{\nu}_{iL}\gamma_\mu l_{iL}.$$

These charged currents are precisely those that entered the Fermi theory of beta decay. The action contains the charge current piece

$$\mathcal{L}_{CC} = \frac{g}{\sqrt{2}}(j_\mu^+ W^{-\mu} + j_\mu^- W^{+\mu}).$$

For energy much less than the mass of the W-boson, the effective theory becomes the current–current interaction of the Fermi theory.

However, gauge invariance now requires that the component W^3 of the gauge field also be coupled to a current that lies in the triplet of SU(2). However, this mixes with the U(1), and another current in that sector is needed. These currents must be uncharged in order to conserve charge. So we require the **neutral currents**

$$j_\mu^3 = \frac{1}{2}(\overline{U}_{iL}\gamma_\mu U_{iL} - \overline{D}_{iL}\gamma_\mu D_{iL} + \overline{\nu}_{iL}\gamma_\mu \nu_{iL} - \overline{l}_{iL}\gamma_\mu l_{iL})$$

$$j_\mu^{em} = \frac{2}{3}\overline{U}_i\gamma_\mu U_i - \frac{1}{3}\overline{D}_i\gamma_\mu D_i - \overline{l}_i\gamma_\mu l_i.$$

The neutral current piece in the Lagrangian is then

$$\mathcal{L}_{NC} = e j_\mu^{em} A^\mu + \frac{g}{\cos\theta_W}(J_\mu^3 - \sin^2\theta_W J_\mu^{em})Z^\mu.$$

28.6 See also

- Overview of Standard Model of particle physics
- Fundamental interaction
- Noncommutative standard model
- Open questions: CP violation, Neutrino masses, Quark matter
- Physics beyond the Standard Model
- Strong interactions: Flavour, Quantum chromodynamics, Quark model
- Weak interactions: Electroweak interaction, Fermi's interaction
- Weinberg angle
- Symmetry in quantum mechanics

28.7   References and external links

[1] In fact, there are mathematical issues regarding quantum field theories still under debate (see e.g. Landau pole), but the predictions extracted from the Standard Model by current methods are all self-consistent. For a further discussion see e.g. R. Mann, chapter 25.

[2] https://fas.org/sgp/othergov/doe/lanl/pubs/00326607.pdf

[3] http://t2k-experiment.org/neutrinos/oscillations-today/

[4] https://fas.org/sgp/othergov/doe/lanl/pubs/00326607.pdf

[5] http://www.mpi-hd.mpg.de/personalhomes/schwetz/tueb-2.pdf

[6] W.-M. Yao *et al.* (Particle Data Group) (2006). "Review of Particle Physics: Quarks" (PDF). *Journal of Physics G* **33**: 1. arXiv:astro-ph/0601168. Bibcode:2006JPhG...33....1Y. doi:10.1088/0954-3899/33/1/001.

[7] Mark Thomson (5 September 2013). *Modern Particle Physics*. Cambridge University Press. pp. 499–500. ISBN 978-1-107-29254-3.

- *An introduction to quantum field theory*, by M.E. Peskin and D.V. Schroeder (HarperCollins, 1995) ISBN 0-201-50397-2.

- *Gauge theory of elementary particle physics*, by T.P. Cheng and L.F. Li (Oxford University Press, 1982) ISBN 0-19-851961-3.

- Standard Model Lagrangian with explicit Higgs terms (T.D. Gutierrez, ca 1999) (PDF, PostScript, and LaTeX version)

- *The quantum theory of fields* (vol 2), by S. Weinberg (Cambridge University Press, 1996) ISBN 0-521-55002-5.

- *Quantum Field Theory in a Nutshell* (Second Edition), by A. Zee (Princeton University Press, 2010) ISBN 978-1-4008-3532-4.

- *An Introduction to Particle Physics and the Standard Model*, by R. Mann (CRC Press, 2010) ISBN 978-1420082982

Chapter 29

Spontaneous symmetry breaking

Spontaneous symmetry breaking[1][2][3] is a mode of realization of symmetry breaking in a physical system, where the underlying laws are invariant under a symmetry transformation, but the system as a whole changes under such transformations, in contrast to explicit symmetry breaking. It is a spontaneous process by which a system in a symmetrical state ends up in an asymmetrical state. It thus describes systems where the equations of motion or the Lagrangian obey certain symmetries, but the lowest-energy solutions do not exhibit that symmetry.

Consider a symmetrical upward dome with a trough circling the bottom. If a ball is put at the very peak of the dome, the system is symmetrical with respect to a rotation around the center axis. But the ball may *spontaneously break* this symmetry by rolling down the dome into the trough, a point of lowest energy. Afterward, the ball has come to a rest at some fixed point on the perimeter. The dome and the ball retain their individual symmetry, but the system does not.[4]

Most simple phases of matter and phase transitions, like crystals, magnets, and conventional superconductors can be simply understood from the viewpoint of spontaneous symmetry breaking. Notable exceptions include topological phases of matter like the fractional quantum Hall effect.

29.1 Spontaneous symmetry breaking in physics

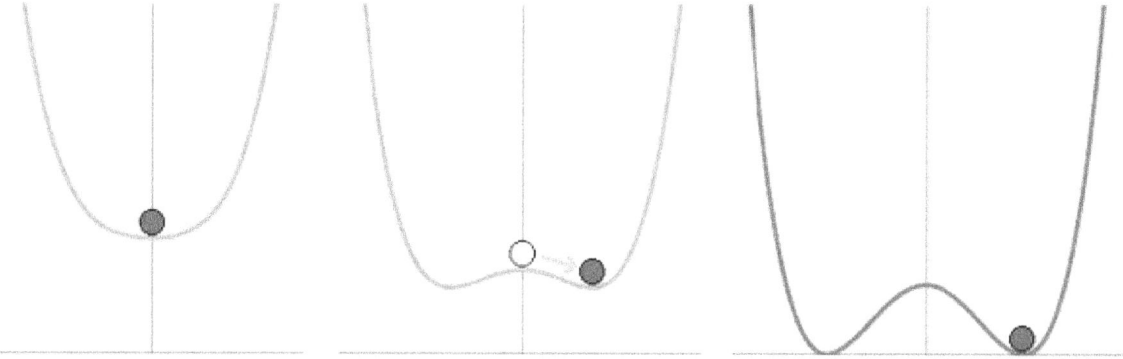

Spontaneous symmetry breaking simplified: – *At high energy levels (left) the ball settles in the center, and the result is symmetrical. At lower energy levels (right), the overall "rules" remain symmetrical, but the "Mexican hat" potential comes into effect: "local" symmetry is inevitably broken since eventually the ball must roll one way (at random) and not another.*

29.1.1   Particle physics

In particle physics the force carrier particles are normally specified by field equations with gauge symmetry; their equations predict that certain measurements will be the same at any point in the field. For instance, field equations might predict that the mass of two quarks is constant. Solving the equations to find the mass of each quark might give two solutions. In one solution, quark A is heavier than quark B. In the second solution, quark B is heavier than quark A *by the same amount*. The symmetry of the equations is not reflected by the individual solutions, but it is reflected by the range of solutions. An actual measurement reflects only one solution, representing a breakdown in the symmetry of the underlying theory. "Hidden" is perhaps a better term than "broken" because the symmetry is always there in these equations. This phenomenon is called *spontaneous* symmetry breaking because *nothing* (that we know) breaks the symmetry in the equations.[5]:194–195

Chiral symmetry

Main article: Chiral symmetry breaking

Chiral symmetry breaking is an example of spontaneous symmetry breaking affecting the chiral symmetry of the strong interactions in particle physics. It is a property of quantum chromodynamics, the quantum field theory describing these interactions, and is responsible for the bulk of the mass (over 99%) of the nucleons, and thus of all common matter, as it converts very light bound quarks into 100 times heavier constituents of baryons. The approximate Nambu–Goldstone bosons in this spontaneous symmetry breaking process are the pions, whose mass is an order of magnitude lighter than the mass of the nucleons. It served as the prototype and significant ingredient of the Higgs mechanism underlying the electroweak symmetry breaking.

Higgs mechanism

Main articles: Brout–Englert–Higgs mechanism and Yukawa interaction

The strong, weak, and electromagnetic forces can all be understood as arising from gauge symmetries. The Higgs mechanism, the spontaneous symmetry breaking of gauge symmetries, is an important component in understanding the superconductivity of metals and the origin of particle masses in the standard model of particle physics. One important consequence of the distinction between true symmetries and *gauge symmetries*, is that the spontaneous breaking of a gauge symmetry does not give rise to characteristic massless Nambu–Goldstone modes, but only massive modes, like the plasma mode in a superconductor, or the Higgs mode observed in particle physics.

In the standard model of particle physics, spontaneous symmetry breaking of the SU(2) × U(1) gauge symmetry associated with the electro-weak force generates masses for several particles, and separates the electromagnetic and weak forces. The W and Z bosons are the elementary particles that mediate the weak interaction, while the photon mediates the electromagnetic interaction. At energies much greater than 100 GeV all these particles behave in a similar manner. The Weinberg–Salam theory predicts that, at lower energies, this symmetry is broken so that the photon and the massive W and Z bosons emerge.[6] In addition, fermions develop mass consistently.

Without spontaneous symmetry breaking, the Standard Model of elementary particle interactions requires the existence of a number of particles. However, some particles (the W and Z bosons) would then be predicted to be massless, when, in reality, they are observed to have mass. To overcome this, spontaneous symmetry breaking is augmented by the Higgs mechanism to give these particles mass. It also suggests the presence of a new particle, the Higgs boson, reported as possibly identifiable with a boson detected in 2012. (If the Higgs boson were not confirmed to have been found, it would mean that the simplest implementation of the Higgs mechanism and spontaneous symmetry breaking *as they are currently formulated* require modification.)

Superconductivity of metals is a condensed-matter analog of the Higgs phenomena, in which a condensate of Cooper pairs of electrons spontaneously breaks the U(1) gauge "symmetry" associated with light and electromagnetism.

29.1.2 Condensed matter physics

Most phases of matter can be understood through the lens of spontaneous symmetry breaking. For example, crystals are periodic arrays of atoms that are not invariant under all translations (only under a small subset of translations by a lattice vector). Magnets have north and south poles that are oriented in a specific direction, breaking rotational symmetry. In addition to these examples, there are a whole host of other symmetry-breaking phases of matter including nematic phases of liquid crystals, charge- and spin-density waves, superfluids and many others.

There are several known examples of matter that cannot be described by spontaneous symmetry breaking, including: topologically ordered phases of matter like fractional quantum Hall liquids, and spin-liquids. These states do not break any symmetry, but are distinct phases of matter. Unlike the case of spontaneous symmetry breaking, there is not a general framework for describing such states.

Continuous symmetry

The ferromagnet is the canonical system which spontaneously breaks the continuous symmetry of the spins below the Curie temperature and at $h = 0$, where h is the external magnetic field. Below the Curie temperature the energy of the system is invariant under inversion of the magnetization $m(\mathbf{x})$ such that $m(\mathbf{x}) = -m(-\mathbf{x})$. The symmetry is spontaneously broken as $h \rightarrow 0$ when the Hamiltonian becomes invariant under the inversion transformation, but the expectation value is not invariant.

Spontaneously, symmetry broken phases of matter are characterized by an order parameter that describes the quantity which breaks the symmetry under consideration. For example, in a magnet, the order parameter is the local magnetization.

Spontaneously breaking of a continuous symmetry is inevitably accompanied by gapless (meaning that these modes do not cost any energy to excite) Nambu–Goldstone modes associated with slow long-wavelength fluctuations of the order parameter. For example, vibrational modes in a crystal, known as phonons, are associated with slow density fluctuations of the crystal's atoms. The associated Goldstone mode for magnets are oscillating waves of spin known as spin-waves. For symmetry-breaking states, whose order parameter is not a conserved quantity, Nambu–Goldstone modes are typically massless and propagate at a constant velocity.

An important theorem, due to Mermin and Wagner, states that, at finite temperature, thermally activated fluctuations of Nambu–Goldstone modes destroy the long-range order, and prevent spontaneous symmetry breaking in one- and two-dimensional systems. Similarly, quantum fluctuations of the order parameter prevent most types of continuous symmetry breaking in one-dimensional systems even at zero temperature (an important exception is ferromagnets, whose order parameter, magnetization, is an exactly conserved quantity and does not have any quantum fluctuations).

Other long-range interacting systems such as cylindrical curved surfaces interacting via the Coulomb potential or Yukawa potential has been shown to break translational and rotational symmetries.[7] It was shown, in the presence of a symmetric Hamiltonian, and in the limit of infinite volume, the system spontaneously adopts a chiral configuration, i.e. breaks mirror plane symmetry.

29.1.3 Dynamical symmetry breaking

Dynamical symmetry breaking (DSB) is a special form of spontaneous symmetry breaking where the ground state of the system has reduced symmetry properties compared to its theoretical description (Lagrangian).

Dynamical breaking of a global symmetry is a spontaneous symmetry breaking, that happens not at the (classical) tree level (i.e. at the level of the bare action), but due to quantum corrections (i.e. at the level of the effective action).

Dynamical breaking of a gauge symmetry is subtler. In the conventional spontaneous gauge symmetry breaking, there exists an unstable Higgs particle in the theory, which drives the vacuum to a symmetry-broken phase (see e.g. Electroweak interaction). In dynamical gauge symmetry breaking, however, no unstable Higgs particle operates in the theory, but the bound states of the system itself provide the unstable fields that render the phase transition. For example, Bardeen, Hill, and Lindner published a paper which attempts to replace the conventional Higgs mechanism in the standard model, by a DSB that is driven by a bound state of top-antitop quarks (such models, where a composite particle plays the role of the Higgs boson, are often referred to as "Composite Higgs models").[8] Dynamical breaking of gauge symmetries is often due to creation of a fermionic condensate; for example the quark condensate, which is connected to the dynamical breaking of chiral symmetry in quantum chromodynamics. Conventional superconductivity is the paradigmatic example from the condensed matter side, where phonon-mediated attractions lead electrons to become bound in pairs and then condense, thereby breaking the electromagnetic gauge symmetry.

29.2   Generalisation and technical usage

For spontaneous symmetry breaking to occur, there must be a system in which there are several equally likely outcomes. The system as a whole is therefore symmetric with respect to these outcomes. (If we consider any two outcomes, the probability is the same. This contrasts sharply to explicit symmetry breaking.) However, if the system is sampled (i.e. if the system is actually used or interacted with in any way), a specific outcome must occur. Though the system as a whole is symmetric, it is never encountered with this symmetry, but only in one specific asymmetric state. Hence, the symmetry is said to be spontaneously broken in that theory. Nevertheless, the fact that each outcome is equally likely is a reflection of the underlying symmetry, which is thus often dubbed "hidden symmetry", and has crucial formal consequences. (See the article on the Goldstone boson).

When a theory is symmetric with respect to a symmetry group, but requires that one element of the group be distinct, then spontaneous symmetry breaking has occurred. The theory must not dictate *which* member is distinct, only that *one is*. From this point on, the theory can be treated as if this element actually is distinct, with the proviso that any results found in this way must be resymmetrized, by taking the average of each of the elements of the group being the distinct one.

The crucial concept in physics theories is the order parameter. If there is a field (often a background field) which acquires an expectation value (not necessarily a *vacuum* expectation value) which is not invariant under the symmetry in question, we say that the system is in the ordered phase, and the symmetry is spontaneously broken. This is because other subsystems interact with the order parameter, which specifies a "frame of reference" to be measured against. In that case, the vacuum state does not obey the initial symmetry (which would keep it invariant, in the linearly realized **Wigner mode** in which it would be a singlet), and, instead changes under the (hidden) symmetry, now implemented in the (nonlinear) **Nambu–Goldstone mode**. Normally, in the absence of the Higgs mechanism, massless Goldstone bosons arise.

The symmetry group can be discrete, such as the space group of a crystal, or continuous (e.g., a Lie group), such as the rotational symmetry of space. However, if the system contains only a single spatial dimension, then only discrete symmetries may be broken in a vacuum state of the full quantum theory, although a classical solution may break a continuous symmetry.

29.3   A pedagogical example: the Mexican hat potential

In the simplest idealized relativistic model, the spontaneously broken symmetry is summarized through an illustrative scalar field theory. The relevant Lagrangian, which essentially dictates how a system behaves, can be split up into kinetic and potential terms,

It is in this potential term $V(\Phi)$ that the symmetry breaking is triggered. An example of a potential, due to Jeffrey Goldstone[9] is illustrated in the graph at the right.

This potential has an infinite number of possible minima (vacuum states) given by

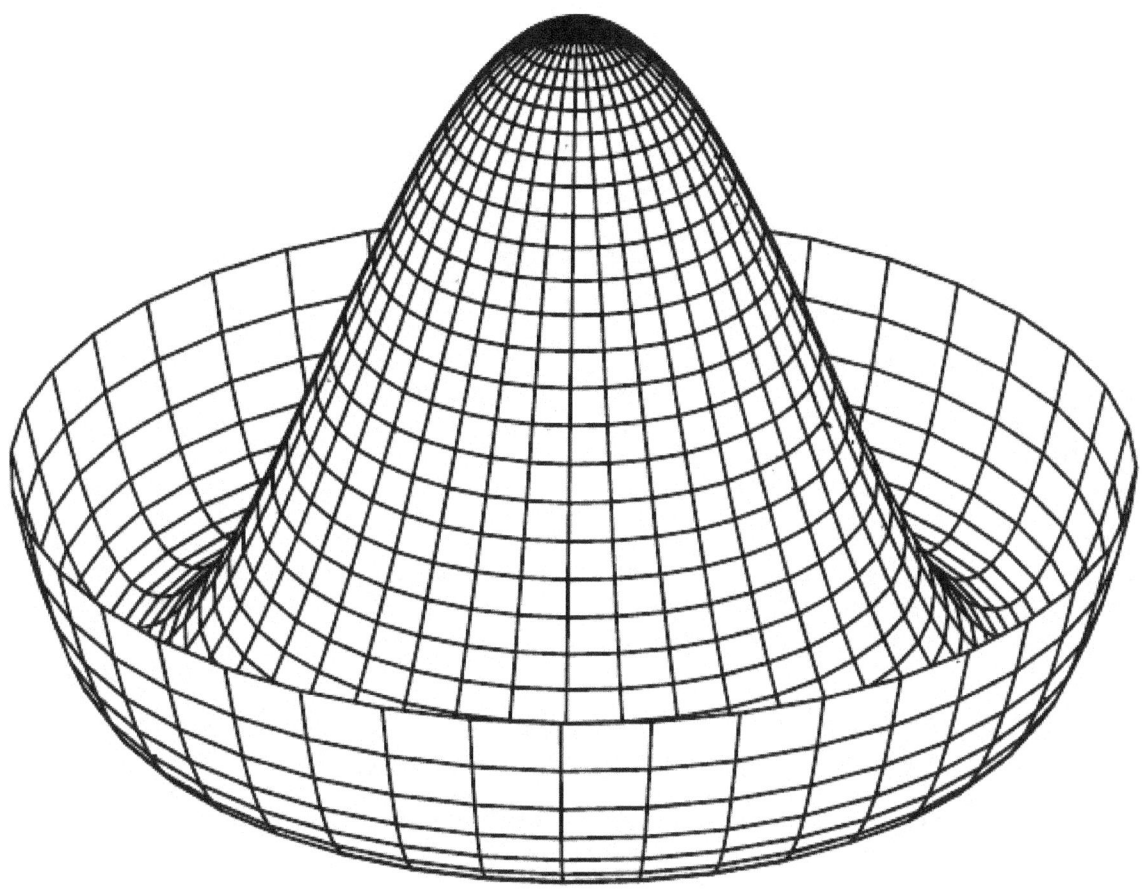

Graph of Goldstone's "Mexican hat" potential function V versus φ.

for any real θ between 0 and 2π. The system also has an unstable vacuum state corresponding to $\Phi = 0$. This state has a U(1) symmetry. However, once the system falls into a specific stable vacuum state (amounting to a choice of θ), this symmetry will appear to be lost, or "spontaneously broken".

In fact, any other choice of θ would have exactly the same energy, implying the existence of a massless Nambu–Goldstone boson, the mode running around the circle at the minimum of this potential, and indicating there is some memory of the original symmetry in the Lagrangian.

29.4 Other examples

- For ferromagnetic materials, the underlying laws are invariant under spatial rotations. Here, the order parameter is the magnetization, which measures the magnetic dipole density. Above the Curie temperature, the order parameter is zero, which is spatially invariant, and there is no symmetry breaking. Below the Curie temperature, however, the magnetization acquires a constant nonvanishing value, which points in a certain direction (in the idealized situation where we have full equilibrium; otherwise, translational symmetry gets broken as well). The residual rotational symmetries which leave the orientation of this vector invariant remain unbroken, unlike the other rotations which do not and are thus spontaneously broken.

- The laws describing a solid are invariant under the full Euclidean group, but the solid itself spontaneously breaks this group down to a space group. The displacement and the orientation are the order parameters.

- General relativity has a Lorentz symmetry, but in FRW cosmological models, the mean 4-velocity field defined by averaging over the velocities of the galaxies (the galaxies act like gas particles at cosmological scales) acts as an order parameter breaking this symmetry. Similar comments can be made about the cosmic microwave background.

- For the electroweak model, as explained earlier, a component of the Higgs field provides the order parameter breaking the electroweak gauge symmetry to the electromagnetic gauge symmetry. Like the ferromagnetic example, there is a phase transition at the electroweak temperature. The same comment about us not tending to notice broken symmetries suggests why it took so long for us to discover electroweak unification.

- In superconductors, there is a condensed-matter collective field ψ, which acts as the order parameter breaking the electromagnetic gauge symmetry.

- Take a thin cylindrical plastic rod and push both ends together. Before buckling, the system is symmetric under rotation, and so visibly cylindrically symmetric. But after buckling, it looks different, and asymmetric. Nevertheless, features of the cylindrical symmetry are still there: ignoring friction, it would take no force to freely spin the rod around, displacing the ground state in time, and amounting to an oscillation of vanishing frequency, unlike the radial oscillations in the direction of the buckle. This spinning mode is effectively the requisite Nambu–Goldstone boson.

- Consider a uniform layer of fluid over an infinite horizontal plane. This system has all the symmetries of the Euclidean plane. But now heat the bottom surface uniformly so that it becomes much hotter than the upper surface. When the temperature gradient becomes large enough, convection cells will form, breaking the Euclidean symmetry.

- Consider a bead on a circular hoop that is rotated about a vertical diameter. As the rotational velocity is increased gradually from rest, the bead will initially stay at its initial equilibrium point at the bottom of the hoop (intuitively stable, lowest gravitational potential). At a certain critical rotational velocity, this point will become unstable and the bead will jump to one of two other newly created equilibria, equidistant from the center. Initially, the system is symmetric with respect to the diameter, yet after passing the critical velocity, the bead ends up in one of the two new equilibrium points, thus breaking the symmetry.

29.5   Nobel Prize

On October 7, 2008, the Royal Swedish Academy of Sciences awarded the 2008 Nobel Prize in Physics to three scientists for their work in subatomic physics symmetry breaking. Yoichiro Nambu, of the University of Chicago, won half of the prize for the discovery of the mechanism of spontaneous broken symmetry in the context of the strong interactions, specifically chiral symmetry breaking. Physicists Makoto Kobayashi and Toshihide Maskawa shared the other half of the prize for discovering the origin of the explicit breaking of CP symmetry in the weak interactions.[10] This origin is ultimately reliant on the Higgs mechanism, but, so far understood as a "just so" feature of Higgs couplings, not a spontaneously broken symmetry phenomenon.

29.6   See also

- Autocatalytic reactions and order creation

- Catastrophe theory

- Chiral symmetry breaking

- CP-violation

- Explicit symmetry breaking

- Gauge gravitation theory

- Goldstone boson

- Grand unified theory

- Higgs mechanism

- Higgs boson

- Higgs field (classical)

- Irreversibility

- Magnetic catalysis of chiral symmetry breaking

- Mermin–Wagner theorem

- Quantum fluctuation

- Sakurai Prize for Theoretical Particle Physics

- Second-order phase transition

- Symmetry breaking

- Tachyon condensation

- Tachyonic field

- Wheeler–Feynman absorber theory

- 1964 PRL symmetry breaking papers

29.7 Notes

- ^ Note that (as in fundamental Higgs driven spontaneous gauge symmetry breaking) the term "symmetry breaking" is a misnomer when applied to gauge symmetries.

29.8 References

[1] *Dynamical Symmetry Breaking in Quantum Field Theories*. By Vladimir A. Miranskij. Pg 15.

[2] Patterns of Symmetry Breaking. Edited by Henryk Arodz, Jacek Dziarmaga, Wojciech Hubert Zurek. Pg 141.

[3] Bubbles, Voids and Bumps in Time: The New Cosmology. Edited by James Cornell. Pg 125.

[4] Gerald M. Edelman, Bright Air, Brilliant Fire: On the Matter of the Mind (New York: BasicBooks, 1992) 203.

[5] Steven Weinberg (20 April 2011). *Dreams of a Final Theory: The Scientist's Search for the Ultimate Laws of Nature*. Knopf Doubleday Publishing Group. ISBN 978-0-307-78786-6.

[6] A Brief History of Time. Stephen Hawking. Bantam: 10th anniversary edition (September 1, 1998). pp. 73–74.

[7] Kohlstedt, K.L.; Vernizzi, G.; Solis, F.J.; Olvera de la Cruz, M. (2007). "Spontaneous Chirality via Long-range Electrostatic Forces".*Physical Review Letters***99**: 030602. arXiv:0704.3435. Bibcode:2007PhRvL..99c0602K.doi:10.1103/Phys.

[8] William A. Bardeen; Christopher T. Hill; Manfred Lindner (1990). "Minimal dynamical symmetry breaking of the standard model". *Physical Review D* **41** (5): 1647–1660. Bibcode:1990PhRvD..41.1647B. doi:10.1103/PhysRevD.41.1647.

[9] Goldstone.J. (1961). "Field theories with"Superconductor"solutions".*Il Nuovo Cimento***19**: 154–164. doi:10.1007/BF.

[10] The Nobel Foundation. "The Nobel Prize in Physics 2008". *nobelprize.org*. Retrieved January 15, 2008.

29.9   External links

- Spontaneous symmetry breaking

- Physical Review Letters – 50th Anniversary Milestone Papers

- In CERN Courier, Steven Weinberg reflects on spontaneous symmetry breaking

- Englert–Brout–Higgs–Guralnik–Hagen–Kibble Mechanism on Scholarpedia

- History of Englert–Brout–Higgs–Guralnik–Hagen–Kibble Mechanism on Scholarpedia

- The History of the Guralnik, Hagen and Kibble development of the Theory of Spontaneous Symmetry Breaking and Gauge Particles

- International Journal of Modern Physics A: The History of the Guralnik, Hagen and Kibble development of the Theory of Spontaneous Symmetry Breaking and Gauge Particles

- Guralnik, G S; Hagen, C R and Kibble, T W B (1967). Broken Symmetries and the Goldstone Theorem. Advances in Physics, vol. 2 Interscience Publishers, New York. pp. 567–708 ISBN 0-470-17057-3

- Spontaneous Symmetry Breaking in Gauge Theories: a Historical Survey

Chapter 30

Lorentz covariance

In physics, **Lorentz symmetry**, named for Hendrik Lorentz, is "the feature of nature that says experimental results are independent of the orientation or the boost velocity of the laboratory through space".[1] **Lorentz covariance**, a related concept, is a key property of spacetime following from the special theory of relativity. Lorentz covariance has two distinct, but closely related meanings:

1. A physical quantity is said to be Lorentz covariant if it transforms under a given representation of the Lorentz group. According to the representation theory of the Lorentz group, these quantities are built out of scalars, four-vectors, four-tensors, and spinors. In particular, a scalar (e.g., the space-time interval) remains the same under Lorentz transformations and is said to be a "Lorentz invariant" (i.e., they transform under the trivial representation).

2. An equation is said to be Lorentz covariant if it can be written in terms of Lorentz covariant quantities (confusingly, some use the term "invariant" here). The key property of such equations is that if they hold in one inertial frame, then they hold in any inertial frame; this follows from the result that if all the components of a tensor vanish in one frame, they vanish in every frame. This condition is a requirement according to the principle of relativity, i.e., all non-gravitational laws must make the same predictions for identical experiments taking place at the same spacetime event in two different inertial frames of reference.

This usage of the term *covariant* should not be confused with the related concept of a *covariant vector*. On manifolds, the words *covariant* and *contravariant* refer to how objects transform under general coordinate transformations. Confusingly, both covariant and contravariant four-vectors can be Lorentz covariant quantities.

Local Lorentz covariance, which follows from general relativity, refers to Lorentz covariance applying only *locally* in an infinitesimal region of spacetime at every point. There is a generalization of this concept to cover Poincaré covariance and Poincaré invariance.

30.1 Examples

In general, the nature of a Lorentz tensor can be identified by its tensor order, which is the number of indices it has. No indices implies it is a scalar, one implies that it is a vector, etc. Furthermore, any number of new scalars, vectors etc. can be made by contracting any kinds of tensors together, but many of these may not have any real physical meaning. Some of those tensors that do have a physical interpretation are listed (by no means exhaustively) below.

Please note, the metric sign convention such that $\eta = \text{diag}\,(1, -1, -1, -1)$ is used throughout the article.

30.1.1 Scalars

Spacetime interval:

$$\Delta s^2 = \Delta x^a \Delta x^b \eta_{ab} = c^2 \Delta t^2 - \Delta x^2 - \Delta y^2 - \Delta z^2$$

Proper time (for timelike intervals):

$$\Delta \tau = \sqrt{\frac{\Delta s^2}{c^2}}, \ \Delta s^2 > 0$$

Proper distance (for spacelike intervals):

$$L = \sqrt{-\Delta s^2}, \ \Delta s^2 < 0$$

Rest mass:

$$m_0^2 c^2 = P^a P^b \eta_{ab} = \frac{E^2}{c^2} - p_x^2 - p_y^2 - p_z^2$$

Electromagnetism invariants:

$$F_{ab} F^{ab} = 2 \left(B^2 - \frac{E^2}{c^2} \right)$$

$$G_{cd} F^{cd} = \frac{1}{2} \epsilon_{abcd} F^{ab} F^{cd} = -\frac{4}{c} \left(\vec{B} \cdot \vec{E} \right)$$

D'Alembertian/wave operator:

$$\Box = \eta^{\mu\nu} \partial_\mu \partial_\nu = \frac{1}{c^2} \frac{\partial^2}{\partial t^2} - \frac{\partial^2}{\partial x^2} - \frac{\partial^2}{\partial y^2} - \frac{\partial^2}{\partial z^2}$$

30.1.2 Four-vectors

4-Displacement:

$$\Delta X^\alpha = (c\Delta t, \vec{\Delta x}) = (c\Delta t, \Delta x, \Delta y, \Delta z)$$

4-Position:

$$X^a = (ct, \vec{x}) = (ct, x, y, z)$$

4-Gradient: with is the 4D Partial derivative:

$$\partial^a = \left(\frac{\partial_t}{c}, -\vec{\nabla} \right) = \left(\frac{1}{c} \frac{\partial}{\partial t}, -\frac{\partial}{\partial x}, -\frac{\partial}{\partial y}, -\frac{\partial}{\partial z} \right)$$

4-Velocity:

$$U^a = \gamma(c, \vec{u}) = \gamma\left(c, \frac{dx}{dt}, \frac{dy}{dt}, \frac{dz}{dt}\right)$$

where $U^a = \frac{dX^a}{d\tau}$

4-Momentum:

$$P^a = (mc, \vec{p}) = \left(\frac{E}{c}, \vec{p}\right) = \left(\frac{E}{c}, p_x, p_y, p_z\right)$$

where $P^a = m_o U^a$

4-Current:

$$J^a = (c\rho, \vec{j}) = (c\rho, j_x, j_y, j_z)$$

where $J^a = \rho_o U^a$

30.1.3 Four-tensors

The Kronecker delta:

$$\delta_b^a = \begin{cases} 1 & \text{if } a = b, \\ 0 & \text{if } a \neq b. \end{cases}$$

The Minkowski metric (the metric of flat space according to general relativity):

$$\eta_{ab} = \eta^{ab} = \begin{cases} 1 & \text{if } a = b = 0, \\ -1 & \text{if } a = b = 1, 2, 3, \\ 0 & \text{if } a \neq b. \end{cases}$$

The Levi-Civita symbol:

$$\epsilon_{abcd} = -\epsilon^{abcd} = \begin{cases} +1 & \text{if } \{abcd\} \text{ is an even permutation of } \{0123\}, \\ -1 & \text{if } \{abcd\} \text{ is an odd permutation of } \{0123\}, \\ 0 & \text{otherwise.} \end{cases}$$

Electromagnetic field tensor (using a metric signature of $+ - - -$):

$$F_{ab} = \begin{bmatrix} 0 & E_x/c & E_y/c & E_z/c \\ -E_x/c & 0 & -B_z & B_y \\ -E_y/c & B_z & 0 & -B_x \\ -E_z/c & -B_y & B_x & 0 \end{bmatrix}$$

Dual electromagnetic field tensor:

$$G_{cd} = \frac{1}{2}\epsilon_{abcd}F^{ab} = \begin{bmatrix} 0 & B_x & B_y & B_z \\ -B_x & 0 & E_z/c & -E_y/c \\ -B_y & -E_z/c & 0 & E_x/c \\ -B_z & E_y/c & -E_x/c & 0 \end{bmatrix}$$

30.2   Lorentz violating models

See also: Modern searches for Lorentz violation

In standard field theory, there are very strict and severe constraints on marginal and relevant Lorentz violating operators within both QED and the Standard Model. Irrelevant Lorentz violating operators may be suppressed by a high cutoff scale, but they typically induce marginal and relevant Lorentz violating operators via radiative corrections. So, we also have very strict and severe constraints on irrelevant Lorentz violating operators.

Since some approaches to quantum gravity lead to violations of Lorentz invariance,[2] these studies are part of Phenomenological Quantum Gravity.

Lorentz violating models typically fall into four classes:

- The laws of physics are exactly Lorentz covariant but this symmetry is spontaneously broken. In special relativistic theories, this leads to phonons, which are the Goldstone bosons. The phonons travel at *less* than the speed of light.

- Similar to the approximate Lorentz symmetry of phonons in a lattice (where the speed of sound plays the role of the critical speed), the Lorentz symmetry of special relativity (with the speed of light as the critical speed in vacuum) is only a low-energy limit of the laws of physics, which involve new phenomena at some fundamental scale. Bare conventional "elementary" particles are not point-like field-theoretical objects at very small distance scales, and a nonzero fundamental length must be taken into account. Lorentz symmetry violation is governed by an energy-dependent parameter which tends to zero as momentum decreases.[3] Such patterns require the existence of a privileged local inertial frame (the "vacuum rest frame"). They can be tested, at least partially, by ultra-high energy cosmic ray experiments like the Pierre Auger Observatory.[4]

- The laws of physics are symmetric under a deformation of the Lorentz or more generally, the Poincaré group, and this deformed symmetry is exact and unbroken. This deformed symmetry is also typically a quantum group symmetry, which is a generalization of a group symmetry. Deformed special relativity is an example of this class of models. It is not accurate to call such models Lorentz-violating as much as Lorentz deformed any more than special relativity can be called a violation of Galilean symmetry rather than a deformation of it. The deformation is scale dependent, meaning that at length scales much larger than the Planck scale, the symmetry looks pretty much like the Poincaré group. Ultra-high energy cosmic ray experiments cannot test such models.

- This is a class of its own; a subgroup of the Lorentz group is sufficient to give us all the standard predictions if CP is an exact symmetry. However, CP isn't exact. This is called Very Special Relativity.

Models belonging to the first two classes can be consistent with experiment if Lorentz breaking happens at Planck scale or beyond it, or even before it in suitable preonic models,[5] and if Lorentz symmetry violation is governed by a suitable energy-dependent parameter. One then has a class of models which deviate from Poincaré symmetry near the Planck scale but still flows towards an exact Poincaré group at very large length scales. This is also true for the third class, which is furthermore protected from radiative corrections as one still has an exact (quantum) symmetry.

Even though there is no evidence of the violation of Lorentz invariance, several experimental searches for such violations have been performed during recent years. A detailed summary of the results of these searches is given in the Data Tables for Lorentz and CPT Violation.[6]

30.3   See also

- Antimatter tests of Lorentz violation

- General covariance

- Lorentz invariance in loop quantum gravity

- Lorentz-violating neutrino oscillations

- Symmetry in physics

30.4 References

[1] "Framing Lorentz symmetry". CERN Courier. 2004-11-24. Retrieved 2013-05-26.

[2] Mattingly,David(2005). "Modern Tests of Lorentz Invariance".*Living Reviews in Relativity* 8. arXiv:gr-qc/0502097. Bibcode. doi:10.12942/lrr-2005-5.

[3] Luis Gonzalez-Mestres (1995-05-25). "Properties of a possible class of particles able to travel faster than light".

[4] Luis Gonzalez-Mestres (1997-05-26). "Absence of Greisen-Zatsepin-Kuzmin Cutoff and Stability of Unstable Particles at Very High Energy, as a Consequence of Lorentz Symmetry Violation".

[5] Luis Gonzalez-Mestres (2014). "Ultra-high energy physics and standard basic principles. Do Planck units really make sense?" (PDF). EPJ Web of Conferences (ICNFP 2013 Conference). doi:10.1051/epjconf/20147100062.

[6] Kostelecky, V.A.; Russell, N. (2010). "Data Tables for Lorentz and CPT Violation". arXiv:0801.0287v3.

- Background information on Lorentz and CPT violation: http://www.physics.indiana.edu/~{}kostelec/faq.html

- Mattingly, David (2005). "Modern Tests of Lorentz Invariance". *Living Reviews in Relativity* 8. arXiv:gr-qc/0502097. Bibcode:2005LRR.....8....5M. doi:10.12942/lrr-2005-5.

- Amelino-Camelia G, Ellis J, Mavromatos N E, Nanopoulos D V, and Sarkar S (June 1998). "Tests of quantum gravity from observations of bold gamma-ray bursts". *Nature* 393 (6687): 763–765. arXiv:astro-ph/9712103. Bibcode:1998Natur.393..763A. doi:10.1038/31647. Retrieved 2007-12-22.

- Jacobson T, Liberati S, and Mattingly D (August 2003). "A strong astrophysical constraint on the violation of special relativity by quantum gravity". *Nature* 424 (6952): 1019–1021. arXiv:astro-ph/0212190. Bibcode:2003Natur. .doi:10.1038/nature01882. PMID 12944959. Retrieved 2007-12-22.

- Carroll S (August 2003). "Quantum gravity: An astrophysical constraint". *Nature* 424 (6952): 1007–1008. Bibcode:2003Natur.424.1007C. doi:10.1038/4241007a. PMID 12944951. Retrieved 2007-12-22.

- Jacobson, T.; Liberati, S.; Mattingly, D. (2003). "Threshold effects and Planck scale Lorentz violation: Combined constraints from high energy astrophysics". *Physical Review D* 67 (12). arXiv:hep-ph/0209264. Bibcode:2003 PhRvD..67l4011J.doi:10.1103/PhysRevD.67.124011.

30.5 External links

Chapter 31

Transformation (function)

For other uses, see Transformation.

In mathematics, particularly in semigroup theory, a **transformation** is any function f mapping a set X to itself, i.e. $f:X{\to}X$.[1][2][3] In other areas of mathematics, a transformation may simply be any function, regardless of domain and codomain.[4] This wider sense shall not be considered in this article; refer instead to the article on function for that sense.

Examples include linear transformations and affine transformations, rotations, reflections and translations. These can be carried out in Euclidean space, particularly in dimensions 2 and 3. They are also operations that can be performed using linear algebra, and described explicitly using matrices.

31.1 Translation

Main article: Translation (geometry)

A **translation**, or **translation operator**, is an affine transformation of Euclidean space which moves every point by a fixed distance in the same direction. It can also be interpreted as the addition of a constant vector to every point, or as shifting the origin of the coordinate system. In other words, if $\mathbf{v}$ is a fixed vector, then the translation $T_\mathbf{v}$ will work as $T_\mathbf{v}(\mathbf{p}) = \mathbf{p} + \mathbf{v}$.

For the purpose of visualization, consider a browser window. This window, if maximized to full dimensions of the screen, is the reference plane. Imagine one of the corners as the reference point or origin $(0, 0)$.

Consider a point $P(x, y)$ in the corresponding plane. Now the axes are shifted from the original axes to a distance (h, k) and this is the corresponding reference axes. Now the origin (previous axes) is (x, y) and the point P is (X, Y) and therefore the equations are:

$X = x - h$ or $x = X + h$ or $h = x - X$ and $Y = y - k$ or $y = Y + k$ or $k = y - Y$.

Replacing these values or using these equations in the respective equation we obtain the transformed equation or new reference axes, old reference axes, point lying on the plane.

31.2 Reflection

Main article: Reflection (mathematics)

A **reflection** is a map that transforms an object into its mirror image with respect to a "mirror", which is a hyperplane of fixed points in the geometry. For example, a reflection of the small Latin letter p with respect to a vertical line would look like a "q". In order to reflect a planar figure one needs the "mirror" to be a line (*axis of reflection* or *axis of symmetry*),

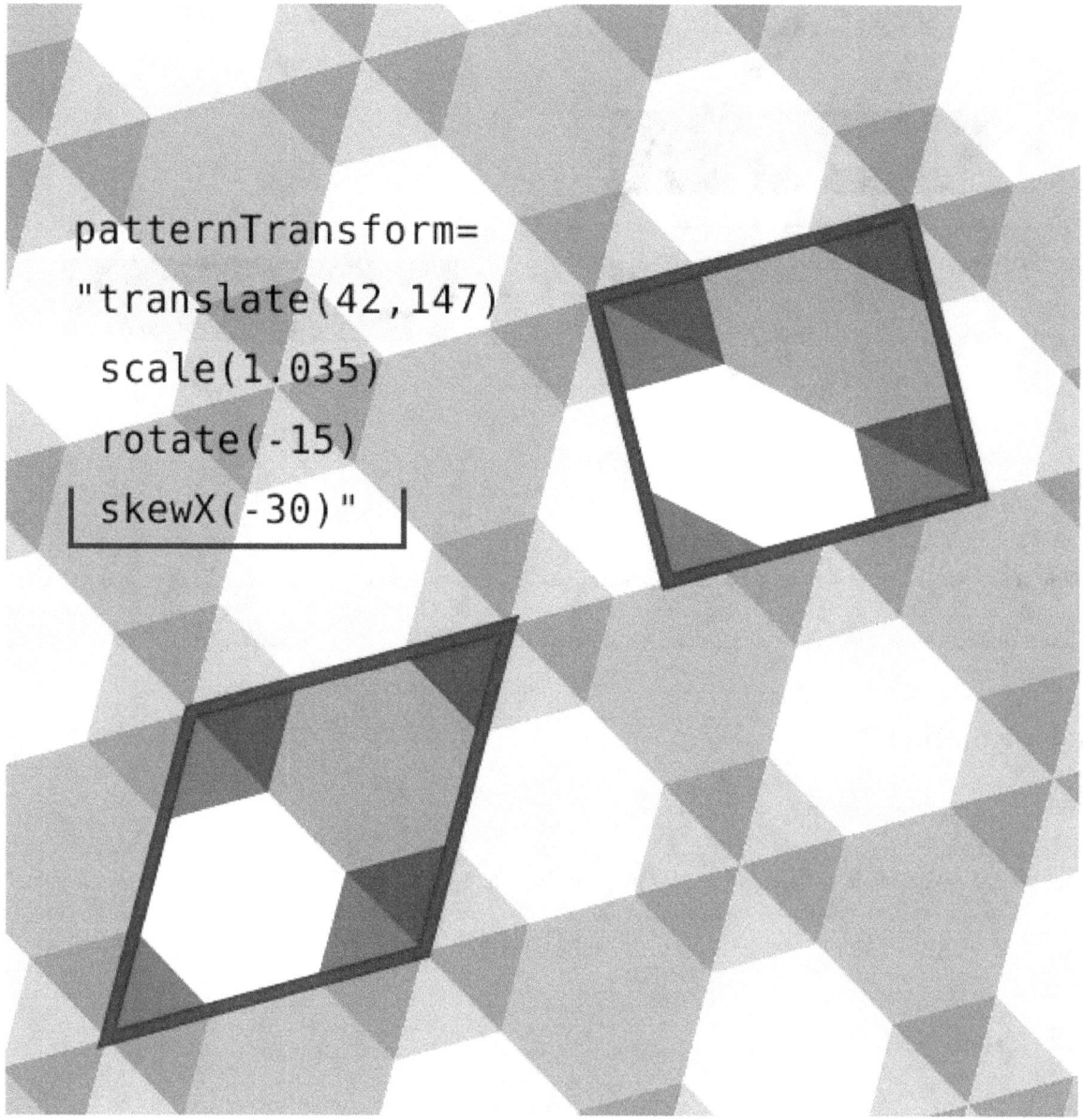

A composition of four mappings coded in SVG,
*which **transforms** a rectangular repetitive pattern*
into a rhombic pattern.

while for reflections in the three-dimensional space one would use a plane (the *plane of reflection* or *symmetry*) for a mirror. Reflection may be considered as the limiting case of inversion as the radius of the reference circle increases without bound.

Reflection is considered to be an *opposite* motion since it changes the orientation of the figures it reflects.

31.3 Glide reflection

Main article: Glide reflection

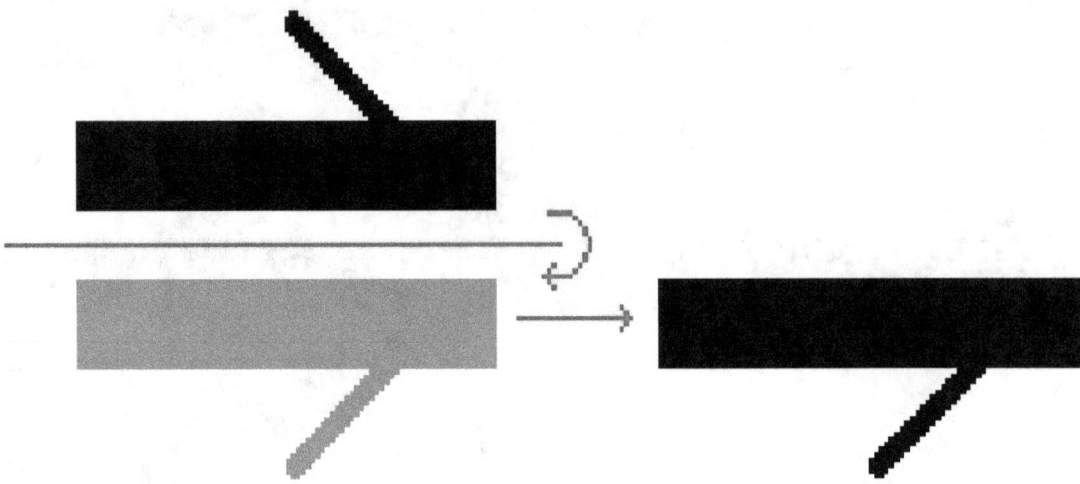

Example of a glide reflection

A **glide reflection** is a type of isometry of the Euclidean plane: the combination of a reflection in a line and a translation along that line. Reversing the order of combining gives the same result. Depending on context, we may consider a simple reflection (without translation) as a special case where the translation vector is the zero vector.

31.4 Rotation

Main article: Rotation (geometry)

A rotation is a transformation that is performed by "spinning" the object around a fixed point known as the center of rotation. You can rotate the object at any degree measure, but 90° and 180° are two of the most common. Rotation by a positive angle rotates the object counterclockwise, whereas rotation by a negative angle rotates the object clockwise.

31.5 Scaling

Main article: Scaling (geometry)

Uniform scaling is a linear transformation that enlarges or diminishes objects; the scale factor is the same in all directions; it is also called a homothety or dilation. The result of uniform scaling is similar (in the geometric sense) to the original.

More general is **scaling** with a separate scale factor for each axis direction; a special case is **directional scaling** (in one direction). Shapes not aligned with the axes may be subject to shear (see below) as a side effect: although the angles between lines parallel to the axes are preserved, other angles are not.

31.6 Shear

Main article: Shear mapping

Shear is a transform that effectively rotates one axis so that the axes are no longer perpendicular. Under shear, a rectangle becomes a parallelogram, and a circle becomes an ellipse. Even if lines parallel to the axes stay the same length, others do not. As a mapping of the plane, it lies in the class of equi-areal mappings.

31.7 More generally

More generally, a **transformation** in mathematics means a mathematical function (synonyms: *map* and *mapping*). A transformation can be an invertible function from a set X to itself, or from X to another set Y. The choice of the term *transformation* may simply flag that a function's more geometric aspects are being considered (for example, with attention paid to invariants).

31.7.1 Partial transformations

The notion of transformation generalized to partial functions. A **partial transformation** is a function $f: A \to B$, where both A and B are subsets of some set X.[5]

31.8 Algebraic structures

The set of all transformations on a given base set together with function composition forms a regular semigroup.

31.9 Combinatorics

For a finite set of cardinality n, there are n^n transformations and $(n+1)^n$ partial transformations.[6]

31.10 See also

- Coordinate transformation
- Data transformation (statistics)
- Infinitesimal transformation
- Linear transformation
- Transformation geometry
- Transformation group
- Transformation matrix

31.11 References

[1] Olexandr Ganyushkin; Volodymyr Mazorchuk (2008). *Classical Finite Transformation Semigroups: An Introduction*. Springer Science & Business Media. p. 1. ISBN 978-1-84800-281-4.

[2] Pierre A. Grillet (1995). *Semigroups: An Introduction to the Structure Theory*. CRC Press. p. 2. ISBN 978-0-8247-9662-4.

[3] Wilkinson, Leland & Graham (2005). *The Grammar of Graphics* (2nd ed.). Springer. p. 29. ISBN 978-0-387-24544-7.

[4] P. R. Halmos (1960). *Naive Set Theory*. Springer Science & Business Media. pp. 30–. ISBN 978-0-387-90092-6.

[5] Christopher Hollings (2014). *Mathematics across the Iron Curtain: A History of the Algebraic Theory of Semigroups*. American Mathematical Society. p. 251. ISBN 978-1-4704-1493-1.

[6] Olexandr Ganyushkin; Volodymyr Mazorchuk (2008). *Classical Finite Transformation Semigroups: An Introduction*. Springer Science & Business Media. p. 2. ISBN 978-1-84800-281-4.

Chapter 32

Circle group

For the jazz group, see Circle (jazz band).

In mathematics, the **circle group**, denoted by **T**, is the multiplicative group of all complex numbers with absolute value 1, i.e., the unit circle in the complex plane or simply the **unit complex numbers**[1]

$$\mathbb{T} = \{z \in \mathbb{C} : |z| = 1\}.$$

The circle group forms a subgroup of $\mathbb{C}^\times$, the multiplicative group of all nonzero complex numbers. Since $\mathbb{C}^\times$ is abelian, it follows that **T** is as well. The circle group is also the group **U(1)** of 1×1 unitary matrices; these act on the complex plane by rotation about the origin. The circle group can be parametrized by the angle θ of rotation by

$$\theta \mapsto z = e^{i\theta} = \cos\theta + i\sin\theta.$$

This is the exponential map for the circle group.

The circle group plays a central role in Pontryagin duality, and in the theory of Lie groups.

The notation **T** for the circle group stems from the fact that, with the standard topology (see below), the circle group is a 1-torus. More generally $\mathbf{T}^n$ (the direct product of **T** with itself n times) is geometrically an n-torus.

32.1 Elementary introduction

One way to think about the circle group is that it describes how to add *angles*, where only angles between 0° and 360° are permitted. For example, the diagram illustrates how to add 150° to 270°. The answer should be 150° + 270° = 420°, but when thinking in terms of the circle group, we need to "forget" the fact that we have wrapped once around the circle. Therefore we adjust our answer by 360° which gives 420° = 60° (mod 360°).

Another description is in terms of ordinary addition, where only numbers between 0 and 1 are allowed (with 1 corresponding to a full rotation). To achieve this, we might need to throw away digits occurring before the decimal point. For example, when we work out 0.784 + 0.925 + 0.446, the answer should be 2.155, but we throw away the leading 2, so the answer (in the circle group) is just 0.155.

32.2 Topological and analytic structure

The circle group is more than just an abstract algebraic object. It has a natural topology when regarded as a subspace of the complex plane. Since multiplication and inversion are continuous functions on $\mathbb{C}^\times$, the circle group has the structure

252

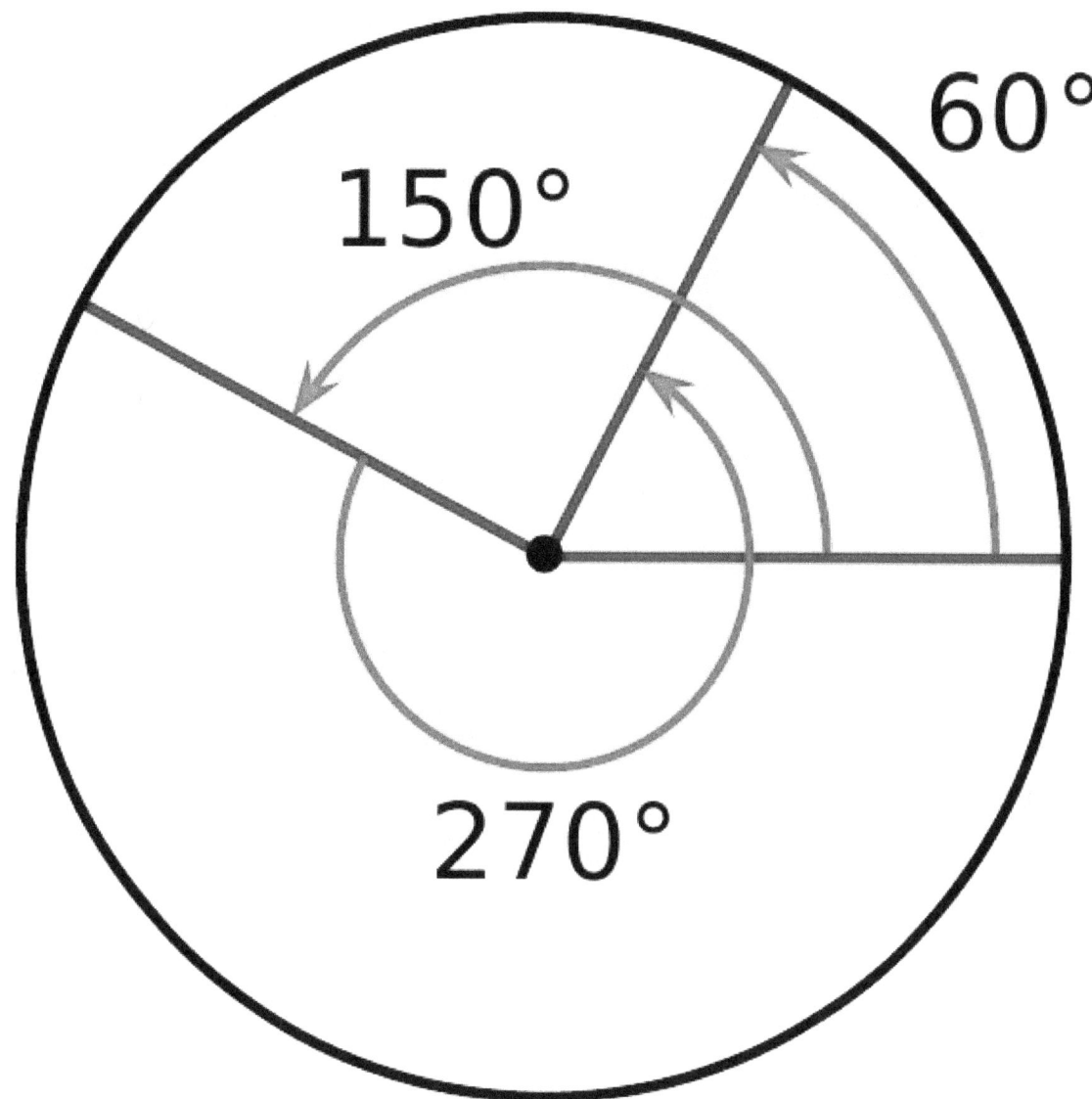

Multiplication on the circle group is equivalent to addition of angles

of a topological group. Moreover, since the unit circle is a closed subset of the complex plane, the circle group is a closed subgroup of $\mathbf{C}^\times$ (itself regarded as a topological group).

One can say even more. The circle is a 1-dimensional real manifold and multiplication and inversion are real-analytic maps on the circle. This gives the circle group the structure of a one-parameter group, an instance of a Lie group. In fact, up to isomorphism, it is the unique 1-dimensional compact, connected Lie group. Moreover, every n-dimensional compact, connected, abelian Lie group is isomorphic to $\mathbf{T}^n$.

32.3   Isomorphisms

The circle group shows up in a variety of forms in mathematics. We list some of the more common forms here. Specifically, we show that

$\mathbb{T} \cong U(1) \cong \mathbb{R}/\mathbb{Z} \cong SO(2)$.

Note that the slash (/) denotes here quotient group.

The set of all 1×1 unitary matrices clearly coincides with the circle group; the unitary condition is equivalent to the condition that its element have absolute value 1. Therefore, the circle group is canonically isomorphic to U(1), the first unitary group.

The exponential function gives rise to a group homomorphism exp : $\mathbb{R} \to \mathbb{T}$ from the additive real numbers $\mathbb{R}$ to the circle group $\mathbb{T}$ via the map

$$\theta \mapsto e^{i\theta} = \cos\theta + i\sin\theta.$$

The last equality is Euler's formula or the complex exponential. The real number θ corresponds to the angle on the unit circle as measured counterclockwise from the positive x-axis. That this map is a homomorphism follows from the fact that the multiplication of unit complex numbers corresponds to addition of angles:

$$e^{i\theta_1} e^{i\theta_2} = e^{i(\theta_1 + \theta_2)}.$$

This exponential map is clearly a surjective function from $\mathbb{R}$ to $\mathbb{T}$. It is not, however, injective. The kernel of this map is the set of all integer multiples of 2π. By the first isomorphism theorem we then have that

$$\mathbb{T} \cong \mathbb{R}/2\pi\mathbb{Z}.$$

After rescaling we can also say that $\mathbf{T}$ is isomorphic to $\mathbf{R/Z}$.

If complex numbers are realized as 2×2 real matrices (see complex number), the unit complex numbers correspond to 2×2 orthogonal matrices with unit determinant. Specifically, we have

$$e^{i\theta} \leftrightarrow \begin{bmatrix} \cos\theta & -\sin\theta \\ \sin\theta & \cos\theta \end{bmatrix}.$$

The circle group is therefore isomorphic to the special orthogonal group SO(2). This has the geometric interpretation that multiplication by a unit complex number is a proper rotation in the complex plane, and every such rotation is of this form.

32.4   Properties

Every compact Lie group G of dimension > 0 has a subgroup isomorphic to the circle group. That means that, thinking in terms of symmetry, a compact symmetry group acting *continuously* can be expected to have one-parameter circle subgroups acting; the consequences in physical systems are seen for example at rotational invariance, and spontaneous symmetry breaking.

The circle group has many subgroups, but its only proper closed subgroups consist of roots of unity: For each integer $n > 0$, the nth roots of unity form a cyclic group of order n, which is unique up to isomorphism.

32.5   Representations

The representations of the circle group are easy to describe. It follows from Schur's lemma that the irreducible complex representations of an abelian group are all 1-dimensional. Since the circle group is compact, any representation $\rho : \mathbf{T} \to$

$GL(1, \mathbf{C}) \cong \mathbf{C}^\times$, must take values in $U(1) \cong \mathbf{T}$. Therefore, the irreducible representations of the circle group are just the homomorphisms from the circle group to itself.

These representations are all inequivalent. The representation φ_{-n} is conjugate to φ_n,

$$\phi_{-n} = \overline{\phi_n}.$$

These representations are just the characters of the circle group. The character group of $\mathbf{T}$ is clearly an infinite cyclic group generated by φ_1:

$$\mathrm{Hom}(\mathbb{T}, \mathbb{T}) \cong \mathbb{Z}.$$

The irreducible real representations of the circle group are the trivial representation (which is 1-dimensional) and the representations

$$\rho_n(e^{i\theta}) = \begin{bmatrix} \cos n\theta & -\sin n\theta \\ \sin n\theta & \cos n\theta \end{bmatrix}, \quad n \in \mathbb{Z}^+,$$

taking values in SO(2). Here we only have positive integers n since the representation ρ_{-n} is equivalent to ρ_n.

32.6 Group structure

In this section we will forget about the topological structure of the circle group and look only at its structure as an abstract group.

The circle group $\mathbf{T}$ is a divisible group. Its torsion subgroup is given by the set of all nth roots of unity for all n, and is isomorphic to $\mathbf{Q/Z}$. The structure theorem for divisible groups and the axiom of choice together tell us that $\mathbf{T}$ is isomorphic to the direct sum of $\mathbf{Q/Z}$ with a number of copies of $\mathbf{Q}$. The number of copies of $\mathbf{Q}$ must be c (the cardinality of the continuum) in order for the cardinality of the direct sum to be correct. But the direct sum of c copies of $\mathbf{Q}$ is isomorphic to $\mathbf{R}$, as $\mathbf{R}$ is a vector space of dimension c over $\mathbf{Q}$. Thus

$$\mathbb{T} \cong \mathbb{R} \oplus (\mathbb{Q}/\mathbb{Z}).$$

The isomorphism

$$\mathbb{C}^\times \cong \mathbb{R} \oplus (\mathbb{Q}/\mathbb{Z})$$

can be proved in the same way, as $\mathbf{C}^\times$ is also a divisible abelian group whose torsion subgroup is the same as the torsion subgroup of $\mathbf{T}$.

32.7 See also

- Rotation number
- Torus
- One-parameter subgroup
- Unitary group

- Orthogonal group

- Group of rational points on the unit circle

- Phase factor (application in quantum-mechanics)

32.8   Notes

[1] "a **unit complex number** is a complex number of unit absolute value" (James & James 1992, p. 436)

32.9   References

- James, Robert C.; James, Glenn (1992), *Mathematics Dictionary* (Fifth ed.), Chapman & Hall

32.10   Further reading

- Hua Luogeng (1981) *Starting with the unit circle*, Springer Verlag, ISBN 0-387-90589-8 .

32.11   External links

- Homeomorphism and the Group Structure on a Circle

Chapter 33

Special unitary group

"SU(5)" redirects here. For the specific grand unification theory, see Georgi–Glashow model.

In mathematics, the **special unitary group** of degree n, denoted SU(n), is the Lie group of $n{\times}n$ unitary matrices with determinant 1 (i.e., real-valued determinant, not complex as for general unitary matrices). The group operation is that of matrix multiplication. The special unitary group is a subgroup of the unitary group U(n), consisting of all $n{\times}n$ unitary matrices. As a compact classical group, U(n) is the group that preserves the standard inner product on $\mathbf{C}^n$.[nb 1] It is itself a subgroup of the general linear group, SU(n) $\subset$ U(n) $\subset$ GL(n, $\mathbf{C}$).

The SU(n) groups find wide application in the Standard Model of particle physics, especially SU(2) in the electroweak interaction and SU(3) in quantum chromodynamics.[1]

The simplest case, SU(1), is the trivial group, having only a single element. The group SU(2) is isomorphic to the group of quaternions of norm 1, and is thus diffeomorphic to the 3-sphere. Since unit quaternions can be used to represent rotations in 3-dimensional space (up to sign), there is a surjective homomorphism from SU(2) to the rotation group SO(3) whose kernel is $\{+I, -I\}$.[nb 2] SU(2) is also identical to one of the symmetry groups of spinors, Spin(3), that enables a spinor presentation of rotations.

33.1   Properties

The special unitary group SU(n) is a real Lie group (though not a complex Lie group). Its dimension as a real manifold is $n^2 - 1$. Topologically, it is compact and simply connected. Algebraically, it is a simple Lie group (meaning its Lie algebra is simple; see below). [2]

The center of SU(n) is isomorphic to the cyclic group Zn, and is composed of the diagonal matrices ζI for ζ an nth root of unity and I the $n{\times}n$ identity matrix.

Its outer automorphism group, for $n \geq 3$, is Z_2, while the outer automorphism group of SU(2) is the trivial group.

A maximal torus, of rank $n - 1$, is given by the set of diagonal matrices with determinant 1. The Weyl group is the symmetric group Sn, which is represented by signed permutation matrices (the signs being necessary to ensure the determinant is 1).

The Lie algebra of SU(n), denoted by **su**(n), can be identified with the set of traceless antihermitian $n{\times}n$ complex matrices, with the regular commutator as Lie bracket. Particle physicists often use a different, equivalent representation: the set of traceless hermitian $n{\times}n$ complex matrices with Lie bracket given by $-i$ times the commutator.

33.2   Infinitesimal generators

The Lie algebra **su**(n) can be generated by n^2 operators $\hat{O}_{ij}$, $i, j= 1, 2, ..., n$, which satisfy the commutator relationships

$$\left[\hat{O}_{ij}, \hat{O}_{k\ell}\right] = \delta_{jk}\hat{O}_{i\ell} - \delta_{i\ell}\hat{O}_{kj}$$

for $i, j, k, \ell = 1, 2, ..., n$, where δ_{jk} denotes the Kronecker delta. Additionally, the operator

$$\hat{N} = \sum_{i=1}^{n} \hat{O}_{ii}$$

satisfies

$$\left[\hat{N}, \hat{O}_{ij}\right] = 0,$$

which implies that the number of *independent* generators of the Lie algebra is $n^2 - 1$.[3]

33.2.1   Fundamental representation

In the defining, or fundamental, representation of **su**(n) the generators Ta are represented by traceless hermitian matrices complex $n \times n$ matrices, where:

$$T_a T_b = \frac{1}{2n}\delta_{ab}I_n + \frac{1}{2}\sum_{c=1}^{n^2-1}(if_{abc} + d_{abc})T_c$$

where the f are the structure constants and are antisymmetric in all indices, while the d-coefficients are symmetric in all indices. As a consequence:

$$[T_a, T_b]_+ = \frac{1}{n}\delta_{ab}I_n + \sum_{c=1}^{n^2-1} d_{abc}T_c$$

$$[T_a, T_b]_- = i\sum_{c=1}^{n^2-1} f_{abc}T_c .$$

We also take

$$\sum_{c,e=1}^{n^2-1} d_{ace}d_{bce} = \frac{n^2-4}{n}\delta_{ab}$$

as a normalization convention.

33.2.2   Adjoint representation

In the $(n^2 - 1)$ -dimensional adjoint representation, the generators are represented by $(n^2 - 1) \times (n^2 - 1)$ matrices, whose elements are defined by the structure constants themselves:

$$(T_a)_{jk} = -if_{ajk}.$$

33.3 $n = 2$

See also: Versor and Pauli matrices

SU(2) is the following group,

$$SU(2) = \left\{ \begin{pmatrix} \alpha & -\overline{\beta} \\ \beta & \overline{\alpha} \end{pmatrix} : \alpha, \beta \in \mathbf{C}, |\alpha|^2 + |\beta|^2 = 1 \right\}.$$

where the overline denotes complex conjugation.

Now, consider the following map,

$$\varphi : \mathbf{C}^2 \to M(2, \mathbf{C})$$
$$\varphi(\alpha, \beta) = \begin{pmatrix} \alpha & -\overline{\beta} \\ \beta & \overline{\alpha} \end{pmatrix}.$$

where $M(2, \mathbf{C})$ denotes the set of 2 by 2 complex matrices. By considering $\mathbf{C}^2$ diffeomorphic to $\mathbf{R}^4$ and $M(2, \mathbf{C})$ diffeomorphic to $\mathbf{R}^8$, we can see that φ is an injective real linear map and hence an embedding. Now, considering the restriction of φ to the 3-sphere (since modulus is 1), denoted S^3, we can see that this is an embedding of the 3-sphere onto a compact submanifold of $M(2, \mathbf{C})$. However, it is also clear that $\varphi(S^3) = SU(2)$.

Therefore, as a manifold S^3 is diffeomorphic to $SU(2)$ and so $SU(2)$ is a compact, connected Lie group.

The Lie algebra of $SU(2)$ is

$$\mathfrak{su}(2) = \left\{ \begin{pmatrix} ia & -\overline{z} \\ z & -ia \end{pmatrix} : a \in \mathbf{R}, z \in \mathbf{C} \right\}.$$

It is easily verified that matrices of this form have trace zero and are antihermitian. The Lie algebra is then generated by the following matrices,

$$u_1 = \begin{pmatrix} 0 & i \\ i & 0 \end{pmatrix} \qquad u_2 = \begin{pmatrix} 0 & -1 \\ 1 & 0 \end{pmatrix} \qquad u_3 = \begin{pmatrix} i & 0 \\ 0 & -i \end{pmatrix}.$$

which are easily seen to have the form of the general element specified above.

These satisfy $u_3 u_2 = -u_2 u_3 = -u_1$ and $u_2 u_1 = -u_1 u_2 = -u_3$. The commutator bracket is therefore specified by

$$[u_3, u_1] = 2u_2, \qquad [u_1, u_2] = 2u_3, \qquad [u_2, u_3] = 2u_1.$$

The above generators are related to the Pauli matrices by $u_1 = i \sigma_1, u_2 = -i \sigma_2$ and $u_3 = i \sigma_3$. This representation is routinely used in quantum mechanics to represent the spin of fundamental particles such as electrons. They also serve as unit vectors for the description of our 3 spatial dimensions in loop quantum gravity.

The Lie algebra serves to work out the representations of $SU(2)$.

See also: Rotation group SO(3) § A note on representations

33.4 $n = 3$

The generators of $\mathfrak{su}(3)$, T, in the defining representation, are:

$$T_a = \frac{\lambda_a}{2}.$$

where λ the Gell-Mann matrices, are the SU(3) analog of the Pauli matrices for SU(2):

These λ_a span all traceless Hermitian matrices H of the Lie algebra, as required.

They obey the relations

$$[T_a, T_b] = i \sum_{c=1}^{8} f_{abc} T_c$$

$$\{T_a, T_b\} = \frac{1}{3} \delta_{ab} + \sum_{c=1}^{8} d_{abc} T_c$$

$$\{\lambda_a, \lambda_b\} = \frac{4}{3} \delta_{ab} + 2 \sum_{c=1}^{8} d_{abc} \lambda_c$$

The f are the structure constants of the Lie algebra, given by:

$$f_{123} = 1$$

$$f_{147} = -f_{156} = f_{246} = f_{257} = f_{345} = -f_{367} = \frac{1}{2}$$

$$f_{458} = f_{678} = \frac{\sqrt{3}}{2},$$

while all other f_{abc} not related to these by permutation are zero.

The symmetric coefficients d take the values:

$$d_{118} = d_{228} = d_{338} = -d_{888} = \frac{1}{\sqrt{3}}$$

$$d_{448} = d_{558} = d_{668} = d_{778} = -\frac{1}{2\sqrt{3}}$$

$$d_{146} = d_{157} = -d_{247} = d_{256} = d_{344} = d_{355} = -d_{366} = -d_{377} = \frac{1}{2}.$$

As a topological space, *SU(3)* is a direct product of a 3-sphere and a 5-sphere, $S^3 \boxtimes S^5$.

A generic *SU(3)* group element generated by a traceless 3×3 hermitian matrix H, normalized as $\text{tr}(H^2) = 2$, is given by[4]

$$\exp(i\theta H) =$$

$$\left[-\frac{1}{3} I \sin(\phi + 2\pi/3) \sin(\phi - 2\pi/3) - \frac{1}{2\sqrt{3}} H \sin(\phi) - \frac{1}{4} H^2 \right] \frac{\exp\left(\frac{2}{\sqrt{3}} i\theta \sin\phi \right)}{\cos(\phi + 2\pi/3)\cos(\phi - 2\pi/3)}$$

$$+ \left[-\tfrac{1}{3} I \sin(\phi) \sin(\phi - 2\pi/3) - \tfrac{1}{2\sqrt{3}} H \sin(\phi + 2\pi/3) - \tfrac{1}{4} H^2 \right] \frac{\exp\left(\tfrac{2}{\sqrt{3}} i\theta \sin(\phi + 2\pi/3) \right)}{\cos(\phi) \cos(\phi - 2\pi/3)}$$

$$+ \left[-\tfrac{1}{3} I \sin(\phi) \sin(\phi + 2\pi/3) - \tfrac{1}{2\sqrt{3}} H \sin(\phi - 2\pi/3) - \tfrac{1}{4} H^2 \right] \frac{\exp\left(\tfrac{2}{\sqrt{3}} i\theta \sin(\phi - 2\pi/3) \right)}{\cos(\phi) \cos(\phi + 2\pi/3)}$$

where

$$\phi \equiv \tfrac{1}{3} \left(\arccos\left(\tfrac{3}{2}\sqrt{3} \det H \right) - \tfrac{\pi}{2} \right)$$

See also: Clebsch–Gordan coefficients for SU(3)

for elementary representation theory facts.

33.5   Lie algebra structure

The above representation bases generalize to $n > 3$, using generalized Pauli matrices.

If we choose an (arbitrary) particular basis, then the subspace of traceless diagonal $n \times n$ matrices with imaginary entries forms an $(n-1)$-dimensional Cartan subalgebra.

Complexify the Lie algebra, so that any traceless $n \times n$ matrix is now allowed. The weight eigenvectors are the Cartan subalgebra itself, as well as the matrices with only one nonzero entry which is off diagonal. Even though the Cartan subalgebra **h** is only $(n-1)$-dimensional, to simplify calculations, it is often convenient to introduce an auxiliary element, the unit matrix which commutes with everything else (which is not an element of the Lie algebra!) for the purpose of computing weights—and that only. So, we have a basis where the i-th basis vector is the matrix with 1 on the i-th diagonal entry and zero elsewhere. Weights would then be given by n coordinates and the sum over all n coordinates has to be zero (because the unit matrix is only auxiliary).

So, SU(n) is of rank $n-1$ and its Dynkin diagram is given by An_{-1}, a chain of $n-1$ vertices, o–o–o–o–⋯–o. Its root system consists of $n(n-1)$ roots spanning a $n-1$ Euclidean space. Here, we use n redundant coordinates instead of $n-1$ to emphasize the symmetries of the root system (the n coordinates have to add up to zero).

In other words, we are embedding this $n-1$ dimensional vector space in an n-dimensional one. Thus, the roots consists of all the $n(n-1)$ permutations of $(1, -1, 0, ..., 0)$. The construction given above explains why. A choice of simple roots is

$(1, -1, 0, ..., 0)$,

$(0, 1, -1, ..., 0)$,

$\cdots$

$(0, 0, 0, ..., 1, -1)$.

Its Cartan matrix is

$$\begin{pmatrix} 2 & -1 & 0 & \ldots & 0 \\ -1 & 2 & -1 & \ldots & 0 \\ 0 & -1 & 2 & \ldots & 0 \\ \vdots & \vdots & \vdots & \ddots & \vdots \\ 0 & 0 & 0 & \ldots & 2 \end{pmatrix}.$$

Its Weyl group or Coxeter group is the symmetric group Sn, the symmetry group of the $(n-1)$-simplex.

33.6 Generalized special unitary group

For a field F, the **generalized special unitary group over** F, SU(p, q; F), is the group of all linear transformations of determinant 1 of a vector space of rank $n = p + q$ over F which leave invariant a nondegenerate, Hermitian form of signature (p, q). This group is often referred to as the **special unitary group of signature** p q **over** F. The field F can be replaced by a commutative ring, in which case the vector space is replaced by a free module.

Specifically, fix a Hermitian matrix A of signature p q in GL(n, $\mathbf{R}$), then all

$$M \in \mathrm{SU}(p, q, R)$$

satisfy

$$M^* A M = A$$
$$\det M = 1.$$

Often one will see the notation SU(p, q) without reference to a ring or field; in this case, the ring or field being referred to is $\mathbf{C}$ and this gives one of the classical Lie groups. The standard choice for A when $F = \mathbf{C}$ is

$$A = \begin{bmatrix} 0 & 0 & i \\ 0 & I_{n-2} & 0 \\ -i & 0 & 0 \end{bmatrix}.$$

However there may be better choices for A for certain dimensions which exhibit more behaviour under restriction to subrings of $\mathbf{C}$.

33.6.1 Example

An important example of this type of group is the Picard modular group SU(2, 1; $\mathbf{Z}[i]$) which acts (projectively) on complex hyperbolic space of degree two, in the same way that SL(2,9;$\mathbf{Z}$) acts (projectively) on real hyperbolic space of dimension two. In 2005 Gábor Francsics and Peter Lax computed an explicit fundamental domain for the action of this group on HC^2.[5]

A further example is SU(1, 1; $\mathbf{C}$), which is isomorphic to SL(2.$\mathbf{R}$).

33.7 Important subgroups

In physics the special unitary group is used to represent bosonic symmetries. In theories of symmetry breaking it is important to be able to find the subgroups of the special unitary group. Subgroups of SU(n) that are important in GUT physics are, for $p > 1$, $n - p > 1$,

$$\mathrm{SU}(n) \supset \mathrm{SU}(p) \times \mathrm{SU}(n - p) \times \mathrm{U}(1)$$

where × denotes the direct product and U(1), known as the circle group, is the multiplicative group of all complex numbers with absolute value 1.

For completeness, there are also the orthogonal and symplectic subgroups,

$$\mathrm{SU}(n) \supset \mathrm{SO}(n),$$
$$\mathrm{SU}(2n) \supset \mathrm{Sp}(n).$$

Since the rank of SU(n) is $n-1$ and of U(1) is 1, a useful check is that the sum of the ranks of the subgroups is less than or equal to the rank of the original group. SU(n) is a subgroup of various other Lie groups,

$$SO(2n) \supset SU(n)$$
$$Sp(n) \supset SU(n)$$
$$Spin(4) = SU(2) \times SU(2)$$
$$E_6 \supset SU(6)$$
$$E_7 \supset SU(8)$$
$$G_2 \supset SU(3)$$

See spin group, and simple Lie groups for E_6, E_7, and G_2.

There are also the accidental isomorphisms: $SU(4) = Spin(6)$, $SU(2) = Spin(3) = Sp(1)$,[nb 3] and $U(1) = Spin(2) = SO(2)$.

One may finally mention that SU(2) is the double covering group of SO(3), a relation that plays an important role in the theory of rotations of 2-spinors in non-relativistic quantum mechanics.

33.8 See also

- Projective special unitary group, PSU(n)

- Generalizations of Pauli matrices

33.9 Remarks

[1] For a characterization of U(n) and hence SU(n) in terms of preservation of the standard inner product on C^n, see Classical group.

[2] For an explicit description of the homomorphism SU(2) → SO(3), see Connection between SO(3) and SU(2).

[3] Sp(n) is the compact real form of Sp(2n, C). It is sometimes denoted USp(2n). The dimension of the Sp(n)-matrices is $2n \times 2n$.

33.10 References

[1] Halzen, Francis; Martin, Alan (1984). *Quarks & Leptons: An Introductory Course in Modern Particle Physics*. John Wiley & Sons. ISBN 0-471-88741-2.

[2] Wybourne, B G (1974). *Classical Groups for Physicists*. Wiley-Interscience. ISBN 0471965057 .

[3] R.R. Puri, *Mathematical Methods of Quantum Optics*, Springer, 2001.

[4] Rosen, S P (1971). "Finite Transformations in Various Representations of SU(3)". *Journal of Mathematical Physics* 12 (4): 673. doi:10.1063/1.1665634. ISSN 0022-2488.; Curtright, T L; Zachos, C K (2015). "Elementary results for the fundamental representation of SU(3)". *Researchgate*. doi:10.13140/RG.2.1.1743.2163.

[5] Francsics, Gabor; Lax, Peter D. "An Explicit Fundamental Domain For The Picard Modular Group In Two Complex Dimensions". arXiv:math/0509708v1.

Chapter 34

Symmetry in quantum mechanics

Symmetries in quantum mechanics describe features of spacetime and particles which are unchanged under some transformation, in the context of quantum mechanics, relativistic quantum mechanics and quantum field theory, with applications in the mathematical formulation of the standard model and condensed matter physics. In general, symmetry in physics, invariance, and conservation laws, are fundamentally important constraints for formulating physical theories and models. In practice; they are powerful methods for solving problems and predicting what could happen. While conservation laws do not always give the answer to the problem directly and alone, they form the correct constraints and the first steps to solving the problem.

This article outlines the connection between the classical form of continuous symmetries as well as their quantum operators, and relates them to the Lie groups, and relativistic transformations in the Lorentz group, and Poincaré group.

34.1 Notation

The notational conventions used in this article are as follows. Boldface indicates vectors, four vectors, matrices, and vectorial operators, while quantum states use bra–ket notation. Wide hats are for operators, narrow hats are for unit vectors (including their components in tensor index notation). The summation convention on the repeated tensor indices is used, unless stated otherwise. The Minkowski metric signature is (+−−−).

34.2 Symmetry transformations on the wavefunction in non-relativistic quantum mechanics

34.2.1 Continuous symmetries

Generally, the correspondence between continuous symmetries and conservation laws is given by Noether's theorem.

The form of the fundamental quantum operators, for example energy as a partial time derivative and momentum as a spatial gradient, becomes clear when one considers the initial state, then changes one parameter of it slightly. This can be done for displacements (lengths), durations (time), and angles (rotations). Additionally, the invariance of certain quantities can be seen by making such changes in lengths and angles, which illustrates conservation of these quantities.

In what follows, transformations on only one-particle wavefunctions in the form:

$$\widehat{\Omega}\psi(\mathbf{r}, t) = \psi(\mathbf{r}', t')$$

are considered, where $\widehat{\Omega}$ denotes a unitary operator. Unitarity is generally required for operators representing transformations of space, time, and spin, since the norm of a state (representing the total probability of finding the particle somewhere with some spin) must be invariant under these transformations. The inverse is the Hermitian conjugate $\hat{\Omega}^{-1} = \hat{\Omega}^{\dagger}$. The results can be extended to many-particle wavefunctions. Written in Dirac notation as standard, the transformations on quantum state vectors are:

$$\hat{\Omega}\,|\mathbf{r}(t)\rangle = |\mathbf{r}'(t')\rangle$$

Now, the action of $\hat{\Omega}$ changes $\psi(\mathbf{r}, t)$ to $\psi(\mathbf{r}', t')$, so the inverse $\hat{\Omega} = \hat{\Omega}^{\dagger}$ changes $\psi(\mathbf{r}', t')$ back to $\psi(\mathbf{r}, t)$, so an operator $\hat{A}$ invariant under $\hat{\Omega}$ satisfies:

$$\hat{A}\psi = \hat{\Omega}^{\dagger}\hat{A}\hat{\Omega}\psi \quad \Rightarrow \quad \hat{\Omega}\hat{A}\psi = \hat{A}\hat{\Omega}\psi$$

and thus:

$$[\hat{\Omega}, \hat{A}]\psi = 0$$

for any state ψ. Quantum operators representing observables are also required to be Hermitian so that their eigenvalues are real numbers, i.e. the operator equals its Hermitian conjugate, $\hat{A} = \hat{A}^{\dagger}$.

34.2.2   Overview of Lie group theory

For a full exposition and details, see Lie group and Generator (mathematics).

Following are the key points of group theory relevant to quantum theory, examples are given throughout the article. For an alternative approach using matrix groups, see the books of Hall[1][2]

Let G be a *Lie group*, which is a group parameterized by a finite number N of real continuously varying parameters ξ_1, ξ_2, ... ξN.

- the *dimension of the group*, N, is the number of parameters it has.

- the *group elements*, g, in G are functions of the parameters:

    $$g = G(\xi_1, \xi_2, \cdots)$$

    and all parameters set to zero returns the *identity element* of the group:

    $$I = G(0, 0 \cdots)$$

    Group elements are often matrices which act on vectors, or transformations acting on functions.

- The *generators of the group* are the partial derivatives of the group elements with respect to the group parameters with the result evaluated when the parameter is set to zero:

    $$X_j = \left.\frac{\partial g}{\partial \xi_j}\right|_{\xi_j = 0}$$

One aspect of generators in theoretical physics is they can be construed themselves as operators corresponding to symmetries, which may be written as matrices, or as differential operators. In quantum theory, for unitary representations of the group, the generators require a factor of i:

$$X_j = i \left. \frac{\partial g}{\partial \xi_j} \right|_{\xi_j = 0}$$

The generators of the group form a vector space, which means linear combinations of generators also form a generator.

- The generators (whether matrices or differential operators) satisfy the *commutator*:

$$[X_a, X_b] = i f_{abc} X_c$$

where f_{abc} are the (basis dependent) *structure constants* of the group. This makes, together with the vector space property, the set of all generators of a group a Lie algebra. Due to the antisymmetry of the bracket, the structure constants of the group are antisymmetric in the first two indices.

- The *representations of the group* are denoted using a capital D and defined by:

$$D[g(\xi_j)] \equiv D(\xi_j) = e^{i \xi_j D(X_j)}$$

without summation on the repeated index j. Representations are linear operators that take in group elements and preserve the composition rule:

$$D(\xi_a) D(\xi_b) = D(\xi_a \xi_b).$$

A representation which cannot be decomposed into a direct sum of other representations, is called *irreducible*. It is conventional to label irreducible representations by a superscripted number n in brackets, as in $D^{(n)}$, or if there is more than one number, we write $D^{(n, m, \dots)}$.

Representations also exist for the generators and the same notation of a capital D is used in this context: $D(X)$. The D in the representation of a generator $D(X)$ is not the same mapping as the D in a representation of a group element, nevertheless this notational abuse of using the same letter to denote two different mappings is used in the literature. An example of this abuse is to be found in the defining equation above.

34.2.3 Momentum and energy as generators of translation and time evolution, and rotation

The space translation operator $\widehat{T}(\Delta \mathbf{r})$ acts on a wavefunction to shift the space coordinates by an infinitesimal displacement $\Delta \mathbf{r}$. The explicit expression $\widehat{T}$ can be quickly determined by a Taylor expansion of $\psi(\mathbf{r} + \Delta \mathbf{r}, t)$ about $\mathbf{r}$, then (keeping the first order term and neglecting second and higher order terms), replace the space derivatives by the momentum operator $\widehat{\mathbf{p}}$. Similarly for the time translation operator acting on the time parameter, the Taylor expansion of $\psi(\mathbf{r}, t + \Delta t)$ is about t, and the time derivative replaced by the energy operator $\widehat{E}$.

The exponential functions arise by definition as those limits, due to Euler, and can be understood physically and mathematically as follows. A net translation can be composed of many small translations, so to obtain the translation operator for a finite increment, replace $\Delta \mathbf{r}$ by $\Delta \mathbf{r}/N$ and Δt by $\Delta t/N$, where N is a positive non-zero integer. Then as N increases, the magnitude of $\Delta \mathbf{r}$ and Δt become even smaller, while leaving the directions unchanged. Acting the infinitesimal operators on the wavefunction N times and taking the limit as N tends to infinity gives the finite operators.

Space and time translations commute, which means the operators and generators commute.

For a time-independent Hamiltonian, energy is conserved in time and quantum states are stationary states: the eigenstates of the Hamiltonian are the energy eigenvalues E:

$$\hat{U}(t) = \exp\left(-\frac{i\Delta t E}{\hbar}\right)$$

and all stationary states have the form

$$\psi(\mathbf{r}, t + t_0) = \hat{U}(t - t_0)\psi(\mathbf{r}, t_0)$$

where t_0 is the initial time, usually set to zero since there is no loss of continuity when the initial time is set. An alternative notation is $\hat{U}(t - t_0) \equiv U(t, t_0)$.

34.2.4 Angular momentum as the generator of rotations

Orbital angular momentum

The rotation operator acts on a wavefunction to rotate the spatial coordinates of a particle by a constant angle $\Delta\theta$:

$$R(\Delta\theta, \hat{\mathbf{a}})\psi(\mathbf{r}, t) = \psi(\mathbf{r}', t)$$

where $\mathbf{r}'$ are the rotated coordinates about an axis defined by a unit vector $\hat{\mathbf{a}} = (a_1, a_2, a_3)$ through an angular increment $\Delta\theta$, given by:

$$\mathbf{r}' = \hat{R}(\Delta\theta, \hat{\mathbf{a}})\mathbf{r}.$$

where $\hat{R}(\Delta\theta, \hat{\mathbf{a}})$ is a rotation matrix dependent on the axis and angle. In group theoretic language, the rotation matrices are group elements, and the angles and axis $\Delta\theta\hat{\mathbf{a}} = \Delta\theta(a_1, a_2, a_3)$ are the parameters, of the three-dimensional special orthogonal group, SO(3). The rotation matrices about the standard Cartesian basis vector $\hat{\mathbf{e}}_x, \hat{\mathbf{e}}_y, \hat{\mathbf{e}}_z$ through angle $\Delta\theta$, and the corresponding generators of rotations $\mathbf{J} = (Jx, Jy, Jz)$, are:

More generally for rotations about an axis defined by $\hat{\mathbf{a}}$, the rotation matrix elements are:[3]

$$[\hat{R}(\theta, \hat{\mathbf{a}})]_{ij} = (\delta_{ij} - a_i a_j)\cos\theta - \varepsilon_{ijk}a_k \sin\theta + a_i a_j$$

where δ_{ij} is the Kronecker delta, and ε_{ijk} is the Levi-Civita symbol.

It is not as obvious how to determine the rotational operator compared to space and time translations. We may consider a special case (rotations about the x, y, or z-axis) then infer the general result, or use the general rotation matrix directly and tensor index notation with δ_{ij} and ε_{ijk}. To derive the infinitesimal rotation operator, which corresponds to small $\Delta\theta$, we use the small angle approximations $\sin(\Delta\theta) \approx \Delta\theta$ and $\cos(\Delta\theta) \approx 1$, then Taylor expand about $\mathbf{r}$ or ri, keep the first order term, and substitute the angular momentum operator components.

The z-component of angular momentum can be replaced by the component along the axis defined by $\hat{\mathbf{a}}$, using the dot product $\hat{\mathbf{a}} \cdot \widehat{\mathbf{L}}$.

Again, a finite rotation can be made from lots of small rotations, replacing $\Delta\theta$ by $\Delta\theta/N$ and taking the limit as N tends to infinity gives the rotation operator for a finite rotation.

Rotations about the *same* axis do commute, for example a rotation through angles θ_1 and θ_2 about axis i can be written

$$R(\theta_1 + \theta_2, \mathbf{e}_i) = R(\theta_1 \mathbf{e}_i)R(\theta_2 \mathbf{e}_i), \quad [R(\theta_1 \mathbf{e}_i), R(\theta_2 \mathbf{e}_i)] = 0.$$

However, rotations about *different* axes do not commute. The general commutation rules are summarized by

$$[L_i, L_j] = i\hbar\varepsilon_{ijk}L_k.$$

In this sense, orbital angular momentum has the common sense properties of rotations. Each of the above commutators can be easily demonstrated by holding an everyday object and rotating it through the same angle about any two different axes in both possible orderings; the final configurations are different.

In quantum mechanics, there is another form of rotation which mathematically appears similar to the orbital case, but has different properties, described next.

Spin angular momentum

All previous quantities have classical definitions. Spin is a quantity possessed by particles in quantum mechanics without any classical analogue, having the units of angular momentum. The spin vector operator is denoted $\widehat{\mathbf{S}} = (\widehat{S_x}, \widehat{S_y}, \widehat{S_z})$. The eigenvalues of its components are the possible outcomes (in units of $\hbar$) of a measurement of the spin projected onto one of the basis directions.

Rotations (of ordinary space) about an axis $\hat{\mathbf{a}}$ through angle θ about the unit vector $\hat{a}$ in space acting on a multicomponent wave function (spinor) at a point in space is represented by:

However, unlike orbital angular momentum in which the z-projection quantum number ℓ can only take positive or negative integer values (including zero), the z-projection spin quantum number s can take all positive and negative half-integer values. There are rotational matrices for each spin quantum number.

Evaluating the exponential for a given z-projection spin quantum number s gives a $(2s + 1)$-dimensional spin matrix. This can be used to define a spinor as a column vector of $2s + 1$ components which transforms to a rotated coordinate system according to the spin matrix at a fixed point in space.

For the simplest non-trivial case of $s = 1/2$, the spin operator is given by

$$\widehat{\mathbf{S}} = \frac{\hbar}{2}\sigma$$

where the Pauli matrices in the standard representation are:

$$\sigma_1 = \sigma_x = \begin{pmatrix} 0 & 1 \\ 1 & 0 \end{pmatrix}, \quad \sigma_2 = \sigma_y = \begin{pmatrix} 0 & -i \\ i & 0 \end{pmatrix}, \quad \sigma_3 = \sigma_z = \begin{pmatrix} 1 & 0 \\ 0 & -1 \end{pmatrix}$$

Total angular momentum

The total angular momentum operator is the sum of the orbital and spin

$$\hat{\mathbf{J}} = \hat{\mathbf{L}} + \hat{\mathbf{S}}$$

and is an important quantity for multi-particle systems, especially in nuclear physics and the quantum chemistry of multi-electron atoms and molecules.

We have a similar rotation matrix:

$$\hat{J}(\theta, \hat{\mathbf{a}}) = \exp\left(-\frac{i}{\hbar}\theta\hat{\mathbf{a}} \cdot \hat{\mathbf{J}}\right)$$

34.3 Lorentz group in relativistic quantum mechanics

Following is an overview of the Lorentz group; a treatment of boosts and rotations in spacetime. Throughout this section, see (for example) T. Ohlsson (2011)[4] and E. Abers (2004).[5]

Lorentz transformations can be parametrized by rapidity φ for a boost in the direction of a three-dimensional unit vector $\hat{\mathbf{n}} = (n_1, n_2, n_3)$, and a rotation angle θ about a three-dimensional unit vector $\hat{\mathbf{a}} = (a_1, a_2, a_3)$ defining an axis, so $\varphi\hat{\mathbf{n}} = \varphi(n_1, n_2, n_3)$ and $\theta\hat{\mathbf{a}} = \theta(a_1, a_2, a_3)$ are together six parameters of the Lorentz group (three for rotations and three for boosts). The Lorentz group is 6-dimensional.

34.3.1 Pure rotations in spacetime

The rotation matrices and rotation generators considered above form the spacelike part of a four-dimensional matrix, representing pure-rotation Lorentz transformations. Three of the Lorentz group elements $\hat{R}_x$, $\hat{R}_y$, $\hat{R}_z$ and generators $\mathbf{J} = (J_1, J_2, J_3)$ for pure rotations are:

The rotation matrices act on any four vector $\mathbf{A} = (A_0, A_1, A_2, A_3)$ and rotate the space-like components according to

$$\mathbf{A}' = \hat{R}(\Delta\theta, \hat{\mathbf{n}})\mathbf{A}$$

leaving the time-like coordinate unchanged. In matrix expressions, $\mathbf{A}$ is treated as a column vector.

34.3.2 Pure boosts in spacetime

A boost with velocity $c\tanh\varphi$ in the x, y, or z directions given by the standard Cartesian basis vector $\hat{\mathbf{e}}_x$, $\hat{\mathbf{e}}_y$, $\hat{\mathbf{e}}_z$, are the boost transformation matrices. These matrices $\hat{B}_x$, $\hat{B}_y$, $\hat{B}_z$ and the corresponding generators $\mathbf{K} = (K_1, K_2, K_3)$ are the remaining three group elements and generators of the Lorentz group:

The boost matrices act on any four vector $\mathbf{A} = (A_0, A_1, A_2, A_3)$ and mix the time-like and the space-like components, according to:

$$\mathbf{A}' = \widehat{B}(\varphi, \hat{\mathbf{n}})\mathbf{A}$$

The term "boost" refers to the relative velocity between two frames, and is not to be conflated with momentum as the *generator of translations*, as explained below.

34.3.3 Combining boosts and rotations

Products of rotations give another rotation (a frequent exemplification of a subgroup), while products of boosts and boosts or of rotations and boosts cannot be expressed as pure boosts or pure rotations. In general, any Lorentz transformation can be expressed as a product of a pure rotation and a pure boost. For more background see (for example) B.R. Durney (2011)[6] and H.L. Berk et al.[7] and references therein.

The boost and rotation generators have representations denoted $D(\mathbf{K})$ and $D(\mathbf{J})$ respectively, the capital D in this context indicates a group representation.

For the Lorentz group, the representations $D(\mathbf{K})$ and $D(\mathbf{J})$ of the generators $\mathbf{K}$ and $\mathbf{J}$ fulfill the following commutation rules.

In all commutators, the boost entities mixed with those for rotations, although rotations alone simply give another rotation. Exponentiating the generators gives the boost and rotation operators which combine into the general Lorentz transformation, under which the spacetime coordinates transform from one rest frame to another boosted and/or rotating frame. Likewise, exponentiating the representations of the generators gives the representations of the boost and rotation operators, under which a particle's spinor field transforms.

In the literature, the boost generators $\mathbf{K}$ and rotation generators $\mathbf{J}$ are sometimes combined into one generator for Lorentz transformations $\mathbf{M}$, an antisymmetric four-dimensional matrix with entries:

$$M^{0a} = -M^{a0} = K_a, \quad M^{ab} = \varepsilon_{abc}J_c.$$

and correspondingly, the boost and rotation parameters are collected into another antisymmetric four-dimensional matrix $\boldsymbol{\omega}$, with entries:

$$\omega_{0a} = -\omega_{a0} = \varphi n_a, \quad \omega_{ab} = \theta\varepsilon_{abc}a_c.$$

The general Lorentz transformation is then:

$$\Lambda(\varphi, \hat{\mathbf{n}}, \theta, \hat{\mathbf{a}}) = \exp\left(-\frac{i}{2}\omega_{\alpha\beta}M^{\alpha\beta}\right) = \exp\left[-\frac{i}{2}\left(\varphi\hat{\mathbf{n}}\cdot\mathbf{K} + \theta\hat{\mathbf{a}}\cdot\mathbf{J}\right)\right]$$

with summation over repeated matrix indices α and β. The Λ matrices act on any four vector $\mathbf{A} = (A_0, A_1, A_2, A_3)$ and mix the time-like and the space-like components, according to:

$$\mathbf{A}' = \Lambda(\varphi, \hat{\mathbf{n}}, \theta, \hat{\mathbf{a}})\mathbf{A}$$

34.3.4 Transformations of spinor wavefunctions in relativistic quantum mechanics

In relativistic quantum mechanics, wavefunctions are no longer single-component scalar fields, but now $2(2s + 1)$ component spinor fields, where s is the spin of the particle. The transformations of these functions in spacetime are given below.

Under a proper orthochronous Lorentz transformation $(\mathbf{r}, t) \to \Lambda(\mathbf{r}, t)$ in Minkowski space, all one-particle quantum states ψ_σ locally transform under some representation D of the Lorentz group:[8] [9]

$$\psi_\sigma(\mathbf{r}, t) \to D(\Lambda)\psi_\sigma(\Lambda^{-1}(\mathbf{r}, t))$$

where $D(\Lambda)$ is a finite-dimensional representation, in other words a $(2s + 1) \times (2s + 1)$ dimensional square matrix, and ψ is thought of as a column vector containing components with the $(2s + 1)$ allowed values of σ:

$$\psi(\mathbf{r}, t) = \begin{bmatrix} \psi_{\sigma=s}(\mathbf{r}, t) \\ \psi_{\sigma=s-1}(\mathbf{r}, t) \\ \vdots \\ \psi_{\sigma=-s+1}(\mathbf{r}, t) \\ \psi_{\sigma=-s}(\mathbf{r}, t) \end{bmatrix} \quad \rightleftharpoons \quad \psi(\mathbf{r}, t)^\dagger = \begin{bmatrix} \psi_{\sigma=s}(\mathbf{r}, t)^* & \psi_{\sigma=s-1}(\mathbf{r}, t)^* & \cdots & \psi_{\sigma=-s+1}(\mathbf{r}, t)^* & \psi_{\sigma=-s}(\mathbf{r}, t)^* \end{bmatrix}$$

34.3.5 Real irreducible representations and spin

The irreducible representations of $D(\mathbf{K})$ and $D(\mathbf{J})$, in short "irreps", can be used to build to spin representations of the Lorentz group. Defining new operators:

$$\mathbf{A} = \frac{\mathbf{J} + i\mathbf{K}}{2}, \quad \mathbf{B} = \frac{\mathbf{J} - i\mathbf{K}}{2},$$

so $\mathbf{A}$ and $\mathbf{B}$ are simply complex conjugates of each other, it follows they satisfy the symmetrically formed commutators:

$$[A_i, A_j] = \varepsilon_{ijk}A_k, \quad [B_i, B_j] = \varepsilon_{ijk}B_k, \quad [A_i, B_j] = 0,$$

and these are essentially the commutators the orbital and spin angular momentum operators satisfy. Therefore $\mathbf{A}$ and $\mathbf{B}$ form operator algebras analogous to angular momentum; same ladder operators, z-projections, etc., independently of each other as each of their components mutually commute. By the analogy to the spin quantum number, we can introduce positive integers or half integers, a, b, with corresponding sets of values $m = a, a - 1, \ldots -a + 1, -a$ and $n = b, b - 1, \ldots -b + 1, -b$. The matrices satisfying the above commutation relations are the same as for spins a and b have components given by multiplying Kronecker delta values with angular momentum matrix elements:

$$(A_x)_{m'n', mn} = \delta_{n'n}\left(J_x^{(m)}\right)_{m'm} \qquad (B_x)_{m'n', mn} = \delta_{m'm}\left(J_x^{(n)}\right)_{n'n}$$

$$(A_y)_{m'n', mn} = \delta_{n'n}\left(J_y^{(m)}\right)_{m'm} \qquad (B_y)_{m'n', mn} = \delta_{m'm}\left(J_y^{(n)}\right)_{n'n}$$

$$(A_z)_{m'n', mn} = \delta_{n'n}\left(J_z^{(m)}\right)_{m'm} \qquad (B_z)_{m'n', mn} = \delta_{m'm}\left(J_z^{(n)}\right)_{n'n}$$

where in each case the row number $m'n'$ and column number mn are separated by a comma, and in turn:

$$\left(J_z^{(m)}\right)_{m'm} = m\delta_{m'm} \qquad \left(J_x^{(m)} \pm iJ_y^{(m)}\right)_{m'm} = m\delta_{a', a\pm 1}\sqrt{(a \mp m)(a \pm m + 1)}$$

and similarly for $\mathbf{J}^{(n)}$.[note 1] The three $\mathbf{J}^{(m)}$ matrices are each $(2m + 1) \times (2m + 1)$ square matrices, and the three $\mathbf{J}^{(n)}$ are each $(2n + 1) \times (2n + 1)$ square matrices. The integers or half-integers m and n numerate all the irreducible representations by, in equivalent notations used by authors: $D^{(m, n)} \equiv (m, n) \equiv D^{(m)} \otimes D^{(n)}$, which are each $[(2m + 1)(2n + 1)] \times [(2m + 1)(2n + 1)]$ square matrices.

Applying this to particles with spin s;

- left-handed $(2s + 1)$-component spinors transform under the real irreps $D^{(s, 0)}$,

- right-handed $(2s + 1)$-component spinors transform under the real irreps $D^{(0, s)}$,

- taking direct sums symbolized by $\oplus$ (see direct sum of matrices for the simpler matrix concept), one obtains the representations under which $2(2s + 1)$-component spinors transform: $D^{(m, n)} \oplus D^{(n, m)}$ where $m + n = s$. These are also real irreps, but as shown above, they split into complex conjugates.

In these cases the D refers to any of $D(\mathbf{J})$, $D(\mathbf{K})$, or a full Lorentz transformation $D(\Lambda)$.

34.3.6   Relativistic wave equations

In the context of the Dirac equation and Weyl equation, the Weyl spinors satisfying the Weyl equation transform under the simplest irreducible spin representations of the Lorentz group, since the spin quantum number in this case is the smallest non-zero number allowed: 1/2. The 2-component left-handed Weyl spinor transforms under $D^{(1/2, 0)}$ and the 2-component right-handed Weyl spinor transforms under $D^{(0, 1/2)}$. Dirac spinors satisfying the Dirac equation transform under the representation $D^{(1/2, 0)} \oplus D^{(0, 1/2)}$, the direct sum of the irreps for the Weyl spinors.

34.4   The Poincaré group in relativistic quantum mechanics and field theory

Space translations, time translations, rotations, and boosts, all taken together, constitute the Poincaré group. The group elements are the three rotation matrices and three boost matrices (as in the Lorentz group), and one for time translations and three for space translations in spacetime. There is a generator for each. Therefore the Poincaré group is 10-dimensional.

In special relativity, space and time can be collected into a four-position vector $\mathbf{X} = (ct, -\mathbf{r})$, and in parallel so can energy and momentum which combine into a four-momentum vector $\mathbf{P} = (E/c, -\mathbf{p})$. With relativistic quantum mechanics in mind, the time duration and spatial displacement parameters (four in total, one for time and three for space) combine into a spacetime displacement $\Delta\mathbf{X} = (c\Delta t, -\Delta\mathbf{r})$, and the energy and momentum operators are inserted in the four-momentum to obtain a four-momentum operator,

$$\widehat{\mathbf{P}} = \left(\frac{\widehat{E}}{c}, -\widehat{\mathbf{p}} \right) = i\hbar \left(\frac{1}{c} \frac{\partial}{\partial t}, \nabla \right) ,$$

which are the generators of spacetime translations (four in total, one time and three space):

$$\widehat{X}(\Delta\mathbf{X}) = \exp\left(-\frac{i}{\hbar} \Delta\mathbf{X} \cdot \widehat{\mathbf{P}} \right) = \exp\left[-\frac{i}{\hbar} \left(\Delta t \widehat{E} + \Delta\mathbf{r} \cdot \widehat{\mathbf{p}} \right) \right] .$$

There are commutation relations between the components four-momentum $\mathbf{P}$ (generators of spacetime translations), and angular momentum $\mathbf{M}$ (generators of Lorentz transformations), that define the Poincaré algebra:[10][11]

- $[P_\mu, P_\nu] = 0$

- $\frac{1}{i}[M_{\mu\nu}, P_\rho] = \eta_{\mu\rho} P_\nu - \eta_{\nu\rho} P_\mu$

- $\frac{1}{i}[M_{\mu\nu}, M_{\rho\sigma}] = \eta_{\mu\rho} M_{\nu\sigma} - \eta_{\mu\sigma} M_{\nu\rho} - \eta_{\nu\rho} M_{\mu\sigma} + \eta_{\nu\sigma} M_{\mu\rho}$

where η is the Minkowski metric tensor. (It is common to drop any hats for the four-momentum operators in the commutation relations). These equations are an expression of the fundamental properties of space and time as far as they are known today. They have a classical counterpart where the commutators are replaced by Poisson brackets.

To describe spin in relativistic quantum mechanics, the Pauli–Lubanski pseudovector

$$W_\mu = \frac{1}{2}\varepsilon_{\mu\nu\rho\sigma} J^{\nu\rho} P^\sigma,$$

a Casimir operator, is the constant spin contribution to the total angular momentum, and there are commutation relations between **P** and **W** and between **M** and **W**:

$$[P^\mu, W^\nu] = 0,$$

$$[J^{\mu\nu}, W^\rho] = i\left(\eta^{\rho\nu} W^\mu - \eta^{\rho\mu} W^\nu\right),$$

$$[W_\mu, W_\nu] = -i\epsilon_{\mu\nu\rho\sigma} W^\rho P^\sigma.$$

Invariants constructed from **W**, instances of Casimir invariants can be used to classify irreducible representations of the Lorentz group.

34.5 Symmetries in quantum field theory and particle physics

34.5.1 Unitary groups in quantum field theory

Group theory is an abstract way of mathematically analyzing symmetries. Unitary operators are paramount to quantum theory, so unitary groups are important in particle physics. The group of N dimensional unitary square matrices is denoted U(N). Unitary operators preserve inner products which means probabilities are also preserved, so the quantum mechanics of the system is invariant under unitary transformations. Let $\hat{U}$ be a unitary operator, so the inverse is the Hermitian adjoint $\hat{U} = \hat{U}^\dagger$, which commutes with the Hamiltonian:

$$\left[\hat{U}, \hat{H}\right] = 0$$

then the observable corresponding to the operator $\hat{U}$ is conserved, and the Hamiltonian is invariant under the transformation $\hat{U}$.

Since the predictions of quantum mechanics should be invariant under the action of a group, physicists look for unitary transformations to represent the group.

Important subgroups of each U(N) are those unitary matrices which have unit determinant (or are "unimodular"): these are called the special unitary groups and are denoted SU(N).

U(1) and SU(1)

The simplest unitary group is U(1), which is just a complex number of modulus 1. This one-dimensional matrix entry is of the form:

$$U = e^{-i\theta}$$

in which θ is the parameter of the group, and the group is Abelian since one-dimensional matrices always commute under matrix multiplication. Lagrangians in quantum field theory for complex scalar fields are often invariant under U(1) transformations. If there is a quantum number a associated with the U(1) symmetry, for example baryon and the three lepton numbers in electromagnetic interactions, we have:

$$U = e^{-ia\theta}$$

U(2) and SU(2)

The general form of an element of a U(2) element is parametrized by two complex numbers a and b:

$$U = \begin{pmatrix} a & b \\ -b^* & a^* \end{pmatrix}$$

and for SU(2), the determinant is restricted to 1:

$$\det(U) = aa^* + bb^* = |a|^2 + |b|^2 = 1$$

In group theoretic language, the Pauli matrices are the generators of the special unitary group in two dimensions, denoted SU(2). Their commutation relation is the same as for orbital angular momentum, aside from a factor of 2:

$$[\sigma_a, \sigma_b] = 2i\hbar\varepsilon_{abc}\sigma_c$$

A group element of SU(2) can be written:

$$U(\theta, \hat{\mathbf{e}}_j) = e^{i\theta\sigma_j/2}$$

where σj is a Pauli matrix, and the group parameters are the angles turned through about an axis.

U(3) and SU(3)

The eight Gell-Mann matrices λn (see article for them and the structure constants) are important for quantum chromodynamics. They originally arose in the theory SU(3) of flavor which is still of practical importance in nuclear physics. They are the generators for the SU(3) group, so an element of SU(3) can be written analogously to an element of SU(2):

$$U(\theta, \hat{\mathbf{e}}_j) = \exp\left(-\frac{i}{2}\sum_{n=1}^{8}\theta_n\lambda_n\right)$$

where θn are eight independent parameters. The λn matrices satisfy the commutator:

$$[\lambda_a, \lambda_b] = 2if_{abc}\lambda_c$$

where the indices a, b, c take the values 1, 2, 3... 8. The structure constants f_{abc} are totally antisymmetric in all indices analogous to those of SU(2). In the standard colour charge basis (r for red, g for green, b for blue):

$$|r\rangle = \begin{pmatrix} 1 \\ 0 \\ 0 \end{pmatrix}, \quad |g\rangle = \begin{pmatrix} 0 \\ 1 \\ 0 \end{pmatrix}, \quad |b\rangle = \begin{pmatrix} 0 \\ 0 \\ 1 \end{pmatrix}$$

the colour states are eigenstates of the λ_3 and λ_8 matrices, while the other matrices mix colour states together.

The eight gluons states (8-dimensional column vectors) are simultaneous eigenstates of the adjoint representation of SU(3), the 8-dimensional representation acting on its own Lie algebra su(3), for the λ_3 and λ_8 matrices. By forming tensor products of representations (the standard representation and its dual) and taking appropriate quotients, protons and neutrons, and other hadrons are eigenstates of various representations of SU(3) of color. The adjoint representation of above is isomorphic to the tensor product of the standard representation and its dual.

34.5.2 Matter and antimatter

In relativistic quantum mechanics, relativistic wave equations predict a remarkable symmetry of nature: that every particle has a corresponding antiparticle. This is mathematically contained in the spinor fields which are the solutions of the relativistic wave equations.

Charge conjugation switches particles and antiparticles. Physical laws and interactions unchanged by this operation have C symmetry.

34.5.3 Discrete spacetime symmetries

- Parity mirrors the orientation of the spatial coordinates from left-handed to right-handed. Informally, space is "reflected" into its mirror image. Physical laws and interactions unchanged by this operation have P symmetry.

- Time reversal negates the time coordinate, which amounts to time running from future to past. A curious property of time, which space does not have, is that it is unidirectional: particles traveling forwards in time are equivalent to antiparticles traveling back in time. Physical laws and interactions unchanged by this operation have T symmetry.

34.5.4 *C, P, T* symmetries

- CPT theorem

- CP violation

- Lorentz violation

34.5.5 Gauge theory

Main article: Gauge theory

In quantum electrodynamics, the symmetry group is U(1) and is abelian. In quantum chromodynamics, the symmetry group is SU(3) and is non-abelian.

The electromagnetic interaction is mediated by photons, which have no electric charge. The electromagnetic tensor has an electromagnetic four-potential field possessing gauge symmetry.

The strong (color) interaction is mediated by gluons, which can have eight color charges. There are eight gluon field strength tensors with corresponding gluon four potentials field, each possessing gauge symmetry.

34.5.6   The strong (color) interaction

Color charge

Analogous to the spin operator, there are color charge operators in terms of the Gell-Mann matrices λ_j:

$$\hat{F}_j = \frac{1}{2}\lambda_j$$

and since color charge is a conserved charge, all color charge operators must commute with the Hamiltonian:

$$\left[\hat{F}_j, \hat{H}\right] = 0$$

Isospin

Isospin is conserved in strong interactions.

34.5.7   The weak and electromagnetic interactions

Duality transformation

Magnetic monopoles can be theoretically realized, although current observations and theory are consistent with them existing or not existing. Electric and magnetic charges can effectively be "rotated into one another" by a duality transformation.

Electroweak symmetry

- Electroweak symmetry

- Electroweak symmetry breaking

34.5.8   Supersymmetry

Main article: Supersymmetry

A Lie superalgebra is an algebra in which (suitable) basis elements either have a commutation relation or have an anti-commutation relation. Symmetries have been proposed to the effect that all fermionic particles have bosonic analogues, and vice versa. These symmetry have theoretical appeal in that no extra assumptions (such as existence of strings) barring symmetries are made. In addition, by assuming supersymmetry, a number puzzling issues can be resolved. These symmetries, which are represented by Lie superalgebras, have not been confirmed experimentally. It is now believed that they are broken symmetries, if they exist. But it has been speculated that dark matter is constitutes gravitinos, a spin 3/2 particle with mass, its supersymmetric partner being the graviton.

34.6   Exchange symmetry

See also: Exchange interaction, Identical particles and Holstein–Herring method

The concept of **exchange symmetry** is derived from a fundamental postulate of quantum statistics, which states that no observable physical quantity should change after exchanging two identical particles. It states that because all observables are proportional to $|\psi|^2$ for a system of identical particles, the wave function ψ must either remain the same or change sign upon such an exchange.

Because the exchange of two identical particles is mathematically equivalent to the rotation of each particle by 180 degrees (and so to the rotation of one particle's frame by 360 degrees),[12] the symmetric nature of the wave function depends on the particle's spin after the rotation operator is applied to it. Integer spin particles do not change the sign of their wave function upon a 360 degree rotation—therefore the sign of the wave function of the entire system does not change. Semi-integer spin particles change the sign of their wave function upon a 360 degree rotation (see more in spin–statistics theorem).

Particles for which the wave function does not change sign upon exchange are called bosons, or particles with a symmetric wave function. The particles for which the wave function of the system changes sign are called fermions, or particles with an antisymmetric wave function.

Fermions therefore obey different statistics (called Fermi–Dirac statistics) than bosons (which obey Bose–Einstein statistics). One of the consequences of Fermi–Dirac statistics is the exclusion principle for fermions—no two identical fermions can share the same quantum state (in other words, the wave function of two identical fermions in the same state is zero). This in turn results in degeneracy pressure for fermions—the strong resistance of fermions to compression into smaller volume. This resistance gives rise to the "stiffness" or "rigidity" of ordinary atomic matter (as atoms contain electrons which are fermions).

34.7 See also

- Casimir operator

- Pauli–Lubanski pseudovector

- Symmetries in general relativity

- Renormalization group

- Center of mass (relativistic)

- Representation of a Lie group

- Representation theory of the Poincaré group

- Representation theory of the Lorentz group

34.8 Footnotes

[1] Sometimes the tuple abbreviations:

$$(\mathbf{A})_{m'n',mn} \equiv \left[(A_x)_{m'n',mn}, (A_y)_{m'n',mn}, (A_z)_{m'n',mn} \right]$$

$$(\mathbf{B})_{m'n',mn} \equiv \left[(B_x)_{m'n',mn}, (B_y)_{m'n',mn}, (B_z)_{m'n',mn} \right]$$

$$\left(\mathbf{J}^{(m)} \right)_{m'm} \equiv \left[\left(J_x^{(m)} \right)_{m'm}, \left(J_y^{(m)} \right)_{m'm}, \left(J_z^{(m)} \right)_{m'm} \right]$$

are used.

34.9   References

[1] B.C. Hall (2003). *Lie groups, Lie algebras, and representations*. Springer.

[2] B.C. Hall (2013). *Quantum Theory for Mathematicians*. Springer.

[3] C.B. Parker (1994). *McGraw Hill Encyclopaedia of Physics* (2nd ed.). McGraw Hill. p. 1333. ISBN 0-07-051400-3.

[4] T. Ohlsson (2011). *Relativistic Quantum Physics: From Advanced Quantum Mechanics to Introductory Quantum*. Cambridge University Press. pp. 7–10. ISBN 1-13950-4320.

[5] E. Abers (2004). *Quantum Mechanics*. Addison Wesley. pp. 11, 104, 105, 410–411. ISBN 978-0-13-146100-0.

[6] B.R. Durney. "Lorentz Transformations". arXiv:1103.0156.

[7] H.L. Berk, K. Chaicherdsakul, T. Udagaw "The Proper Homogeneous Lorentz Transformation Operator$L = e^{-\omega S - \xi K}$, Where's It Going, What's the Twist" (PDF). Texas, Austin.

[8] Weinberg, S. (1964). "Feynman Rules *for Any* spin" (PDF). *Phys. Rev.* **133** (5B): B1318–B1332. Bibcode:1964PhRv..133 W. doi:10.1103/PhysRev.133.B1318.; Weinberg, S. (1964). "Feynman Rules *for Any* spin. II. Massless Particles" (PDF). *Phys. Rev.* **134** (4B): B882–B896. Bibcode:1964PhRv..134..882W. doi:10.1103/PhysRev.134.B882.; Weinberg, S. (1969). "Feynman Rules *for Any* spin. III" (PDF). *Phys. Rev.* **181** (5): 1893–1899. Bibcode:1969PhRv..181.1893W. doi:10.1103/PhysRev.181.1893.

[9] K. Masakatsu (2012). "Superradiance Problem of Bosons and Fermions for Rotating Black Holes in Bargmann–Wigner Formulation". Nara, Japan. arXiv:1208.0644.

[10] N.N. Bogolubov (1989). *General Principles of Quantum Field Theory* (2nd ed.). Springer. p. 272. ISBN 0-7923-0540-X.

[11] T. Ohlsson (2011). *Relativistic Quantum Physics: From Advanced Quantum Mechanics to Introductory Quantum Field Theory*. Cambridge University Press. p. 10. ISBN 1-13950-4320.

[12] Feynman, Richard. *The 1986 Dirac Memorial Lectures*. Cambridge University Press. p. 57. ISBN 978-0-521-65862-1.

- D. McMahon (2008). *Quantum Field Theory*. Mc Graw Hill. ISBN 978-0-07-154382-8.

- B. R. Martin, G. Shaw. *Particle Physics* (3rd ed.). Manchester Physics Series, John Wiley & Sons. p. 3. ISBN   978-0-470-03294-7.

- M. Chaichian, R. Hagedorn (1998). *Symmetry in quantum mechanics: From angular momentum to supersymmetry*. Graduate student series in physics. Institute of physic s (Bristol and Philadelphia). ISBN 0-7503-0408-1.

- W. Ludwig, C. Falter (1996). *Symmetries in physics*. Solid state science (2nd ed.). Springer. ISBN 3-540-60284-4.

- M. F. C. Ladd (1989). *Symmetry in molecules and crystals*. Solid state science. Ellis Horwood series in physical chemistry. ISBN 0-85312-255-5.

- K. J. Barnes (2010). *Group theory for the standard model and beyond*. Series in high energy physics, cosmology, and gravitation. Taylor & Francis. ISBN 142-007-874-7.

- S. Haywood (2011). *Symmetries and Conservation Laws in Particle Physics: An Introduction to Group Theory for Particle Physicists*. World Scientific. ISBN 184-816-703-2.

34.10   External links

- (2010) *Irreducible Tensor Operators and the Wigner-Eckart Theorem*

- R.D. Reece (2006) *A Derivation of the Quantum Mechanical Momentum Operator in the Position Representation*

- D. E. Soper (2011) *Position and momentum in quantum mechanics*

- *Lie groups*

- F. Porter (2009) *Lie Groups and Lie Algebras*

- *Continuous Groups, Lie Groups, and Lie Algebras*

- P.J. Mulders (2011) *Quantum field theory*

- arXiv:math-ph/0005032v1 B.C. Hall (2000) *An Elementary Introduction to Groups and Representations*

Chapter 35

Noether's theorem

This article is about Emmy Noether's first theorem, which derives conserved quantities from symmetries. For other uses, see Noether's theorem (disambiguation).

Noether's (first)[1] **theorem** states that every differentiable symmetry of the action of a physical system has a corresponding conservation law. The theorem was proven by German mathematician Emmy Noether in 1915 and published in 1918.[2] The action of a physical system is the integral over time of a Lagrangian function (which may or may not be an integral over space of a Lagrangian density function), from which the system's behavior can be determined by the principle of least action.

Noether's theorem has become a fundamental tool of modern theoretical physics and the calculus of variations. A generalization of the seminal formulations on constants of motion in Lagrangian and Hamiltonian mechanics (developed in 1788 and 1833, respectively), it does not apply to systems that cannot be modeled with a Lagrangian alone (e.g. systems with a Rayleigh dissipation function). In particular, dissipative systems with continuous symmetries need not have a corresponding conservation law.

35.1   Basic illustrations and background

As an illustration, if a physical system behaves the same regardless of how it is oriented in space, its Lagrangian is rotationally symmetric: from this symmetry, Noether's theorem dictates that the angular momentum of the system be conserved, as a consequence of its laws of motion. The physical system itself need not be symmetric; a jagged asteroid tumbling in space conserves angular momentum despite its asymmetry — it is *the laws of its motion* that are symmetric.

As another example, if a physical process exhibits the same outcomes regardless of place or time, then its Lagrangian is symmetric under continuous translations in space and time: by Noether's theorem, these symmetries account for the conservation laws of linear momentum and energy within this system, respectively.

Noether's theorem is important, both because of the insight it gives into conservation laws, and also as a practical calculational tool. It allows investigators to determine the conserved quantities (invariants) from the observed symmetries of a physical system. Conversely, it allows researchers to consider whole classes of hypothetical Lagrangians with given invariants, to describe a physical system. As an illustration, suppose that a physical theory is proposed which conserves a quantity X. A researcher can calculate the types of Lagrangians that conserve X through a continuous symmetry. Due to Noether's theorem, the properties of these Lagrangians provide further criteria to understand the implications and judge the fitness of the new theory.

There are numerous versions of Noether's theorem, with varying degrees of generality. The original version only applied to ordinary differential equations (particles) and not partial differential equations (fields). The original versions also assume that the Lagrangian only depends upon the first derivative, while later versions generalize the theorem to Lagrangians depending on the nth derivative. There are natural quantum counterparts of this theorem, expressed in the

Ward–Takahashi identities. Generalizations of Noether's theorem to superspaces are also available.

35.2 Informal statement of the theorem

All fine technical points aside, Noether's theorem can be stated informally

> If a system has a continuous symmetry property, then there are corresponding quantities whose values are conserved in time.[3]

A more sophisticated version of the theorem involving fields states that:

> To every differentiable symmetry generated by local actions, there corresponds a conserved current.

The word "symmetry" in the above statement refers more precisely to the covariance of the form that a physical law takes with respect to a one-dimensional Lie group of transformations satisfying certain technical criteria. The conservation law of a physical quantity is usually expressed as a continuity equation.

The formal proof of the theorem utilizes the condition of invariance to derive an expression for a current associated with a conserved physical quantity. In modern (since ca. 1980[4]) terminology, the conserved quantity is called the *Noether charge*, while the flow carrying that charge is called the *Noether current*. The Noether current is defined up to a solenoidal (divergenceless) vector field.

In the context of gravitation, Felix Klein's statement of Noether's theorem for action *I* stipulates for the invariants:[5]

> If an integral I is invariant under a continuous group G_ρ with ρ parameters, then ρ linearly independent combinations of the Lagrangian expressions are divergences.

35.3 Historical context

Main articles: Constant of motion, conservation law and conserved current

A conservation law states that some quantity X in the mathematical description of a system's evolution remains constant throughout its motion — it is an invariant. Mathematically, the rate of change of X (its derivative with respect to time) vanishes,

$$\frac{dX}{dt} = 0 .$$

Such quantities are said to be conserved; they are often called constants of motion (although motion *per se* need not be involved, just evolution in time). For example, if the energy of a system is conserved, its energy is invariant at all times, which imposes a constraint on the system's motion and may help in solving for it. Aside from insights that such constants of motion give into the nature of a system, they are a useful calculational tool; for example, an approximate solution can be corrected by finding the nearest state that satisfies the suitable conservation laws.

The earliest constants of motion discovered were momentum and energy, which were proposed in the 17th century by René Descartes and Gottfried Leibniz on the basis of collision experiments, and refined by subsequent researchers. Isaac Newton was the first to enunciate the conservation of momentum in its modern form, and showed that it was a consequence of Newton's third law. According to general relativity, the conservation laws of linear momentum, energy and angular momentum are only exactly true globally when expressed in terms of the sum of the stress–energy tensor (non-gravitational stress–energy) and the Landau–Lifshitz stress–energy–momentum pseudotensor (gravitational stress–energy). The local conservation of non-gravitational linear momentum and energy in a free-falling reference frame is expressed by the vanishing of the covariant divergence of the stress–energy tensor. Another important conserved quantity, discovered in studies of the celestial mechanics of astronomical bodies, is the Laplace–Runge–Lenz vector.

In the late 18th and early 19th centuries, physicists developed more systematic methods for discovering invariants. A major advance came in 1788 with the development of Lagrangian mechanics, which is related to the principle of least action. In this approach, the state of the system can be described by any type of generalized coordinates $\mathbf{q}$; the laws of motion need not be expressed in a Cartesian coordinate system, as was customary in Newtonian mechanics. The action is defined as the time integral I of a function known as the Lagrangian L

$$ I = \int L(\mathbf{q}, \dot{\mathbf{q}}, t) \, dt \, . $$

where the dot over $\mathbf{q}$ signifies the rate of change of the coordinates $\mathbf{q}$,

$$ \dot{\mathbf{q}} = \frac{d\mathbf{q}}{dt} \, . $$

Hamilton's principle states that the physical path $\mathbf{q}(t)$—the one actually taken by the system—is a path for which infinitesimal variations in that path cause no change in I, at least up to first order. This principle results in the Euler–Lagrange equations,

$$ \frac{d}{dt} \left(\frac{\partial L}{\partial \dot{\mathbf{q}}} \right) = \frac{\partial L}{\partial \mathbf{q}} \, . $$

Thus, if one of the coordinates, say qk, does not appear in the Lagrangian, the right-hand side of the equation is zero, and the left-hand side requires that

$$ \frac{d}{dt} \left(\frac{\partial L}{\partial \dot{q}_k} \right) = \frac{dp_k}{dt} = 0 \, . $$

where the momentum

$$ p_k = \frac{\partial L}{\partial \dot{q}_k} $$

is conserved throughout the motion (on the physical path).

Thus, the absence of the **ignorable** coordinate qk from the Lagrangian implies that the Lagrangian is unaffected by changes or transformations of qk; the Lagrangian is invariant, and is said to exhibit a symmetry under such transformations. This is the seed idea generalized in Noether's theorem.

Several alternative methods for finding conserved quantities were developed in the 19th century, especially by William Rowan Hamilton. For example, he developed a theory of canonical transformations which allowed changing coordinates so that some coordinates disappeared from the Lagrangian, as above, resulting in conserved canonical momenta. Another approach, and perhaps the most efficient for finding conserved quantities, is the Hamilton–Jacobi equation.

35.4   Mathematical expression

See also: Perturbation theory

35.4.1 Simple form using perturbations

The essence of Noether's theorem is generalizing the ignorable coordinates outlined.

Imagine that the action I defined above is invariant under small perturbations (warpings) of the time variable t and the generalized coordinates $\mathbf{q}$; in a notation commonly used in physics,

$$t \to t' = t + \delta t$$

$$\mathbf{q} \to \mathbf{q}' = \mathbf{q} + \delta\mathbf{q} .$$

where the perturbations δt and $\delta\mathbf{q}$ are both small, but variable. For generality, assume there are (say) N such symmetry transformations of the action, i.e. transformations leaving the action unchanged; labelled by an index $r = 1, 2, 3, \ldots, N$.

Then the resultant perturbation can be written as a linear sum of the individual types of perturbations,

$$\delta t = \sum_r \varepsilon_r T_r$$

$$\delta\mathbf{q} = \sum_r \varepsilon_r \mathbf{Q}_r .$$

where εr are infinitesimal parameter coefficients corresponding to each:

- generator Tr of time evolution, and

- generator $\mathbf{Q}r$ of the generalized coordinates.

For translations, $\mathbf{Q}r$ is a constant with units of length; for rotations, it is an expression linear in the components of $\mathbf{q}$, and the parameters make up an angle.

Using these definitions, Noether showed that the N quantities

$$\left(\frac{\partial L}{\partial \dot{\mathbf{q}}} \cdot \dot{\mathbf{q}} - L \right) T_r - \frac{\partial L}{\partial \dot{\mathbf{q}}} \cdot \mathbf{Q}_r$$

(which have the dimensions of [energy]·[time] + [momentum]·[length] = [action]) are conserved (constants of motion).

Examples

Time invariance

For illustration, consider a Lagrangian that does not depend on time, i.e., that is invariant (symmetric) under changes $t \to t + \delta t$, without any change in the coordinates $\mathbf{q}$. In this case, $N = 1$, $T = 1$ and $\mathbf{Q} = 0$; the corresponding conserved quantity is the total energy H[6]

$$H = \frac{\partial L}{\partial \dot{\mathbf{q}}} \cdot \dot{\mathbf{q}} - L.$$

Translational invariance

Consider a Lagrangian which does not depend on an ("ignorable", as above) coordinate qk; so it is invariant (symmetric) under changes $qk \rightarrow qk + \delta qk$. In that case, $N = 1$, $T = 0$, and $Qk = 1$; the conserved quantity is the corresponding momentum pk[7]

$$p_k = \frac{\partial L}{\partial \dot{q}_k}.$$

In special and general relativity, these apparently separate conservation laws are aspects of a single conservation law, that of the stress–energy tensor,[8] that is derived in the next section.

Rotational invariance

The conservation of the angular momentum $\mathbf{L} = \mathbf{r} \times \mathbf{p}$ is analogous to its linear momentum counterpart.[9] It is assumed that the symmetry of the Lagrangian is rotational, i.e., that the Lagrangian does not depend on the absolute orientation of the physical system in space. For concreteness, assume that the Lagrangian does not change under small rotations of an angle $\delta\theta$ about an axis $\mathbf{n}$; such a rotation transforms the Cartesian coordinates by the equation

$$\mathbf{r} \rightarrow \mathbf{r} + \delta\theta \mathbf{n} \times \mathbf{r}.$$

Since time is not being transformed, $T=0$. Taking $\delta\theta$ as the ε parameter and the Cartesian coordinates $\mathbf{r}$ as the generalized coordinates $\mathbf{q}$, the corresponding $\mathbf{Q}$ variables are given by

$$\mathbf{Q} = \mathbf{n} \times \mathbf{r}.$$

Then Noether's theorem states that the following quantity is conserved,

$$\frac{\partial L}{\partial \dot{\mathbf{q}}} \cdot \mathbf{Q}_r = \mathbf{p} \cdot (\mathbf{n} \times \mathbf{r}) = \mathbf{n} \cdot (\mathbf{r} \times \mathbf{p}) = \mathbf{n} \cdot \mathbf{L}.$$

In other words, the component of the angular momentum $\mathbf{L}$ along the $\mathbf{n}$ axis is conserved.

If $\mathbf{n}$ is arbitrary, i.e., if the system is insensitive to any rotation, then every component of $\mathbf{L}$ is conserved; in short, angular momentum is conserved.

35.4.2   Field theory version

Although useful in its own right, the version of Noether's theorem just given is a special case of the general version derived in 1915. To give the flavor of the general theorem, a version of the Noether theorem for continuous fields in four-dimensional space–time is now given. Since field theory problems are more common in modern physics than mechanics problems, this field theory version is the most commonly used version (or most often implemented) of Noether's theorem.

Let there be a set of differentiable fields φ defined over all space and time; for example, the temperature $T(\mathbf{x}, t)$ would be representative of such a field, being a number defined at every place and time. The principle of least action can be applied to such fields, but the action is now an integral over space and time

$$I = \int L\left(\phi, \partial_\mu \phi, x^\mu\right) d^4 x$$

(the theorem can actually be further generalized to the case where the Lagrangian depends on up to the nth derivative using jet bundles)

Let the action be invariant under certain transformations of the space–time coordinates x^μ and the fields φ

$$x^\mu \to x^\mu + \delta x^\mu$$

$$\phi \to \phi + \delta\phi$$

where the transformations can be indexed by $r = 1, 2, 3, \ldots, N$

$$\delta x^\mu = \varepsilon_r X_r^\mu$$

$$\delta\phi = \varepsilon_r \Psi_r \ .$$

For such systems, Noether's theorem states that there are N conserved current densities

$$j_r^\nu = -\left(\frac{\partial L}{\partial\phi_{,\nu}}\right) \cdot \Psi_r + \left[\left(\frac{\partial L}{\partial\phi_{,\nu}}\right) \cdot \phi_{,\sigma} - L\delta_\sigma^\nu\right] X_r^\sigma$$

In such cases, the conservation law is expressed in a four-dimensional way

$$\partial_\mu j^\mu = 0$$

which expresses the idea that the amount of a conserved quantity within a sphere cannot change unless some of it flows out of the sphere. For example, electric charge is conserved; the amount of charge within a sphere cannot change unless some of the charge leaves the sphere.

For illustration, consider a physical system of fields that behaves the same under translations in time and space, as considered above; in other words, $L\left(\phi, \partial_\mu\phi, x^\mu\right)$ is constant in its third argument. In that case, $N = 4$, one for each dimension of space and time. Since only the positions in space–time are being warped, not the fields, the Ψ are all zero and the $X\mu^\nu$ equal the Kronecker delta $\delta\mu^\nu$, where we have used μ instead of r for the index. In that case, Noether's theorem corresponds to the conservation law for the stress–energy tensor $T\mu^\nu$[8]

$$T_\mu^{\ \nu} = \left[\left(\frac{\partial L}{\partial\phi_{,\nu}}\right) \cdot \phi_{,\sigma} - L\,\delta_\sigma^\nu\right]\delta_\mu^\sigma = \left(\frac{\partial L}{\partial\phi_{,\nu}}\right) \cdot \phi_{,\mu} - L\,\delta_\mu^\nu$$

The conservation of electric charge, by contrast, can be derived by considering zero $X\mu^\nu = 0$ and Ψ linear in the fields φ themselves.[10] In quantum mechanics, the probability amplitude $\psi(\mathbf{x})$ of finding a particle at a point $\mathbf{x}$ is a complex field φ, because it ascribes a complex number to every point in space and time. The probability amplitude itself is physically unmeasurable; only the probability $p = |\psi|^2$ can be inferred from a set of measurements. Therefore, the system is invariant under transformations of the ψ field and its complex conjugate field ψ^* that leave $|\psi|^2$ unchanged, such as

$$\psi \to e^{i\theta}\psi \ , \quad \psi^* \to e^{-i\theta}\psi^* \ .$$

a complex rotation. In the limit when the phase θ becomes infinitesimally small, $\delta\theta$, it may be taken as the parameter ε, while the Ψ are equal to $i\psi$ and $-i\psi^*$, respectively. A specific example is the Klein–Gordon equation, the relativistically correct version of the Schrödinger equation for spinless particles, which has the Lagrangian density

$$L = \psi_{,\nu}\psi_{,\mu}^*\eta^{\nu\mu} + m^2\psi\psi^* \ .$$

In this case, Noether's theorem states that the conserved ($\partial \cdot j = 0$) current equals

$$j^{\nu} = i \left(\frac{\partial \psi}{\partial x^{\mu}} \psi^* - \frac{\partial \psi^*}{\partial x^{\mu}} \psi \right) \eta^{\nu\mu} ,$$

which, when multiplied by the charge on that species of particle, equals the electric current density due to that type of particle. This "gauge invariance" was first noted by Hermann Weyl, and is one of the prototype gauge symmetries of physics.

35.5 Derivations

35.5.1 One independent variable

Consider the simplest case, a system with one independent variable, time. Suppose the dependent variables $\mathbf{q}$ are such that the action integral

$$I = \int_{t_1}^{t_2} L[\mathbf{q}[t], \dot{\mathbf{q}}[t], t] \, dt$$

is invariant under brief infinitesimal variations in the dependent variables. In other words, they satisfy the Euler–Lagrange equations

$$\frac{d}{dt} \frac{\partial L}{\partial \dot{\mathbf{q}}}[t] = \frac{\partial L}{\partial \mathbf{q}}[t].$$

And suppose that the integral is invariant under a continuous symmetry. Mathematically such a symmetry is represented as a flow, $\boldsymbol{\varphi}$, which acts on the variables as follows

$$t \to t' = t + \varepsilon T$$

$$\mathbf{q}[t] \to \mathbf{q}'[t'] = \phi[\mathbf{q}[t], \varepsilon] = \phi[\mathbf{q}[t' - \varepsilon T], \varepsilon]$$

where ε is a real variable indicating the amount of flow, and T is a real constant (which could be zero) indicating how much the flow shifts time.

$$\dot{\mathbf{q}}[t] \to \dot{\mathbf{q}}'[t'] = \frac{d}{dt} \phi[\mathbf{q}[t], \varepsilon] = \frac{\partial \phi}{\partial \mathbf{q}}[\mathbf{q}[t' - \varepsilon T], \varepsilon] \dot{\mathbf{q}}[t' - \varepsilon T].$$

The action integral flows to

$$I'[\varepsilon] = \int_{t_1 + \varepsilon T}^{t_2 + \varepsilon T} L[\mathbf{q}'[t'], \dot{\mathbf{q}}'[t'], t'] \, dt'$$

$$= \int_{t_1 + \varepsilon T}^{t_2 + \varepsilon T} L[\phi[\mathbf{q}[t' - \varepsilon T], \varepsilon], \frac{\partial \phi}{\partial \mathbf{q}}[\mathbf{q}[t' - \varepsilon T], \varepsilon] \dot{\mathbf{q}}[t' - \varepsilon T], t'] \, dt'$$

which may be regarded as a function of ε. Calculating the derivative at $\varepsilon = 0$ and using the symmetry, we get

$$0 = \frac{dI'}{d\varepsilon}[0] = L[\mathbf{q}[t_2], \dot{\mathbf{q}}[t_2], t_2]T - L[\mathbf{q}[t_1], \dot{\mathbf{q}}[t_1], t_1]T$$

$$+ \int_{t_1}^{t_2} \frac{\partial L}{\partial \mathbf{q}} \left(-\frac{\partial \phi}{\partial \mathbf{q}} \dot{\mathbf{q}} T + \frac{\partial \phi}{\partial \varepsilon} \right) + \frac{\partial L}{\partial \dot{\mathbf{q}}} \left(-\frac{\partial^2 \phi}{(\partial \mathbf{q})^2} \dot{\mathbf{q}}^2 T + \frac{\partial^2 \phi}{\partial \varepsilon \partial \mathbf{q}} \dot{\mathbf{q}} - \frac{\partial \phi}{\partial \mathbf{q}} \ddot{\mathbf{q}} T \right) \, dt.$$

Notice that the Euler–Lagrange equations imply

$$
\frac{d}{dt}\left(\frac{\partial L}{\partial \dot{\mathbf{q}}}\frac{\partial \phi}{\partial \mathbf{q}}\dot{\mathbf{q}}T\right) = \left(\frac{d}{dt}\frac{\partial L}{\partial \dot{\mathbf{q}}}\right)\frac{\partial \phi}{\partial \mathbf{q}}\dot{\mathbf{q}}T + \frac{\partial L}{\partial \dot{\mathbf{q}}}\left(\frac{d}{dt}\frac{\partial \phi}{\partial \mathbf{q}}\right)\dot{\mathbf{q}}T + \frac{\partial L}{\partial \dot{\mathbf{q}}}\frac{\partial \phi}{\partial \mathbf{q}}\ddot{\mathbf{q}}T
$$

$$
= \frac{\partial L}{\partial \mathbf{q}}\frac{\partial \phi}{\partial \mathbf{q}}\dot{\mathbf{q}}T + \frac{\partial L}{\partial \dot{\mathbf{q}}}\left(\frac{\partial^2 \phi}{(\partial \mathbf{q})^2}\dot{\mathbf{q}}\right)\dot{\mathbf{q}}T + \frac{\partial L}{\partial \dot{\mathbf{q}}}\frac{\partial \phi}{\partial \mathbf{q}}\ddot{\mathbf{q}}\,T.
$$

Substituting this into the previous equation, one gets

$$
0 = \frac{dI'}{d\varepsilon}[0] = L[\mathbf{q}[t_2],\dot{\mathbf{q}}[t_2],t_2]T - L[\mathbf{q}[t_1],\dot{\mathbf{q}}[t_1],t_1]T - \frac{\partial L}{\partial \dot{\mathbf{q}}}\frac{\partial \phi}{\partial \mathbf{q}}\dot{\mathbf{q}}[t_2]T + \frac{\partial L}{\partial \dot{\mathbf{q}}}\frac{\partial \phi}{\partial \mathbf{q}}\dot{\mathbf{q}}[t_1]T
$$

$$
+ \int_{t_1}^{t_2}\frac{\partial L}{\partial \mathbf{q}}\frac{\partial \phi}{\partial \varepsilon} + \frac{\partial L}{\partial \dot{\mathbf{q}}}\frac{\partial^2 \phi}{\partial \varepsilon \partial \mathbf{q}}\dot{\mathbf{q}}\,dt.
$$

Again using the Euler–Lagrange equations we get

$$
\frac{d}{dt}\left(\frac{\partial L}{\partial \dot{\mathbf{q}}}\frac{\partial \phi}{\partial \varepsilon}\right) = \left(\frac{d}{dt}\frac{\partial L}{\partial \dot{\mathbf{q}}}\right)\frac{\partial \phi}{\partial \varepsilon} + \frac{\partial L}{\partial \dot{\mathbf{q}}}\frac{\partial^2 \phi}{\partial \varepsilon \partial \mathbf{q}}\dot{\mathbf{q}} = \frac{\partial L}{\partial \mathbf{q}}\frac{\partial \phi}{\partial \varepsilon} + \frac{\partial L}{\partial \dot{\mathbf{q}}}\frac{\partial^2 \phi}{\partial \varepsilon \partial \mathbf{q}}\dot{\mathbf{q}}.
$$

Substituting this into the previous equation, one gets

$$
0 = L[\mathbf{q}[t_2],\dot{\mathbf{q}}[t_2],t_2]T - L[\mathbf{q}[t_1],\dot{\mathbf{q}}[t_1],t_1]T - \frac{\partial L}{\partial \dot{\mathbf{q}}}\frac{\partial \phi}{\partial \mathbf{q}}\dot{\mathbf{q}}[t_2]T + \frac{\partial L}{\partial \dot{\mathbf{q}}}\frac{\partial \phi}{\partial \mathbf{q}}\dot{\mathbf{q}}[t_1]T
$$

$$
+ \frac{\partial L}{\partial \dot{\mathbf{q}}}\frac{\partial \phi}{\partial \varepsilon}[t_2] - \frac{\partial L}{\partial \dot{\mathbf{q}}}\frac{\partial \phi}{\partial \varepsilon}[t_1].
$$

From which one can see that

$$
\left(\frac{\partial L}{\partial \dot{\mathbf{q}}}\frac{\partial \phi}{\partial \mathbf{q}}\dot{\mathbf{q}} - L\right)T - \frac{\partial L}{\partial \dot{\mathbf{q}}}\frac{\partial \phi}{\partial \varepsilon}
$$

is a constant of the motion, i.e., it is a conserved quantity. Since $\varphi[\mathbf{q},0] = \mathbf{q}$, we get $\frac{\partial \phi}{\partial \mathbf{q}} = 1$ and so the conserved quantity simplifies to

$$
\left(\frac{\partial L}{\partial \dot{\mathbf{q}}}\dot{\mathbf{q}} - L\right)T - \frac{\partial L}{\partial \dot{\mathbf{q}}}\frac{\partial \phi}{\partial \varepsilon}.
$$

To avoid excessive complication of the formulas, this derivation assumed that the flow does not change as time passes. The same result can be obtained in the more general case.

35.5.2   Field-theoretic derivation

Noether's theorem may also be derived for tensor fields φ^A where the index A ranges over the various components of the various tensor fields. These field quantities are functions defined over a four-dimensional space whose points are labeled by coordinates x^μ where the index μ ranges over time ($\mu = 0$) and three spatial dimensions ($\mu = 1, 2, 3$). These four coordinates are the independent variables; and the values of the fields at each event are the dependent variables. Under an infinitesimal transformation, the variation in the coordinates is written

$$x^\mu \rightarrow \xi^\mu = x^\mu + \delta x^\mu$$

whereas the transformation of the field variables is expressed as

$$\phi^A \rightarrow \alpha^A(\xi^\mu) = \phi^A(x^\mu) + \delta\phi^A(x^\mu).$$

By this definition, the field variations $\delta\varphi^A$ result from two factors: intrinsic changes in the field themselves and changes in coordinates, since the transformed field α^A depends on the transformed coordinates ξ^μ. To isolate the intrinsic changes, the field variation at a single point x^μ may be defined

$$\alpha^A(x^\mu) = \phi^A(x^\mu) + \bar\delta\phi^A(x^\mu).$$

If the coordinates are changed, the boundary of the region of space–time over which the Lagrangian is being integrated also changes; the original boundary and its transformed version are denoted as Ω and Ω', respectively.

Noether's theorem begins with the assumption that a specific transformation of the coordinates and field variables does not change the action, which is defined as the integral of the Lagrangian density over the given region of spacetime. Expressed mathematically, this assumption may be written as

$$\int_{\Omega'} L\left(\alpha^A, \alpha^A_{,\nu}, \xi^\mu\right) d^4\xi - \int_\Omega L\left(\phi^A, \phi^A_{,\nu}, x^\mu\right) d^4x = 0$$

where the comma subscript indicates a partial derivative with respect to the coordinate(s) that follows the comma, e.g.

$$\phi^A_{,\sigma} = \frac{\partial\phi^A}{\partial x^\sigma}.$$

Since ξ is a dummy variable of integration, and since the change in the boundary Ω is infinitesimal by assumption, the two integrals may be combined using the four-dimensional version of the divergence theorem into the following form

$$\int_\Omega \left\{ \left[L\left(\alpha^A, \alpha^A_{,\nu}, x^\mu\right) - L\left(\phi^A, \phi^A_{,\nu}, x^\mu\right) \right] + \frac{\partial}{\partial x^\sigma}\left[L\left(\phi^A, \phi^A_{,\nu}, x^\mu\right) \delta x^\sigma \right] \right\} d^4x = 0.$$

The difference in Lagrangians can be written to first-order in the infinitesimal variations as

$$\left[L\left(\alpha^A, \alpha^A_{,\nu}, x^\mu\right) - L\left(\phi^A, \phi^A_{,\nu}, x^\mu\right) \right] = \frac{\partial L}{\partial\phi^A}\bar\delta\phi^A + \frac{\partial L}{\partial\phi^A_{,\sigma}}\bar\delta\phi^A_{,\sigma}.$$

However, because the variations are defined at the same point as described above, the variation and the derivative can be done in reverse order; they commute

$$\bar\delta\phi^A_{,\sigma} = \bar\delta\frac{\partial\phi^A}{\partial x^\sigma} = \frac{\partial}{\partial x^\sigma}(\bar\delta\phi^A).$$

Using the Euler–Lagrange field equations

$$\frac{\partial}{\partial x^\sigma}\left(\frac{\partial L}{\partial\phi^A_{,\sigma}} \right) = \frac{\partial L}{\partial\phi^A}$$

the difference in Lagrangians can be written neatly as

$$\left[L\left(a^A, a^A{}_{,\nu}, x^\mu \right) - L\left(\phi^A, \phi^A{}_{,\nu}, x^\mu \right) \right] = \frac{\partial}{\partial x^\sigma} \left(\frac{\partial L}{\partial \phi^A{}_{,\sigma}} \right) \bar\delta \phi^A + \frac{\partial L}{\partial \phi^A{}_{,\sigma}} \bar\delta \phi^A{}_{,\sigma} = \frac{\partial}{\partial x^\sigma} \left(\frac{\partial L}{\partial \phi^A{}_{,\sigma}} \bar\delta \phi^A \right) .$$

Thus, the change in the action can be written as

$$\int_\Omega \frac{\partial}{\partial x^\sigma} \left\{ \frac{\partial L}{\partial \phi^A{}_{,\sigma}} \bar\delta \phi^A + L\left(\phi^A, \phi^A{}_{,\nu}, x^\mu \right) \delta x^\sigma \right\} d^4x = 0 .$$

Since this holds for any region Ω, the integrand must be zero

$$\frac{\partial}{\partial x^\sigma} \left\{ \frac{\partial L}{\partial \phi^A{}_{,\sigma}} \bar\delta \phi^A + L\left(\phi^A, \phi^A{}_{,\nu}, x^\mu \right) \delta x^\sigma \right\} = 0 .$$

For any combination of the various symmetry transformations, the perturbation can be written

$$\delta x^\mu = \varepsilon X^\mu$$

$$\delta \phi^A = \varepsilon \Psi^A = \bar\delta \phi^A + \varepsilon \mathcal{L}_X \phi^A$$

where $\mathcal{L}_X \phi^A$ is the Lie derivative of φ^A in the X^μ direction. When φ^A is a scalar or $X^\mu{}_{,\nu} = 0$,

$$\mathcal{L}_X \phi^A = \frac{\partial \phi^A}{\partial x^\mu} X^\mu .$$

These equations imply that the field variation taken at one point equals

$$\bar\delta \phi^A = \varepsilon \Psi^A - \varepsilon \mathcal{L}_X \phi^A .$$

Differentiating the above divergence with respect to ε at $\varepsilon = 0$ and changing the sign yields the conservation law

$$\frac{\partial}{\partial x^\sigma} j^\sigma = 0$$

where the conserved current equals

$$j^\sigma = \left[\frac{\partial L}{\partial \phi^A{}_{,\sigma}} \mathcal{L}_X \phi^A - L X^\sigma \right] - \left(\frac{\partial L}{\partial \phi^A{}_{,\sigma}} \right) \Psi^A .$$

35.5.3 Manifold/fiber bundle derivation

Suppose we have an n-dimensional oriented Riemannian manifold, M and a target manifold T. Let C be the configuration space of smooth functions from M to T. (More generally, we can have smooth sections of a fiber bundle over M.)

Examples of this M in physics include:

- In classical mechanics, in the Hamiltonian formulation, M is the one-dimensional manifold **R**, representing time and the target space is the cotangent bundle of space of generalized positions.

- In field theory, M is the spacetime manifold and the target space is the set of values the fields can take at any given point. For example, if there are m real-valued scalar fields, $\phi_1, \ldots, \phi_m$, then the target manifold is $\mathbf{R}^m$. If the field is a real vector field, then the target manifold is isomorphic to $\mathbf{R}^3$.

Now suppose there is a functional

$$S : \mathcal{C} \to \mathbf{R},$$

called the action. (Note that it takes values into $\mathbf{R}$, rather than $\mathbf{C}$; this is for physical reasons, and doesn't really matter for this proof.)

To get to the usual version of Noether's theorem, we need additional restrictions on the action. We assume $S[\phi]$ is the integral over M of a function

$$\mathcal{L}(\phi, \partial_\mu \phi, x)$$

called the Lagrangian density, depending on φ, its derivative and the position. In other words, for φ in $\mathcal{C}$

$$S[\phi] = \int_M \mathcal{L}[\phi(x), \partial_\mu \phi(x), x] \mathrm{d}^n x.$$

Suppose we are given boundary conditions, i.e., a specification of the value of φ at the boundary if M is compact, or some limit on φ as x approaches ∞. Then the subspace of $\mathcal{C}$ consisting of functions φ such that all functional derivatives of S at φ are zero, that is:

$$\frac{\delta S[\phi]}{\delta \phi(x)} \approx 0$$

and that φ satisfies the given boundary conditions, is the subspace of on shell solutions. (See principle of stationary action)

Now, suppose we have an infinitesimal transformation on $\mathcal{C}$, generated by a functional derivation, Q such that

$$Q\left[\int_N \mathcal{L} \, \mathrm{d}^n x \right] \approx \int_{\partial N} f^\mu [\phi(x), \partial \phi, \partial \partial \phi, \ldots] \, \mathrm{d}s_\mu$$

for all compact submanifolds N or in other words,

$$Q[\mathcal{L}(x)] \approx \partial_\mu f^\mu(x)$$

for all x, where we set

$$\mathcal{L}(x) = \mathcal{L}[\phi(x), \partial_\mu \phi(x), x].$$

If this holds on shell and off shell, we say Q generates an off-shell symmetry. If this only holds on shell, we say Q generates an on-shell symmetry. Then, we say Q is a generator of a one parameter symmetry Lie group.

Now, for any N, because of the Euler–Lagrange theorem, on shell (and only on-shell), we have

Since this is true for any N, we have

$$\partial_\mu \left[\frac{\partial \mathcal{L}}{\partial(\partial_\mu \phi)} Q[\phi] - f^\mu \right] \approx 0.$$

But this is the continuity equation for the current J^μ defined by:[11]

$$J^\mu = \frac{\partial \mathcal{L}}{\partial(\partial_\mu \phi)} Q[\phi] - f^\mu,$$

which is called the **Noether current** associated with the symmetry. The continuity equation tells us that if we integrate this current over a space-like slice, we get a conserved quantity called the Noether charge (provided, of course, if M is noncompact, the currents fall off sufficiently fast at infinity).

35.5.4   Comments

Noether's theorem is an on shell theorem: it relies on use of the equations of motion—the classical path. It reflects the relation between the boundary conditions and the variational principle. Assuming no boundary terms in the action, Noether's theorem implies that

$$\int_{\partial N} J^\mu \mathrm{d}s_\mu \approx 0 \,.$$

The quantum analogs of Noether's theorem involving expectation values, e.g. $\langle \int d^4x\, \partial \cdot J \rangle = 0$, probing off shell quantities as well are the Ward–Takahashi identities.

35.5.5   Generalization to Lie algebras

Suppose say we have two symmetry derivations Q_1 and Q_2. Then, $[Q_1, Q_2]$ is also a symmetry derivation. Let's see this explicitly. Let's say

$$Q_1[\mathcal{L}] \approx \partial_\mu f_1^\mu$$

and

$$Q_2[\mathcal{L}] \approx \partial_\mu f_2^\mu$$

Then,

$$[Q_1, Q_2][\mathcal{L}] = Q_1[Q_2[\mathcal{L}]] - Q_2[Q_1[\mathcal{L}]] \approx \partial_\mu f_{12}^\mu$$

where $f_{12} = Q_1[f_2^\mu] - Q_2[f_1^\mu]$. So,

$$j_{12}^\mu = \left(\frac{\partial}{\partial(\partial_\mu \phi)} \mathcal{L} \right) (Q_1[Q_2[\phi]] - Q_2[Q_1[\phi]]) - f_{12}^\mu.$$

This shows we can extend Noether's theorem to larger Lie algebras in a natural way.

35.5.6   Generalization of the proof

This applies to *any* local symmetry derivation Q satisfying $QS \approx 0$, and also to more general local functional differentiable actions, including ones where the Lagrangian depends on higher derivatives of the fields. Let ε be any arbitrary smooth function of the spacetime (or time) manifold such that the closure of its support is disjoint from the boundary. ε is a test function. Then, because of the variational principle (which does *not* apply to the boundary, by the way), the derivation distribution q generated by $q[\varepsilon][\Phi(x)] = \varepsilon(x)Q[\Phi(x)]$ satisfies $q[\varepsilon][S] \approx 0$ for every ε, or more compactly, $q(x)[S] \approx 0$ for all x not on the boundary (but remember that $q(x)$ is a shorthand for a derivation *distribution*, not a derivation parametrized by x in general). This is the generalization of Noether's theorem.

To see how the generalization is related to the version given above, assume that the action is the spacetime integral of a Lagrangian that only depends on φ and its first derivatives. Also, assume

$$Q[\mathcal{L}] \approx \partial_\mu f^\mu$$

Then,

$$\begin{aligned}
q[\varepsilon][\mathcal{S}] &= \int q[\varepsilon][\mathcal{L}]\,\mathrm{d}^n x \\
&= \int \left\{ \left(\frac{\partial}{\partial\phi}\mathcal{L}\right)\varepsilon Q[\phi] + \left[\frac{\partial}{\partial(\partial_\mu\phi)}\mathcal{L}\right]\partial_\mu(\varepsilon Q[\phi]) \right\}\,\mathrm{d}^n x \\
&= \int \left\{ \varepsilon Q[\mathcal{L}] + \partial_\mu\varepsilon\left[\frac{\partial}{\partial(\partial_\mu\phi)}\mathcal{L}\right]Q[\phi] \right\}\,\mathrm{d}^n x \\
&\approx \int \varepsilon\partial_\mu\left\{ f^\mu - \left[\frac{\partial}{\partial(\partial_\mu\phi)}\mathcal{L}\right]Q[\phi] \right\}\,\mathrm{d}^n x
\end{aligned}$$

for all ε.

More generally, if the Lagrangian depends on higher derivatives, then

$$\partial_\mu\left[f^\mu - \left[\frac{\partial}{\partial(\partial_\mu\phi)}\mathcal{L}\right]Q[\phi] - 2\left[\frac{\partial}{\partial(\partial_\mu\partial_\nu\phi)}\mathcal{L}\right]\partial_\nu Q[\phi] + \partial_\nu\left[\left[\frac{\partial}{\partial(\partial_\mu\partial_\nu\phi)}\mathcal{L}\right]Q[\phi]\right] - \cdots \right] \approx 0.$$

35.6   Examples

35.6.1   Example 1: Conservation of energy

Looking at the specific case of a Newtonian particle of mass m, coordinate x, moving under the influence of a potential V, coordinatized by time t. The action, S, is:

$$\begin{aligned}
\mathcal{S}[x] &= \int L[x(t), \dot{x}(t)]\,dt \\
&= \int \left(\frac{m}{2}\sum_{i=1}^{3}\dot{x}_i^2 - V(x(t)) \right)\,dt.
\end{aligned}$$

The first term in the brackets is the kinetic energy of the particle, whilst the second is its potential energy. Consider the generator of time translations $Q = \partial/\partial t$. In other words, $Q[x(t)] = \dot{x}(t)$. Note that x has an explicit dependence on time, whilst V does not; consequently:

$$Q[L] = m \sum_i \dot{x}_i \ddot{x}_i - \sum_i \frac{\partial V(x)}{\partial x_i} \dot{x}_i = \frac{d}{dt} \left[\frac{m}{2} \sum_i \dot{x}_i^2 - V(x) \right]$$

so we can set

$$f = \frac{m}{2} \sum_i \dot{x}_i^2 - V(x).$$

Then,

$$
\begin{aligned}
j &= \sum_{i=1}^{3} \frac{\partial L}{\partial \dot{x}_i} Q[x_i] - f \\
&= m \sum_i \dot{x}_i^2 - \left[\frac{m}{2} \sum_i \dot{x}_i^2 - V(x) \right] \\
&= \frac{m}{2} \sum_i \dot{x}_i^2 + V(x).
\end{aligned}
$$

The right hand side is the energy, and Noether's theorem states that $\dot{j} = 0$ (i.e. the principle of conservation of energy is a consequence of invariance under time translations).

More generally, if the Lagrangian does not depend explicitly on time, the quantity

$$\sum_{i=1}^{3} \frac{\partial L}{\partial \dot{x}_i} \dot{x}_i - L$$

(called the Hamiltonian) is conserved.

35.6.2 Example 2: Conservation of center of momentum

Still considering 1-dimensional time, let

$$
\begin{aligned}
S[x] &= \int \mathcal{L}[\vec{x}(t), \dot{\vec{x}}(t)] \, dt \\
&= \int \left[\sum_{\alpha=1}^{N} \frac{m_\alpha}{2} (\dot{\vec{x}}_\alpha)^2 - \sum_{\alpha < \beta} V_{\alpha\beta}(\vec{x}_\beta - \vec{x}_\alpha) \right] \, dt
\end{aligned}
$$

i.e. N Newtonian particles where the potential only depends pairwise upon the relative displacement.

For $\vec{Q}$, let's consider the generator of Galilean transformations (i.e. a change in the frame of reference). In other words,

$$Q_i[x_\alpha^j(t)] = t \delta_i^j.$$

Note that

$$Q_i[\mathcal{L}] = \sum_\alpha m_\alpha \dot{x}^i_\alpha - \sum_{\alpha < \beta} \partial_i V_{\alpha\beta}(\vec{x}_\beta - \vec{x}_\alpha)(t - t)$$

$$= \sum_\alpha m_\alpha \dot{x}^i_\alpha.$$

This has the form of $\frac{\mathrm{d}}{\mathrm{d}t} \sum_\alpha m_\alpha x^i_\alpha$ so we can set

$$\vec{f} = \sum_\alpha m_\alpha \vec{x}_\alpha.$$

Then,

$$\vec{j} = \sum_\alpha \left(\frac{\partial}{\partial \dot{\vec{x}}_\alpha} \mathcal{L} \right) \cdot \vec{Q}[\vec{x}_\alpha] - \vec{f}$$

$$= \sum_\alpha (m_\alpha \dot{\vec{x}}_\alpha t - m_\alpha \vec{x}_\alpha)$$

$$= \vec{P}t - M\vec{x}_{CM}$$

where $\vec{P}$ is the total momentum, M is the total mass and $\vec{x}_{CM}$ is the center of mass. Noether's theorem states:

$$\dot{\vec{j}} = 0 \Rightarrow \vec{P} - M\dot{\vec{x}}_{CM} = 0.$$

35.6.3   Example 3: Conformal transformation

Both examples 1 and 2 are over a 1-dimensional manifold (time). An example involving spacetime is a conformal transformation of a massless real scalar field with a quartic potential in $(3 + 1)$-Minkowski spacetime.

For Q, consider the generator of a spacetime rescaling. In other words,

$$Q[\phi(x)] = x^\mu \partial_\mu \phi(x) + \phi(x).$$

The second term on the right hand side is due to the "conformal weight" of φ. Note that

$$Q[\mathcal{L}] = \partial^\mu \phi \left(\partial_\mu \phi + x^\nu \partial_\mu \partial_\nu \phi + \partial_\mu \phi \right) - 4\lambda \phi^3 \left(x^\mu \partial_\mu \phi + \phi \right).$$

This has the form of

$$\partial_\mu \left[\frac{1}{2} x^\mu \partial^\nu \phi \partial_\nu \phi - \lambda x^\mu \phi^4 \right] = \partial_\mu \left(x^\mu \mathcal{L} \right)$$

(where we have performed a change of dummy indices) so set

$$f^\mu = x^\mu \mathcal{L}.$$

Then,

$$j^\mu = \left[\frac{\partial}{\partial(\partial_\mu \phi)}\mathcal{L}\right] Q[\phi] - f^\mu$$

$$= \partial^\mu \phi \left(x^\nu \partial_\nu \phi + \phi\right) - x^\mu \left(\frac{1}{2}\partial^\nu \phi \partial_\nu \phi - \lambda \phi^4\right).$$

Noether's theorem states that $\partial_\mu j^\mu = 0$ (as one may explicitly check by substituting the Euler–Lagrange equations into the left hand side).

(Aside: If one tries to find the Ward–Takahashi analog of this equation, one runs into a problem because of anomalies.)

35.7 Applications

Application of Noether's theorem allows physicists to gain powerful insights into any general theory in physics, by just analyzing the various transformations that would make the form of the laws involved invariant. For example:

- the invariance of physical systems with respect to spatial translation (in other words, that the laws of physics do not vary with locations in space) gives the law of conservation of linear momentum;

- invariance with respect to rotation gives the law of conservation of angular momentum;

- invariance with respect to time translation gives the well-known law of conservation of energy

In quantum field theory, the analog to Noether's theorem, the Ward–Takahashi identity, yields further conservation laws, such as the conservation of electric charge from the invariance with respect to a change in the phase factor of the complex field of the charged particle and the associated gauge of the electric potential and vector potential.

The Noether charge is also used in calculating the entropy of stationary black holes.[12]

35.8 See also

- Charge (physics)

- Gauge symmetry

- Gauge symmetry (mathematics)

- Invariant (physics)

- Goldstone boson

- Symmetry in physics

35.9   Notes

[1] See also Noether's second theorem.

[2] Noether E (1918). "Invariante Variationsprobleme". *Nachr. D. König. Gesellsch. D. Wiss. Zu Göttingen, Math-phys. Klasse* **1918**: 235–257.

[3] Thompson, W.J. (1994). *Angular Momentum: an illustrated guide to rotational symmetries for physical systems* **1**. Wiley. p. 5. ISBN 0-471-55264-X.

[4] The term "Noether charge" occurs in Seligman, *Group theory and its applications in physics, 1980: Latin American School of Physics, Mexico City*, American Institute of Physics, 1981. It comes enters wider use during the 1980s, e.g. by G. Takeda in: Errol Gotsman, Gerald Tauber (eds.) *From SU(3) to Gravity: Festschrift in Honor of Yuval Ne'eman*, 1985, p. 196.

[5] Nina Byers (1998) "E. Noether's Discovery of the Deep Connection Between Symmetries and Conservation Laws." in Proceedings of a Symposium on the Heritage of Emmy Noether, held on 2–4 December 1996, at the Bar-Ilan University, Israel, Appendix B.

[6] Lanczos 1970, pp. 401–3

[7] Lanczos 1970, pp. 403–4

[8] Goldstein 1980, pp. 592–3

[9] Lanczos 1970, pp. 404–5

[10] Goldstein 1980, pp. 593–4

[11] Michael E. Peskin, Daniel V. Schroeder (1995). *An Introduction to Quantum Field Theory*. Basic Books. p. 18. ISBN 0-201-50397-2.

[12] Vivek Iyer; Wald (1995). "A comparison of Noether charge and Euclidean methods for Computing the Entropy of Stationary Black Holes".*Physical Review D***52**(8): 4430–9. arXiv:gr-qc/9503052. Bibcode:1995PhRvD..52.4430I.doi:10.1103/.

35.10   References

- Goldstein, Herbert (1980). *Classical Mechanics* (2nd ed.). Reading, MA: Addison-Wesley. pp. 588–596. ISBN 0-201-02918-9.

- Kosmann-Schwarzbach, Yvette (2010). *The Noether theorems: Invariance and conservation laws in the twentieth century*. Sources and Studies in the History of Mathematics and Physical Sciences. Springer-Verlag. ISBN 978-0-387-87867-6

- Lanczos, C. (1970). *The Variational Principles of Mechanics* (4th ed.). New York: Dover Publications. pp. 401–5. ISBN 0-486-65067-7.

- Olver, Peter (1993). *Applications of Lie groups to differential equations*. Graduate Texts in Mathematics **107** (2nd ed.). Springer-Verlag. ISBN 0-387-95000-1

35.11   External links

- Emmy Noether; Mort Tavel (translator) (1971). "Invariant Variation Problems". *Transport Theory and Statistical Physics* **1** (3): 186–207. arXiv:physics/0503066. Bibcode:1971TTSP....1..186N. doi:10.1080/00411457108231823. (Original in*Gott.Nachr.* 1918:235–257)

- Emmy Noether (1918). "Invariante Variationenprobleme" (in German).

- *Emmy Noether and The Fabric of Reality* (video) on YouTube

- Byers, Nina (1998). "E. Noether's Discovery of the Deep Connection Between Symmetries and Conservation Laws". arXiv:physics/9807044 [physics.hist-ph].

- John Baez (2002) "Noether's Theorem in a Nutshell."

- Hanca, J.; Tulejab, S.; Hancova, M. (2004). "Symmetries and conservation laws: Consequences of Noether's theorem". *American Journal of Physics* **72** (4): 428–35. Bibcode:2004AmJPh..72..428H. doi:10.1119/1.1591764.

- Merced Montesinos; Ernesto Flores (2006). "Symmetric energy–momentum tensor in Maxwell, Yang–Mills, and Proca theories obtained using only Noether's theorem" (PDF). *Revista Mexicana de Física* **52**: 29–36. arXiv:hep-th/0602190. Bibcode:2006RMxF...52...29M.

- Vladimir Cuesta; Merced Montesinos; José David Vergara (2007). "Gauge invariance of the action principle for gauge systems with noncanonical symplectic structures". *Physical Review D* **76**: 025025. Bibcode:2007PhRvD .doi:10.1103/PhysRevD.76.025025.

- Sardanashvily (2009). "Gauge conservation laws in a general setting. Superpotential". *International Journal of Geometric Methods in Modern Physics* **6** (06): 1047. arXiv:0906.1732. Bibcode:2009arXiv0906.1732S. doi:10.1142/S0219887809003862.
- Neuenschwander, Dwight E. (2010). *Emmy Noether's Wonderful Theorem*. Johns Hopkins University Press. ISBN 978-0-8018-9694-1.

- Noether's Theorem at MathPages.

Emmy Noether was an influential German mathematician known for her groundbreaking contributions to abstract algebra and theoretical physics.

Chapter 36

Electroweak interaction

In particle physics, the **electroweak interaction** is the unified description of two of the four known fundamental interactions of nature: electromagnetism and the weak interaction. Although these two forces appear very different at everyday low energies, the theory models them as two different aspects of the same force. Above the unification energy, on the order of 100 GeV, they would merge into a single **electroweak force**. Thus, if the universe is hot enough (approximately 10^{15} K, a temperature exceeded until shortly after the Big Bang), then the electromagnetic force and weak force merge into a combined electroweak force. During the electroweak epoch, the electroweak force separated from the strong force. During the quark epoch, the electroweak force split into the electromagnetic and weak force.

Sheldon Glashow, Abdus Salam, and Steven Weinberg were awarded the 1979 Nobel Prize in Physics for their contributions to the unification of the weak and electromagnetic interaction between elementary particles.[1][2] The existence of the electroweak interactions was experimentally established in two stages, the first being the discovery of neutral currents in neutrino scattering by the Gargamelle collaboration in 1973, and the second in 1983 by the UA1 and the UA2 collaborations that involved the discovery of the W and Z gauge bosons in proton–antiproton collisions at the converted Super Proton Synchrotron. In 1999, Gerardus 't Hooft and Martinus Veltman were awarded the Nobel prize for showing that the electroweak theory is renormalizable.

36.1 Formulation

Mathematically, the unification is accomplished under an $SU(2) \times U(1)$ gauge group. The corresponding gauge bosons are the **three** W bosons of weak isospin from SU(2) (W_1, W_2, and W_3, and the B boson of weak hypercharge from U(1), respectively, all of which are massless.

In the Standard Model, the W± and Z0 bosons, and the photon, are produced by the spontaneous symmetry breaking of the **electroweak symmetry** from $SU(2) \times U(1)Y$ to $U(1)_{em}$, caused by the Higgs mechanism (see also Higgs boson).[3][4][5][6] $U(1)Y$ and $U(1)_{em}$ are different copies of $U(1)$; the generator of $U(1)_{em}$ is given by $Q = Y/2 + I_3$, where Y is the generator of $U(1)Y$ (called the weak hypercharge), and I_3 is one of the $SU(2)$ generators (a component of weak isospin).

The spontaneous symmetry breaking causes the W_3 and B bosons to coalesce together into two different bosons – the Z0 boson, and the photon (γ) as follows:

$$
\begin{pmatrix} \gamma \\ Z^0 \end{pmatrix} = \begin{pmatrix} \cos\theta_W & \sin\theta_W \\ -\sin\theta_W & \cos\theta_W \end{pmatrix} \begin{pmatrix} B \\ W_3 \end{pmatrix}
$$

Where 0W is the *weak mixing angle*. The axes representing the particles have essentially just been rotated, in the (W_3, B) plane, by the angle 0W. This also introduces a discrepancy between the mass of the Z0 and the mass of the W± particles (denoted as MZ and MW, respectively):

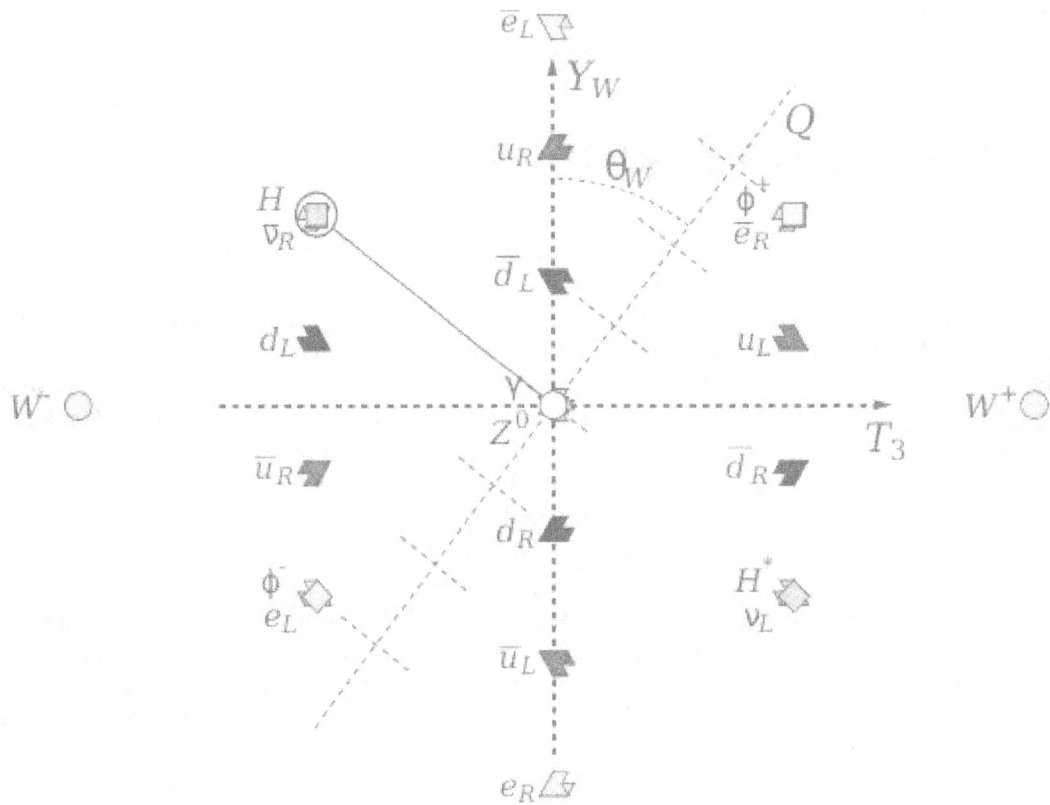

The pattern of weak isospin, T_3, and weak hypercharge, YW, of the known elementary particles, showing electric charge, Q, along the weak mixing angle. The neutral Higgs field (circled) breaks the electroweak symmetry and interacts with other particles to give them mass. Three components of the Higgs field become part of the massive W and Z bosons.

$$M_Z = \frac{M_W}{\cos \theta_W}$$

W_1 and W_2 bosons, in turn, combine to give massive charged bosons

$$W^\pm = \frac{1}{\sqrt{2}}(W_1 \mp iW_2)$$

The distinction between electromagnetism and the weak force arises because there is a (nontrivial) linear combination of Y and I_3 that vanishes for the Higgs boson (it is an eigenstate of both Y and I_3, so the coefficients may be taken as $-I_3$ and

Y): $U(1)_{em}$ is defined to be the group generated by this linear combination, and is unbroken because it does not interact with the Higgs.

36.2 Lagrangian

36.2.1 Before electroweak symmetry breaking

The Lagrangian for the electroweak interactions is divided into four parts before electroweak symmetry breaking

$$\mathcal{L}_{EW} = \mathcal{L}_g + \mathcal{L}_f + \mathcal{L}_h + \mathcal{L}_y.$$

The $\mathcal{L}_g$ term describes the interaction between the three W particles and the B particle.

$$\mathcal{L}_g = -\frac{1}{4} W_a^{\mu\nu} W_{\mu\nu}^a - \frac{1}{4} B^{\mu\nu} B_{\mu\nu}$$

where $W^{a\mu\nu}$ ($a = 1, 2, 3$) and $B^{\mu\nu}$ are the field strength tensors for the weak isospin and weak hypercharge fields.

$\mathcal{L}_f$ is the kinetic term for the Standard Model fermions. The interaction of the gauge bosons and the fermions are through the gauge covariant derivative.

$$\mathcal{L}_f = \overline{Q}_i i \slashed{D} Q_i + \overline{u}_i i \slashed{D} u_i + \overline{d}_i i \slashed{D} d_i + \overline{L}_i i \slashed{D} L_i + \overline{e}_i i \slashed{D} e_i$$

where the subscript i runs over the three generations of fermions, Q, u, and d are the left-handed doublet, right-handed singlet up, and right handed singlet down quark fields, and L and e are the left-handed doublet and right-handed singlet electron fields.

The h term describes the Higgs field F.

$$\mathcal{L}_h = |D_\mu h|^2 - \lambda \left(|h|^2 - \frac{v^2}{2} \right)^2$$

The y term gives the Yukawa interaction that generates the fermion masses after the Higgs acquires a vacuum expectation value.

$$\mathcal{L}_y = -y_{u\,ij} \epsilon^{ab} h_b^\dagger \overline{Q}_{ia} u_j^c - y_{d\,ij} h \overline{Q}_i d_j^c - y_{e\,ij} h \overline{L}_i e_j^c + h.c.$$

36.2.2 After electroweak symmetry breaking

The Lagrangian reorganizes itself after the Higgs boson acquires a vacuum expectation value. Due to its complexity, this Lagrangian is best described by breaking it up into several parts as follows.

$$\mathcal{L}_{EW} = \mathcal{L}_K + \mathcal{L}_N + \mathcal{L}_C + \mathcal{L}_H + \mathcal{L}_{HV} + \mathcal{L}_{WWV} + \mathcal{L}_{WWVV} + \mathcal{L}_Y$$

The kinetic term $\mathcal{L}_K$ contains all the quadratic terms of the Lagrangian, which include the dynamic terms (the partial derivatives) and the mass terms (conspicuously absent from the Lagrangian before symmetry breaking)

$$\mathcal{L}_K = \sum_f \overline{f}(i\partial\!\!\!/ - m_f)f - \frac{1}{4}A_{\mu\nu}A^{\mu\nu} - \frac{1}{2}W_{\mu\nu}^+W^{-\mu\nu} + m_W^2 W_\mu^+ W^{-\mu}$$

$$-\frac{1}{4}Z_{\mu\nu}Z^{\mu\nu} + \frac{1}{2}m_Z^2 Z_\mu Z^\mu + \frac{1}{2}(\partial^\mu H)(\partial_\mu H) - \frac{1}{2}m_H^2 H^2$$

where the sum runs over all the fermions of the theory (quarks and leptons), and the fields $A_{\mu\nu}$, $Z_{\mu\nu}$, $W_{\mu\nu}^-$, and $W_{\mu\nu}^+ \equiv (W_{\mu\nu}^-)^\dagger$ are given as

$$X_{\mu\nu} = \partial_\mu X_\nu - \partial_\nu X_\mu + g f^{abc} X_\mu^b X_\nu^c \,,\text{(replace X by the relevant field, and } f^{abc} \text{ with the structure constants}$$
for the gauge group).

The neutral current $\mathcal{L}_N$ and charged current $\mathcal{L}_C$ components of the Lagrangian contain the interactions between the fermions and gauge bosons.

$$\mathcal{L}_N = e J_\mu^{em} A^\mu + \frac{g}{\cos\theta_W}(J_\mu^3 - \sin^2\theta_W J_\mu^{em})Z^\mu$$

where the electromagnetic current J_μ^{em} and the neutral weak current J_μ^3 are

$$J_\mu^{em} = \sum_f q_f \overline{f}\gamma_\mu f$$

and

$$J_\mu^3 = \sum_f I_f^3 \overline{f}\gamma_\mu \frac{1-\gamma^5}{2}f$$

q_f and I_f^3 are the fermions' electric charges and weak isospin.

The charged current part of the Lagrangian is given by

$$\mathcal{L}_C = -\frac{g}{\sqrt{2}}\left[\overline{u}_i\gamma^\mu \frac{1-\gamma^5}{2}M_{ij}^{CKM}d_j + \overline{\nu}_i\gamma^\mu \frac{1-\gamma^5}{2}e_i\right]W_\mu^+ + h.c.$$

$\mathcal{L}_H$ contains the Higgs three-point and four-point self interaction terms.

$$\mathcal{L}_H = -\frac{gm_H^2}{4m_W}H^3 - \frac{g^2 m_H^2}{32m_W^2}H^4$$

$\mathcal{L}_{HV}$ contains the Higgs interactions with gauge vector bosons.

$$\mathcal{L}_{HV} = \left(gm_W H + \frac{g^2}{4}H^2\right)\left(W_\mu^+ W^{-\mu} + \frac{1}{2\cos^2\theta_W}Z_\mu Z^\mu\right)$$

$\mathcal{L}_{WWV}$ contains the gauge three-point self interactions.

$$\mathcal{L}_{WWV} = -ig[(W_{\mu\nu}^+ W^{-\mu} - W^{+\mu}W_{\mu\nu}^-)(A^\nu \sin\theta_W - Z^\nu \cos\theta_W) + W_\nu^- W_\mu^+ (A^{\mu\nu}\sin\theta_W - Z^{\mu\nu}\cos\theta_W)]$$

$\mathcal{L}_{WWVV}$ contains the gauge four-point self interactions

$$\mathcal{L}_{WWVV} = -\frac{g^2}{4}\left\{ [2W_\mu^+ W^{-\mu} + (A_\mu \sin\theta_W - Z_\mu \cos\theta_W)^2]^2 \right.$$
$$\left. - [W_\mu^+ W_\nu^- + W_\nu^+ W_\mu^- + (A_\mu \sin\theta_W - Z_\mu \cos\theta_W)(A_\nu \sin\theta_W - Z_\nu \cos\theta_W)]^2 \right\}$$

and $\mathcal{L}_Y$ contains the Yukawa interactions between the fermions and the Higgs field.

$$\mathcal{L}_Y = -\sum_f \frac{gm_f}{2m_W}\bar{f}fH$$

Note the $\frac{1-\gamma^5}{2}$ factors in the weak couplings: these factors project out the left handed components of the spinor fields. This is why electroweak theory (after symmetry breaking) is commonly said to be a chiral theory.

36.3 See also

- Fundamental forces

- Formulation of the standard model

- Weinberg angle

- Unitarity gauge

36.4 References

[1] S. Bais (2005). *The Equations: Icons of knowledge*. p. 84. ISBN 0-674-01967-9.

[2] "The Nobel Prize in Physics 1979". The Nobel Foundation. Retrieved 2008-12-16.

[3] F. Englert, R. Brout (1964). "Broken Symmetry and the Mass of Gauge Vector Mesons". *Physical Review Letters* **13** (9): 321–323. Bibcode:1964PhRvL..13..321E. doi:10.1103/PhysRevLett.13.321.

[4] P.W. Higgs (1964). "Broken Symmetries and the Masses of Gauge Bosons". *Physical Review Letters* **13** (16): 508–509. Bibcode:1964PhRvL..13..508H. doi:10.1103/PhysRevLett.13.508.

[5] G.S. Guralnik, C.R. Hagen, T.W.B. Kibble (1964). "Global Conservation Laws and Massless Particles". *Physical Review Letters* **13** (20): 585–587. Bibcode:1964PhRvL..13..585G. doi:10.1103/PhysRevLett.13.585.

[6] G.S. Guralnik (2009). "The History of the Guralnik, Hagen and Kibble development of the Theory of Spontaneous Symmetry Breaking and Gauge Particles". *International Journal of Modern Physics A* **24** (14): 2601–2627. arXiv:0907.3466. Bibcode:2009IJMPA..24.2601G. doi:10.1142/S0217751X09045431.

36.4.1 General readers

- B. A. Schumm (2004). *Deep Down Things: The Breathtaking Beauty of Particle Physics*. Johns Hopkins University Press. ISBN 0-8018-7971-X. Conveys much of the Standard Model with no formal mathematics. Very thorough on the weak interaction.

36.4.2   Texts

- D. J. Griffiths (1987). *Introduction to Elementary Particles*. John Wiley & Sons. ISBN 0-471-60386-4.

- W. Greiner, B. Müller (2000). *Gauge Theory of Weak Interactions*. Springer. ISBN 3-540-67672-4.

- G. L. Kane (1987). *Modern Elementary Particle Physics*. Perseus Books. ISBN 0-201-11749-5.

36.4.3   Articles

- E. S. Abers, B. W. Lee (1973). "Gauge theories". *Physics Reports* **9**: 1–141. Bibcode:1973PhR.....9....1A. doi:10.1016/0370-1573(73)90027-6.

- Y. Hayato; et al. (1999). "Search for Proton Decay through $p \rightarrow \nu K^+$ in a Large Water Cherenkov Detector". *Physical Review Letters* **83**(8): 1529. arXiv:hep-ex/9904020. Bibcode:1999PhRvL..83.1529H.doi:10.1103/Phys.

- J. Hucks (1991). "Global structure of the standard model, anomalies, and charge quantization". *Physical Review D* **43** (8): 2709–2717. Bibcode:1991PhRvD..43.2709H. doi:10.1103/PhysRevD.43.2709.

- S. F. Novaes (2000). "Standard Model: An Introduction". arXiv:hep-ph/0001283 [hep-ph].

- D. P. Roy (1999). "Basic Constituents of Matter and their Interactions — A Progress Report". arXiv:hep-ph/9912523 [hep-ph].

Chapter 37

Higgs mechanism

In the Standard Model of particle physics, the **Higgs mechanism** is essential to explain the generation mechanism of the property "mass" for gauge bosons. Without the Higgs mechanism, or some other effect like it, all bosons (a type of fundamental particle) would be massless, but measurements show that the W^+, W^-, and Z bosons actually have relatively large masses of around 80 GeV/c^2. The Higgs field resolves this conundrum. The simplest description of the mechanism adds a quantum field (the Higgs field) that permeates all space, to the Standard Model. Below some extremely high temperature, the field causes spontaneous symmetry breaking during interactions. The breaking of symmetry triggers the Higgs mechanism, causing the bosons it interacts with to have mass. In the Standard Model, the phrase "Higgs mechanism" refers specifically to the generation of masses for the $W^\pm$, and Z weak gauge bosons through electroweak symmetry breaking.[11] The Large Hadron Collider at CERN announced results consistent with the Higgs particle on March 14, 2013, making it extremely likely that the field, or one like it, exists, and explaining how the Higgs mechanism takes place in nature.

The mechanism was proposed in 1962 by Philip Warren Anderson,[2] following work in the late 1950s on symmetry breaking in superconductivity and a 1960 paper by Yoichiro Nambu that discussed its application within particle physics. A theory able to finally explain mass generation without "breaking" gauge theory was published almost simultaneously by three independent groups in 1964: by Robert Brout and François Englert;[3] by Peter Higgs;[4] and by Gerald Guralnik, C. R. Hagen, and Tom Kibble.[5][6][7] The Higgs mechanism is therefore also called the **Brout–Englert–Higgs mechanism** or **Englert–Brout–Higgs–Guralnik–Hagen–Kibble mechanism**,[8] **Anderson–Higgs mechanism**,[9] **Anderson–Higgs–Kibble mechanism**,[10] **Higgs–Kibble mechanism** by Abdus Salam[11] and **ABEGHHK'tH mechanism** [for Anderson, Brout, Englert, Guralnik, Hagen, Higgs, Kibble and 't Hooft] by Peter Higgs.[11]

On October 8, 2013, following the discovery at CERN's Large Hadron Collider of a new particle that appeared to be the long-sought Higgs boson predicted by the theory, it was announced that Peter Higgs and François Englert had been awarded the 2013 Nobel Prize in Physics (Englert's co-author Robert Brout had died in 2011 and the Nobel Prize is not usually awarded posthumously).[12]

37.1   Standard model

The Higgs mechanism was incorporated into modern particle physics by Steven Weinberg and Abdus Salam, and is an essential part of the standard model.

In the standard model, at temperatures high enough that electroweak symmetry is unbroken, all elementary particles are massless. At a critical temperature the Higgs field becomes tachyonic, the symmetry is spontaneously broken by condensation, and the W and Z bosons acquire masses. (EWSB, ElectroWeak Symmetry Breaking, is an abbreviation used for this.)

Fermions, such as the leptons and quarks in the Standard Model, can also acquire mass as a result of their interaction with the Higgs field, but not in the same way as the gauge bosons.

37.1.1   Structure of the Higgs field

In the standard model, the Higgs field is an **SU**(2) doublet, a complex scalar with four real components (or equivalently with two complex components). Its (weak hypercharge) **U**(1) charge is 1. That means that it transforms as a spinor under **SU**(2). Under **U**(1) rotations, it is multiplied by a phase, which thus mixes the real and imaginary parts of the complex spinor into each other—so this is *not the same* as two complex spinors mixing under **U**(1) (which would have eight real components between them), but instead is the spinor representation of the group **U**(2).

The Higgs field, through the interactions specified (summarized, represented, or even simulated) by its potential, induces spontaneous breaking of three out of the four generators ("directions") of the gauge group **SU**(2) × **U**(1): three out of its four components would ordinarily amount to Goldstone bosons, if they were not coupled to gauge fields.

However, after symmetry breaking, these three of the four degrees of freedom in the Higgs field mix with the three W and Z bosons (W+, W− and Z), and are only observable as spin components of these weak bosons, which are now massive; while the one remaining degree of freedom becomes the Higgs boson—a new scalar particle.

37.1.2   The photon as the part that remains massless

The gauge group of the electroweak part of the standard model is **SU**(2) × **U**(1). The group **SU**(2) is the group of all 2-by-2 unitary matrices with unit determinant; all the orthonormal changes of coordinates in a complex two dimensional vector space.

Rotating the coordinates so that the second basis vector points in the direction of the Higgs boson makes the vacuum expectation value of H the spinor $(0, v)$. The generators for rotations about the x, y, and z axes are by half the Pauli matrices σx, σy, and σz, so that a rotation of angle θ about the z-axis takes the vacuum to

$$(0, ve^{-i\theta/2}).$$

While the T_x and T_y generators mix up the top and bottom components of the spinor, the T_z rotations only multiply each by opposite phases. This phase can be undone by a **U**(1) rotation of angle $1/2\theta$. Consequently, under both an **SU**(2) T_z-rotation and a **U**(1) rotation by an amount $1/2\theta$, *the vacuum is invariant.*

This combination of generators

$$Q = T_z + \frac{Y}{2}$$

defines the unbroken part of the gauge group, where Q is the electric charge, T_z is the generator of rotations around the z-axis in the **SU**(2) and Y is the hypercharge generator of the **U**(1). This combination of generators (a z rotation in the **SU**(2) and a simultaneous **U**(1) rotation by half the angle) preserves the vacuum, and defines the unbroken gauge group in the standard model, namely *the electric charge* group. The part of the gauge field in this direction stays massless, and amounts to the physical photon.

37.1.3   Consequences for fermions

In spite of the introduction of spontaneous symmetry breaking, the mass terms oppose the chiral gauge invariance. For these fields the mass terms should always be replaced by a gauge-invariant "Higgs" mechanism. One possibility is some kind of "Yukawa coupling" (see below) between the fermion field ψ and the Higgs field Φ, with unknown couplings $G\psi$, which after symmetry breaking (more precisely: after expansion of the Lagrange density around a suitable ground state) again results in the original mass terms, which are now, however (i.e. by introduction of the Higgs field) written in a gauge-invariant way. The Lagrange density for the "Yukawa" interaction of a fermion field ψ and the Higgs field Φ is

$$\mathcal{L}_{\text{Fermion}}(\phi, A, \psi) = \overline{\psi}\gamma^\mu D_\mu \psi + G_\psi \overline{\psi}\phi\psi,$$

where again the gauge field A only enters $D\mu$ (i.e., it is only indirectly visible). The quantities γ^μ are the Dirac matrices, and $G\psi$ is the already-mentioned "Yukawa" coupling parameter. Already now the mass-generation follows the same principle as above, namely from the existence of a finite expectation value $|\langle\phi\rangle|$, as described above. Again, this is crucial for the existence of the property "mass".

37.2 History of research

37.2.1 Background

Spontaneous symmetry breaking offered a framework to introduce bosons into relativistic quantum field theories. However, according to Goldstone's theorem, these bosons should be massless.[13] The only observed particles which could be approximately interpreted as Goldstone bosons were the pions, which Yoichiro Nambu related to chiral symmetry breaking.

A similar problem arises with Yang–Mills theory (also known as non-abelian gauge theory), which predicts massless spin−1 gauge bosons. Massless weakly interacting gauge bosons lead to long-range forces, which are only observed for electromagnetism and the corresponding massless photon. Gauge theories of the weak force needed a way to describe massive gauge bosons in order to be consistent.

37.2.2 Discovery

The mechanism was proposed in 1962 by Philip Warren Anderson,[2] who discussed its consequences for particle physics but did not work out an explicit relativistic model. The relativistic model was developed in 1964 by three independent groups – Robert Brout and François Englert;[3] Peter Higgs;[4] and Gerald Guralnik, Carl Richard Hagen, and Tom Kibble.[5][6][7] Slightly later, in 1965, but independently from the other publications[14][15][16][17][18][19] the mechanism was also proposed by Alexander Migdal and Alexander Polyakov,[20] at that time Soviet undergraduate students. However, the paper was delayed by the Editorial Office of JETP, and was published only in 1966.

The mechanism is closely analogous to phenomena previously discovered by Yoichiro Nambu involving the "vacuum structure" of quantum fields in superconductivity.[21] A similar but distinct effect (involving an affine realization of what is now recognized as the Higgs field), known as the Stueckelberg mechanism, had previously been studied by Ernst Stueckelberg.

These physicists discovered that when a gauge theory is combined with an additional field that spontaneously breaks the symmetry group, the gauge bosons can consistently acquire a nonzero mass. In spite of the large values involved (see below) this permits a gauge theory description of the weak force, which was independently developed by Steven Weinberg and Abdus Salam in 1967. Higgs's original article presenting the model was rejected by Physics Letters. When revising the article before resubmitting it to Physical Review Letters, he added a sentence at the end,[22] mentioning that it implies the existence of one or more new, massive scalar bosons, which do not form complete representations of the symmetry group; these are the Higgs bosons.

The three papers by Brout and Englert; Higgs; and Guralnik, Hagen, and Kibble were each recognized as "milestone letters" by Physical Review Letters in 2008.[23] While each of these seminal papers took similar approaches, the contributions and differences among the 1964 PRL symmetry breaking papers are noteworthy. All six physicists were jointly awarded the 2010 J. J. Sakurai Prize for Theoretical Particle Physics for this work.[24]

Benjamin W. Lee is often credited with first naming the "Higgs-like" mechanism, although there is debate around when this first occurred.[25][26][27] One of the first times the *Higgs* name appeared in print was in 1972 when Gerardus 't Hooft and Martinus J. G. Veltman referred to it as the "Higgs–Kibble mechanism" in their Nobel winning paper.[28][29]

Philip W. Anderson, the first to propose the mechanism in 1962.

37.3   Examples

The Higgs mechanism occurs whenever a charged field has a vacuum expectation value. In the nonrelativistic context, this is the Landau model of a charged Bose–Einstein condensate, also known as a superconductor. In the relativistic condensate, the condensate is a scalar field, and is relativistically invariant.

Five of the six 2010 APS Sakurai Prize Winners – (L to R) Tom Kibble, Gerald Guralnik, Carl Richard Hagen, François Englert, and Robert Brout

37.3.1 Landau model

The Higgs mechanism is a type of superconductivity which occurs in the vacuum. It occurs when all of space is filled with a sea of particles which are charged, or, in field language, when a charged field has a nonzero vacuum expectation value. Interaction with the quantum fluid filling the space prevents certain forces from propagating over long distances (as it does in a superconducting medium; e.g., in the Ginzburg–Landau theory).

A superconductor expels all magnetic fields from its interior, a phenomenon known as the Meissner effect. This was mysterious for a long time, because it implies that electromagnetic forces somehow become short-range inside the superconductor. Contrast this with the behavior of an ordinary metal. In a metal, the conductivity shields electric fields by rearranging charges on the surface until the total field cancels in the interior. But magnetic fields can penetrate to any distance, and if a magnetic monopole (an isolated magnetic pole) is surrounded by a metal the field can escape without collimating into a string. In a superconductor, however, electric charges move with no dissipation, and this allows for permanent surface currents, not just surface charges. When magnetic fields are introduced at the boundary of a superconductor, they produce surface currents which exactly neutralize them. The Meissner effect is due to currents in a thin surface layer, whose thickness, the London penetration depth, can be calculated from a simple model (the Ginzburg–Landau theory).

This simple model treats superconductivity as a charged Bose–Einstein condensate. Suppose that a superconductor contains bosons with charge q. The wavefunction of the bosons can be described by introducing a quantum field, ψ, which obeys the Schrödinger equation as a field equation (in units where the reduced Planck constant, $\hbar$, is set to 1):

$$i\frac{\partial}{\partial t}\psi = \frac{(\nabla - iqA)^2}{2m}\psi.$$

The operator $\psi(x)$ annihilates a boson at the point x, while its adjoint $\psi^\dagger$ creates a new boson at the same point. The

Number six: Peter Higgs 2009

wavefunction of the Bose–Einstein condensate is then the expectation value ψ of $\psi(x)$, which is a classical function that obeys the same equation. The interpretation of the expectation value is that it is the phase that one should give to a newly created boson so that it will coherently superpose with all the other bosons already in the condensate.

When there is a charged condensate, the electromagnetic interactions are screened. To see this, consider the effect of a gauge transformation on the field. A gauge transformation rotates the phase of the condensate by an amount which changes from point to point, and shifts the vector potential by a gradient:

$$\psi \to e^{iq\phi(x)}\psi$$
$$A \to A + \nabla\phi.$$

When there is no condensate, this transformation only changes the definition of the phase of ψ at every point. But when there is a condensate, the phase of the condensate defines a preferred choice of phase.

The condensate wave function can be written as

$$\psi(x) = \rho(x)\,e^{i\theta(x)},$$

where ρ is real amplitude, which determines the local density of the condensate. If the condensate were neutral, the flow would be along the gradients of θ, the direction in which the phase of the Schrödinger field changes. If the phase θ changes slowly, the flow is slow and has very little energy. But now θ can be made equal to zero just by making a gauge transformation to rotate the phase of the field.

The energy of slow changes of phase can be calculated from the Schrödinger kinetic energy.

$$H = \frac{1}{2m} |(qA + \nabla)\psi|^2.$$

and taking the density of the condensate ρ to be constant,

$$H \approx \frac{\rho^2}{2m}(qA + \nabla\theta)^2.$$

Fixing the choice of gauge so that the condensate has the same phase everywhere, the electromagnetic field energy has an extra term,

$$\frac{q^2\rho^2}{2m}A^2.$$

When this term is present, electromagnetic interactions become short-ranged. Every field mode, no matter how long the wavelength, oscillates with a nonzero frequency. The lowest frequency can be read off from the energy of a long wavelength A mode,

$$E \approx \frac{\dot{A}^2}{2} + \frac{q^2\rho^2}{2m}A^2.$$

This is a harmonic oscillator with frequency

$$\sqrt{\frac{1}{m}q^2\rho^2}.$$

The quantity $|\psi|^2$ ($=\rho^2$) is the density of the condensate of superconducting particles.

In an actual superconductor, the charged particles are electrons, which are fermions not bosons. So in order to have superconductivity, the electrons need to somehow bind into Cooper pairs. The charge of the condensate q is therefore twice the electron charge e. The pairing in a normal superconductor is due to lattice vibrations, and is in fact very weak: this means that the pairs are very loosely bound. The description of a Bose–Einstein condensate of loosely bound pairs is actually more difficult than the description of a condensate of elementary particles, and was only worked out in 1957 by Bardeen, Cooper and Schrieffer in the famous BCS theory.

37.3.2   Abelian Higgs mechanism

Gauge invariance means that certain transformations of the gauge field do not change the energy at all. If an arbitrary gradient is added to A, the energy of the field is exactly the same. This makes it difficult to add a mass term, because a mass term tends to push the field toward the value zero. But the zero value of the vector potential is not a gauge invariant idea. What is zero in one gauge is nonzero in another.

So in order to give mass to a gauge theory, the gauge invariance must be broken by a condensate. The condensate will then define a preferred phase, and the phase of the condensate will define the zero value of the field in a gauge-invariant way. The gauge-invariant definition is that a gauge field is zero when the phase change along any path from parallel transport is equal to the phase difference in the condensate wavefunction.

The condensate value is described by a quantum field with an expectation value, just as in the Ginzburg-Landau model.

In order for the phase of the vacuum to define a gauge, the field must have a phase (also referred to as 'to be charged'). In order for a scalar field Φ to have a phase, it must be complex, or (equivalently) it should contain two fields with a symmetry which rotates them into each other. The vector potential changes the phase of the quanta produced by the field when they move from point to point. In terms of fields, it defines how much to rotate the real and imaginary parts of the fields into each other when comparing field values at nearby points.

The only renormalizable model where a complex scalar field Φ acquires a nonzero value is the Mexican-hat model, where the field energy has a minimum away from zero. The action for this model is

$$S(\phi) = \int \frac{1}{2}|\partial\phi|^2 - \lambda\left(|\phi|^2 - \Phi^2\right)^2,$$

which results in the Hamiltonian

$$H(\phi) = \frac{1}{2}|\dot{\phi}|^2 + |\nabla\phi|^2 + V(|\phi|).$$

The first term is the kinetic energy of the field. The second term is the extra potential energy when the field varies from point to point. The third term is the potential energy when the field has any given magnitude.

This potential energy, $V(z, \Phi) = \lambda(|z|^2 - \Phi^2)^2$,[30] has a graph which looks like a Mexican hat, which gives the model its name. In particular, the minimum energy value is not at $z = 0$, but on the circle of points where the magnitude of z is Φ.

When the field $\Phi(x)$ is not coupled to electromagnetism, the Mexican-hat potential has flat directions. Starting in any one of the circle of vacua and changing the phase of the field from point to point costs very little energy. Mathematically, if

$$\phi(x) = \Phi e^{i\theta(x)}$$

with a constant prefactor, then the action for the field $\theta(x)$, i.e., the "phase" of the Higgs field $\Phi(x)$, has only derivative terms. This is not a surprise. Adding a constant to $\theta(x)$ is a symmetry of the original theory, so different values of $\theta(x)$ cannot have different energies. This is an example of Goldstone's theorem: spontaneously broken continuous symmetries normally produce massless excitations.

The Abelian Higgs model is the Mexican-hat model coupled to electromagnetism:

$$S(\phi, A) = \int -\frac{1}{4}F^{\mu\nu}F_{\mu\nu} + |(\partial - iqA)\phi|^2 - \lambda(|\phi|^2 - \Phi^2)^2.$$

The classical vacuum is again at the minimum of the potential, where the magnitude of the complex field φ is equal to Φ. But now the phase of the field is arbitrary, because gauge transformations change it. This means that the field $\theta(x)$ can be set to zero by a gauge transformation, and does not represent any actual degrees of freedom at all.

Furthermore, choosing a gauge where the phase of the vacuum is fixed, the potential energy for fluctuations of the vector field is nonzero. So in the abelian Higgs model, the gauge field acquires a mass. To calculate the magnitude of the mass, consider a constant value of the vector potential A in the x direction in the gauge where the condensate has constant phase. This is the same as a sinusoidally varying condensate in the gauge where the vector potential is zero. In the gauge where A is zero, the potential energy density in the condensate is the scalar gradient energy:

$$E = \frac{1}{2}\left|\partial\left(\Phi e^{iqAx}\right)\right|^2 = \frac{1}{2}q^2\Phi^2 A^2.$$

This energy is the same as a mass term $1/2m^2A^2$ where $m = q\Phi$.

37.3.3 Nonabelian Higgs mechanism

The Nonabelian Higgs model has the following action:

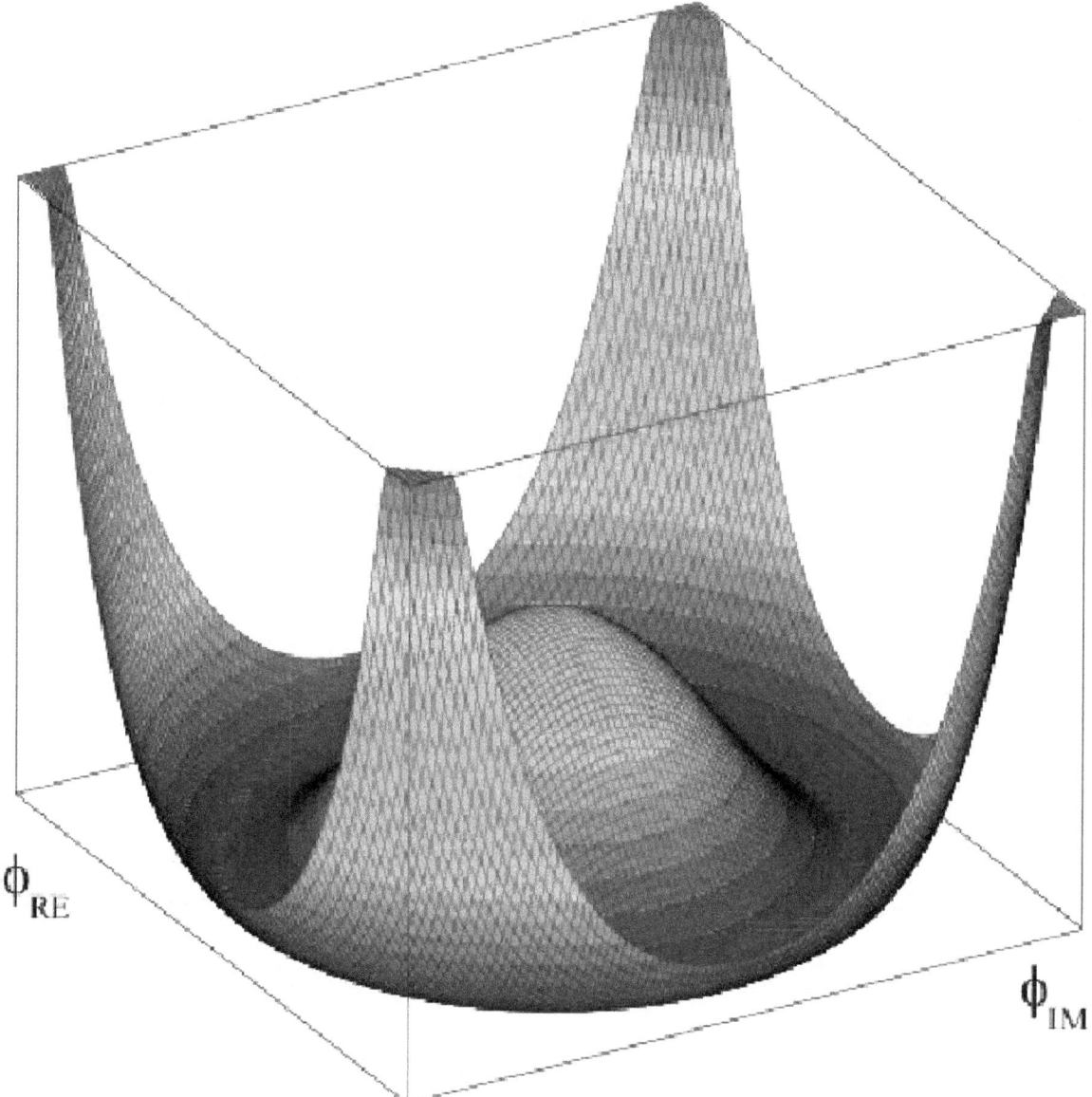

ϕ_{RE}

ϕ_{IM}

Higgs potential V. *For a fixed value of* λ *the potential is presented upwards against the real and imaginary parts of* Φ. *The* Mexican-hat *or* champagne-bottle *profile at the ground should be noted.*

$$S(\phi, \mathbf{A}) = \int \frac{1}{4g^2} \mathrm{tr}(F^{\mu\nu} F_{\mu\nu}) + |D\phi|^2 + V(|\phi|)$$

where now the nonabelian field $\mathbf{A}$ is contained in D and in the tensor components $F^{\mu\nu}$ and $F_{\mu\nu}$ (the relation between $\mathbf{A}$ and those components is well-known from the Yang–Mills theory).

It is exactly analogous to the Abelian Higgs model. Now the field φ is in a representation of the gauge group, and the gauge covariant derivative is defined by the rate of change of the field minus the rate of change from parallel transport using the gauge field A as a connection.

$$D\phi = \partial\phi - iA^k t_k \phi$$

Again, the expectation value of Φ defines a preferred gauge where the vacuum is constant, and fixing this gauge, fluctuations in the gauge field A come with a nonzero energy cost.

Depending on the representation of the scalar field, not every gauge field acquires a mass. A simple example is in the renormalizable version of an early electroweak model due to Julian Schwinger. In this model, the gauge group is **SO**(3) (or **SU**(2) – there are no spinor representations in the model), and the gauge invariance is broken down to **U**(1) or **SO**(2) at long distances. To make a consistent renormalizable version using the Higgs mechanism, introduce a scalar field φ^a which transforms as a vector (a triplet) of **SO**(3). If this field has a vacuum expectation value, it points in some direction in field space. Without loss of generality, one can choose the z-axis in field space to be the direction that φ is pointing, and then the vacuum expectation value of φ is $(0, 0, A)$, where A is a constant with dimensions of mass ($c = \hbar = 1$).

Rotations around the z-axis form a **U**(1) subgroup of **SO**(3) which preserves the vacuum expectation value of φ, and this is the unbroken gauge group. Rotations around the x and y-axis do not preserve the vacuum, and the components of the **SO**(3) gauge field which generate these rotations become massive vector mesons. There are two massive W mesons in the Schwinger model, with a mass set by the mass scale A, and one massless **U**(1) gauge boson, similar to the photon.

The Schwinger model predicts magnetic monopoles at the electroweak unification scale, and does not predict the Z meson. It doesn't break electroweak symmetry properly as in nature. But historically, a model similar to this (but not using the Higgs mechanism) was the first in which the weak force and the electromagnetic force were unified.

37.3.4   Affine Higgs mechanism

Ernst Stueckelberg discovered[311] a version of the Higgs mechanism by analyzing the theory of quantum electrodynamics with a massive photon. Effectively, Stueckelberg's model is a limit of the regular Mexican hat Abelian Higgs model, where the vacuum expectation value H goes to infinity and the charge of the Higgs field goes to zero in such a way that their product stays fixed. The mass of the Higgs boson is proportional to H, so the Higgs boson becomes infinitely massive and decouples, so is not present in the discussion. The vector meson mass, however, equals to the product eH, and stays finite.

The interpretation is that when a **U**(1) gauge field does not require quantized charges, it is possible to keep only the angular part of the Higgs oscillations, and discard the radial part. The angular part of the Higgs field θ has the following gauge transformation law:

$$\theta \to \theta + e\alpha$$

$$A \to A + \alpha.$$

The gauge covariant derivative for the angle (which is actually gauge invariant) is:

$$D\theta = \partial\theta - eA.$$

In order to keep θ fluctuations finite and nonzero in this limit, θ should be rescaled by H, so that its kinetic term in the action stays normalized. The action for the theta field is read off from the Mexican hat action by substituting $\phi = He^{\frac{1}{H}i\theta}$

$$S = \int \frac{1}{4}F^2 + \frac{1}{2}(D\theta)^2 = \int \frac{1}{4}F^2 + \frac{1}{2}(\partial\theta - HeA)^2 = \int \frac{1}{4}F^2 + \frac{1}{2}(\partial\theta - mA)^2$$

since eH is the gauge boson mass. By making a gauge transformation to set $\theta = 0$, the gauge freedom in the action is eliminated, and the action becomes that of a massive vector field:

$$S = \int \frac{1}{4}F^2 + \frac{1}{2}m^2A^2.$$

To have arbitrarily small charges requires that the **U**(1) is not the circle of unit complex numbers under multiplication, but the real numbers **R** under addition, which is only different in the global topology. Such a **U**(1) group is *non-compact*. The field 0 transforms as an affine representation of the gauge group. Among the allowed gauge groups, only non-compact **U**(1) admits affine representations, and the **U**(1) of electromagnetism is experimentally known to be compact, since charge quantization holds to extremely high accuracy.

The Higgs condensate in this model has infinitesimal charge, so interactions with the Higgs boson do not violate charge conservation. The theory of quantum electrodynamics with a massive photon is still a renormalizable theory, one in which electric charge is still conserved, but magnetic monopoles are not allowed. For nonabelian gauge theory, there is no affine limit, and the Higgs oscillations cannot be too much more massive than the vectors.

37.4 See also

- Electromagnetic mass
- Higgs bundle
- Mass generation
- QCD vacuum
- Quantum triviality
- Top quark condensate
- Yang–Mills–Higgs equations

37.5 References

[1] G. Bernardi, M. Carena, and T. Junk: "Higgs bosons: theory and searches", Reviews of Particle Data Group: Hypothetical particles and Concepts, 2007, http://pdg.lbl.gov/2008/reviews/higgs_s055.pdf

[2] P.W.Anderson(1962). "Plasmons,Gauge Invariance,and Mass".*Physical Review* **130**(1): 439–442. Bibcode:1963PhRv.. doi:10.1103/PhysRev.130.439.

[3] F. Englert and R. Brout (1964). "Broken Symmetry and the Mass of Gauge Vector Mesons". *Physical Review Letters* **13** (9): 321–323. Bibcode:1964PhRvL..13..321E. doi:10.1103/PhysRevLett.13.321.

[4] Peter W. Higgs (1964). "Broken Symmetries and the Masses of Gauge Bosons". *Physical Review Letters* **13** (16): 508–509. Bibcode:1964PhRvL..13..508H. doi:10.1103/PhysRevLett.13.508.

[5] G. S. Guralnik, C. R. Hagen, and T. W. B. Kibble (1964). "Global Conservation Laws and Massless Particles". *Physical Review Letters* **13** (20): 585–587. Bibcode:1964PhRvL..13..585G. doi:10.1103/PhysRevLett.13.585.

[6] Gerald S. Guralnik (2009). "The History of the Guralnik, Hagen and Kibble development of the Theory of Spontaneous Symmetry Breaking and Gauge Particles". *International Journal of Modern Physics* **A24** (14): 2601–2627. arXiv:0907.3466. Bibcode:2009IJMPA..24.2601G. doi:10.1142/S0217751X09045431.

[7] History of Englert–Brout–Higgs–Guralnik–Hagen–Kibble Mechanism. Scholarpedia.

[8] "Englert–Brout–Higgs–Guralnik–Hagen–Kibble Mechanism". Scholarpedia. Retrieved 2012-06-16.

[9] Liu, G. Z.; Cheng, G. (2002). "Extension of the Anderson-Higgs mechanism". *Physical Review B* **65** (13): 132513. arXiv:cond-mat/0106070. Bibcode:2002PhRvB..65m2513L. doi:10.1103/PhysRevB.65.132513.

[10] Matsumoto, H.; Papastamatiou, N. J.; Umezawa, H.; Vitiello, G. (1975). "Dynamical rearrangement in the Anderson-Higgs-Kibble mechanism". *Nuclear Physics B* **97**: 61. doi:10.1016/0550-3213(75)90215-1.

[11] Close, Frank (2011). *The Infinity Puzzle: Quantum Field Theory and the Hunt for an Orderly Universe*. Oxford: Oxford University Press. ISBN 978-0-19-959350-7.

[12] "Press release from Royal Swedish Academy of Sciences" (PDF). 8 October 2013. Retrieved 8 October 2013.

[13] "Guralnik, G S; Hagen, C R and Kibble, T W B (1967). Broken Symmetries and the Goldstone Theorem. Advances in Physics, vol. 2" (PDF).

[14] A.M. Polyakov, A View From The Island, 1992

[15] Farhi, E., & Jackiw, R. W. (1982). *Dynamical Gauge Symmetry Breaking: A Collection Of Reprints*. Singapore: World Scientific Pub. Co.

[16] Frank Close. "The Infinity Puzzle." 2011, p.158

[17] Norman Dombey, "Higgs Boson: Credit Where It's Due". The Guardian, July 6, 2012

[18] Cern Courier, Mar 1, 2006

[19] Sean Carrol, "The Particle At The End Of The Universe: The Hunt For The Higgs And The Discovery Of A New World", 2012, p.228

[20] A. A. Migdal and A. M. Polyakov, "Spontaneous Breakdown of Strong Interaction Symmetry and Absence of Massless Particles", *JETP* **51**, 135, July 1966 (English translation: *Soviet Physics JETP*, **24**, 1, January 1967)

[21] Nambu, Y (1960). "Quasiparticles and Gauge Invariance in the Theory of Superconductivity". *Physical Review* **117** (3): 648–663. Bibcode:1960PhRv..117..648N. doi:10.1103/PhysRev.117.648.

[22] Higgs, Peter (2007). "Prehistory of the Higgs boson". *Comptes Rendus Physique* **8** (9): 970–972. Bibcode:2007CRPhy...8..970H. doi:10.1016/j.crhy.2006.12.006.

[23] "Physical Review Letters – 50th Anniversary Milestone Papers". Prl.aps.org. Retrieved 2012-06-16.

[24] "American Physical Society – J. J. Sakurai Prize Winners". Aps.org. Retrieved 2012-06-16.

[25] Department of Physics and Astronomy. "Rochester's Hagen Sakurai Prize Announcement". Pas.rochester.edu. Retrieved 2012-06-16.

[26] FermiFred (2010-02-15). "C.R. Hagen discusses naming of Higgs Boson in 2010 Sakurai Prize Talk". Youtube.com. Retrieved 2012-06-16.

[27] Sample, Ian (2009-05-29). "Anything but the God particle by Ian Sample". Guardian. Retrieved 2012-06-16.

[28] G. 't Hooft and M. Veltman (1972). "Regularization and Renormalization of Gauge Fields". *Nuclear Physics B* **44** (1): 189–219. Bibcode:1972NuPhB..44..189T. doi:10.1016/0550-3213(72)90279-9.

[29] "Regularization and Renormalization of Gauge Fields by t'Hooft and Veltman (PDF)" (PDF). Retrieved 2012-06-16.

[30] Goldstone,J. (1961). "Field theories with "Superconductor"solutions".*Il Nuovo Cimento***19**: 154–164. doi:10.1007/BF028.

[31] Stueckelberg, E. C. G. (1938), "Die Wechselwirkungskräfte in der Elektrodynamik und in der Feldtheorie der Kräfte", *Helv. Phys. Acta*. **11**: 225

37.6   Further reading

- Schumm, Bruce A. (2004) *Deep Down Things*. Johns Hopkins Univ. Press. Chpt. 9.

- Englert-Brout-Higgs-Guralnik-Hagen-Kibble mechanism Tom W B Kibble Scholarpedia, 4(1):6441. doi:10.4249/scholarpedia.6441

37.7 External links

- Guralnik, G.S.; Hagen, C.R.; Kibble, T.W.B. (1964). "Global Conservation Laws and Massless Particles". *Physical Review Letters* **13** (20): 585–87. Bibcode:1964PhRvL..13..585G. doi:10.1103/PhysRevLett.13.585.

- Mark D. Roberts (1999) "A Generalized Higgs Model"

- 2010 Sakurai Prize - All Events - YouTube

- From BCS to the LHC - CERN Courier Jan 21, 2008, Steven Weinberg, University of Texas at Austin.

- Higgs, dark matter and supersymmetry: What the Large Hadron Collider will tell us (Steven Weinberg) - YouTube on YouTube 06-11-2009

- Gerry Guralnik speaks at Brown University about the 1964 PRL papers

- Guralnik, Gerald (2013). "Heretical Ideas that Provided the Cornerstone for the Standard Model of Particle Physics". SPG MITTEILUNGEN March 2013, No. 39, (p. 14)

- Steven Weinberg Praises Teams for Higgs Boson Theory

- Physical Review Letters – 50th Anniversary Milestone Papers

- Imperial College London on PRL 50th Anniversary Milestone Papers

- Englert–Brout–Higgs–Guralnik–Hagen–Kibble Mechanism on Scholarpedia

- History of Englert–Brout–Higgs–Guralnik–Hagen–Kibble Mechanism on Scholarpedia

- The Hunt for the Higgs at Tevatron

- The Mystery of Empty Space on YouTube. A lecture with UCSD physicist Kim Griest (43 minutes)

Chapter 38

Symmetry in mathematics

For other uses, see Symmetry (disambiguation) and Bilateral (disambiguation).

Symmetry occurs not only in geometry, but also in other branches of mathematics. Symmetry is a type of invariance: the property that something does not change under a set of transformations.

Given a structured object X of any sort, a symmetry is a mapping of the object onto itself which preserves the structure. This occurs in many cases; for example, if X is a set with no additional structure, a symmetry is a bijective map from the set to itself, giving rise to permutation groups. If the object X is a set of points in the plane with its metric structure or any other metric space, a symmetry is a bijection of the set to itself which preserves the distance between each pair of points (an isometry).

In general, every kind of structure in mathematics will have its own kind of symmetry, many of which are listed in the given points mentioned above.

38.1 Symmetry in geometry

Main article: Symmetry (geometry)

The types of symmetry considered in basic geometry (like reflection and rotation symmetry) are described more fully in the main article on symmetry.

38.2 Symmetry in calculus

38.2.1 Even and odd functions

Main article: Even and odd functions

Even functions

Let $f(x)$ be a real-valued function of a real variable. Then f is **even** if the following equation holds for all x and $-x$ in the domain of f:

$$f(x) = f(-x).$$

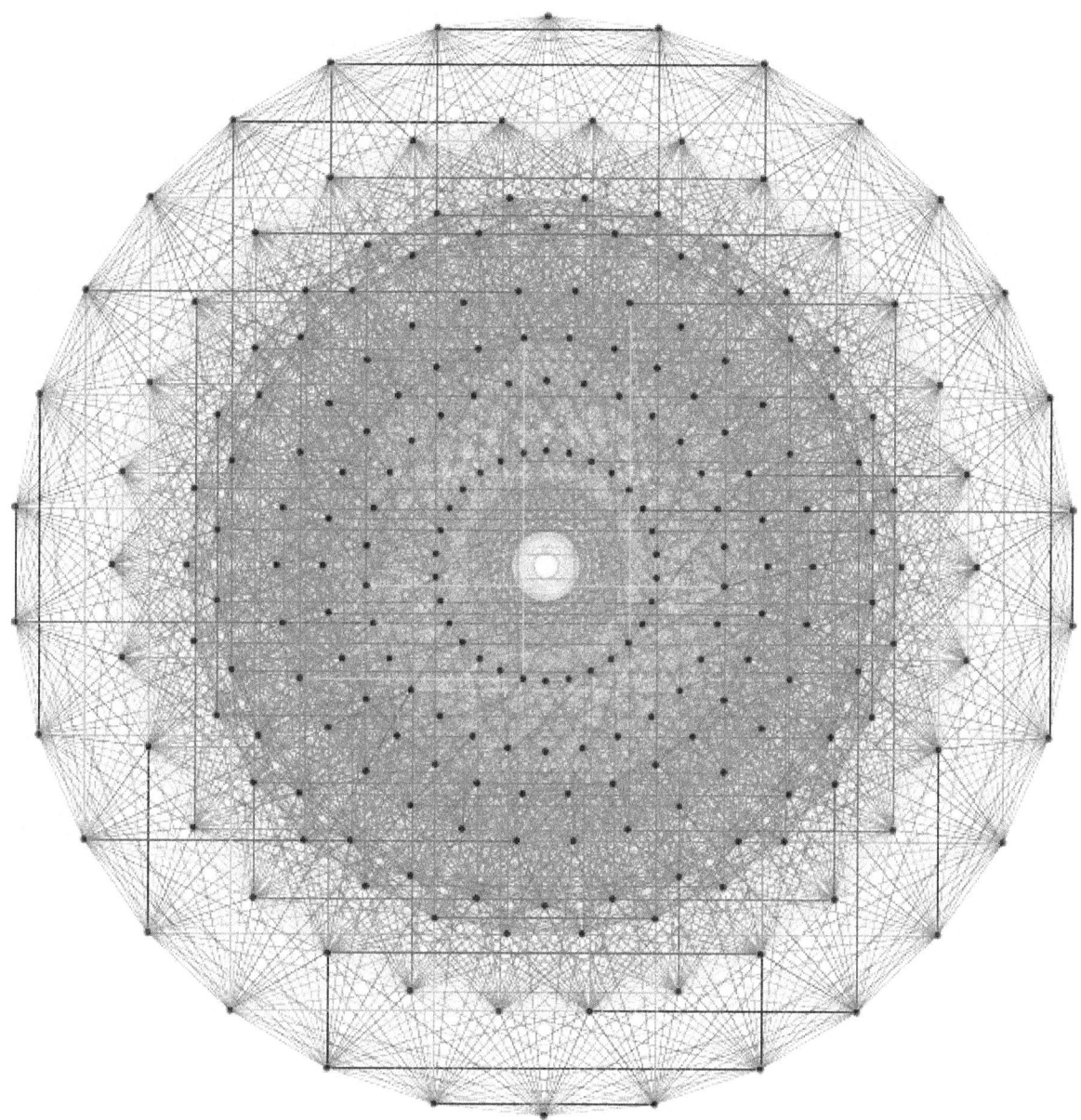

The root system of the exceptional Lie group E₈. Lie groups have many symmetries.

Geometrically speaking, the graph face of an even function is symmetric with respect to the y-axis, meaning that its graph remains unchanged after reflection about the y-axis.

Examples of even functions are x, x^2, x^4, cos(x), and cosh(x).

Odd functions

Again, let $f(x)$ be a real-valued function of a real variable. Then f is **odd** if the following equation holds for all x and $-x$ in the domain of f:

$$-f(x) = f(-x).$$

or

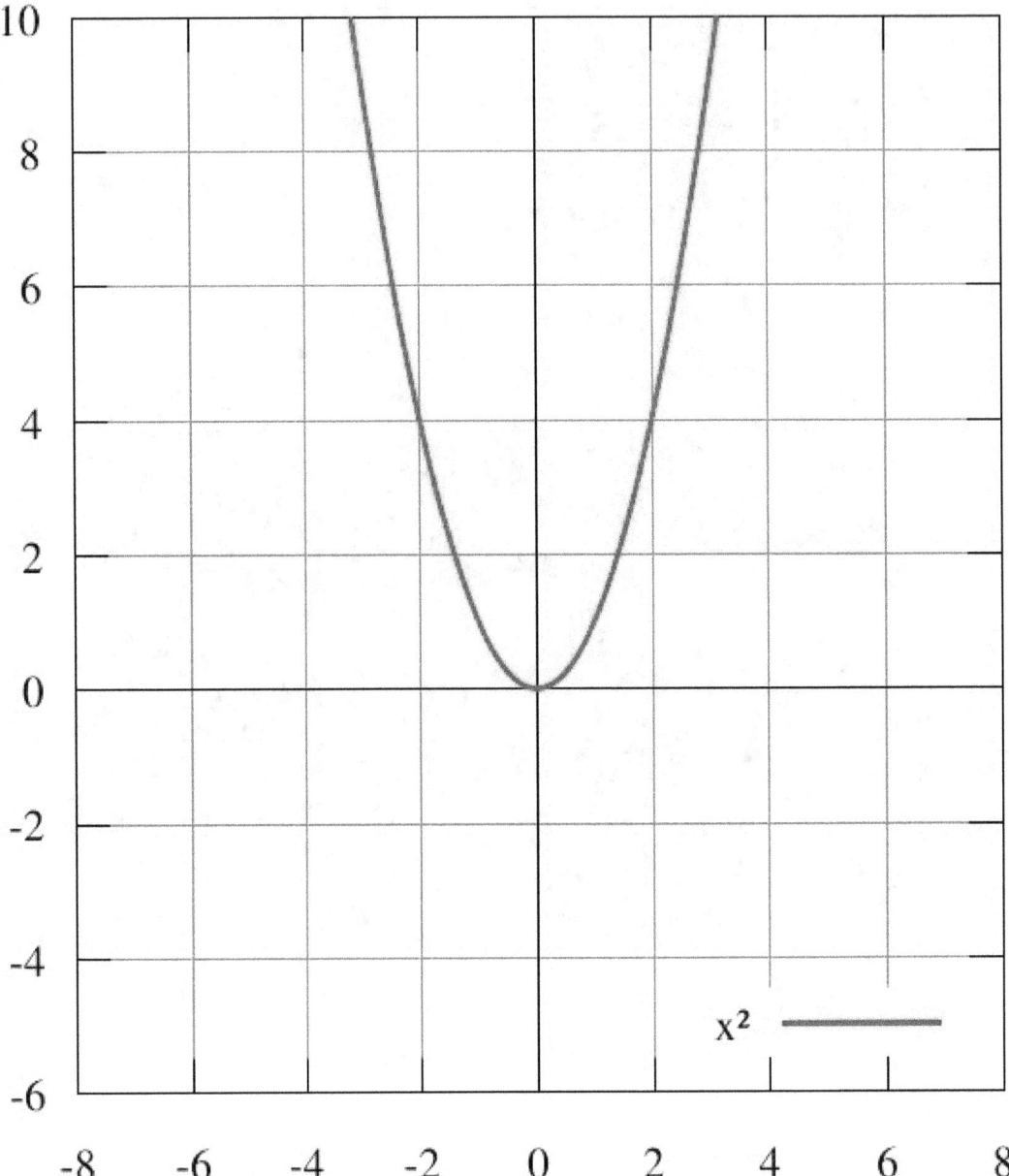

$f(x) = x^2$ *is an example of an even function.*

$$f(x) + f(-x) = 0.$$

Geometrically, the graph of an odd function has rotational symmetry with respect to the origin, meaning that its graph remains unchanged after rotation of 180 degrees about the origin.

Examples of odd functions are x, x^3, $\sin(x)$, $\sinh(x)$, and $\operatorname{erf}(x)$.

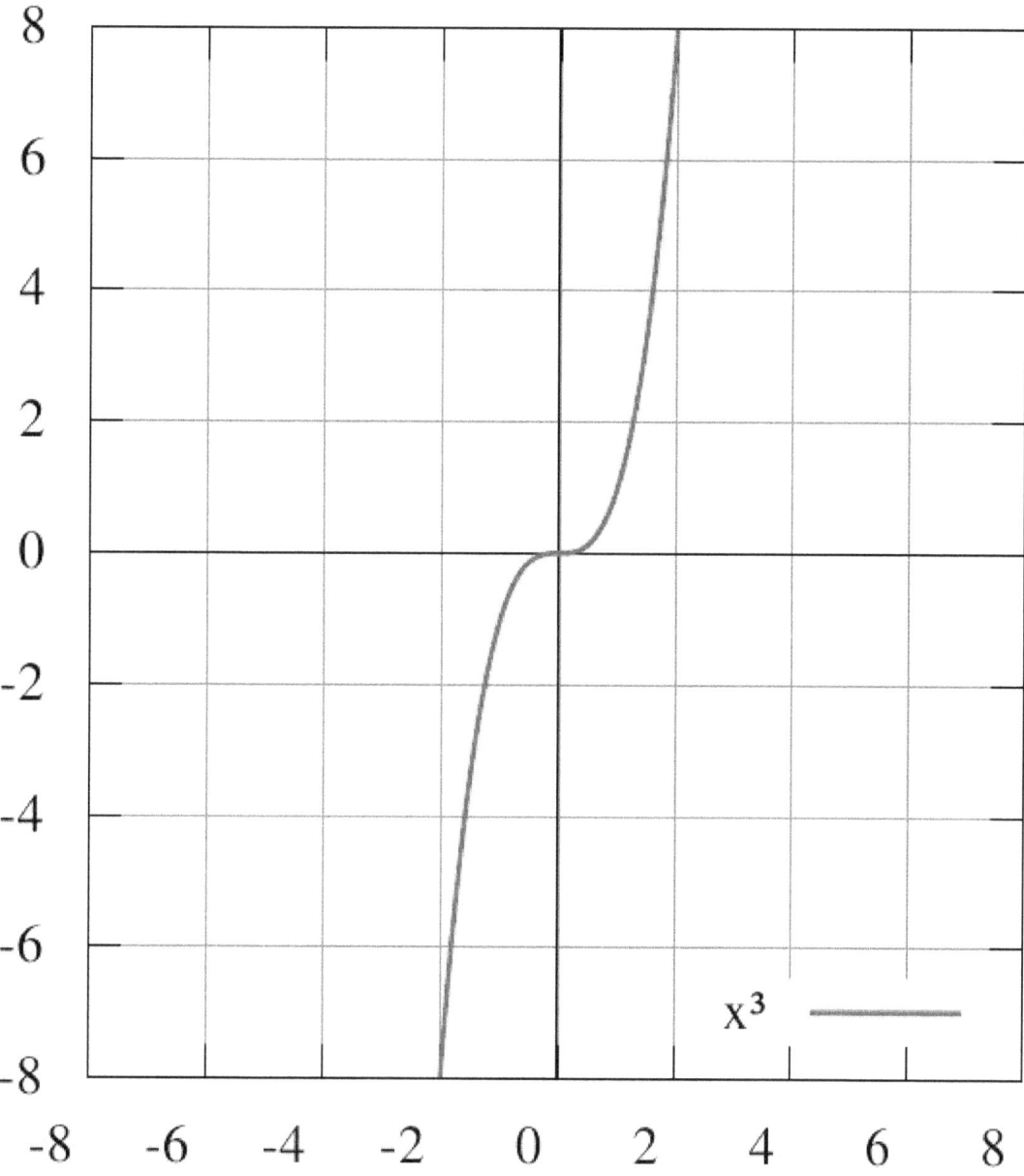

$f(x) = x^3$ *is an example of an odd function.*

38.2.2 Integrating

The integral of an odd function from $-A$ to $+A$ is zero (where A is finite, and the function has no vertical asymptotes between $-A$ and A).

The integral of an even function from $-A$ to $+A$ is twice the integral from 0 to $+A$ (where A is finite, and the function has no vertical asymptotes between $-A$ and A. This also holds true when A is infinite, but only if the integral converges).

38.2.3   Series

- The Maclaurin series of an even function includes only even powers.

- The Maclaurin series of an odd function includes only odd powers.

- The Fourier series of a periodic even function includes only cosine terms.

- The Fourier series of a periodic odd function includes only sine terms.

38.3   Symmetry in linear algebra

38.3.1   Symmetry in matrices

In linear algebra, a **symmetric matrix** is a square matrix that is equal to its transpose. Formally, matrix A is symmetric if

$$A = A^\top$$

and, because the definition of matrix equality demands equality of their dimensions, only square matrices can be symmetric.

The entries of a symmetric matrix are symmetric with respect to the main diagonal. So if the entries are written as $A = (a_{ij})$, then $a_{ij} = a_{ji}$, for all indices i and j.

The following 3×3 matrix is symmetric:

$$\begin{bmatrix} 1 & 7 & 3 \\ 7 & 4 & -5 \\ 3 & -5 & 6 \end{bmatrix}.$$

Every square diagonal matrix is symmetric, since all off-diagonal entries are zero. Similarly, each diagonal element of a skew-symmetric matrix must be zero, since each is its own negative.

In linear algebra, a real symmetric matrix represents a self-adjoint operator over a real inner product space. The corresponding object for a complex inner product space is a Hermitian matrix with complex-valued entries, which is equal to its conjugate transpose. Therefore, in linear algebra over the complex numbers, it is often assumed that a symmetric matrix refers to one which has real-valued entries. Symmetric matrices appear naturally in a variety of applications, and typical numerical linear algebra software makes special accommodations for them.

38.4   Symmetry in abstract algebra

38.4.1   Symmetric groups

Main article: Symmetric group

The **symmetric group** S_n on a finite set of n symbols is the group whose elements are all the permutations of the n symbols, and whose group operation is the composition of such permutations, which are treated as bijective functions from the set of symbols to itself.[1] Since there are $n!$ (n factorial) possible permutations of a set of n symbols, it follows that the order (the number of elements) of the symmetric group S_n is $n!$.

38.4.2 Symmetric polynomials

Main article: Symmetric polynomial

A **symmetric polynomial** is a polynomial $P(X_1, X_2, \ldots, Xn)$ in n variables, such that if any of the variables are interchanged, one obtains the same polynomial. Formally, P is a *symmetric polynomial*, if for any permutation σ of the subscripts 1, 2, ..., n one has $P(X\sigma_{(1)}, X\sigma_{(2)}, \ldots, X\sigma_{(n)}) = P(X_1, X_2, \ldots, Xn)$.

Symmetric polynomials arise naturally in the study of the relation between the roots of a polynomial in one variable and its coefficients, since the coefficients can be given by polynomial expressions in the roots, and all roots play a similar role in this setting. From this point of view the elementary symmetric polynomials are the most fundamental symmetric polynomials. A theorem states that any symmetric polynomial can be expressed in terms of elementary symmetric polynomials, which implies that every *symmetric* polynomial expression in the roots of a monic polynomial can alternatively be given as a polynomial expression in the coefficients of the polynomial.

Examples

In two variables X_1, X_2 one has symmetric polynomials like

- $X_1^3 + X_2^3 - 7$
- $4X_1^2 X_2^2 + X_1^3 X_2 + X_1 X_2^3 + (X_1 + X_2)^4$

and in three variables X_1, X_2, X_3 one has for instance

- $X_1 X_2 X_3 - 2X_1 X_2 - 2X_1 X_3 - 2X_2 X_3$

38.4.3 Symmetric tensors

Main article: Symmetric tensor

In mathematics, a **symmetric tensor** is tensor that is invariant under a permutation of its vector arguments:

$$T(v_1, v_2, \ldots, v_r) = T(v_{\sigma 1}, v_{\sigma 2}, \ldots, v_{\sigma r})$$

for every permutation σ of the symbols $\{1, 2, \ldots, r\}$. Alternatively, an rth order symmetric tensor represented in coordinates as a quantity with r indices satisfies

$$T_{i_1 i_2 \ldots i_r} = T_{i_{\sigma 1} i_{\sigma 2} \ldots i_{\sigma r}}.$$

The space of symmetric tensors of rank r on a finite-dimensional vector space is naturally isomorphic to the dual of the space of homogeneous polynomials of degree r on V. Over fields of characteristic zero, the graded vector space of all symmetric tensors can be naturally identified with the symmetric algebra on V. A related concept is that of the antisymmetric tensor or alternating form. Symmetric tensors occur widely in engineering, physics and mathematics.

38.4.4 Galois theory

Main article: Galois theory

Given a polynomial, it may be that some of the roots are connected by various algebraic equations. For example, it may be that for two of the roots, say A and B, that $A^2 + 5B^3 = 7$. The central idea of Galois theory is to consider those permutations (or rearrangements) of the roots having the property that *any* algebraic equation satisfied by the roots is *still satisfied* after the roots have been permuted. An important proviso is that we restrict ourselves to algebraic equations whose coefficients are rational numbers. Thus, Galois theory studies the symmetries inherent in algebraic equations.

38.4.5 Automorphisms of algebraic objects

Main article: Automorphism

In abstract algebra, an **automorphism** is an isomorphism from a mathematical object to itself. It is, in some sense, a symmetry of the object, and a way of mapping the object to itself while preserving all of its structure. The set of all automorphisms of an object forms a group, called the **automorphism group**. It is, loosely speaking, the symmetry group of the object.

Examples

- In set theory, an arbitrary permutation of the elements of a set X is an automorphism. The automorphism group of X is also called the symmetric group on X.

- In elementary arithmetic, the set of integers, $\mathbf{Z}$, considered as a group under addition, has a unique nontrivial automorphism: negation. Considered as a ring, however, it has only the trivial automorphism. Generally speaking, negation is an automorphism of any abelian group, but not of a ring or field.

- A group automorphism is a group isomorphism from a group to itself. Informally, it is a permutation of the group elements such that the structure remains unchanged. For every group G there is a natural group homomorphism $G \to \text{Aut}(G)$ whose image is the group $\text{Inn}(G)$ of inner automorphisms and whose kernel is the center of G. Thus, if G has trivial center it can be embedded into its own automorphism group.[2]

- In linear algebra, an endomorphism of a vector space V is a linear operator $V \to V$. An automorphism is an invertible linear operator on V. When the vector space is finite-dimensional, the automorphism group of V is the same as the general linear group, $\text{GL}(V)$.

- A field automorphism is a bijective ring homomorphism from a field to itself. In the cases of the rational numbers ($\mathbf{Q}$) and the real numbers ($\mathbf{R}$) there are no nontrivial field automorphisms. Some subfields of $\mathbf{R}$ have nontrivial field automorphisms, which however do not extend to all of $\mathbf{R}$ (because they cannot preserve the property of a number having a square root in $\mathbf{R}$). In the case of the complex numbers, $\mathbf{C}$, there is a unique nontrivial automorphism that sends $\mathbf{R}$ into $\mathbf{R}$: complex conjugation, but there are infinitely (uncountably) many "wild" automorphisms (assuming the axiom of choice).[3] Field automorphisms are important to the theory of field extensions, in particular Galois extensions. In the case of a Galois extension L/K the subgroup of all automorphisms of L fixing K pointwise is called the Galois group of the extension.

38.5 Symmetry in representation theory

38.5.1 Symmetry in quantum mechanics: bosons and fermions

In quantum mechanics, bosons have representatives that are symmetric under permutation operators, and fermions have antisymmetric representatives.

This implies the Pauli exclusion principle for fermions. In fact, the Pauli exclusion principle with a single-valued many-particle wavefunction is equivalent to requiring the wavefunction to be antisymmetric. An antisymmetric two-particle state is represented as a sum of states in which one particle is in state $|x\rangle$ and the other in state $|y\rangle$:

$$|\psi\rangle = \sum_{x,y} A(x,y)|x,y\rangle$$

and antisymmetry under exchange means that $A(x,y) = -A(y,x)$. This implies that $A(x,x) = 0$, which is Pauli exclusion. It is true in any basis, since unitary changes of basis keep antisymmetric matrices antisymmetric, although strictly speaking, the quantity $A(x,y)$ is not a matrix but an antisymmetric rank-two tensor.

Conversely, if the diagonal quantities $A(x,x)$ are zero *in every basis*, then the wavefunction component:

$$A(x,y) = \langle \psi | x,y \rangle = \langle \psi | (|x\rangle \otimes |y\rangle)$$

is necessarily antisymmetric. To prove it, consider the matrix element:

$$\langle \psi | ((|x\rangle + |y\rangle) \otimes (|x\rangle + |y\rangle))$$

This is zero, because the two particles have zero probability to both be in the superposition state $|x\rangle + |y\rangle$. But this is equal to

$$\langle \psi | x,x \rangle + \langle \psi | x,y \rangle + \langle \psi | y,x \rangle + \langle \psi | y,y \rangle$$

The first and last terms on the right hand side are diagonal elements and are zero, and the whole sum is equal to zero. So the wavefunction matrix elements obey:

$$\langle \psi | x,y \rangle + \langle \psi | y,x \rangle = 0$$

or

$$A(x,y) = -A(y,x)$$

38.6 Symmetry in set theory

38.6.1 Symmetric relation

Main article: Symmetric relation

We call a relation symmetric if every time the relation stands from A to B, it stands too from B to A. Note that symmetry is not the exact opposite of antisymmetry.

38.7 Symmetry in metric spaces

38.7.1 Isometries of a space

Main article: Isometry

An **isometry** is a distance-preserving map between metric spaces. Given a metric space, or a set and scheme for assigning distances between elements of the set, an isometry is a transformation which maps elements to another metric space such that the distance between the elements in the new metric space is equal to the distance between the elements in the original metric space. In a two-dimensional or three-dimensional space, two geometric figures are congruent if they are related by an isometry: related by either a rigid motion, or a composition of a rigid motion and a reflection. Up to a relation by a rigid motion, they are equal if related by a direct isometry.

Isometries were used to build a unfying symmetry definition working in geometry and for functions, probability distributions, matrices, strings, graphs, etc.[4]

38.8   Symmetries of differential equations

A symmetry of a differential equation is a transformation that leaves the differential equation invariant. Knowledge of such symmetries may help solve the differential equation.

A Line symmetry of a system of differential equations is a continuous symmetry of the system of differential equations. Knowledge of a Line symmetry can be used to simplify an ordinary differential equation through reduction of order.[5]

For ordinary differential equations, knowledge of an appropriate set of Lie symmetries allows one to explicitly calculate a set of first integrals, yielding a complete solution without integration.

Symmetries may be found by solving a related set of ordinary differential equations.[5] Solving these equations is often much simpler than solving the original differential equations.

38.9   Symmetry in probability

In the case of a finite number of possible outcomes, symmetry with respect to permutations (relabelings) implies a discrete uniform distribution.

In the case of a real interval of possible outcomes, symmetry with respect to interchanging sub-intervals of equal length corresponds to a continuous uniform distribution.

In other cases, such as "taking a random integer" or "taking a random real number", there are no probability distributions at all symmetric with respect to relabellings or to exchange of equally long subintervals. Other reasonable symmetries do not single out one particular distribution, or in other words, there is not a unique probability distribution providing maximum symmetry.

There is one type of isometry in one dimension that may leave the probability distribution unchanged, that is reflection in a point, for example zero.

A possible symmetry for randomness with positive outcomes is that the former applies for the logarithm, i.e., the outcome and its reciprocal have the same distribution. However this symmetry does not single out any particular distribution uniquely.

For a "random point" in a plane or in space, one can choose an origin, and consider a probability distribution with circular or spherical symmetry, respectively.

38.10   See also

- Use of symmetry in integration

- Invariance (mathematics)

38.11 References

[1] Jacobson (2009), p. 31.

[2] PJ Pahl, R Damrath (2001). "§7.5.5 Automorphisms". *Mathematical foundations of computational engineering* (Felix Pahl translation ed.). Springer. p. 376. ISBN 3-540-67995-2.

[3] Yale, Paul B. (May 1966). "Automorphisms of the Complex Numbers" (PDF). *Mathematics Magazine* **39** (3): 135–141. doi:10.2307/2689301. JSTOR 2689301.

[4] Petitjean, Michel (2007). "A definition of symmetry" (PDF). *Symmetry: Culture and Science* **18** (2-3): 99–119. Zbl 1274.58003.

[5] Olver, Peter J. (1986). *Applications of Lie Groups to Differential Equations*. New York: Springer Verlag. ISBN 978-0-387-95000-6.

38.12 Bibliography

- Hermann Weyl, *Symmetry*. Reprint of the 1952 original. Princeton Science Library. Princeton University Press, Princeton, NJ, 1989. viii+168 pp. ISBN 0-691-02374-3

- Mark Ronan, *Symmetry and the Monster*, Oxford University Press, 2006. ISBN 978-0-19-280723-6 (Concise introduction for lay reader)

- Marcus du Sautoy, *Finding Moonshine: a Mathematician's Journey through Symmetry*, Fourth Estate, 2009

Chapter 39

Lie algebra

"Lie bracket" redirects here. For the operation on vector fields, see Lie bracket of vector fields.

In mathematics, a **Lie algebra** (/ˈliː/, not /ˈlaɪ/) is a vector space together with a non-associative multiplication called "Lie bracket" $[x, y]$. It was introduced to study the concept of infinitesimal transformations. Hermann Weyl introduced the term "Lie algebra" (after Sophus Lie) in the 1930s. In older texts, the name "**infinitesimal group**" is used.

Lie algebras are closely related to Lie groups which are groups that are also smooth manifolds, with the property that the group operations of multiplication and inversion are smooth maps. Any Lie group gives rise to a Lie algebra. Conversely, to any finite-dimensional Lie algebra over real or complex numbers, there is a corresponding connected Lie group unique up to covering (Lie's third theorem). This correspondence between Lie groups and Lie algebras allows one to study Lie groups in terms of Lie algebras.

39.1   Definitions

A **Lie algebra** is a vector space $\mathfrak{g}$ over some field F together with a binary operation $[\cdot, \cdot] : \mathfrak{g} \times \mathfrak{g} \to \mathfrak{g}$ called the **Lie bracket**, which satisfies the following axioms:

- Bilinearity,

  $$[ax + by, z] = a[x, z] + b[y, z], \quad [z, ax + by] = a[z, x] + b[z, y]$$

  for all scalars a, b in F and all elements x, y, z in $\mathfrak{g}$.

- Alternativity on $\mathfrak{g}$,

  $$[x, x] = 0$$

  for all x in $\mathfrak{g}$.

- The Jacobi identity,

  $$[x, [y, z]] + [z, [x, y]] + [y, [z, x]] = 0$$

  for all x, y, z in $\mathfrak{g}$.

Using bilinearity to expand the Lie bracket $[x + y, x + y]$ and using alternativity shows that $[x, y] + [y, x] = 0$ for all elements x, y in $\mathfrak{g}$, showing that bilinearity and alternativity together imply

- Anticommutativity,

  $$[x, y] = -[y, x],$$

  for all elements x, y in $\mathfrak{g}$. Anticommutativity only implies the alternating property if the field's characteristic is not 2.[1]

It is customary to express a Lie algebra in lower-case fraktur, like $\mathfrak{g}$. If a Lie algebra is associated with a Lie group, then the spelling of the Lie algebra is the same as that Lie group. For example, the Lie algebra of $SU(n)$ is written as $\mathfrak{su}(n)$.

39.1.1 Generators and dimension

Elements of a Lie algebra $\mathfrak{g}$ are said to be **generators** of the Lie algebra if the smallest subalgebra of $\mathfrak{g}$ containing them is $\mathfrak{g}$ itself. The **dimension** of a Lie algebra is its dimension as a vector space over F. The cardinality of a minimal generating set of a Lie algebra is always less than or equal to its dimension.

39.1.2 Subalgebras, ideals and homomorphisms

The Lie bracket is not associative in general, meaning that $[[x, y], z]$ need not equal $[x, [y, z]]$. Nonetheless, much of the terminology that was developed in the theory of associative rings or associative algebras is commonly applied to Lie algebras. A subspace $\mathfrak{h} \subseteq \mathfrak{g}$ that is closed under the Lie bracket is called a **Lie subalgebra**. If a subspace $I \subseteq \mathfrak{g}$ satisfies a stronger condition that

$$[\mathfrak{g}, I] \subseteq I,$$

then I is called an **ideal** in the Lie algebra $\mathfrak{g}$.[2] A **homomorphism** between two Lie algebras (over the same base field) is a linear map that is compatible with the respective Lie brackets:

$$f : \mathfrak{g} \to \mathfrak{g}', \quad f([x, y]) = [f(x), f(y)],$$

for all elements x and y in $\mathfrak{g}$. As in the theory of associative rings, ideals are precisely the kernels of homomorphisms, given a Lie algebra $\mathfrak{g}$ and an ideal I in it, one constructs the **factor algebra** $\mathfrak{g}/I$, and the first isomorphism theorem holds for Lie algebras.

Let S be a subset of $\mathfrak{g}$. The set of elements x such that $[x, s] = 0$ for all s in S forms a subalgebra called the centralizer of S. The centralizer of $\mathfrak{g}$ itself is called the center of $\mathfrak{g}$. Similar to centralizers, if S is a subspace,[3] then the set of x such that $[x, s]$ is in S for all s in S forms a subalgebra called the normalizer of S.

39.1.3 Direct sum and semidirect product

Given two Lie algebras $\mathfrak{g}$ and $\mathfrak{g}'$, their direct sum is the Lie algebra consisting of the vector space $\mathfrak{g} \oplus \mathfrak{g}'$, of the pairs (x, x'), $x \in \mathfrak{g}$, $x' \in \mathfrak{g}'$, with the operation

$$[(x, x'), (y, y')] = ([x, y], [x', y']), \quad x, y \in \mathfrak{g}, \ x', y' \in \mathfrak{g}'.$$

Let $\mathfrak{g}$ be a Lie algebra and $\mathfrak{i}$ its ideal. If the canonical map $\mathfrak{g} \to \mathfrak{g}/\mathfrak{i}$ splits (i.e., admits a section), then $\mathfrak{g}$ is said to be a semidirect product of $\mathfrak{i}$ and $\mathfrak{g}/\mathfrak{i}$.

Levi's theorem says that a finite-dimensional Lie algebra is a semidirect product of its radical and the complementary subalgebra (Levi subalgebra).

39.2   Properties

39.2.1   Admits an enveloping algebra

See also: Universal enveloping algebra

For any associative algebra A with multiplication $*$, one can construct a Lie algebra $L(A)$. As a vector space, $L(A)$ is the same as A. The Lie bracket of two elements of $L(A)$ is defined to be their commutator in A:

$$[a, b] = a * b - b * a.$$

The associativity of the multiplication $*$ in A implies the Jacobi identity of the commutator in $L(A)$. For example, the associative algebra of $n \times n$ matrices over a field F gives rise to the general linear Lie algebra $\mathfrak{gl}_n(F)$. The associative algebra A is called an **enveloping algebra** of the Lie algebra $L(A)$. Every Lie algebra can be embedded into one that arises from an associative algebra in this fashion; see universal enveloping algebra.

39.2.2   Representation

Given a vector space V, let $\mathfrak{gl}(V)$ denote the Lie algebra enveloped by the associative algebra of all linear endomorphisms of V. A representation of a Lie algebra $\mathfrak{g}$ on V is a Lie algebra homomorphism

$$\pi : \mathfrak{g} \to \mathfrak{gl}(V).$$

A representation is said to be faithful if its kernel is trivial. Every finite-dimensional Lie algebra has a faithful representation on a finite-dimensional vector space (Ado's theorem).[4]

For example,

$$\mathrm{ad} : \mathfrak{g} \to \mathfrak{gl}(\mathfrak{g})$$

given by $\mathrm{ad}(x)(y) = [x, y]$ is a representation of $\mathfrak{g}$ on the vector space $\mathfrak{g}$ called the adjoint representation. A derivation on the Lie algebra $\mathfrak{g}$ (in fact on any non-associative algebra) is a linear map $\delta : \mathfrak{g} \to \mathfrak{g}$ that obeys the Leibniz' law, that is,

$$\delta([x, y]) = [\delta(x), y] + [x, \delta(y)]$$

for all x and y in the algebra. For any x, $\mathrm{ad}(x)$ is a derivation; a consequence of the Jacobi identity. Thus, the image of ad lies in the subalgebra of $\mathfrak{gl}(\mathfrak{g})$ consisting of derivations on $\mathfrak{g}$. A derivation that happens to be in the image of ad is called an inner derivation. If $\mathfrak{g}$ is semisimple, every derivation on $\mathfrak{g}$ is inner.

39.3   Examples

39.3.1   Vector spaces

- Any vector space V endowed with the identically zero Lie bracket becomes a Lie algebra. Such Lie algebras are called abelian, cf. below. Any one-dimensional Lie algebra over a field is abelian, by the antisymmetry of the Lie bracket.

- The real vector space of all $n \times n$ skew-hermitian matrices is closed under the commutator and forms a real Lie algebra denoted $\mathfrak{u}(n)$. This is the Lie algebra of the unitary group $U(n)$.

39.3.2 Subspaces

- The subspace of the general linear Lie algebra $\mathfrak{gl}_n(F)$ consisting of matrices of trace zero is a subalgebra,[5] the special linear Lie algebra, denoted $\mathfrak{sl}_n(F)$.

39.3.3 Real matrix groups

- Any Lie group G defines an associated real Lie algebra $\mathfrak{g}$ =Lie(G). The definition in general is somewhat technical, but in the case of real matrix groups, it can be formulated via the exponential map, or the matrix exponent. The Lie algebra $\mathfrak{g}$ consists of those matrices X for which $\exp(tX) \in G$, ∀ real numbers t.

  The Lie bracket of $\mathfrak{g}$ is given by the commutator of matrices. As a concrete example, consider the special linear group SL(n,**R**), consisting of all $n \times n$ matrices with real entries and determinant 1. This is a matrix Lie group, and its Lie algebra consists of all $n \times n$ matrices with real entries and trace 0.

39.3.4 Three dimensions

- The three-dimensional Euclidean space $\mathbf{R}^3$ with the Lie bracket given by the cross product of vectors becomes a three-dimensional Lie algebra.

- The Heisenberg algebra $H_3(\mathrm{R})$ is a three-dimensional Lie algebra generated by elements x, y and z with Lie brackets

$$[x,y] = z, \quad [x,z] = 0, \quad [y,z] = 0.$$

It is explicitly realized as the space of 3×3 strictly upper-triangular matrices, with the Lie bracket given by the matrix commutator,

$$x = \begin{pmatrix} 0 & 1 & 0 \\ 0 & 0 & 0 \\ 0 & 0 & 0 \end{pmatrix}, \quad y = \begin{pmatrix} 0 & 0 & 0 \\ 0 & 0 & 1 \\ 0 & 0 & 0 \end{pmatrix}, \quad z = \begin{pmatrix} 0 & 0 & 1 \\ 0 & 0 & 0 \\ 0 & 0 & 0 \end{pmatrix}.$$

Any element of the Heisenberg group is thus representable as a product of group generators, i.e., matrix exponentials of these Lie algebra generators,

$$\begin{pmatrix} 1 & a & c \\ 0 & 1 & b \\ 0 & 0 & 1 \end{pmatrix} = e^{by} e^{cz} e^{ax}.$$

- The commutation relations between the x, y, and z components of the angular momentum operator in quantum mechanics are the same as those of $\mathfrak{su}(2)$ and $\mathfrak{so}(3)$.

$$[L_x, L_y] = i\hbar L_z$$

$$[L_y, L_z] = i\hbar L_x$$

$$[L_z, L_x] = i\hbar L_y$$

(The physicist convention for Lie algebras is used in the above equations, hence the factor of i.) The Lie algebra formed by these operators have, in fact, representations of all finite dimensions.

39.3.5 Infinite dimensions

- An important class of infinite-dimensional real Lie algebras arises in differential topology. The space of smooth vector fields on a differentiable manifold M forms a Lie algebra, where the Lie bracket is defined to be the commutator of vector fields. One way of expressing the Lie bracket is through the formalism of Lie derivatives, which identifies a vector field X with a first order partial differential operator LX acting on smooth functions by letting $LX(f)$ be the directional derivative of the function f in the direction of X. The Lie bracket $[X,Y]$ of two vector fields is the vector field defined through its action on functions by the formula:

$$L_{[X,Y]}f = L_X(L_Y f) - L_Y(L_X f).$$

- A Kac–Moody algebra is an example of an infinite-dimensional Lie algebra.

- The Moyal algebra is an infinite-dimensional Lie algebra which contains all classical Lie algebras as subalgebras.

39.4 Structure theory and classification

Lie algebras can be classified to some extent. In particular, this has an application to the classification of Lie groups.

39.4.1 Abelian, nilpotent, and solvable

Analogously to abelian, nilpotent, and solvable groups, defined in terms of the derived subgroups, one can define abelian, nilpotent, and solvable Lie algebras.

A Lie algebra $\mathfrak{g}$ is **abelian** if the Lie bracket vanishes, i.e. $[x,y] = 0$, for all x and y in $\mathfrak{g}$. Abelian Lie algebras correspond to commutative (or abelian) connected Lie groups such as vector spaces K^n or tori T^n, and are all of the form $\mathfrak{k}^n$, meaning an n-dimensional vector space with the trivial Lie bracket.

A more general class of Lie algebras is defined by the vanishing of all commutators of given length. A Lie algebra $\mathfrak{g}$ is **nilpotent** if the lower central series

$$\mathfrak{g} > [\mathfrak{g},\mathfrak{g}] > [[\mathfrak{g},\mathfrak{g}],\mathfrak{g}] > [[[\mathfrak{g},\mathfrak{g}],\mathfrak{g}],\mathfrak{g}] > \cdots$$

becomes zero eventually. By Engel's theorem, a Lie algebra is nilpotent if and only if for every u in $\mathfrak{g}$ the adjoint endomorphism

$$\mathrm{ad}(u) : \mathfrak{g} \to \mathfrak{g}, \quad \mathrm{ad}(u)v = [u,v]$$

is nilpotent.

More generally still, a Lie algebra $\mathfrak{g}$ is said to be **solvable** if the derived series:

$$\mathfrak{g} > [\mathfrak{g},\mathfrak{g}] > [[\mathfrak{g},\mathfrak{g}],[\mathfrak{g},\mathfrak{g}]] > [[[\mathfrak{g},\mathfrak{g}],[\mathfrak{g},\mathfrak{g}]],[[\mathfrak{g},\mathfrak{g}],[\mathfrak{g},\mathfrak{g}]]] > \cdots$$

becomes zero eventually.

Every finite-dimensional Lie algebra has a unique maximal solvable ideal, called its radical. Under the Lie correspondence, nilpotent (respectively, solvable) connected Lie groups correspond to nilpotent (respectively, solvable) Lie algebras.

39.4.2 Simple and semisimple

A Lie algebra is "simple" if it has no non-trivial ideals and is not abelian. A Lie algebra $\mathfrak{g}$ is called **semisimple** if its radical is zero. Equivalently, $\mathfrak{g}$ is semisimple if it does not contain any non-zero abelian ideals. In particular, a simple Lie algebra is semisimple. Conversely, it can be proven that any semisimple Lie algebra is the direct sum of its minimal ideals, which are canonically determined simple Lie algebras.

The concept of semisimplicity for Lie algebras is closely related with the complete reducibility (semisimplicity) of their representations. When the ground field F has characteristic zero, any finite-dimensional representation of a semisimple Lie algebra is semisimple (i.e., direct sum of irreducible representations.) In general, a Lie algebra is called reductive if the adjoint representation is semisimple. Thus, a semisimple Lie algebra is reductive.

39.4.3 Cartan's criterion

Cartan's criterion gives conditions for a Lie algebra to be nilpotent, solvable, or semisimple. It is based on the notion of the Killing form, a symmetric bilinear form on $\mathfrak{g}$ defined by the formula

$$K(u, v) = \mathrm{tr}(\mathrm{ad}(u)\,\mathrm{ad}(v)).$$

where tr denotes the trace of a linear operator. A Lie algebra $\mathfrak{g}$ is semisimple if and only if the Killing form is nondegenerate. A Lie algebra $\mathfrak{g}$ is solvable if and only if $K(\mathfrak{g}, [\mathfrak{g}, \mathfrak{g}]) = 0$.

39.4.4 Classification

The Levi decomposition expresses an arbitrary Lie algebra as a semidirect sum of its solvable radical and a semisimple Lie algebra, almost in a canonical way. Furthermore, semisimple Lie algebras over an algebraically closed field have been completely classified through their root systems. However, the classification of solvable Lie algebras is a 'wild' problem, and cannot be accomplished in general.

39.5 Relation to Lie groups

See also: Lie group–Lie algebra correspondence

Although Lie algebras are often studied in their own right, historically they arose as a means to study Lie groups.

Lie's fundamental theorems describe a relation between Lie groups and Lie algebras. In particular, any Lie group gives rise to a canonically determined Lie algebra (concretely, *the tangent space at the identity*); and, conversely, for any Lie algebra there is a corresponding connected Lie group (Lie's third theorem; see the Baker–Campbell–Hausdorff formula). This Lie group is not determined uniquely; however, any two connected Lie groups with the same Lie algebra are *locally isomorphic*, and in particular, have the same universal cover. For instance, the special orthogonal group SO(3) and the special unitary group SU(2) give rise to the same Lie algebra, which is isomorphic to $\mathbf{R}^3$ with the cross-product, while SU(2) is a simply-connected twofold cover of SO(3).

Given a Lie group, a Lie algebra can be associated to it either by endowing the tangent space to the identity with the differential of the adjoint map, or by considering the left-invariant vector fields as mentioned in the examples. In the case of real matrix groups, the Lie algebra $\mathfrak{g}$ consists of those matrices X for which $\exp(tX) \in G$ for all real numbers t, where exp is the exponential map.

Some examples of Lie algebras corresponding to Lie groups are the following:

- The Lie algebra $\mathfrak{gl}_n(\mathbb{C})$ for the group $GL_n(\mathbb{C})$ is the algebra of complex $n{\times}n$ matrices

- The Lie algebra $\mathfrak{sl}_n(\mathbb{C})$ for the group $SL_n(\mathbb{C})$ is the algebra of complex $n \times n$ matrices with trace 0

- The Lie algebras $\mathfrak{o}_n(\mathbb{R})$ for the group $O_n(\mathbb{R})$ and $\mathfrak{so}_n(\mathbb{R})$ for $SO_n(\mathbb{R})$ are both the algebra of real anti-symmetric $n \times n$ matrices (See Antisymmetric matrix: Infinitesimal rotations for a discussion)

- The Lie algebra $\mathfrak{u}_n(\mathbb{C})$ for the group $U_n(\mathbb{C})$ is the algebra of skew-Hermitian complex $n \times n$ matrices while the Lie algebra $\mathfrak{su}_n(\mathbb{C})$ for $SU_n(\mathbb{C})$ is the algebra of skew-Hermitian, traceless complex $n \times n$ matrices.

In the above examples, the Lie bracket $[X, Y]$ (for X and Y matrices in the Lie algebra) is defined as $[X, Y] = XY - YX$.

Given a set of generators T^a, the **structure constants** f^{abc} express the Lie brackets of pairs of generators as linear combinations of generators from the set, i.e., $[T^a, T^b] = f^{abc} T^c$. The structure constants determine the Lie brackets of elements of the Lie algebra, and consequently nearly completely determine the group structure of the Lie group. The structure of the Lie group near the identity element is displayed explicitly by the Baker–Campbell–Hausdorff formula, an expansion in Lie algebra elements X, Y and their Lie brackets, all nested together within a single exponent, $\exp(tX)\exp(tY) = \exp(tX + tY + \frac{1}{2} t^2 [X, Y] + O(t^3))$.

The mapping from Lie groups to Lie algebras is functorial, which implies that homomorphisms of Lie groups lift to homomorphisms of Lie algebras, and various properties are satisfied by this lifting: it commutes with composition, it maps Lie subgroups, kernels, quotients and cokernels of Lie groups to subalgebras, kernels, quotients and cokernels of Lie algebras, respectively.

The functor L which takes each Lie group to its Lie algebra and each homomorphism to its differential is faithful and exact. It is however not an equivalence of categories: different Lie groups may have isomorphic Lie algebras (for example $SO(3)$ and $SU(2)$), and there are (infinite dimensional) Lie algebras that are not associated to any Lie group.[6]

However, when the Lie algebra $\mathfrak{g}$ is finite-dimensional, one can associate to it a simply connected Lie group having $\mathfrak{g}$ as its Lie algebra. More precisely, the Lie algebra functor L has a left adjoint functor Γ from finite-dimensional (real) Lie algebras to Lie groups, factoring through the full subcategory of simply connected Lie groups.[7] In other words, there is a natural isomorphism of bifunctors

$$\operatorname{Hom}(\Gamma(\mathfrak{g}), H) \cong \operatorname{Hom}(\mathfrak{g}, L(H)).$$

The adjunction $\mathfrak{g} \to L(\Gamma(\mathfrak{g}))$ (corresponding to the identity on $\Gamma(\mathfrak{g})$) is an isomorphism, and the other adjunction $\Gamma(L(H)) \to H$ is the projection homomorphism from the universal cover group of the identity component of H to H. It follows immediately that if G is simply connected, then the Lie algebra functor establishes a bijective correspondence between Lie group homomorphisms $G \to H$ and Lie algebra homomorphisms $L(G) \to L(H)$.

The universal cover group above can be constructed as the image of the Lie algebra under the exponential map. More generally, we have that the Lie algebra is homeomorphic to a neighborhood of the identity. But globally, if the Lie group is compact, the exponential will not be injective, and if the Lie group is not connected, simply connected or compact, the exponential map need not be surjective.

If the Lie algebra is infinite-dimensional, the issue is more subtle. In many instances, the exponential map is not even locally a homeomorphism (for example, in $\operatorname{Diff}(S^1)$, one may find diffeomorphisms arbitrarily close to the identity that are not in the image of exp). Furthermore, some infinite-dimensional Lie algebras are not the Lie algebra of any group.

The correspondence between Lie algebras and Lie groups is used in several ways, including in the classification of Lie groups and the related matter of the representation theory of Lie groups. Every representation of a Lie algebra lifts uniquely to a representation of the corresponding connected, simply connected Lie group, and conversely every representation of any Lie group induces a representation of the group's Lie algebra; the representations are in one to one correspondence. Therefore, knowing the representations of a Lie algebra settles the question of representations of the group.

As for classification, it can be shown that any connected Lie group with a given Lie algebra is isomorphic to the universal cover mod a discrete central subgroup. So classifying Lie groups becomes simply a matter of counting the discrete subgroups of the center, once the classification of Lie algebras is known (solved by Cartan et al. in the semisimple case).

39.6 Category theoretic definition

Using the language of category theory, a **Lie algebra** can be defined as an object A in **Vec**k, the category of vector spaces over a field k of characteristic not 2, together with a morphism $[...]: A \otimes A \to A$, where $\otimes$ refers to the monoidal product of **Vec**k, such that

- $[\cdot, \cdot] \circ (\mathrm{id} + \tau_{A,A}) = 0$

- $[\cdot, \cdot] \circ ([\cdot, \cdot] \otimes \mathrm{id}) \circ (\mathrm{id} + \sigma + \sigma^2) = 0$

where $\tau (a \otimes b) := b \otimes a$ and σ is the cyclic permutation braiding $(\mathrm{id} \otimes \tau_{A,A}) \circ (\tau_{A,A} \otimes \mathrm{id})$. In diagrammatic form:

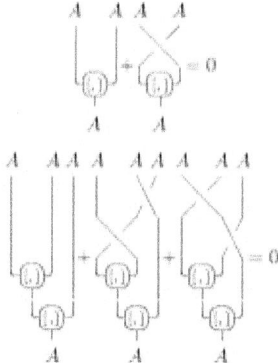

39.7 Lie ring

A **Lie ring** arises as a generalisation of Lie algebras, or through the study of the lower central series of groups. A **Lie ring** is defined as a nonassociative ring with multiplication that is anticommutative and satisfies the Jacobi identity. More specifically we can define a Lie ring L to be an abelian group with an operation $[\cdot, \cdot]$ that has the following properties:

- Bilinearity:

$$[x + y, z] = [x, z] + [y, z], \quad [z, x + y] = [z, x] + [z, y]$$

for all $x, y, z \in L$.

- The *Jacobi identity*:

$$[x, [y, z]] + [y, [z, x]] + [z, [x, y]] = 0$$

for all x, y, z in L.

- For all x in L:

$$[x, x] = 0$$

Lie rings need not be Lie groups under addition. Any Lie algebra is an example of a Lie ring. Any associative ring can be made into a Lie ring by defining a bracket operator $[x, y] = xy - yx$. Conversely to any Lie algebra there is a corresponding ring, called the universal enveloping algebra.

Lie rings are used in the study of finite p-groups through the Lazard correspondence. The lower central factors of a p-group are finite abelian p-groups, so modules over $\mathbf{Z}/p\mathbf{Z}$. The direct sum of the lower central factors is given the structure of a Lie ring by defining the bracket to be the commutator of two coset representatives. The Lie ring structure is enriched with another module homomorphism, then pth power map, making the associated Lie ring a so-called restricted Lie ring.

Lie rings are also useful in the definition of a p-adic analytic groups and their endomorphisms by studying Lie algebras over rings of integers such as the p-adic integers. The definition of finite groups of Lie type due to Chevalley involves restricting from a Lie algebra over the complex numbers to a Lie algebra over the integers, and the reducing modulo p to get a Lie algebra over a finite field.

39.7.1   Examples

- Any Lie algebra over a general ring instead of a field is an example of a Lie ring. Lie rings are *not* Lie groups under addition, despite the name.

- Any associative ring can be made into a Lie ring by defining a bracket operator $[x, y] = xy - yx$.

- For an example of a Lie ring arising from the study of groups, let G be a group with $(x, y) = x^{-1}y^{-1}xy$ the commutator operation, and let $G = G_0 \supseteq G_1 \supseteq G_2 \supseteq \cdots \supseteq G_n \supseteq \cdots$ be a central series in G — that is the commutator subgroup (G_i, G_j) is contained in G_{i+j} for any i, j. Then

$$L = \bigoplus G_i/G_{i+1}$$

  is a Lie ring with addition supplied by the group operation (which will be commutative in each homogeneous part), and the bracket operation given by

$$[xG_i, yG_j] = (x, y)G_{i+j}$$

(x, y)

39.8   See also

39.9   Notes

[1] Humphreys p. 1

[2] Due to the anticommutativity of the commutator, the notions of a left and right ideal in a Lie algebra coincide.

[3] Jacobson 1962, pg. 28

[4] Jacobson 1962, Ch. VI

[5] Humphreys p.2

[6] Beltita 2005, pg. 75

[7] Adjoint property is discussed in more general context in Hofman & Morris (2007) (e.g., page 130) but is a straightforward consequence of, e.g., Bourbaki (1989) Theorem 1 of page 305 and Theorem 3 of page 310.

39.10 References

- Beltita, Daniel. *Smooth Homogeneous Structures in Operator Theory*, CRC Press, 2005. ISBN 978-1-4200-3480-6

- Boza, Luis; Fedriani, Eugenio M. & Núñez, Juan. *A new method for classifying complex filiform Lie algebras*, Applied Mathematics and Computation, 121 (2-3): 169–175, 2001

- Bourbaki, Nicolas. "Lie Groups and Lie Algebras - Chapters 1-3", Springer, 1989. ISBN 3-540-64242-0

- Erdmann, Karin & Wildon, Mark. *Introduction to Lie Algebras*, 1st edition, Springer, 2006. ISBN 1-84628-040-0

- Hall, Brian C. *Lie Groups, Lie Algebras, and Representations: An Elementary Introduction*, Springer, 2003. ISBN 0-387-40122-9

- Hofman, Karl & Morris, Sidney. "The Lie Theory of Connected Pro-Lie Groups", European Mathematical Society, 2007. ISBN 978-3-03719-032-6

- Humphreys, James E. *Introduction to Lie Algebras and Representation Theory*. Second printing, revised. Graduate Texts in Mathematics, 9. Springer-Verlag, New York, 1978. ISBN 0-387-90053-5

- Jacobson, Nathan. *Lie algebras*. Republication of the 1962 original. Dover Publications, Inc., New York, 1979. ISBN 0-486-63832-4

- Kac, Victor G. et al. *Course notes for MIT 18.745: Introduction to Lie Algebras*, math.mit.edu

- O'Connor, J.J. & Robertson, E.F. Biography of Sophus Lie, MacTutor History of Mathematics Archive, www-history.mcs.st-andrews.ac.uk

- O'Connor, J.J. & Robertson, E.F. Biography of Wilhelm Killing, MacTutor History of Mathematics Archive, www-history.mcs.st-andrews.ac.uk

- Serre, Jean-Pierre. "Lie Algebras and Lie Groups", 2nd edition, Springer, 2006. ISBN 3-540-55008-9

- Steeb, W.-H. *Continuous Symmetries, Lie Algebras, Differential Equations and Computer Algebra*, second edition, World Scientific, 2007, ISBN 978-981-270-809-0

- Varadarajan, V.S. *Lie Groups, Lie Algebras, and Their Representations*, 1st edition, Springer, 2004. ISBN 0-387-90969-9.

39.11 External links

- Hazewinkel, Michiel, ed. (2001), "Lie algebra", *Encyclopedia of Mathematics*, Springer, ISBN 978-1-55608-010-4

- McKenzie, Douglas, (2015), "An Elementary Introduction to Lie Algebras for Physicists"

Chapter 40

Diffeomorphism

In mathematics, a **diffeomorphism** is an isomorphism of smooth manifolds. It is an invertible function that maps one differentiable manifold to another such that both the function and its inverse are smooth.

40.1 Definition

Given two manifolds M and N, a differentiable map $f : M \to N$ is called a **diffeomorphism** if it is a bijection and its inverse $f^{-1} : N \to M$ is differentiable as well. If these functions are r times continuously differentiable, f is called a C^r-**diffeomorphism**.

Two manifolds M and N are **diffeomorphic** (symbol usually being $\simeq$) if there is a diffeomorphism f from M to N. They are C^r **diffeomorphic** if there is an r times continuously differentiable bijective map between them whose inverse is also r times continuously differentiable.

40.2 Diffeomorphisms of subsets of manifolds

Given a subset X of a manifold M and a subset Y of a manifold N, a function $f : X \to Y$ is said to be smooth if for all p in X there is a neighborhood $U \subset M$ of p and a smooth function $g : U \to N$ such that the restrictions agree $g_{|U \cap X} = f_{|U \cap X}$ (note that g is an extension of f). f is said to be a diffeomorphism if it is bijective, smooth and its inverse is smooth.

40.3 Local description

Model example

If U, V are connected open subsets of $\mathbf{R}^n$ such that V is simply connected, a differentiable map $f : U \to V$ is a **diffeomorphism** if it is proper and if the differential $Df_x : \mathbf{R}^n \to \mathbf{R}^n$ is bijective at each point x in U.

> **First remark**
> It is essential for V to be simply connected for the function f to be globally invertible (under the sole condition that its derivative is a bijective map at each point). For example, consider the "realification" of the complex square function
>
> $$\begin{cases} f : \mathbf{R}^2 \setminus \{(0,0)\} \to \mathbf{R}^2 \setminus \{(0,0)\} \\ (x,y) \mapsto (x^2 - y^2, 2xy) \end{cases}$$

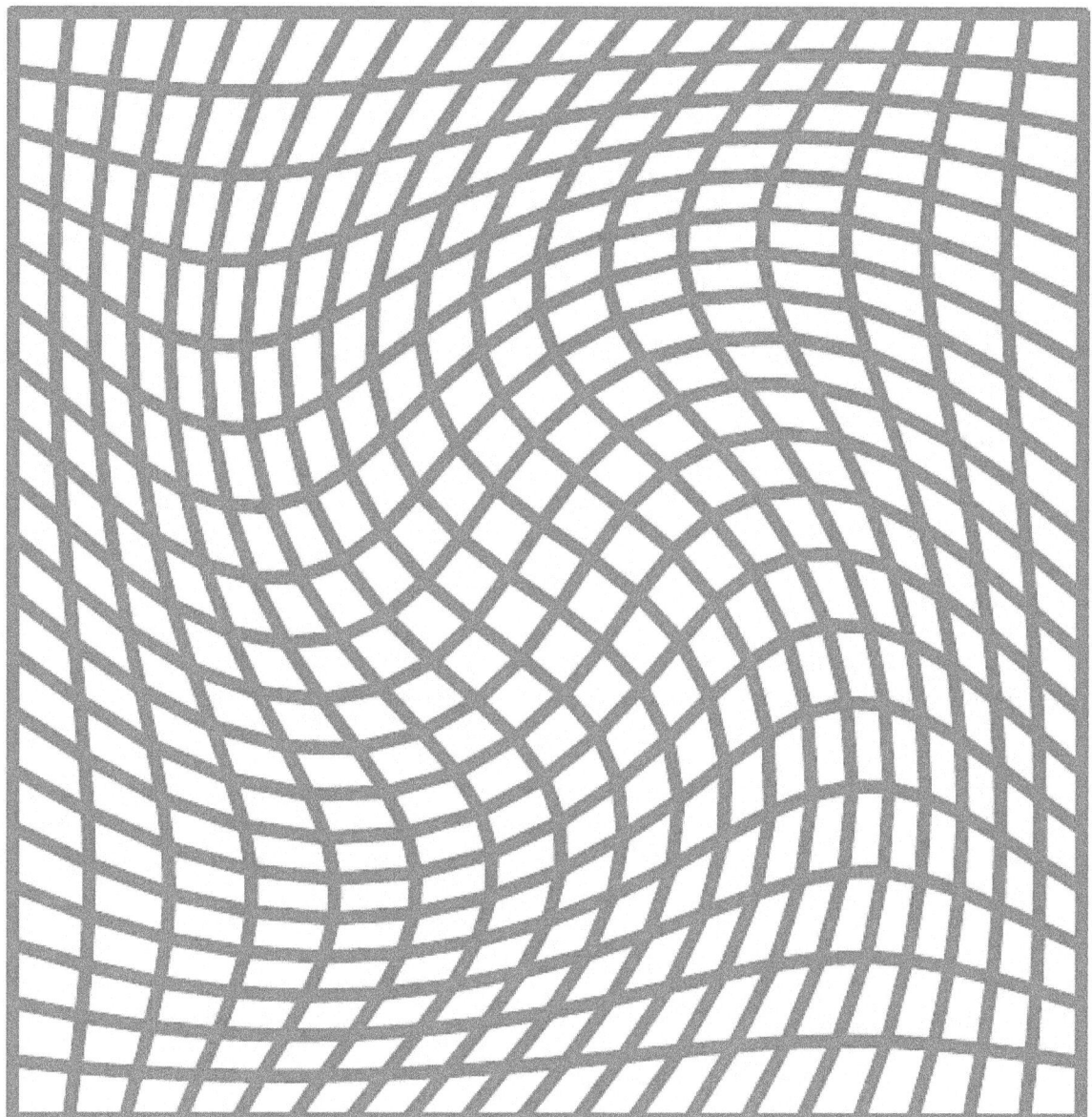

The image of a rectangular grid on a square under a diffeomorphism from the square onto itself.

Then f is surjective and it satisfies

$$\det D f_x = 4(x^2 + y^2) \neq 0$$

Thus, though Dfx is bijective at each point, f is not invertible because it fails to be injective (e.g. $f(1,0) = (1,0) = f(-1,0)$).

Second remark
Since the differential at a point (for a differentiable function)

$$D f_x : T_x U \to T_{f(x)} V$$

is a linear map, it has a well-defined inverse if and only if $Df x$ is a bijection. The matrix representation of $Df x$ is the $n \times n$ matrix of first-order partial derivatives whose entry in the i-th row and j-th column is $\partial f_i / \partial x_j$. This so-called Jacobian matrix is often used for explicit computations.

Third remark

Diffeomorphisms are necessarily between manifolds of the same dimension. Imagine f going from dimension n to dimension k. If $n < k$ then $Df x$ could never be surjective; and if $n > k$ then $Df x$ could never be injective. In both cases, therefore, $Df x$ fails to be a bijection.

Fourth remark

If $Df x$ is a bijection at x then f is said to be a local diffeomorphism (since, by continuity, $Df y$ will also be bijective for all y sufficiently close to x).

Fifth remark

Given a smooth map from dimension n to dimension k, if Df (or, locally, $Df x$) is surjective, f is said to be a submersion (or, locally, a "local submersion"); and if Df (or, locally, $Df x$) is injective, f is said to be an immersion (or, locally, a "local immersion").

Sixth remark

A differentiable bijection is *not* necessarily a diffeomorphism. $f(x) = x^3$, for example, is not a diffeomorphism from **R** to itself because its derivative vanishes at 0 (and hence its inverse is not differentiable at 0). This is an example of a homeomorphism that is not a diffeomorphism.

Seventh remark

When f is a map between *differentiable* manifolds, a diffeomorphic f is a stronger condition than a homeomorphic f. For a diffeomorphism, f and its inverse need to be differentiable; for a homeomorphism, f and its inverse need only be continuous. Every diffeomorphism is a homeomorphism, but not every homeomorphism is a diffeomorphism.

$f : M \to N$ is called a **diffeomorphism** if, in coordinate charts, it satisfies the definition above. More precisely: Pick any cover of M by compatible coordinate charts and do the same for N. Let φ and ψ be charts on, respectively, M and N, with U and V as, respectively, the images of φ and ψ. The map $\psi f \varphi^{-1} : U \to V$ is then a diffeomorphism as in the definition above, whenever $f(\varphi^{-1}(U)) \subset \psi^{-1}(V)$.

40.4   Examples

Since any manifold can be locally parametrised, we can consider some explicit maps from $\mathbf{R}^2$ into $\mathbf{R}^2$.

- Let

$$f(x, y) = \left(x^2 + y^3, x^2 - y^3 \right).$$

We can calculate the Jacobian matrix:

$$J_f = \begin{pmatrix} 2x & 3y^2 \\ 2x & -3y^2 \end{pmatrix}.$$

The Jacobian matrix has zero determinant if, and only if $xy = 0$. We see that f is a diffeomorphism away from the x-axis and the y-axis.

- Let

$$g(x,y) = (a_0 + a_{1,0}x + a_{0,1}y + \cdots, b_0 + b_{1,0}x + b_{0,1}y + \cdots)$$

where the $a_{i,j}$ and $b_{i,j}$ are arbitrary real numbers, and the omitted terms are of degree at least two in x and y. We can calculate the Jacobian matrix at **0**:

$$J_g(0,0) = \begin{pmatrix} a_{1,0} & a_{0,1} \\ b_{1,0} & b_{0,1} \end{pmatrix}.$$

We see that g is a local diffeomorphism at **0** if, and only if,

$$a_{1,0}b_{0,1} - a_{0,1}b_{1,0} \neq 0,$$

i.e. the linear terms in the components of g are linearly independent as polynomials.

- Let

$$h(x,y) = \left(\sin(x^2 + y^2), \cos(x^2 + y^2) \right).$$

We can calculate the Jacobian matrix:

$$J_h = \begin{pmatrix} 2x\cos(x^2 + y^2) & 2y\cos(x^2 + y^2) \\ -2x\sin(x^2 + y^2) & -2y\sin(x^2 + y^2) \end{pmatrix}.$$

The Jacobian matrix has zero determinant everywhere! In fact we see that the image of h is the unit circle.

40.4.1 Surface deformations

In mechanics, a stress-induced transformation is called a deformation and may be described by a diffeomorphism. A diffeomorphism $f : U \to V$ between two surfaces U and V has a Jacobian matrix Df that is an invertible matrix. In fact, it is required that for p in U, there is a neighborhood of p in which the Jacobian Df stays non-singular. Since the Jacobian is a 2×2 real matrix, Df can be read as one of three types of complex number: ordinary complex, split complex number, or dual number. Suppose that in a chart of the surface, $f(x,y) = (u,v)$.

The total differential of u is

$$du = \frac{\partial u}{\partial x}dx + \frac{\partial u}{\partial y}dy., \text{ and similarly for } v.$$

Then the image $(du, dv) = (dx, dy)Df$ is a linear transformation, fixing the origin, and expressible as the action of a complex number of a particular type. When (dx, dy) is also interpreted as that type of complex number, the action is of complex multiplication in the appropriate complex number plane. As such, there is a type of angle (Euclidean, hyperbolic, or slope) that is preserved in such a multiplication. Due to Df being invertible, the type of complex number is uniform over the surface.

Consequently, a surface deformation or diffeomorphism of surfaces has the **conformal property** of preserving (the appropriate type of) angles.

40.5 Diffeomorphism group

Let M be a differentiable manifold that is second-countable and Hausdorff. The **diffeomorphism group** of M is the group of all C^r diffeomorphisms of M to itself, denoted by $\mathrm{Diff}^r(M)$ or, when r is understood, $\mathrm{Diff}(M)$. This is a "large" group, in the sense that – provided M is not zero-dimensional – it is not locally compact.

40.5.1 Topology

The diffeomorphism group has two natural topologies: *weak* and *strong* (Hirsch 1997). When the manifold is compact, these two topologies agree. The weak topology is always metrizable. When the manifold is not compact, the strong topology captures the behavior of functions "at infinity" and is not metrizable. It is, however, still Baire.

Fixing a Riemannian metric on M, the weak topology is the topology induced by the family of metrics

$$d_K(f,g) = \sup_{x \in K} d(f(x), g(x)) + \sum_{1 \le p \le r} \sup_{x \in K} \|D^p f(x) - D^p g(x)\|$$

as K varies over compact subsets of M. Indeed, since M is σ-compact, there is a sequence of compact subsets Kn whose union is M. Then:

$$d(f,g) = \sum_n 2^{-n} \frac{d_{K_n}(f,g)}{1 + d_{K_n}(f,g)}.$$

The diffeomorphism group equipped with its weak topology is locally homeomorphic to the space of C^r vector fields (Leslie 1967). Over a compact subset of M, this follows by fixing a Riemannian metric on M and using the exponential map for that metric. If r is finite and the manifold is compact, the space of vector fields is a Banach space. Moreover, the transition maps from one chart of this atlas to another are smooth, making the diffeomorphism group into a Banach manifold with smooth right translations; left translations and inversion are only continuous. If $r = \infty$, the space of vector fields is a Fréchet space. Moreover, the transition maps are smooth, making the diffeomorphism group into a Fréchet manifold and even into a regular Fréchet Lie group.

If the manifold is σ-compact and not compact the full diffeomorphism group is not locally contractible for any of the two topologies. One has to restrict the group by controlling the deviation from the identity near infinity to obtain a diffeomorphism group which is a manifold; see (Michor & Mumford 2013).

40.5.2 Lie algebra

The Lie algebra of the diffeomorphism group of M consists of all vector fields on M equipped with the Lie bracket of vector fields. Somewhat formally, this is seen by making a small change to the coordinate x at each point in space:

$$x^\mu \to x^\mu + \varepsilon h^\mu(x)$$

so the infinitesimal generators are the vector fields

$$L_h = h^\mu(x) \frac{\partial}{\partial x_\mu}.$$

40.5.3 Examples

- When $M = G$ is a Lie group, there is a natural inclusion of G in its own diffeomorphism group via left-translation. Let Diff(G) denote the diffeomorphism group of G, then there is a splitting Diff$(G) \simeq G \times$ Diff(G, e), where Diff(G, e) is the subgroup of Diff(G) that fixes the identity element of the group.

- The diffeomorphism group of Euclidean space $\mathbf{R}^n$ consists of two components, consisting of the orientation preserving and orientation reversing diffeomorphisms. In fact, the general linear group is a deformation retract of subgroup Diff$(\mathbf{R}^n, 0)$ of diffeomorphisms fixing the origin under the map $f(x) \mapsto f(tx)/t$, $t \in \& (0,1]$. In particular, the general linear group is also a deformation retract of the full diffeomorphism group.

- For a finite set of points, the diffeomorphism group is simply the symmetric group. Similarly, if M is any manifold there is a group extension $0 \to \mathrm{Diff}_0(M) \to \mathrm{Diff}(M) \to \Sigma(\pi_0(M))$. Here $\mathrm{Diff}_0(M)$ is the subgroup of $\mathrm{Diff}(M)$ that preserves all the components of M, and $\Sigma(\pi_0(M))$ is the permutation group of the set $\pi_0(M)$ (the components of M). Moreover, the image of the map $\mathrm{Diff}(M) \to \Sigma(\pi_0(M))$ is the bijections of $\pi_0(M)$ that preserve diffeomorphism classes.

40.5.4 Transitivity

For a connected manifold M, the diffeomorphism group acts transitively on M. More generally, the diffeomorphism group acts transitively on the configuration space CkM. If M is at least two-dimensional, the diffeomorphism group acts transitively on the configuration space FkM and the action on M is multiply transitive (Banyaga 1997, p. 29).

40.5.5 Extensions of diffeomorphisms

In 1926, Tibor Radó asked whether the harmonic extension of any homeomorphism or diffeomorphism of the unit circle to the unit disc yields a diffeomorphism on the open disc. An elegant proof was provided shortly afterwards by Hellmuth Kneser. In 1945, Gustave Choquet, apparently unaware of this result, produced a completely different proof.

The (orientation-preserving) diffeomorphism group of the circle is pathwise connected. This can be seen by noting that any such diffeomorphism can be lifted to a diffeomorphism f of the reals satisfying $[f(x+1) = f(x) + 1]$; this space is convex and hence path-connected. A smooth, eventually constant path to the identity gives a second more elementary way of extending a diffeomorphism from the circle to the open unit disc (a special case of the Alexander trick). Moreover, the diffeomorphism group of the circle has the homotopy-type of the orthogonal group $O(2)$.

The corresponding extension problem for diffeomorphisms of higher-dimensional spheres S^{n-1} was much studied in the 1950s and 1960s, with notable contributions from René Thom, John Milnor and Stephen Smale. An obstruction to such extensions is given by the finite abelian group Γn, the "group of twisted spheres", defined as the quotient of the abelian component group of the diffeomorphism group by the subgroup of classes extending to diffeomorphisms of the ball B^n.

40.5.6 Connectedness

For manifolds, the diffeomorphism group is usually not connected. Its component group is called the mapping class group. In dimension 2 (i.e. surfaces), the mapping class group is a finitely presented group generated by Dehn twists (Dehn, Lickorish, Hatcher). Max Dehn and Jakob Nielsen showed that it can be identified with the outer automorphism group of the fundamental group of the surface.

William Thurston refined this analysis by classifying elements of the mapping class group into three types: those equivalent to a periodic diffeomorphism; those equivalent to a diffeomorphism leaving a simple closed curve invariant; and those equivalent to pseudo-Anosov diffeomorphisms. In the case of the torus $S^1 \times S^1 = R^2/Z^2$, the mapping class group is simply the modular group $SL(2, Z)$ and the classification becomes classical in terms of elliptic, parabolic and hyperbolic matrices. Thurston accomplished his classification by observing that the mapping class group acted naturally on a compactification of Teichmüller space; as this enlarged space was homeomorphic to a closed ball, the Brouwer fixed-point theorem became applicable.

Smale conjectured that if M is an oriented smooth closed manifold, the identity component of the group of orientation-preserving diffeomorphisms is simple. This had first been proved for a product of circles by Michel Herman; it was proved in full generality by Thurston.

40.5.7 Homotopy types

- The diffeomorphism group of S^2 has the homotopy-type of the subgroup $O(3)$. This was proved by Steve Smale.[1]

- The diffeomorphism group of the torus has the homotopy-type of its linear automorphisms: $S^1 \times S^1 \times GL(2, Z)$.

- The diffeomorphism groups of orientable surfaces of genus $g > 1$ have the homotopy-type of their mapping class groups (i.e. the components are contractible).

- The homotopy-type of the diffeomorphism groups of 3-manifolds are fairly well-understood via the work of Ivanov, Hatcher, Gabai and Rubinstein, although there are a few outstanding open cases (primarily 3-manifolds with finite fundamental groups).

- The homotopy-type of diffeomorphism groups of n-manifolds for $n > 3$ are poorly undersood. For example, it is an open problem whether or not $\mathrm{Diff}(\mathbf{S}^4)$ has more than two components. Via Milnor, Kahn and Antonelli, however, it is known that provided $n > 6$, $\mathrm{Diff}(\mathbf{S}^n)$ does not have the homotopy-type of a finite CW-complex.

40.6   Homeomorphism and diffeomorphism

Unlike non-diffeomorphic homeomorphisms, it is relatively difficult to find a pair of homeomorphic manifolds that are not diffeomorphic. In dimensions 1, 2, 3, any pair of homeomorphic smooth manifolds are diffeomorphic. In dimension 4 or greater, examples of homeomorphic but not diffeomorphic pairs have been found. The first such example was constructed by John Milnor in dimension 7. He constructed a smooth 7-dimensional manifold (called now Milnor's sphere) that is homeomorphic to the standard 7-sphere but not diffeomorphic to it. There are, in fact, 28 oriented diffeomorphism classes of manifolds homeomorphic to the 7-sphere (each of them is the total space of a fiber bundle over the 4-sphere with the 3-sphere as the fiber).

More unusual phenomena occur for 4-manifolds. In the early 1980s, a combination of results due to Simon Donaldson and Michael Freedman led to the discovery of exotic R4s: there are uncountably many pairwise non-diffeomorphic open subsets of $\mathbf{R}^4$ each of which is homeomorphic to $\mathbf{R}^4$, and also there are uncountably many pairwise non-diffeomorphic differentiable manifolds homeomorphic to $\mathbf{R}^4$ that do not embed smoothly in $\mathbf{R}^4$.

40.7   See also

- Étale morphism

- Large diffeomorphism

- Local diffeomorphism

- Superdiffeomorphism

40.8   Notes

[1] Smale, "Diffeomorphisms of the 2-sphere", *Proc. Amer. Math. Soc.* 10 (1959), pp. 621–626.

40.9   References

- Chaudhuri, Shyamoli, Hakuru Kawai and S.-H Henry Tye. "Path-integral formulation of closed strings", *Phys. Rev. D*, 36: 1148 (1987).

- Banyaga, Augustin (1997), *The structure of classical diffeomorphism groups*, Mathematics and its Applications, 400, Kluwer Academic, ISBN 0-7923-4475-8

- Duren, Peter L. (2004), *Harmonic Mappings in the Plane*, Cambridge Mathematical Tracts, 156, Cambridge University Press, ISBN 0-521-64121-7

- Hazewinkel, Michiel, ed. (2001), "Diffeomorphism", *Encyclopedia of Mathematics*, Springer, ISBN 978-1-55608-010-4

- Hirsch, Morris (1997), *Differential Topology*, Berlin, New York: Springer-Verlag, ISBN 978-0-387-90148-0

- Kriegl, Andreas; Michor, Peter (1997), *The convenient setting of global analysis*, Mathematical Surveys and Monographs, 53, American Mathematical Society, ISBN 0-8218-0780-3

- Leslie, J. A. (1967), "On a differential structure for the group of diffeomorphisms", *Topology* **6** (2): 263–271, doi:10.1016/0040-9383(67)90038-9, ISSN 0040-9383, MR 0210147

- Michor, Peter W.; Mumford, David (2013), "A zoo of diffeomorphism groups on $\mathbf{R}^n$.", *Annals of Global Analysis and Geometry* **44** (4): 529–540, doi:10.1007/s10455-013-9380-2 (arXiv:1211.5704)

- Milnor, John W. (2007), *Collected Works Vol. III, Differential Topology*, American Mathematical Society, ISBN 0-8218-4230-7

- Omori, Hideki (1997), *Infinite-dimensional Lie groups*, Translations of Mathematical Monographs, 158, American Mathematical Society, ISBN 0-8218-4575-6

- Kneser, Hellmuth (1926), "Lösung der Aufgabe 41.", *Jahresbericht der Deutschen Mathematiker-Vereinigung* (in German) **35** (2): 123

Chapter 41

Symmetry breaking

This article is about the concept in physics. For the concept in biology, see Symmetry breaking and cortical rotation. For the concept in mathematics, see Symmetry-breaking constraints.

In physics, **symmetry breaking** is a phenomenon in which (infinitesimally) small fluctuations acting on a system crossing a critical point decide the system's fate, by determining which branch of a bifurcation is taken. To an outside observer unaware of the fluctuations (or "noise"), the choice will appear arbitrary. This process is called symmetry "breaking", because such transitions usually bring the system from a symmetric but disorderly state into one or more definite states. Symmetry breaking is supposed to play a major role in pattern formation.

In 1972, Nobel laureate P.W. Anderson used the idea of symmetry breaking to show some of the drawbacks of reductionism in his paper titled "More is different" in *Science*.[1]

Symmetry breaking can be distinguished into two types, explicit symmetry breaking and spontaneous symmetry breaking, characterized by whether the equations of motion fail to be invariant or the ground state fails to be invariant.

41.1 Explicit symmetry breaking

Main article: Explicit symmetry breaking

In explicit symmetry breaking, the equations of motion describing a system are not invariant under the broken symmetry.

41.2 Spontaneous symmetry breaking

Main article: Spontaneous symmetry breaking

In spontaneous symmetry breaking, the equations of motion of the system are invariant, but the system is not because the background (spacetime) of the system, its vacuum, is non-invariant. Such a symmetry breaking is parametrized by an order parameter. A special case of this type of symmetry breaking is dynamical symmetry breaking.

41.3 Examples

Symmetry breaking can cover any of the following scenarios:[2]

- The breaking of an exact symmetry of the underlying laws of physics by the random formation of some structure;

- A situation in physics in which a minimal energy state has less symmetry than the system itself;

- Situations where the actual state of the system does not reflect the underlying symmetries of the dynamics because the manifestly symmetric state is unstable (stability is gained at the cost of local asymmetry);

- Situations where the equations of a theory may have certain symmetries, though their solutions may not (the symmetries are "hidden").

One of the first cases of broken symmetry discussed in the physics literature is related to the form taken by a uniformly rotating body of incompressible fluid in gravitational and hydrostatic equilibrium. Jacobi[3] and soon later Liouville,[4] in 1834, discussed the fact that a tri-axial ellipsoid was an equilibrium solution for this problem when the kinetic energy compared to the gravitational energy of the rotating body exceeded a certain critical value. The axial symmetry presented by the McLaurin spheroids is broken at this bifurcation point. Furthermore, above this bifurcation point, and for constant angular momentum, the solutions that minimize the kinetic energy are the *non*-axially symmetric Jacobi ellipsoids instead of the Maclaurin spheroids.

41.4 See also

- Anomalous symmetry breaking

- Higgs mechanism

- QCD vacuum

- Goldstone boson

- 1964 PRL symmetry breaking papers

- J. J. Sakurai Prize for Theoretical Particle Physics

41.5 References

[1] Anderson,P.W. (1972). "More is Different"(PDF).*Science***177**(4047): 393–396. Bibcode:1972Sci...177..393A.doi:1.393. PMID 17796623.

[2] http://www.angelfire.com/stars5/astroinfo/gloss/s.html

[3] Jacobi, C.G.J. (1834). "Über die figur des gleichgewichts". *Annalen der Physik und Chemie* (33): 229–238.

[4] Liouville, J. (1834). "Sur la figure d'une masse fluide homogène, en équilibre et douée d'un mouvement de rotation". *Journal de l'École Polytechnique* (14): 289–296.

Chapter 42

Wheeler–Feynman absorber theory

The **Wheeler–Feynman absorber theory** (also called the **Wheeler–Feynman time-symmetric theory**), named after its originators, the physicists Richard Feynman and John Archibald Wheeler, is an interpretation of electrodynamics derived from the assumption that the solutions of the electromagnetic field equations must be invariant under time-reversal transformation, as are the field equations themselves. Indeed, there is no apparent reason for the time-reversal symmetry breaking which singles out a preferential time direction and thus makes a distinction between past and future. A time-reversal invariant theory is more logical and elegant. Another key principle, resulting from this interpretation and reminiscent of Mach's principle due to Tetrode, is that elementary particles are not self-interacting. This immediately removes the problem of self-energies.

42.1   T-symmetry and causality

The requirement of time reversal symmetry, in general, is difficult to conjugate with the principle of causality. Maxwell's equations and the equations for electromagnetic waves have, in general, two possible solutions: a retarded (delayed) solution and an advanced one. Accordingly, any charged particle generates waves, say at time $t_0 = 0$ and point $x_0 = 0$, which will arrive at point x_1 at the instant $t_1 = x_1/c$ (here c is the speed of light) after the emission (retarded solution), and other waves which will arrive at the same place at the instant $t_2 = x_1/c$ before the emission (advanced solution). The latter, however, violates the causality principle: advanced waves could be detected before their emission. Thus the advanced solutions are usually discarded in the interpretation of electromagnetic waves. In the absorber theory, instead charged particles are considered as both emitters and absorbers, and the emission process is connected with the absorption process as follows: Both the retarded waves from emitter to absorber and the advanced waves from absorber to emitter are considered. The sum of the two, however, results in *causal waves*, although the anti-causal (advanced) solutions are not discarded *a priori*.

Feynman and Wheeler obtained this result in a very simple and elegant way. They considered all the charged particles (emitters) present in our universe, and assumed all of them to generate time-reversal symmetric waves. The resulting field is

$$E_{\text{tot}}(\mathbf{x}, t) = \sum_n \frac{E_n^{\text{ret}}(\mathbf{x}, t) + E_n^{\text{adv}}(\mathbf{x}, t)}{2}.$$

Then they observed that, if the relation

$$E_{\text{free}}(\mathbf{x}, t) = \sum_n \frac{E_n^{\text{ret}}(\mathbf{x}, t) - E_n^{\text{adv}}(\mathbf{x}, t)}{2} = 0$$

holds, E_{free}, being a solution of the homogeneous Maxwell equation, can be used to obtain the total field

$$E_{tot}(\mathbf{x}, t) = \sum_n \frac{E_n^{ret}(\mathbf{x}, t) + E_n^{adv}(\mathbf{x}, t)}{2} + \sum_n \frac{E_n^{ret}(\mathbf{x}, t) - E_n^{adv}(\mathbf{x}, t)}{2} = \sum_n E_n^{ret}(\mathbf{x}, t).$$

The total field is retarded and causality is not violated.

The assumption that the *free field* is identically zero is the core of the absorber idea. It means that the radiation emitted by each particle is completely absorbed by all other particles present in the universe. To better understand this point, it may be useful to consider how the absorption mechanism works in common materials. At the microscopic scale, it results from the sum of the incoming electromagnetic wave and the waves generated from the electrons of the material, which react to the external perturbation. If the incoming wave is absorbed, the result is a zero outcoming field. In the absorber theory the same concept is used, however in presence of both retarded and advanced waves.

The resulting wave appears to have a preferred time direction, because it respects causality. However, this is only an illusion. Indeed it is always possible to reverse the time direction by simply exchanging the labels *emitter* and *absorber*. Thus, the apparently preferred time direction results from the arbitrary labelling.

42.2 T-symmetry and self-interaction

One of the major results of the absorber theory is the elegant and clear interpretation of the electromagnetic radiation process. A charged particle which experiences acceleration is known to emit electromagnetic waves, i.e., to lose energy. Thus, the Newtonian equation for the particle ($F = ma$) must contain a dissipative force (damping term), which takes into account this energy loss. In the causal interpretation of electromagnetism, Lorentz and Abraham proposed that such a force, later called Abraham–Lorentz force, is due to the retarded self-interaction of the particle with its own field. This first interpretation, however, is not completely satisfactory, as it leads to divergences in the theory and needs some assumptions on the structure of charge distribution of the particle. Dirac generalized the formula to make it relativistically invariant. While doing so, he also suggested a different interpretation. He showed that the damping term can be expressed in terms of a free field acting on the particle at its own position.

$$E^{damping}(\mathbf{x}_j, t) = \frac{E_j^{ret}(\mathbf{x}_j, t) - E_j^{adv}(\mathbf{x}_j, t)}{2}$$

However Dirac did not propose any physical explanation of this interpretation.

A clear and simple explanation can instead be obtained in the framework of absorber theory, starting from the simple idea that each particle does not interact with itself. This is actually the opposite of the first Abraham–Lorentz proposal. The field acting on the particle j at its own position (the point x_j) is then:

$$E^{tot}(\mathbf{x}_j, t) = \sum_{n \neq j} \frac{E_n^{ret}(\mathbf{x}_j, t) + E_n^{adv}(\mathbf{x}_j, t)}{2}.$$

If we sum the *free field term* of this expression we obtain

$$E^{tot}(\mathbf{x}_j, t) = \sum_{n \neq j} \frac{E_n^{ret}(\mathbf{x}_j, t) + E_n^{adv}(\mathbf{x}_j, t)}{2} + \sum_n \frac{E_n^{ret}(\mathbf{x}_j, t) - E_n^{adv}(\mathbf{x}_j, t)}{2}$$

and, thanks to Dirac's result,

$$E^{tot}(\mathbf{x}_j, t) = \sum_{n \neq j} E_n^{ret}(\mathbf{x}_j, t) + E^{damping}(\mathbf{x}_j, t).$$

Thus, the damping force is obtained without the need for self-interaction, which is known to lead to divergences, and also giving a physical justification to the expression derived by Dirac.

42.3   Criticism

The Abraham–Lorentz force is, however, not free of problems. Written in the non-relativistic limit, it gives:

$$E^{\text{damping}}(\mathbf{x}_j, t) = \frac{e}{6\pi c^3} \frac{\mathrm{d}^3}{\mathrm{d}t^3} x$$

Since the third derivative with respect to the time (also called the "jerk" or "jolt") enters in the equation of motion, to derive a solution one needs not only the initial position and velocity of the particle, but also its initial acceleration. This apparent problem however can be solved in the absorber theory, by observing that the equation of motion for the particle has to be solved together with the Maxwell equations for the field. In this case, instead of the initial acceleration, one only needs to specify the initial field and the boundary condition. This interpretation restores the coherence of the physical interpretation of the theory.

Other difficulties may arise trying to solve the equation of motion for a charged particle in the presence of this damping force. It is commonly stated that the Maxwell equations are classical and cannot correctly account for microscopic phenomena, such as the behavior of a point-like particle, where quantum mechanical effects should appear. Nevertheless with absorber theory, Wheeler and Feynman were able to create a coherent classical approach to the problem (see also the "paradoxes" section in the Abraham–Lorentz force).

Also, the time-symmetric interpretation of the electromagnetic waves appears to be in contrast with the experimental evidence that time flows in a given direction and, thus, that the T-symmetry is broken in our world. It is commonly believed, however, that this symmetry breaking appears only in the thermodynamical limit (see, for example, the arrow of time). Wheeler himself accepted that the expansion of the universe is not time symmetric in the thermodynamic limit . This however does not imply that the T-symmetry must be broken also at the microscopic level.

Finally, the main drawback of the theory turned out to be the result that particles are not self-interacting. Indeed, as demonstrated by Hans Bethe, the Lamb shift necessitated a self-energy term to be explained. Feynman and Bethe had an intense discussion over that issue and eventually Feynman himself stated that self-interaction is needed to correctly account for this effect.

42.4   Developments since original formulation

42.4.1   Gravity theory

Main article: Hoyle–Narlikar theory of gravity

Inspired by the Machian nature of the Wheeler–Feynman absorber theory for electrodynamics, Fred Hoyle and Jayant Narliker proposed their own theory of gravity[1][2][3] in the context of general relativity. This model still exists in spite of recent astronomical observations that have challenged the theory.

42.4.2   Transactional interpretation of quantum mechanics

Main article: Transactional interpretation

Again inspired by the Wheeler–Feynman absorber theory, the transactional interpretation of quantum mechanics (TIQM) first proposed in 1986 by John G. Cramer[4] describes quantum interactions in terms of a standing wave formed by retarded (forward-in-time) and advanced (backward-in-time) waves. J. Cramer claims it avoids the philosophical problems with the Copenhagen interpretation and the role of the observer, and resolves various quantum paradoxes, such as quantum nonlocality, quantum entanglement and retrocausality.[5]

42.4.3 Attempted resolution of causality

T. C. Scott and R. A. Moore demonstrated that the apparent acausality suggested by the presence of advanced Liénard–Wiechert potentials could be removed by recasting the theory in terms of retarded potentials only, without the complications of the absorber idea.[6][7] The Lagrangian describing a particle (p_1) under the influence of the time-symmetric potential generated by another particle (p_1) is:

$$L_1 = T_1 - \frac{1}{2}\left((V_R)_1^2 + (V_A)_1^2\right)$$

where T_i is the relativistic kinetic energy functional of particle p_i, and, $(V_R)_i^j$ and $(V_A)_i^j$ are respectively the retarded and advanced Liénard–Wiechert potentials acting on particle p_i and generated by particle p_j. The corresponding Lagrangian for particle p_1 is:

$$L_2 = T_2 - \frac{1}{2}\left((V_R)_2^1 + (V_A)_2^1\right).$$

It was originally demonstrated with computer algebra[8] and then proven analytically[9] that:

$$(V_R)_j^i - (V_A)_i^j$$

is a total time derivative, i.e. a *divergence* in the calculus of variations, and thus it gives no contribution to the Euler–Lagrange equations. Thanks to this result the advanced potentials can be eliminated; here the total derivative plays the same role as the *free field*. The Lagrangian for the N-body system is therefore:

$$L = \sum_{i=1}^{N} T_i - \frac{1}{2}\sum_{i \neq j}^{N} (V_R)_j^i$$

The resulting lagrangian is symmetric under the exchange of p_i with p_j. For $N = 2$ this Lagrangian will generate *exactly* the same equations of motion of L_1 and L_2. Therefore, from the point of view of an *outside* observer, everything is causal. Only if we isolate the forces acting on a particular body do the advanced potentials make their appearance. This recasting of the problem comes at a price: the N-body Lagrangian depends on all the time derivatives of the curves traced by all particles i.e. the Lagrangian is infinite order. However, much progress was made in examining the unresolved issue of quantizing the theory.[10][11] Also, this formulation recovers the Darwin Lagrangian from which the Breit equation was originally derived, but without the dissipative terms.[9] This ensures agreement with theory and experiment, up to but not including the Lamb shift. Numerical solutions for the classical problem were also found.[12] Finally, Moore and Scott[6] showed that the radiation reaction can be alternatively derived using the notion that, on average, the net dipole moment is zero for a collection of charged particles, thereby avoiding the complications of the absorber theory. An important bonus from their approach is the formulation of a total preserved canonical generalized momentum, as presented in a comprehensive review article in the light of quantum nonlocality[13]

This apparent acausality may be viewed as merely apparent, & this entire problem goes away. For the opposing view as held by Einstein see.[14] This result in no way determines this debate in favor of Ritz.

42.4.4 Alternative Lamb shift calculation

As mentioned previously, a serious criticism against the absorber theory is that its Machian assumption that point particles do not act on themselves does not allow (infinite) self-energies and consequently an explanation for the Lamb shift according to Quantum electrodynamics (QED). Ed Jaynes proposed an alternate model where the Lamb-like shift is due instead to the interaction with *other particles* very much along the same notions of the Wheeler–Feynman absorber theory itself. One simple model is to calculate the motion of an oscillator coupled directly with many other oscillators. Jaynes has shown that it is easy to get both spontaneous emission and Lamb shift behavior in classical mechanics.[15] Furthermore, Jayne's alternatives provides a solution to the process of "addition and subtraction of infinities" associated with renormalization.[13][16]

This model leads to essentially the same type of Bethe Logarithm an essential part of the Lamb shift calculation vindicating Jaynes' claim that two different physical models can be mathematically isomorphic to each other and therefore yield the same results, a point also apparently made by Scott and Moore on the issue of causality.

42.5 Conclusions

This universal absorber theory is mentioned in the chapter titled "Monster Minds" in Feynman's autobiographical work *Surely You're Joking, Mr. Feynman!* as well as in Vol. II of the Feynman Lectures on Physics. It led to the formulation of a framework of quantum mechanics using a Lagrangian and action as starting points, rather than a Hamiltonian, namely the formulation using Feynman path integrals which proved useful in Feynman's earliest calculations in quantum electrodynamics and quantum field theory in general. Both retarded and advanced fields appear respectively as retarded and advanced propagators, and also, in the Feynman propagator and the Dyson propagator. In hindsight, the relationship between retarded and advanced potentials shown here is not so surprising in view of the fact that, in field theory, the advanced propagator can be obtained from the retarded propagator by exchanging the roles of field source and test particle (usually within the kernel of a Green's function formalism). In field theory, advanced as well as retarded fields are simply viewed as *mathematical* solutions of Maxwell's equations whose combinations are decided by the boundary conditions.

42.6 See also

- Causality

- Symmetry in physics and T-symmetry

- Transactional interpretation

- Abraham–Lorentz force

- Retrocausality

- Two-state vector formalism

- Paradox of a charge in a gravitational field

- Ni Guangjiong

42.7 Notes

[1] F.Hoyle and J.V.Narlikar(1964). "A New Theory of Gravitation".*Proceedings of the Royal Society A*.Bibcode:1964R91H. doi:10.1098/rspa.1964.0227.

[2] "Cosmology: Math Plus Mach Equals Far-Out Gravity". Time. June 26, 1964. Retrieved 7 August 2010.

[3] Hoyle, F.; Narlikar, J. V. (1995). "Cosmology and action-at-a-distance electrodynamics". *Reviews of Modern Physics* **67** (1): 113–155. Bibcode:1995RvMP...67..113H. doi:10.1103/RevModPhys.67.113.

[4] The Transactional Interpretation of Quantum Mechanics by John Cramer. *Reviews of Modern Physics* 58, 647-688, July (1986)

[5] John G. Cramer, "Quantum Entanglement, Nonlocality, Back-in-Time Messages, *(April 3, 2010)*.

[6] Moore, R. A.; Scott, T. C.; Monagan, M. B. (1987). "Relativistic, many-particle Lagrangean for electromagnetic interactions". *Phys. Rev. Lett.* **59** (5): 525–527. Bibcode:1987PhRvL..59..525M. doi:10.1103/PhysRevLett.59.525.

[7] Moore, R. A.; Scott, T. C.; Monagan, M. B. (1988). "A Model for a Relativistic Many-Particle Lagrangian with Electromagnetic Interactions". *Can. J. Phys.* **66** (3): 206–211. Bibcode:1988CaJPh..66..206M. doi:10.1139/p88-032.

[8] Scott, T. C.; Moore, R. A.; Monagan, M. B. (1989). "Resolution of Many Particle Electrodynamics by Symbolic Manipulation". *Comput. Phys. Commun.* **52** (2): 261–281. Bibcode:1989CoPhC..52..261S. doi:10.1016/0010-4655(89)90009-X.

[9] Scott, T. C. (1986). "Relativistic Classical and Quantum Mechanical Treatment of the Two-body Problem". *MMath thesis* (U. of Waterloo, Canada).

[10] Scott, T. C.; Moore, R. A. (1989). "Quantization of Hamiltonians from High-Order Lagrangians". *Nucl. Phys. B* **6** (Proc. Suppl.): 455–457. Bibcode:1989NuPhS...6..455S. doi:10.1016/0920-5632(89)90498-2.

[11] Moore, R. A.; Scott, T. C. (1991). "Quantization of Second-Order Lagrangians: Model Problem". *Phys. Rev. A* **44** (3): 1477–1484. Bibcode:1991PhRvA..44.1477M. doi:10.1103/PhysRevA.44.1477.

[12] Moore, R. A.; Qi, D.; Scott, T. C. (1992). "Causality of Relativistic Many-Particle Classical Dynamics Theories". *Can. J. Phys.* **70** (9): 772–781. Bibcode:1992CaJPh..70..772M. doi:10.1139/p92-122.

[13] Scott, T. C.; Andrae, D. (2015). "Quantum Nonlocality and Conservation of momentum". *Phys. Essays* **28** (3): 374–385.

[14] http://www.datasync.com/~{}rsf1/rtzein.htm

[15] E.T. Jaynes, "The Lamb Shift in Classical Mechanics in "*Probability in Quantum Theory*, pp. 13-15. (1996) Jaynes' analysis of Lamb shift.

[16] E.T. Jaynes, "Classical Subtraction Physics in "Probability in Quantum Theory, *pp. 15-18. (1996)* Jaynes' analysis of handing the infinities of the Lamb shift calculation.

42.8 Key papers

- Wheeler, J. A.; Feynman, R. P. (1945). "Interaction with the Absorber as the Mechanism of Radiation". *Reviews of Modern Physics* **17** (2–3): 157–161. Bibcode:1945RvMP...17..157W. doi:10.1103/RevModPhys.17.157.

- Wheeler, J. A.; Feynman, R. P. (1949). "Classical Electrodynamics in Terms of Direct Interparticle Action". *Reviews of Modern Physics* **21** (3): 425–433. Bibcode:1949RvMP...21..425W. doi:10.1103/RevModPhys.21.425.

42.9 External links

- J. A. Wheeler and R. P. Feynman, "Interaction with the Absorber as the Mechanism of Radiation" Caltech Library of Authors

Chapter 43

Invariant (mathematics)

In mathematics, an **invariant** is a property, held by a class of mathematical objects, which remains unchanged when transformations of a certain type are applied to the objects. The particular class of objects and type of transformations are usually indicated by the context in which the term is used. For example, the area of a triangle is an invariant with respect to isometries of the Euclidean plane. The phrases "invariant under" and "invariant to" a transformation are both used. More generally, an invariant with respect to an equivalence relation is a property that is constant on each equivalence class.

Invariants are used in diverse areas of mathematics such as geometry, topology and algebra. Some important classes of transformations are defined by an invariant they leave unchanged, for example conformal maps are defined as transformations of the plane that preserve angles. The discovery of invariants is an important step in the process of classifying mathematical objects.

43.1   Simple examples

The most fundamental example of invariance is expressed in our ability to count. For a finite collection of objects of any kind, there appears to be a number to which we invariably arrive, regardless of how we count the objects in the set. The quantity—a cardinal number—is associated with the set, and is invariant under the process of counting.

An identity is an equation that remains true for all values of its variables. There are also inequalities that remain true when the values of their variables change.

Another simple example of invariance is that the distance between two points on a number line is not changed by adding the same quantity to both numbers. On the other hand, multiplication does not have this property, so distance is not invariant under multiplication.

Angles and ratios of distances are invariant under scalings, rotations, translations and reflections. These transformations produce similar shapes, which is the basis of trigonometry. All circles are similar. Therefore, they can be transformed into each other and the ratio of the circumference to the diameter is invariant and equal to pi.

43.2   More advanced examples

Some more complicated examples:

- The real part and the absolute value of a complex number are invariant under complex conjugation.
- The degree of a polynomial is invariant under linear change of variables.
- The dimension and homology groups of a topological object are invariant under homeomorphism.[1]
- The number of fixed points of a dynamical system is invariant under many mathematical operations.

- Euclidean distance is invariant under orthogonal transformations.

- Euclidean area is invariant under a linear map with determinant 1 (see Equi-areal maps).

- Some invariants of projective transformations: collinearity of three or more points, concurrency of three or more lines, conic sections, the cross-ratio.[2]

- The determinant, trace, and eigenvectors and eigenvalues of a square matrix are invariant under changes of basis. In a word, the spectrum of a matrix is invariant to the change of basis.

- Invariants of tensors.

- The singular values of a matrix are invariant under orthogonal transformations.

- Lebesgue measure is invariant under translations.

- The variance of a probability distribution is invariant under translations of the real line; hence the variance of a random variable is unchanged by the addition of a constant to it.

- The fixed points of a transformation are the elements in the domain invariant under the transformation. They may, depending on the application, be called symmetric with respect to that transformation. For example, objects with translational symmetry are invariant under certain translations.

- The integral $\int_M K \, d\mu$ of the Gaussian curvature K of a 2-dimensional Riemannian manifold (M,g) is invariant under changes of the Riemannian metric g. This is the Gauss-Bonnet Theorem.

43.3   Invariant set

A subset S of the domain U of a mapping $T: U \to U$ is an **invariant set** under the mapping when $x \in S \Rightarrow T(x) \in S$. Note that the elements of S are not fixed, but rather the set S is fixed in the power set of U. (Some authors use the terminology *setwise invariant*[3] vs. *pointwise invariant*[4] to distinguish between these cases.) For example, a circle is an invariant subset of the plane under a rotation about the circle's center. Further, a conical surface is invariant as a set under a homothety of space.

An invariant set of an operation T is also said to be **stable under** T. For example, the normal subgroups that are so important in group theory are those subgroups that are stable under the inner automorphisms of the ambient group.[5][6][7] Other examples occur in linear algebra. Suppose a linear transformation T has an eigenvector $\mathbf{v}$. Then the line through 0 and $\mathbf{v}$ is an invariant set under T. The eigenvectors span an invariant subspace which is stable under T.

When T is a screw displacement, the screw axis is an invariant line, though if the pitch is non-zero, T has no fixed points.

43.4   Formal statement

The notion of invariance is formalized in three different ways in mathematics: via group actions, presentations, and deformation.

43.4.1   Unchanged under group action

Firstly, if one has a group G acting on a mathematical object (or set of objects) X, then one may ask which points x are unchanged, "invariant" under the group action, or under an element g of the group.

Very frequently one will have a group acting on a set X and ask which objects in an *associated* set $F(X)$ are invariant. For example, rotation in the plane about a point leaves the point about which it rotates invariant, while translation in the plane does not leave any points invariant, but does leave all lines parallel to the direction of translation invariant as lines.

Formally, define the set of lines in the plane P as $L(P)$; then a rigid motion of the plane takes lines to lines – the group of rigid motions acts on the set of lines – and one may ask which lines are unchanged by an action.

More importantly, one may define a *function* on a set, such as "radius of a circle in the plane" and then ask if this function is invariant under a group action, such as rigid motions.

Dual to the notion of invariants are *coinvariants*, also known as *orbits*, which formalizes the notion of congruence: objects which can be taken to each other by a group action. For example, under the group of rigid motions of the plane, the perimeter of a triangle is an invariant, while the set of triangles congruent to a given triangle is a coinvariant.

These are connected as follows: invariants are constant on coinvariants (for example, congruent triangles have the same perimeter), while two objects which agree in the value of one invariant may or may not be congruent (two triangles with the same perimeter need not be congruent). In classification problems, one seeks to find a complete set of invariants, such that if two objects have the same values for this set of invariants, they are congruent. For example, triangles such that all three sides are equal are congruent, via SSS congruence, and thus the length of all three sides forms a complete set of invariants for triangles.

43.4.2   Independent of presentation

Secondly, a function may be defined in terms of some presentation or decomposition of a mathematical object; for instance, the Euler characteristic of a cell complex is defined as the alternating sum of the number of cells in each dimension. One may forget the cell complex structure and look only at the underlying topological space (the manifold) – as different cell complexes give the same underlying manifold, one may ask if the function is *independent* of choice of *presentation*, in which case it is an *intrinsically* defined invariant. This is the case for the Euler characteristic, and a general method for defining and computing invariants is to define them for a given presentation and then show that they are independent of the choice of presentation. Note that there is no notion of a group action in this sense.

The most common examples are:

- The presentation of a manifold in terms of coordinate charts – invariants must be unchanged under change of coordinates.

- Various manifold decompositions, as discussed for Euler characteristic.

- Invariants of a presentation of a group.

43.4.3   Unchanged under perturbation

Thirdly, if one is studying an object which varies in a family, as is common in algebraic geometry and differential geometry, one may ask if the property is unchanged under perturbation – if an object is constant on families or invariant under change of metric, for instance.

43.5   See also

- Erlangen program

- Invariant (physics)

- Invariant estimator in statistics

- Invariant theory

- Symmetry in mathematics

- Topological invariant

- Invariant differential operator

- Invariant measure

- Mathematical constant

- Mathematical constants and functions

43.6 Notes

[1] Fraleigh (1976, pp. 166–167)

[2] Kay (1969, pp. 219)

[3] Barry Simon. *Representations of Finite and Compact Groups*. American Mathematical Soc. p. 16. ISBN 978-0-8218-7196-6.

[4] Judith Cederberg (1989). *A Course in Modern Geometries*. Springer. p. 174. ISBN 978-1-4757-3831-5.

[5] Fraleigh (1976, p. 103)

[6] Herstein (1964, p. 42)

[7] McCoy (1968, p. 183)

43.7 References

- Fraleigh, John B. (1976), *A First Course In Abstract Algebra* (2nd ed.), Reading: Addison-Wesley, ISBN 0-201-01984-1

- Herstein, I. N. (1964), *Topics In Algebra*, Waltham: Blaisdell Publishing Company, ISBN 978-1114541016

- Kay, David C. (1969), *College Geometry*, New York: Holt, Rinehart and Winston, LCCN 69-12075

- McCoy, Neal H. (1968), *Introduction To Modern Algebra. Revised Edition*, Boston: Allyn and Bacon, LCCN 68-15225

- Weisstein, Eric W., "Invariant", *MathWorld*.

- Popov, V.L. (2001), "Invariant", in Hazewinkel, Michiel, *Encyclopedia of Mathematics*, Springer, ISBN 978-1-55608-010-4

Chapter 44

Symmetric relation

In mathematics and other areas, a binary relation R over a set X is **symmetric** if it holds for all a and b in X that if a is related to b then b is related to a.

In mathematical notation, this is:

$$\forall a, b \in X, \ aRb \Rightarrow \ bRa.$$

44.1 Examples

44.1.1 In mathematics

- "is equal to" (equality) (whereas "is less than" is not symmetric)

- "is comparable to", for elements of a partially ordered set

- "... and ... are odd":

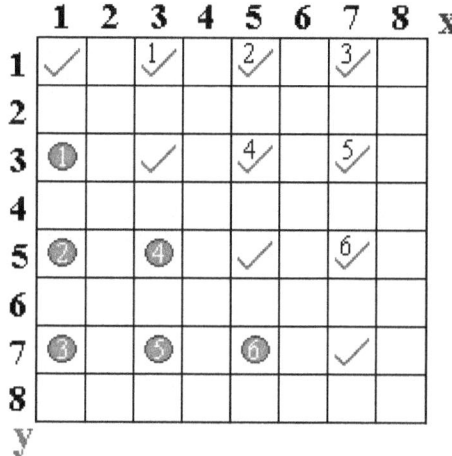

X and y are odd

Is true for this case (need not be true for all cases)

Must be true if the check mark with the same number (z) is true for it to be a symmetric relation

z Is true for this case and requires the circle with the same number (z) to also be true for it to be a symmetric relation

44.1.2   Outside mathematics

- "is married to" (in most legal systems)

- "is a fully biological sibling of"

- "is a homophone of"

44.2   Relationship to asymmetric and antisymmetric relations

By definition, a relation cannot be both symmetric and asymmetric (where if a is related to b, then b cannot be related to a (in the same way)). However, a relation can be neither symmetric nor asymmetric, which is the case for "is less than or equal to" and "preys on").

Symmetric and antisymmetric (where the only way a can be related to b and b be related to a is if $a = b$) are actually independent of each other, as these examples show.

44.3   Additional aspects

A symmetric relation that is also transitive and reflexive is an equivalence relation.

One way to conceptualize a symmetric relation in graph theory is that a symmetric relation is an edge, with the edge's two vertices being the two entities so related. Thus, symmetric relations and undirected graphs are combinatorially equivalent objects.

44.4   See also

- Symmetry in mathematics
- Symmetry
- Asymmetric relation
- Antisymmetric relation

Chapter 45

Symmetry (geometry)

A drawing of a butterfly with bilateral symmetry, with left and right sides as mirror images of each other.

A geometric object has *symmetry* if there is an "operation" or "transformation" (technically, an isometry or affine map) that maps the figure/object onto itself; i.e., it is said that the object has an invariance under the transform.[1] For instance, a circle rotated about its center will have the same shape and size as the original circle—all points before and after the transform would be indistinguishable. A circle is said to be *symmetric under rotation* or to have *rotational symmetry*. If the isometry is the reflection of a plane figure, the figure is said to have reflectional symmetry or line symmetry; moreover, it is possible for a figure/object to have more than one line of symmetry.[2]

The types of symmetries that are possible for a geometric object depend on the set of geometric transforms available, and on what object properties should remain unchanged after a transform. Because the composition of two transforms is also a transform and every transform has an inverse transform that undoes it, the set of transforms under which an object is symmetric form a mathematical group.[3]

45.1   Symmetries in general

The most common group of transforms applied to objects are termed the Euclidean group of "isometries," which are distance-preserving transformations in space commonly referred to as two-dimensional or three-dimensional (i.e., in plane geometry or solid geometry Euclidean spaces). These isometries consist of reflections, rotations, translations, and combinations of these basic operations.[4] Under an isometric transformation, a geometric object is said to be symmetric if, after transformation, the object is indistinguishable from the object before the transformation, i.e., if the transformed object is congruent to the original.[5] A geometric object is typically symmetric only under a subset or "subgroup" of all isometries. The kinds of isometry subgroups are described below, followed by other kinds of transform groups and by the types of object invariance that are possible in geometry.

Hyperbolic space has another transformation, called **striation**, **parabolic transformation**, or **pararotation**, named after the geological term striation for parallel grooves, and is similar to a Euclidean rotation except the center of rotation is seen at infinity on the ideal limit. Two generating mirrors of a striation create infinitely many virtual copies following a horocycle.

45.2   Reflectional symmetry

Main article: reflectional symmetry

Reflectional symmetry, mirror symmetry, mirror-image symmetry, or bilateral symmetry is symmetry with respect to reflection.[6]

In one dimension, there is a point of symmetry about which reflection takes place; in two dimensions there is an axis of symmetry, and in three dimensions there is a plane of symmetry.[7] An object or figure which is indistinguishable from its transformed image is called mirror symmetric (see mirror image).

The axis of symmetry of a two-dimensional figure is a line such that, if a perpendicular is constructed, any two points lying on the perpendicular at equal distances from the axis of symmetry are identical. Another way to think about it is that if the shape were to be folded in half over the axis, the two halves would be identical: the two halves are each other's mirror image. Thus a square has four axes of symmetry, because there are four different ways to fold it and have the edges all match. A circle has infinitely many axes of symmetry passing through its center, for the same reason.[8]

If the letter T is reflected along a vertical axis, it appears the same. This is sometimes called vertical symmetry. One can better use an unambiguous formulation; e.g., "T has a vertical symmetry axis" or "T has left-right symmetry".

The triangles with reflection symmetry are isosceles, the quadrilaterals with this symmetry are the kites and the isosceles trapezoids.[9]

For each line or plane of reflection, the symmetry group is isomorphic with C_s (see point groups in three dimensions), one of the three types of order two (involutions), hence algebraically isomorphic to C_2. The fundamental domain is a half-plane or half-space.[10]

45.3   Point reflection and other involutive isometries

Main article: Point reflection

Reflection symmetry can be generalized to other isometries of m-dimensional space which are involutions, such as

$$(x_1, ..., xm) \mapsto (-x_1, ..., -xk, xk_{+1}, ..., xm)$$

in a certain system of Cartesian coordinates. This reflects the space along an $(m-k)$-dimensional affine subspace.[11] If k = m, then such a transformation is known as a point reflection, or an *inversion through a point*. On the plane (m = 2)

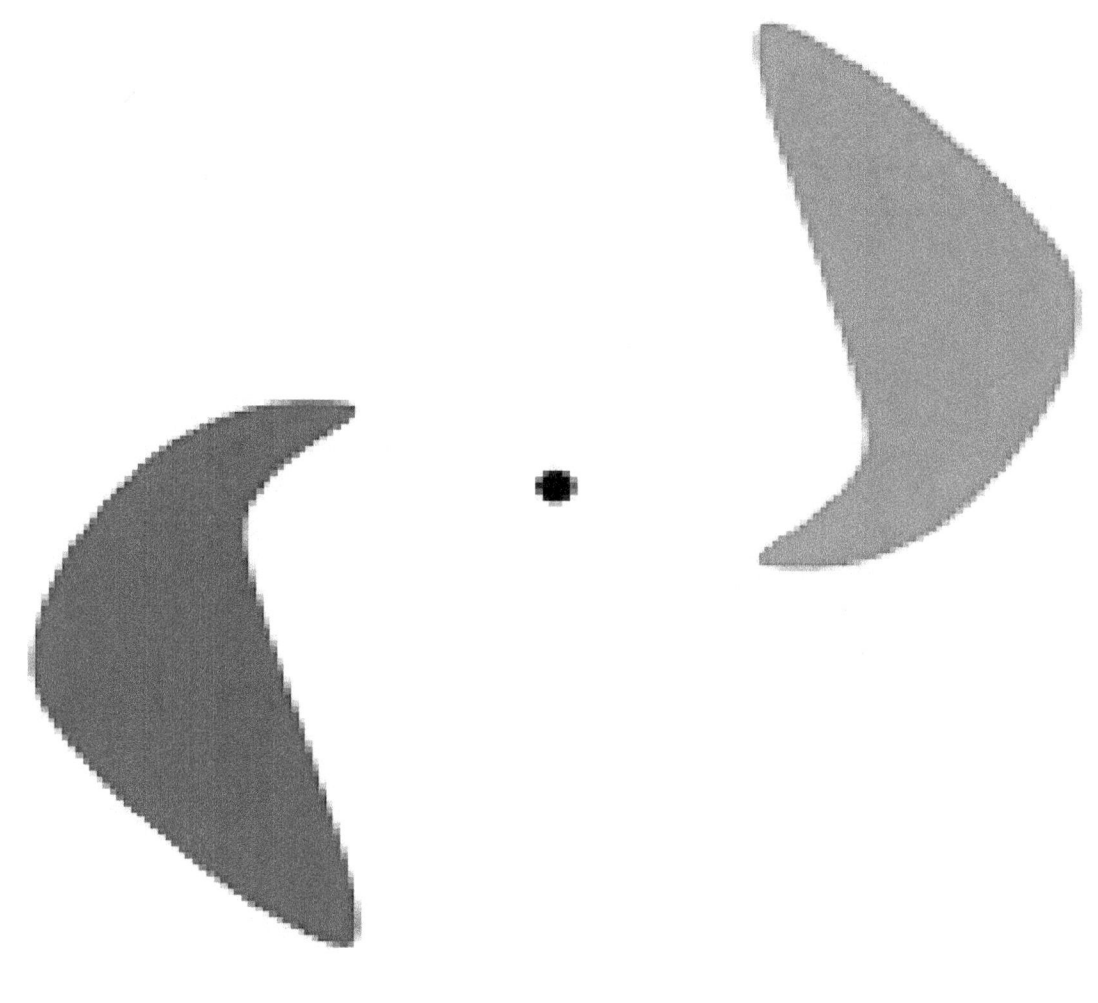

In 2 dimensions, a point reflection is a 180 degree rotation.

a point reflection is the same as a half-turn (180°) rotation; see below. *Antipodal symmetry* is an alternative name for a point reflection symmetry through the origin.[12]

Such a "reflection" preserves orientation if and only if k is an even number.[13] This implies that for m = 3 (as well as for other odd m) a point reflection changes the orientation of the space, like a mirror-image symmetry. That is why in physics the term *P-symmetry* is used for both point reflection and mirror symmetry (P stands for parity). As a point reflection in three dimensions changes a left-handed coordinate system into a right-handed coordinate system, symmetry under a point reflection is also called a left-right symmetry.[14]

45.4   Rotational symmetry

Main article: Rotational symmetry

Rotational symmetry is symmetry with respect to some or all rotations in m-dimensional Euclidean space. Rotations

The triskelion has 3-fold rotational symmetry.

are direct isometries; i.e., isometries preserving orientation.[15] Therefore a symmetry group of rotational symmetry is a subgroup of the special Euclidean group $E^+(m)$.

Symmetry with respect to all rotations about all points implies translational symmetry with respect to all translations (because translations are compositions of rotations about distinct points),[16] and the symmetry group is the whole $E^+(m)$. This does not apply for objects because it makes space homogeneous, but it may apply for physical laws.

For symmetry with respect to rotations about a point we can take that point as origin. These rotations form the special orthogonal group SO(m), which can be represented by the group of $m \times m$ orthogonal matrices with determinant 1. For m = 3 this is the rotation group SO(3).[17]

In another meaning of the word, the rotation group of an object is the symmetry group within $E^+(m)$, the group of rigid motions;[18] in other words, the intersection of the full symmetry group and the group of rigid motions. For chiral objects it is the same as the full symmetry group.

Laws of physics are SO(3)-invariant if they do not distinguish different directions in space. Because of Noether's theorem, rotational symmetry of a physical system is equivalent to the angular momentum conservation law.[19] See also rotational invariance.

45.5 Translational symmetry

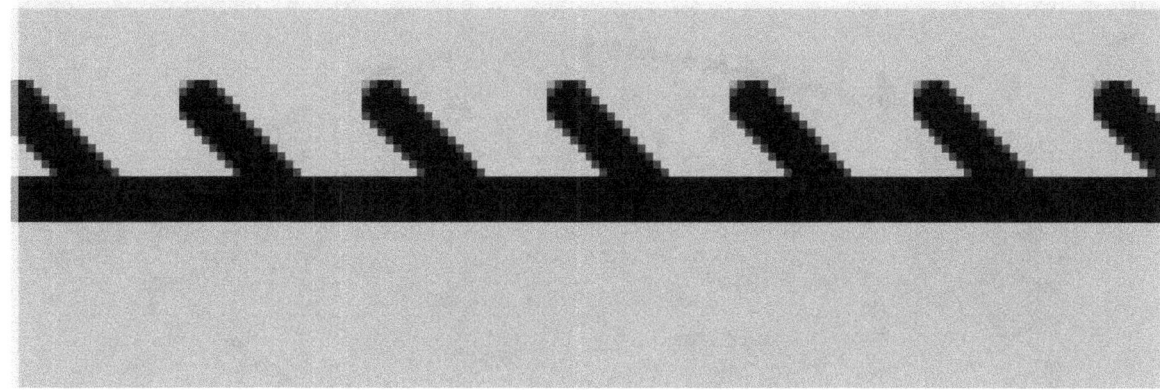

A frieze pattern with translational symmetry

Main article: Translational symmetry

Translational symmetry leaves an object invariant under a discrete or continuous group of translations $T_a(p) = p + a$.[20] The illustration on the right shows four congruent triangles generated by translations along the arrow. If the line of triangles extended to infinity in both directions, they would have a discrete translational symmetry; any translation that mapped one triangle onto another would leave the whole line unchanged.

45.6 Glide reflection symmetry

A frieze pattern with glide reflection symmetry

Main article: glide reflection

In 2D, a **glide reflection** symmetry (in 3D it is called a glide plane symmetry, and a **transflection** in general) means that a reflection in a line or plane combined with a translation along the line / in the plane, results in the same object.[21] The composition of two glide reflections results in a translation symmetry with twice the translation vector. The symmetry group comprising glide reflections and associated translations is the frieze group **p11g** and is isomorphic with the infinite cyclic group **Z**.

45.6.1   Rotoreflection symmetry

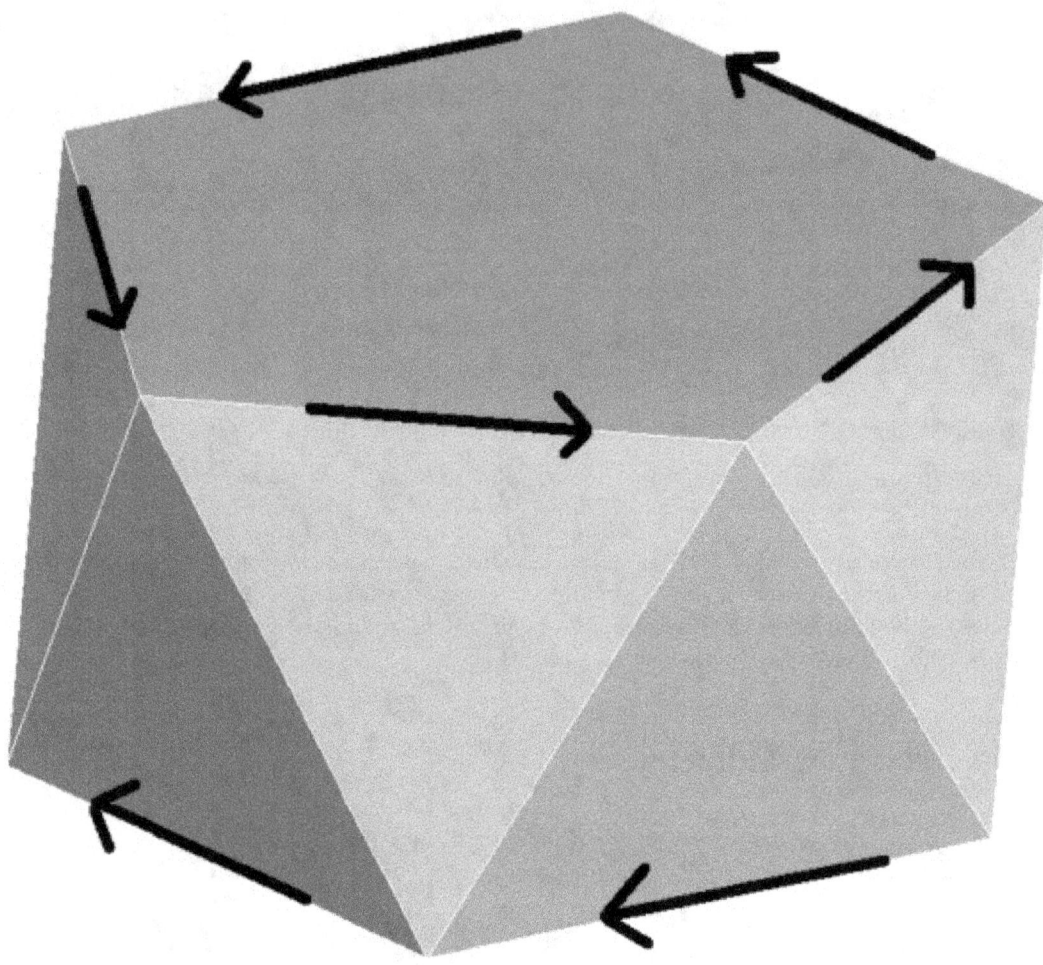

A pentagonal antiprism with marked edges shows rotoreflectional symmetry, with an order of 10.

Main article: improper rotation

In 3D, a **rotary reflection**, **rotoreflection** or **improper rotation** is a rotation about an axis combined with reflection in a plane perpendicular to that axis.[22] The symmetry groups associated with rotoreflections include:

- if the rotation angle has no common divisor with 360°, the symmetry group is not discrete

- if the rotoreflection has a 2n-fold rotation angle (angle of 180°/n), the symmetry group is S_2n of order 2n (not to be confused with symmetric groups, for which the same notation is used; the abstract group is $C2n$). A special case is $n = 1$, an inversion, because it does not depend on the axis and the plane, it is characterized by just the point of inversion.

- the group *Cnh* (angle of 360°/*n*); for odd *n* this is generated by a single symmetry, and the abstract group is C_{2n}, for even *n* this is not a basic symmetry but a combination.
  See also: point groups in three dimensions

45.7   Helical symmetry

See also: Screw axis

In 3D geometry and higher, a screw axis (or rotary translation) is a combination of a rotation and a translation along the rotation axis.[23]

Helical symmetry is the kind of symmetry seen in such everyday objects as springs, Slinky toys, drill bits, and augers. The concept of helical symmetry can be visualized as the tracing in three-dimensional space that results from rotating an object at a constant angular speed while simultaneously translating at a constant linear speed along its axis of rotation. At any one point in time, these two motions combine to give a *coiling angle* that helps define the properties of the traced helix.[24] When the tracing object rotates quickly and translates slowly, the coiling angle will be close to 0°. Conversely, if the rotation is slow and the translation is speedy, the coiling angle will approach 90°.

Three main classes of helical symmetry can be distinguished based on the interplay of the angle of coiling and translation symmetries along the axis:

- **Infinite helical symmetry**: If there are no distinguishing features along the length of a helix or helix-like object, the object will have infinite symmetry much like that of a circle, but with the additional requirement of translation along the long axis of the object to return it to its original appearance.[25] A helix-like object is one that has at every point the regular angle of coiling of a helix, but which can also have a cross section of indefinitely high complexity, provided only that precisely the same cross section exists (usually after a rotation) at every point along the length of the object. Simple examples include evenly coiled springs, slinkies, drill bits, and augers. Stated more precisely, an object has infinite helical symmetries if for any small rotation of the object around its central axis there exists a point nearby (the translation distance) on that axis at which the object will appear exactly as it did before. It is this infinite helical symmetry that gives rise to the curious illusion of movement along the length of an auger or screw bit that is being rotated. It also provides the mechanically useful ability of such devices to move materials along their length, provided that they are combined with a force such as gravity or friction that allows the materials to resist simply rotating along with the drill or auger.

- ***n*-fold helical symmetry**: If the requirement that every cross section of the helical object be identical is relaxed, additional lesser helical symmetries become possible. For example, the cross section of the helical object may change, but still repeats itself in a regular fashion along the axis of the helical object. Consequently, objects of this type will exhibit a symmetry after a rotation by some fixed angle θ and a translation by some fixed distance, but will not in general be invariant for any rotation angle. If the angle (rotation) at which the symmetry occurs divides evenly into a full circle (360°), the result is the helical equivalent of a regular polygon. This case is called *n-fold helical symmetry*, where *n* = 360°; for example, a double helix. This concept can be further generalized to include cases where *mθ* is a multiple of 360° – that is, the cycle does eventually repeat, but only after more than one full rotation of the helical object.

- **Non-repeating helical symmetry**: This is the case in which the angle of rotation θ required to observe the symmetry is irrational. The angle of rotation never repeats exactly no matter how many times the helix is rotated. Such symmetries are created by using a non-repeating point group in two dimensions. DNA, with approximately 10.5 base pairs per turn, is an example of this type of non-repeating helical symmetry.[26]

45.8 Double rotation symmetry

See also: Rotations_in_4-dimensional_Euclidean_space § Double_rotations

In 4D, a double rotation symmetry can be generated as the composite of two orthogonal rotations.[27] It is similar to 3D screw axis which is the composite of a rotation and an orthogonal translation.

45.9 Non-isometric symmetries

A wider definition of geometric symmetry allows operations from a larger group than the Euclidean group of isometries. Examples of larger geometric symmetry groups are:

- The group of similarity transformations;[28] i.e., affine transformations represented by a matrix A that is a scalar times an orthogonal matrix. Thus homothety is added, self-similarity is considered a symmetry.

- The group of affine transformations represented by a matrix A with determinant 1 or −1; i.e., the transformations which preserve area.[29]

    This adds, e.g., oblique reflection symmetry.

- The group of all bijective affine transformations.

- The group of Möbius transformations which preserve cross-ratios.

    This adds, e.g., inversive reflections such as circle reflection on the plane.

In Felix Klein's Erlangen program, each possible group of symmetries defines a geometry in which objects that are related by a member of the symmetry group are considered to be equivalent.[30] For example, the Euclidean group defines Euclidean geometry, whereas the group of Möbius transformations defines projective geometry.

45.10 Scale symmetry and fractals

Scale symmetry means that if an object is expanded or reduced in size, the new object has the same properties as the original.[31] This is *not* true of most physical systems, as witness the difference in the shape of the legs of an elephant and a mouse (so-called allometric scaling). Similarly, if a soft wax candle were enlarged to the size of a tall tree, it would immediately collapse under its own weight.

A more subtle form of scale symmetry is demonstrated by fractals. As conceived by Benoît Mandelbrot, fractals are a mathematical concept in which the structure of a complex form looks similar at any degree of magnification,[32] well seen in the Mandelbrot set. A coast is an example of a naturally occurring fractal, since it retains similar-appearing complexity at every level from the view of a satellite to a microscopic examination of how the water laps up against individual grains of sand. The branching of trees, which enables children to use small twigs as stand-ins for full trees in dioramas, is another example.

Because fractals can generate the appearance of patterns in nature, they have a beauty and familiarity not typically seen with mathematically generated functions. Fractals have also found a place in computer-generated movie effects, where their ability to create complex curves with fractal symmetries results in more realistic virtual worlds.

45.11 Abstract symmetry

45.11.1 Klein's view

With every geometry, Felix Klein associated an underlying group of symmetries. The hierarchy of geometries is thus mathematically represented as a hierarchy of these groups, and hierarchy of their invariants. For example, lengths, angles and areas are preserved with respect to the Euclidean group of symmetries, while only the incidence structure and the cross-ratio are preserved under the most general projective transformations. A concept of parallelism, which is preserved in affine geometry, is not meaningful in projective geometry. Then, by abstracting the underlying groups of symmetries from the geometries, the relationships between them can be re-established at the group level. Since the group of affine geometry is a subgroup of the group of projective geometry, any notion invariant in projective geometry is *a priori* meaningful in affine geometry; but not the other way round. If you add required symmetries, you have a more powerful theory but fewer concepts and theorems (which will be deeper and more general).

45.11.2 Thurston's view

William Thurston introduced a similar version of symmetries in geometry. A **model geometry** is a simply connected smooth manifold X together with a transitive action of a Lie group G on X with compact stabilizers. The Lie group can be thought of as the group of symmetries of the geometry.

A model geometry is called **maximal** if G is maximal among groups acting smoothly and transitively on X with compact stabilizers, i.e. if it is the maximal group of symmetries. Sometimes this condition is included in the definition of a model geometry.

A **geometric structure** on a manifold M is a diffeomorphism from M to X/Γ for some model geometry X, where Γ is a discrete subgroup of G acting freely on X. If a given manifold admits a geometric structure, then it admits one whose model is maximal.

A 3-dimensional model geometry X is relevant to the geometrization conjecture if it is maximal and if there is at least one compact manifold with a geometric structure modelled on X. Thurston classified the 8 model geometries satisfying these conditions; they are listed below and are sometimes called **Thurston geometries**. (There are also uncountably many model geometries without compact quotients.)

45.12 External links

- Calotta: A World of Symmetry

- Dutch: Symmetry Around a Point in the Plane

45.13 References

[1] Martin, G. (1996). *Transformation Geometry: An Introduction to Symmetry*. Springer. p. 28.

[2] Freitag, Mark (2013). *Mathematics for Elementary School Teachers: A Process Approach*. Cengage Learning. p. 721.

[3] Miller, Willard Jr. (1972). *Symmetry Groups and Their Applications*. New York: Academic Press. OCLC 589081. Retrieved 2009-09-28.

[4] Higher dimensional group theory "Higher Dimensional Group Theory". Retrieved 2013-04-16.

[5] "geometric congruence". PlanetMath.org. Retrieved 29 May 2013.

[6] Weyl, Hermann (1982) [1952]. *Symmetry*. Princeton: Princeton University Press. ISBN 0-691-02374-3.

[7] Cowin, Stephen C., Doty, Stephen B. (2007). *Tissue Mechanics*. Springer. p. 152.

[8] Caldecott, Stratford (2009). *Beauty for Truth's Sake: On the Re-enchantment of Education*. Brazos Press. p. 70.

[9] Bassarear, Tom (2011). *Mathematics for Elementary School Teachers* (5 ed.). Cengage Learning. p. 499.

[10] Johnson, N. W. Johnson (2015). "11: Finite symmetry groups". *Geometries and Transformations*.

[11] Hertrich-Jeromin, Udo (2003). *Introduction to Möbius Differential Geometry*. Cambridge University Press.

[12] Dieck, Tammo (2008). *Algebraic Topology*. European Mathematical Society. p. 261. ISBN 9783037190487.

[13] William H. Barker, Roger Howe *Continuous Symmetry: From Euclid to Klein (Google eBook)* American Mathematical Soc

[14] W.M. Gibson and B.R. Pollard (1980). *Symmetry principles in elementary particle physics*. Cambridge University Press. pp. 120–122. ISBN 0 521 29964 0.

[15] Vladimir G. Ivancevic, Tijana T. Ivancevic (2005) *Natural Biodynamics* World Scientific

[16] Singer, David A. (1998). *Geometry: Plane and Fancy*. Springer Science & Business Media.

[17] Joshi, A. W. (2007). *Elements of Group Theory for Physicists*. New Age International. pp. 111ff.

[18] Hartshorne, Robin (2000). *Geometry: Euclid and Beyond*. Springer Science & Business Media.

[19] Kosmann-Schwarzbach, Yvette (2010). *The Noether theorems: Invariance and conservation laws in the twentieth century*. Sources and Studies in the History of Mathematics and Physical Sciences. Springer-Verlag. ISBN 978-0-387-87867-6.

[20] Stenger, Victor J. (2000) and Mahou Shiro (2007). *Timeless Reality*. Prometheus Books. Especially chapter 12. Nontechnical.

[21] Martin, George E. (1982). *Transformation Geometry: An Introduction to Symmetry*, Undergraduate Texts in Mathematics. Springer. p. 64. ISBN 9780387906362.

[22] Robert O. Gould, Steffen Borchardt-Ott (2011)*Crystallography: An Introduction* Springer Science & Business Media

[23] Bottema, O. and B. Roth, *Theoretical Kinematics*, Dover Publications (September 1990)

[24] George R. McGhee (2006) *The Geometry of Evolution: Adaptive Landscapes and Theoretical Morphospaces* Cambridge University Press p.64

[25] Anna Ursyn(2012) *Biologically-inspired Computing for the Arts: Scientific Data Through Graphics* IGI Global Snippet p.209

[26] Sinden, Richard R. (1994). *DNA structure and function*. Gulf Professional Publishing. p. 101. ISBN 9780126457506.

[27] Charles Howard Hinton (1906) *The Fourth Dimension (Google eBook)* S. Sonnenschein & Company p.223

[28] H.S.M. Coxeter (1961,9) *Introduction to Geometry*, §5 Similarity in the Euclidean Plane, pp. 67–76, §7 Isometry and Similarity in Euclidean Space, pp 96–104, John Wiley & Sons.

[29] William Thurston. *Three-dimensional geometry and topology. Vol. 1*. Edited by Silvio Levy. Princeton Mathematical Series, 35. Princeton University Press, Princeton, NJ, 1997. x+311 pp. ISBN 0-691-08304-5

[30] Klein, Felix, 1872. "Vergleichende Betrachtungen über neuere geometrische Forschungen" ('A comparative review of recent researches in geometry'). Mathematische Annalen, 43 (1893) pp. 63–100 (Also: Gesammelte Abh. Vol. 1, Springer, 1921, pp. 460–497).

An English translation by Mellen Haskell appeared in *Bull. N. Y. Math. Soc* 2 (1892–1893): 215–249.

[31] Tian Yu Cao *Conceptual Foundations of Quantum Field Theory* Cambridge University Press p.154-155

[32] Gouyet, Jean-François (1996). *Physics and fractal structures*. Paris/New York: Masson Springer. ISBN 978-0-387-94153-0.

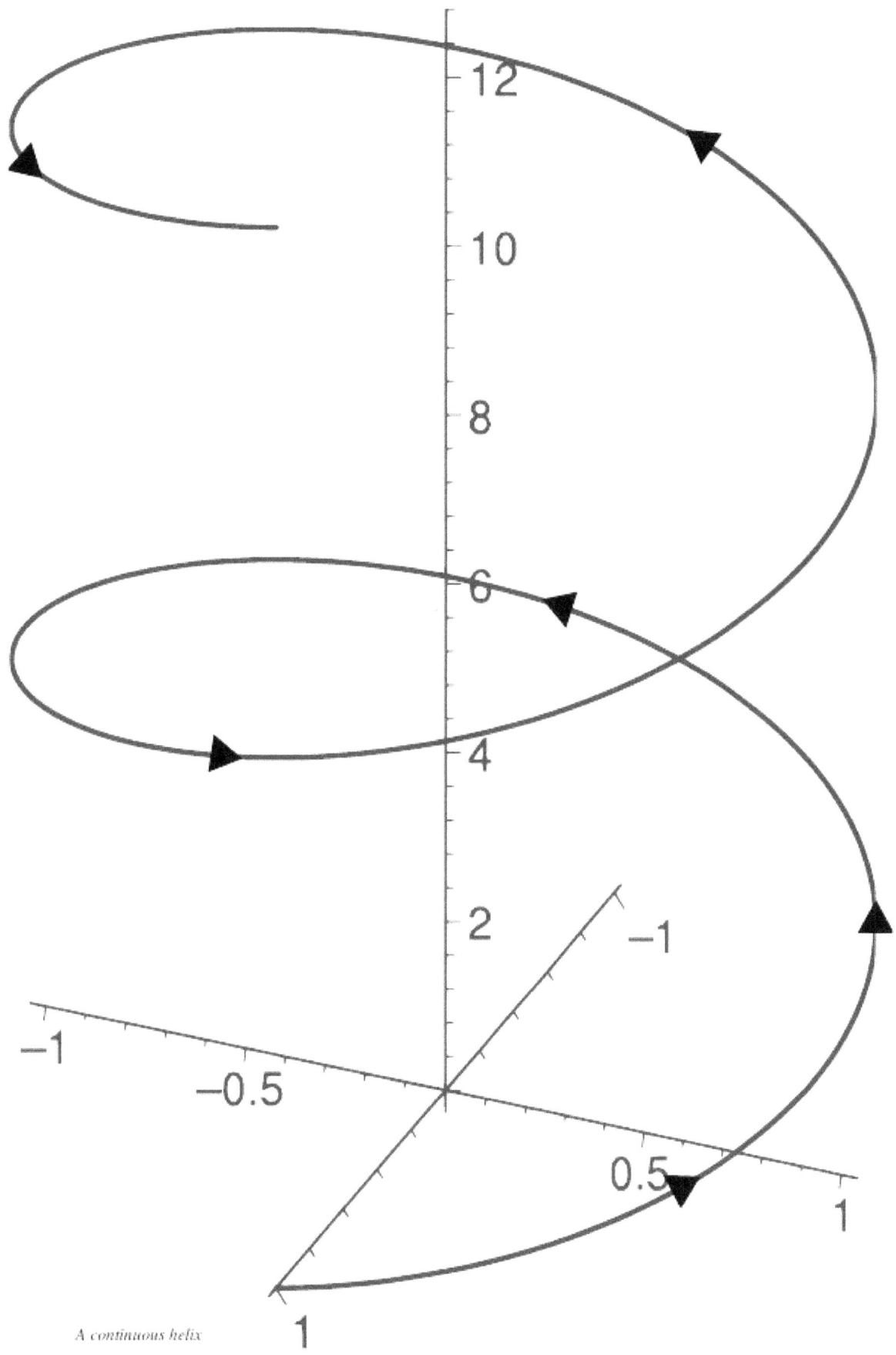

A continuous helix

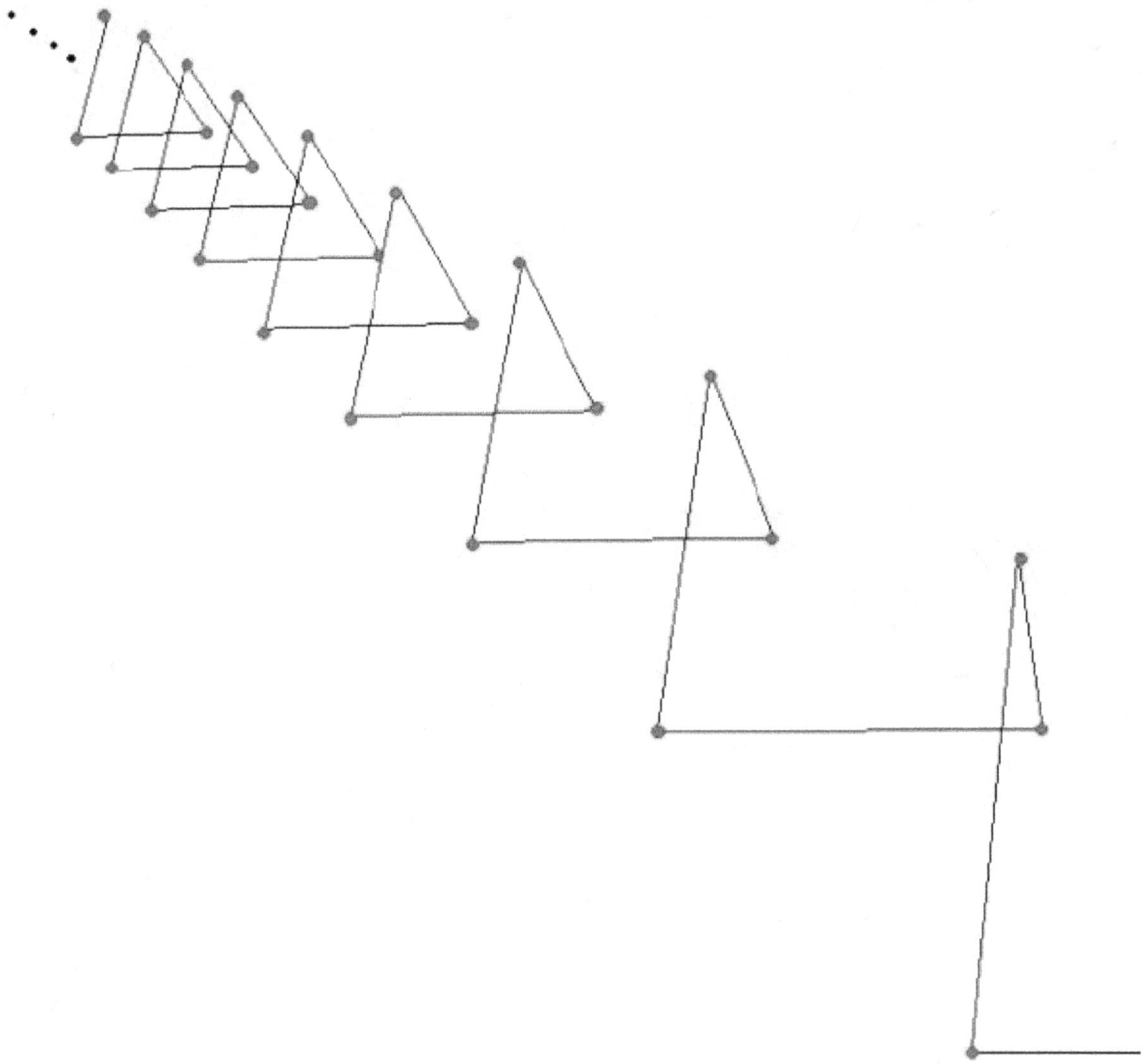

A regular skew-apeirogon has a discrete (3-fold here) screw-axis symmetry, drawn in perspective.

The Boerdijk–Coxeter helix, constructed by augmented regular tetrahedra, is an example of a screw axis symmetry that is nonperiodic.

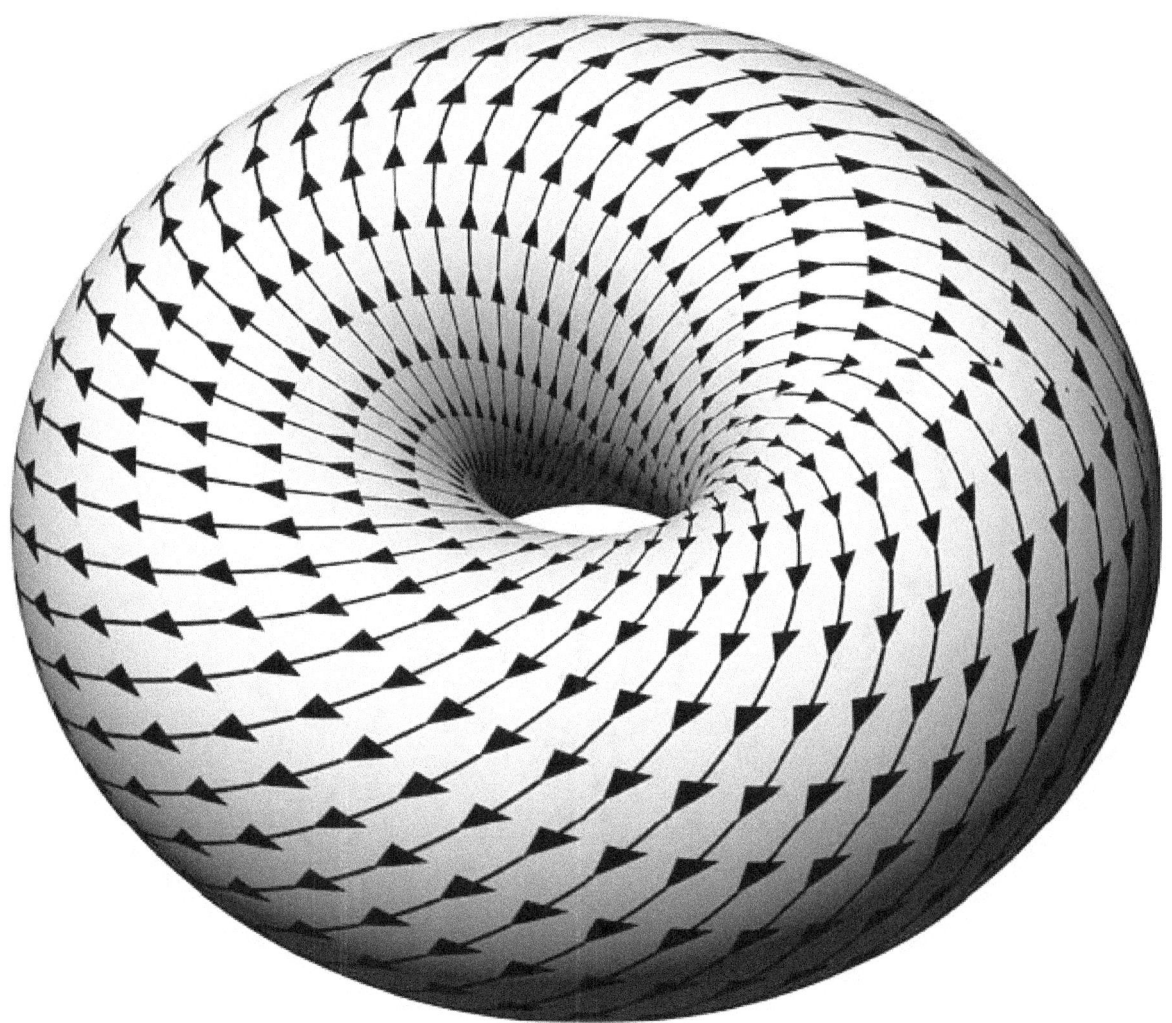

A 4D clifford torus, stereographically projected into 3D, looks like a torus. A double rotation can be seen as a helical path.

A *Julia set has scale symmetry*

Chapter 46

Symmetric polynomial

This article is about individual symmetric polynomials. For the ring of symmetric polynomials, see ring of symmetric functions.

In mathematics, a **symmetric polynomial** is a polynomial $P(X_1, X_2, \ldots, Xn)$ in n variables, such that if any of the variables are interchanged, one obtains the same polynomial. Formally, P is a *symmetric polynomial*, if for any permutation σ of the subscripts 1, 2, ..., n one has $P(X\sigma_{(1)}, X\sigma_{(2)}, \ldots, X\sigma_{(n)}) = P(X_1, X_2, \ldots, Xn)$.

Symmetric polynomials arise naturally in the study of the relation between the roots of a polynomial in one variable and its coefficients, since the coefficients can be given by polynomial expressions in the roots, and all roots play a similar role in this setting. From this point of view the elementary symmetric polynomials are the most fundamental symmetric polynomials. A theorem states that any symmetric polynomial can be expressed in terms of elementary symmetric polynomials, which implies that every *symmetric* polynomial expression in the roots of a monic polynomial can alternatively be given as a polynomial expression in the coefficients of the polynomial.

Symmetric polynomials also form an interesting structure by themselves, independently of any relation to the roots of a polynomial. In this context other collections of specific symmetric polynomials, such as complete homogeneous, power sum, and Schur polynomials play important roles alongside the elementary ones. The resulting structures, and in particular the ring of symmetric functions, are of great importance in combinatorics and in representation theory.

46.1 Examples

Symmetric polynomials in two variables X_1, X_2:

- $X_1^3 + X_2^3 - 7$

- $4X_1^2 X_2^2 + X_1^3 X_2 + X_1 X_2^3 + (X_1 + X_2)^4$

and in three variables X_1, X_2, X_3:

- $X_1 X_2 X_3 - 2X_1 X_2 - 2X_1 X_3 - 2X_2 X_3$

There are many ways to make specific symmetric polynomials in any number of variables, see the various types below. An example of a somewhat different flavor is

- $\prod_{1 \leq i < j \leq n} (X_i - X_j)^2$

where first a polynomial is constructed that changes sign under every exchange of variables, and taking the square renders it completely symmetric (if the variables represent the roots of a monic polynomial, this polynomial gives its discriminant).

On the other hand, the polynomial in two variables

- $X_1 - X_2$

is not symmetric, since if one exchanges X_1 and X_2 one gets a different polynomial, $X_2 - X_1$. Similarly in three variables

- $X_1^4 X_2^2 X_3 + X_1 X_2^4 X_3^2 + X_1^2 X_2 X_3^4$

has only symmetry under cyclic permutations of the three variables, which is not sufficient to be a symmetric polynomial. However, the following is symmetric:

- $X_1^4 X_2^2 X_3 + X_1 X_2^4 X_3^2 + X_1^2 X_2 X_3^4 + X_1^4 X_2 X_3^2 + X_1 X_2^2 X_3^4 + X_1^2 X_2^4 X_3$

46.2   Applications

46.2.1   Galois theory

Main article: Galois theory

One context in which symmetric polynomial functions occur is in the study of monic univariate polynomials of degree n having n roots in a given field. These n roots determine the polynomial, and when they are considered as independent variables, the coefficients of the polynomial are symmetric polynomial functions of the roots. Moreover the fundamental theorem of symmetric polynomials implies that a polynomial function f of the n roots can be expressed as (another) polynomial function of the coefficients of the polynomial determined by the roots if and only if f is given by a symmetric polynomial.

This yields the approach to solving polynomial equations by inverting this map, "breaking" the symmetry – given the coefficients of the polynomial (the elementary symmetric polynomials in the roots), how can one recover the roots? This leads to studying solutions of polynomials using the permutation group of the roots, originally in the form of Lagrange resolvents, later developed in Galois theory.

46.3   Relation with the roots of a monic univariate polynomial

Consider a monic polynomial in t of degree n

$$P = t^n + a_{n-1}t^{n-1} + \cdots + a_2 t^2 + a_1 t + a_0$$

with coefficients a_i in some field k. There exist n roots $x_1, \ldots, x_n$ of P in some possibly larger field (for instance if k is the field of real numbers, the roots will exist in the field of complex numbers); some of the roots might be equal, but the fact that one has *all* roots is expressed by the relation

$$P = t^n + a_{n-1}t^{n-1} + \cdots + a_2 t^2 + a_1 t + a_0 = (t - x_1)(t - x_2) \cdots (t - x_n).$$

By comparison of the coefficients one finds that

$$a_{n-1} = -x_1 - x_2 - \cdots - x_n$$

$$a_{n-2} = x_1 x_2 + x_1 x_3 + \cdots + x_2 x_3 + \cdots + x_{n-1} x_n = \sum_{1 \le i < j \le n} x_i x_j$$

$$\vdots$$

$$a_1 = (-1)^{n-1}(x_2 x_3 \cdots x_n + x_1 x_3 x_4 \cdots x_n + \cdots + x_1 x_2 \cdots x_{n-2} x_n + x_1 x_2 \cdots x_{n-1}) = (-1)^{n-1} \sum_{i=1}^{n} \prod_{j \ne i} x_j$$

$$a_0 = (-1)^n x_1 x_2 \cdots x_n.$$

These are in fact just instances of Viète's formulas. They show that all coefficients of the polynomial are given in terms of the roots by a symmetric polynomial expression: although for a given polynomial P there may be qualitative differences between the roots (like lying in the base field k or not, being simple or multiple roots), none of this affects the way the roots occur in these expressions.

Now one may change the point of view, by taking the roots rather than the coefficients as basic parameters for describing P, and considering them as indeterminates rather than as constants in an appropriate field; the coefficients ai then become just the particular symmetric polynomials given by the above equations. Those polynomials, without the sign $(-1)^{n-i}$, are known as the elementary symmetric polynomials in $x_1, \ldots, x_n$. A basic fact, known as the **fundamental theorem of symmetric polynomials** states that *any* symmetric polynomial in n variables can be given by a polynomial expression in terms of these elementary symmetric polynomials. It follows that any symmetric polynomial expression in the roots of a monic polynomial can be expressed as a polynomial in the *coefficients* of the polynomial, and in particular that its value lies in the base field k that contains those coefficients. Thus, when working only with such symmetric polynomial expressions in the roots, it is unnecessary to know anything particular about those roots, or to compute in any larger field than k in which those roots may lie. In fact the values of the roots themselves become rather irrelevant, and the necessary relations between coefficients and symmetric polynomial expressions can be found by computations in terms of symmetric polynomials only. An example of such relations are Newton's identities, which express the sum of any fixed power of the roots in terms of the elementary symmetric polynomials.

46.4 Special kinds of symmetric polynomials

There are a few types of symmetric polynomials in the variables $X_1, X_2, \ldots, Xn$ that are fundamental.

46.4.1 Elementary symmetric polynomials

Main article: elementary symmetric polynomial

For each nonnegative integer k, the elementary symmetric polynomial $ek(X_1, \ldots, Xn)$ is the sum of all distinct products of k distinct variables. (Some authors denote it by σk instead.) For $k = 0$ there is only the empty product so $e_0(X_1, \ldots, Xn) = 1$, while for $k > n$, no products at all can be formed, so $ek(X_1, X_2, \ldots, Xn) = 0$ in these cases. The remaining n elementary symmetric polynomials are building blocks for all symmetric polynomials in these variables: as mentioned above, any symmetric polynomial in the variables considered can be obtained from these elementary symmetric polynomials using multiplications and additions only. In fact one has the following more detailed facts:

- any symmetric polynomial P in $X_1, \ldots, Xn$ can be written as a polynomial expression in the polynomials $ek(X_1, \ldots, Xn)$ with $1 \le k \le n$;

- this expression is unique up to equivalence of polynomial expressions;

- if P has *integral* coefficients, then the polynomial expression also has integral coefficients.

For example, for $n = 2$, the relevant elementary symmetric polynomials are $e_1(X_1, X_2) = X_1 + X_2$, and $e_2(X_1, X_2) = X_1 X_2$. The first polynomial in the list of examples above can then be written as

$$X_1^3 + X_2^3 - 7 = e_1(X_1, X_2)^3 - 3e_2(X_1, X_2)e_1(X_1, X_2) - 7$$

(for a proof that this is always possible see the fundamental theorem of symmetric polynomials).

46.4.2   Monomial symmetric polynomials

Powers and products of elementary symmetric polynomials work out to rather complicated expressions. If one seeks basic *additive* building blocks for symmetric polynomials, a more natural choice is to take those symmetric polynomials that contain only one type of monomial, with only those copies required to obtain symmetry. Any monomial in $X_1, ..., Xn$ can be written as $X_1^{\alpha_1} ... Xn^{\alpha n}$ where the exponents αi are natural numbers (possibly zero); writing $\alpha = (\alpha_1,...,\alpha n)$ this can be abbreviated to X^α. The **monomial symmetric polynomial** $m\alpha(X_1, ..., Xn)$ is defined as the sum of all monomials x^β where β ranges over all *distinct* permutations of $(\alpha_1,...,\alpha n)$. For instance one has

$$m_{(3,1,1)}(X_1, X_2, X_3) = X_1^3 X_2 X_3 + X_1 X_2^3 X_3 + X_1 X_2 X_3^3$$

$$m_{(3,2,1)}(X_1, X_2, X_3) = X_1^3 X_2^2 X_3 + X_1^3 X_2 X_3^2 + X_1^2 X_2^3 X_3 + X_1^2 X_2 X_3^3 + X_1 X_2^3 X_3^2 + X_1 X_2^2 X_3^3.$$

Clearly $m\alpha = m_\beta$ when β is a permutation of α, so one usually considers only those $m\alpha$ for which $\alpha_1 \geq \alpha_2 \geq ... \geq \alpha n$, in other words for which α is a partition of an integer. These monomial symmetric polynomials form a vector space basis: every symmetric polynomial P can be written as a linear combination of the monomial symmetric polynomials. To do this it suffices to separate the different types of monomial occurring in P. In particular if P has integer coefficients, then so will the linear combination.

The elementary symmetric polynomials are particular cases of monomial symmetric polynomials: for $0 \leq k \leq n$ one has

$$e_k(X_1, ..., X_n) = m_\alpha(X_1, ..., X_n) \text{ where } \alpha \text{ is the partition of } k \text{ into } k \text{ parts } 1 \text{ (followed by } n - k \text{ zeros).}$$

46.4.3   Power-sum symmetric polynomials

Main article: power sum symmetric polynomial

For each integer $k \geq 1$, the monomial symmetric polynomial $m_{(k,0,...,0)}(X_1, ..., Xn)$ is of special interest. It is the power sum symmetric polynomial, defined as

$$p_k(X_1, ..., X_n) = X_1^k + X_2^k + \cdots + X_n^k.$$

All symmetric polynomials can be obtained from the first n power sum symmetric polynomials by additions and multiplications, possibly involving rational coefficients. More precisely,

> Any symmetric polynomial in $X_1, ..., Xn$ can be expressed as a polynomial expression with rational coefficients in the power sum symmetric polynomials $p_1(X_1, ..., Xn), ..., pn(X_1, ..., Xn)$.

In particular, the remaining power sum polynomials $pk(X_1, ..., Xn)$ for $k > n$ can be so expressed in the first n power sum polynomials; for example

$$p_3(X_1, X_2) = \tfrac{3}{2}p_2(X_1, X_2)p_1(X_1, X_2) - \tfrac{1}{2}p_1(X_1, X_2)^3.$$

In contrast to the situation for the elementary and complete homogeneous polynomials, a symmetric polynomial in n variables with *integral* coefficients need not be a polynomial function with integral coefficients of the power sum symmetric polynomials. For an example, for $n = 2$, the symmetric polynomial

$$m_{(2,1)}(X_1, X_2) = X_1^2 X_2 + X_1 X_2^2$$

has the expression

$$m_{(2,1)}(X_1, X_2) = \tfrac{1}{2} p_1(X_1, X_2)^3 - \tfrac{1}{2} p_2(X_1, X_2) p_1(X_1, X_2).$$

Using three variables one gets a different expression

$$m_{(2,1)}(X_1, X_2, X_3) = X_1^2 X_2 + X_1 X_2^2 + X_1^2 X_3 + X_1 X_3^2 + X_2^2 X_3 + X_2 X_3^2$$
$$= p_1(X_1, X_2, X_3) p_2(X_1, X_2, X_3) - p_3(X_1, X_2, X_3).$$

The corresponding expression was valid for two variables as well (it suffices to set X_3 to zero), but since it involves p_3, it could not be used to illustrate the statement for $n = 2$. The example shows that whether or not the expression for a given monomial symmetric polynomial in terms of the first n power sum polynomials involves rational coefficients may depend on n. But rational coefficients are *always* needed to express elementary symmetric polynomials (except the constant ones, and e_1 which coincides with the first power sum) in terms of power sum polynomials. The Newton identities provide an explicit method to do this; it involves division by integers up to n, which explains the rational coefficients. Because of these divisions, the mentioned statement fails in general when coefficients are taken in a field of finite characteristic; however it is valid with coefficients in any ring containing the rational numbers.

46.4.4 Complete homogeneous symmetric polynomials

Main article: complete homogeneous symmetric polynomial

For each nonnegative integer k, the complete homogeneous symmetric polynomial $hk(X_1, \ldots, Xn)$ is the sum of all distinct monomials of degree k in the variables $X_1, \ldots, Xn$. For instance

$$h_3(X_1, X_2, X_3) = X_1^3 + X_1^2 X_2 + X_1^2 X_3 + X_1 X_2^2 + X_1 X_2 X_3 + X_1 X_3^2 + X_2^3 + X_2^2 X_3 + X_2 X_3^2 + X_3^3.$$

The polynomial $hk(X_1, \ldots, Xn)$ is also the sum of all distinct monomial symmetric polynomials of degree k in $X_1, \ldots, Xn$, for instance for the given example

$$h_3(X_1, X_2, X_3) = m_{(3)}(X_1, X_2, X_3) + m_{(2,1)}(X_1, X_2, X_3) + m_{(1,1,1)}(X_1, X_2, X_3)$$
$$= (X_1^3 + X_2^3 + X_3^3) + (X_1^2 X_2 + X_1^2 X_3 + X_1 X_2^2 + X_1 X_3^2 + X_2^2 X_3 + X_2 X_3^2) + (X_1 X_2 X_3).$$

All symmetric polynomials in these variables can be built up from complete homogeneous ones: any symmetric polynomial in $X_1, \ldots, Xn$ can be obtained from the complete homogeneous symmetric polynomials $h_1(X_1, \ldots, Xn), \ldots, hn(X_1, \ldots, Xn)$ via multiplications and additions. More precisely:

> Any symmetric polynomial P in $X_1, \ldots, Xn$ can be written as a polynomial expression in the polynomials $hk(X_1, \ldots, Xn)$ with $1 \le k \le n$.
>
> If P has *integral* coefficients, then the polynomial expression also has *integral* coefficients.

For example, for $n = 2$, the relevant complete homogeneous symmetric polynomials are $h_1(X_1, X_2) = X_1 + X_2$, and $h_2(X_1, X_2) = X_1^2 + X_1 X_2 + X_2^2$. The first polynomial in the list of examples above can then be written as

$$X_1^3 + X_2^3 - 7 = -2 h_1(X_1, X_2)^3 + 3 h_1(X_1, X_2) h_2(X_1, X_2) - 7.$$

As in the case of power sums, the given statement applies in particular to the complete homogeneous symmetric polynomials beyond $hn(X_1, \ldots, Xn)$, allowing them to be expressed in terms of the ones up to that point; again the resulting identities become invalid when the number of variables is increased.

An important aspect of complete homogeneous symmetric polynomials is their relation to elementary symmetric polynomials, which can be expressed as the identities

$$\sum_{i=0}^{k}(-1)^i e_i(X_1, \ldots, X_n)h_{k-i}(X_1, \ldots, X_n) = 0 \text{ , for all } k > 0 \text{, and any number of variables } n.$$

Since $e_0(X_1, \ldots, Xn)$ and $h_0(X_1, \ldots, Xn)$ are both equal to 1, one can isolate either the first or the last term of these summations; the former gives a set of equations that allows one to recursively express the successive complete homogeneous symmetric polynomials in terms of the elementary symmetric polynomials, and the latter gives a set of equations that allows doing the inverse. This implicitly shows that any symmetric polynomial can be expressed in terms of the $hk(X_1, \ldots, Xn)$ with $1 \le k \le n$: one first expresses the symmetric polynomial in terms of the elementary symmetric polynomials, and then expresses those in terms of the mentioned complete homogeneous ones.

46.4.5 Schur polynomials

Main article: Schur polynomial

Another class of symmetric polynomials is that of the Schur polynomials, which are of fundamental importance in the applications of symmetric polynomials to representation theory. They are however not as easy to describe as the other kinds of special symmetric polynomials; see the main article for details.

46.5 Symmetric polynomials in algebra

Symmetric polynomials are important to linear algebra, representation theory, and Galois theory. They are also important in combinatorics, where they are mostly studied through the ring of symmetric functions, which avoids having to carry around a fixed number of variables all the time.

46.6 Alternating polynomials

Main article: Alternating polynomials

Analogous to symmetric polynomials are alternating polynomials: polynomials that, rather than being *invariant* under permutation of the entries, change according to the sign of the permutation.

These are all products of the Vandermonde polynomial and a symmetric polynomial, and form a quadratic extension of the ring of symmetric polynomials: the Vandermonde polynomial is a square root of the discriminant.

46.7 See also

- Symmetric function

- Newton's identities

- Stanley symmetric function

- Muirhead's inequality

46.8 References

- Lang, Serge (2002), *Algebra*, Graduate Texts in Mathematics **211** (Revised third ed.), New York: Springer-Verlag, ISBN 978-0-387-95385-4, Zbl 0984.00001, MR 1878556

- Macdonald, I.G. (1979), *Symmetric Functions and Hall Polynomials*, Oxford Mathematical Monographs, Oxford: Clarendon Press.

- I.G. Macdonald (1995), *Symmetric Functions and Hall Polynomials*, second ed. Oxford: Clarendon Press. ISBN 0-19-850450-0 (paperback, 1998).

- Richard P. Stanley (1999), *Enumerative Combinatorics*, Vol. 2. Cambridge: Cambridge University Press. ISBN 0-521-56069-1

Chapter 47

Symmetric tensor

In mathematics, a **symmetric tensor** is a tensor that is invariant under a permutation of its vector arguments:

$$T(v_1, v_2, \ldots, v_r) = T(v_{\sigma 1}, v_{\sigma 2}, \ldots, v_{\sigma r})$$

for every permutation σ of the symbols $\{1,2,\ldots,r\}$. Alternatively, a symmetric tensor of order r represented in coordinates as a quantity with r indices satisfies

$$T_{i_1 i_2 \ldots i_r} = T_{i_{\sigma 1} i_{\sigma 2} \ldots i_{\sigma r}}.$$

The space of symmetric tensors of order r on a finite-dimensional vector space is naturally isomorphic to the dual of the space of homogeneous polynomials of degree r on V. Over fields of characteristic zero, the graded vector space of all symmetric tensors can be naturally identified with the symmetric algebra on V. A related concept is that of the antisymmetric tensor or alternating form. Symmetric tensors occur widely in engineering, physics and mathematics.

47.1   Definition

Let V be a vector space and

$$T \in V^{\otimes k}$$

a tensor of order k. Then T is a symmetric tensor if

$$\tau_\sigma T = T$$

for the braiding maps associated to every permutation σ on the symbols $\{1,2,\ldots,k\}$ (or equivalently for every transposition on these symbols).

Given a basis $\{e^i\}$ of V, any symmetric tensor T of rank k can be written as

$$T = \sum_{i_1,\ldots,i_k=1}^{N} T_{i_1 i_2 \ldots i_k} e^{i_1} \otimes e^{i_2} \otimes \cdots \otimes e^{i_k}$$

for some unique list of coefficients $T_{i_1 i_2 \ldots i_k}$ (the *components* of the tensor in the basis) that are symmetric on the indices. That is to say

$$T_{i_{\sigma 1} i_{\sigma 2} \ldots i_{\sigma k}} = T_{i_1 i_2 \ldots i_k}$$

for every permutation σ.

The space of all symmetric tensors of order k defined on V is often denoted by $S^k(V)$ or $\mathrm{Sym}^k(V)$. It is itself a vector space, and if V has dimension N then the dimension of $\mathrm{Sym}^k(V)$ is the binomial coefficient

$$\dim \mathrm{Sym}^k(V) = \binom{N + k - 1}{k}.$$

We then construct $\mathrm{Sym}(V)$ as the direct sum of $\mathrm{Sym}^k(V)$ for $k = 0, 1, 2, \ldots$

$$\mathrm{Sym}(V) = \bigoplus_{k=0}^{\infty} \mathrm{Sym}^k(V).$$

47.2 Examples

There are many examples of symmetric tensors. Some include, the metric tensor, $g_{\mu\nu}$, the Einstein tensor, $G_{\mu\nu}$ and the Ricci tensor, $R_{\mu\nu}$.

Many material properties and fields used in physics and engineering can be represented as symmetric tensor fields; for example: stress, strain, and anisotropic conductivity. Also, in diffusion MRI one often uses symmetric tensors to describe diffusion in the brain or other parts of the body.

Ellipsoids are examples of algebraic varieties; and so, for general rank, symmetric tensors, in the guise of homogeneous polynomials, are used to define projective varieties, and are often studied as such.

47.3 Symmetric part of a tensor

Suppose V is a vector space over a field of characteristic 0. If $T \in V^{\otimes k}$ is a tensor of order k, then the symmetric part of T is the symmetric tensor defined by

$$\mathrm{Sym}\, T = \frac{1}{k!} \sum_{\sigma \in \mathfrak{S}_k} \tau_\sigma T,$$

the summation extending over the symmetric group on k symbols. In terms of a basis, and employing the Einstein summation convention, if

$$T = T_{i_1 i_2 \ldots i_k} e^{i_1} \otimes e^{i_2} \otimes \cdots \otimes e^{i_k},$$

then

$$\mathrm{Sym}\, T = \frac{1}{k!} \sum_{\sigma \in \mathfrak{S}_k} T_{i_{\sigma 1} i_{\sigma 2} \ldots i_{\sigma k}} e^{i_1} \otimes e^{i_2} \otimes \cdots \otimes e^{i_k}.$$

The components of the tensor appearing on the right are often denoted by

$$T_{(i_1 i_2 \ldots i_k)} = \frac{1}{k!} \sum_{\sigma \in \mathfrak{S}_k} T_{i_{\sigma 1} i_{\sigma 2} \ldots i_{\sigma k}}$$

with parentheses around the indices which have been symmetrized. [Square brackets are used to indicate anti-symmetrization.]

47.4 Symmetric product

If T is a simple tensor, given as a pure tensor product

$$T = v_1 \otimes v_2 \otimes \cdots \otimes v_r$$

then the symmetric part of T is the symmetric product of the factors:

$$v_1 \odot v_2 \odot \cdots \odot v_r := \frac{1}{r!} \sum_{\sigma \in \mathfrak{S}_r} v_{\sigma 1} \otimes v_{\sigma 2} \otimes \cdots \otimes v_{\sigma r}.$$

In general we can turn Sym(V) into an algebra by defining the commutative and associative product ' $\odot$ '.[1] Given two tensors $T_1 \in \text{Sym}^{k_1}(V)$ and $T_2 \in \text{Sym}^{k_2}(V)$, we use the symmetrization operator to define:

$$T_1 \odot T_2 = \text{Sym}(T_1 \otimes T_2) \quad \left(\in \text{Sym}^{k_1 + k_2}(V) \right).$$

It can be verified (as is done by Kostrikin and Manin[1]) that the resulting product is in fact commutative and associative. In some cases the operator is not written at all: $T_1 T_2 = T_1 \odot T_2$.

In some cases an exponential notation is used:

$$v^{\odot k} = \underbrace{v \odot v \odot \cdots \odot v}_{k\text{-times}} = \underbrace{v \otimes v \otimes \cdots \otimes v}_{k\text{-times}} = v^{\otimes k}.$$

Where v is a vector. Again, in some cases the ' $\odot$ ' is left out:

$$v^k = \underbrace{v\, v \cdots v}_{k\text{-times}} = \underbrace{v \odot v \odot \cdots \odot v}_{k\text{-times}}.$$

47.5 Decomposition

In analogy with the theory of symmetric matrices, a (real) symmetric tensor of order 2 can be "diagonalized". More precisely, for any tensor $T \in \text{Sym}^2(V)$, there are an integer r, non-zero unit vectors $v_1, \ldots, v_r \in V$ and weights $\lambda_1, \ldots, \lambda_r$ such that

$$T = \sum_{i=1}^{r} \lambda_i\, v_i \otimes v_i.$$

The minimum number r for which such a decomposition is possible is the (symmetric) rank of T. The vectors appearing in this minimal expression are the *principal axes* of the tensor, and generally have an important physical meaning. For example, the principal axes of the inertia tensor define the Poinsot's ellipsoid representing the moment of inertia. Also see Sylvester's law of inertia.

For symmetric tensors of arbitrary order k, decompositions

$$T = \sum_{i=1}^{r} \lambda_i \, v_i^{\otimes k}$$

are also possible. The minimum number r for which such a decomposition is possible is the *symmetric* rank of T.[2] This minimal decomposition is called a Waring decomposition; it is a symmetric form of the tensor rank decomposition. For second-order tensors this corresponds to the rank of the matrix representing the tensor in any basis, and it is well known that the maximum rank is equal to the dimension of the underlying vector space. However, for higher orders this need not hold: the rank can be higher than the number of dimensions in the underlying vector space.

47.6 See also

- antisymmetric tensor
- Ricci calculus
- Schur polynomial
- symmetric polynomial
- transpose
- Young symmetrizer

47.7 Notes

[1] Kostrikin, Alexei I.; Manin, Iurii Ivanovich (1997). *Linear algebra and geometry*. Algebra, Logic and Applications **1**. Gordon and Breach. pp. 276–279. ISBN 9056990497.

[2] Comon, P.; Golub, G.; Lim, L. H.; Mourrain, B. (2008). "Symmetric Tensors and Symmetric Tensor Rank". *SIAM Journal on Matrix Analysis and Applications* **30** (3): 1254. doi:10.1137/060661569.

47.8 References

- Bourbaki, Nicolas (1989), *Elements of mathematics, Algebra I*, Springer-Verlag, ISBN 3-540-64243-9.
- Bourbaki, Nicolas (1990), *Elements of mathematics, Algebra II*, Springer-Verlag, ISBN 3-540-19375-8.
- Greub, Werner Hildbert (1967), *Multilinear algebra*, Die Grundlehren der Mathematischen Wissenschaften, Band 136, Springer-Verlag New York, Inc., New York, MR 0224623.
- Sternberg, Shlomo (1983), *Lectures on differential geometry*, New York: Chelsea, ISBN 978-0-8284-0316-0.

47.9 External links

- Cesar O. Aguilar, *The Dimension of Symmetric k-tensors*

Chapter 48

Automorphism

In mathematics, an **automorphism** is an isomorphism from a mathematical object to itself. It is, in some sense, a symmetry of the object, and a way of mapping the object to itself while preserving all of its structure. The set of all automorphisms of an object forms a group, called the **automorphism group**. It is, loosely speaking, the symmetry group of the object.

48.1 Definition

The exact definition of an automorphism depends on the type of "mathematical object" in question and what, precisely, constitutes an "isomorphism" of that object. The most general setting in which these words have meaning is an abstract branch of mathematics called category theory. Category theory deals with abstract objects and morphisms between those objects.

In category theory, an automorphism is an endomorphism (i.e. a morphism from an object to itself) which is also an isomorphism (in the categorical sense of the word).

This is a very abstract definition since, in category theory, morphisms aren't necessarily functions and objects aren't necessarily sets. In most concrete settings, however, the objects will be sets with some additional structure and the morphisms will be functions preserving that structure.

In the context of abstract algebra, for example, a mathematical object is an algebraic structure such as a group, ring, or vector space. An isomorphism is simply a bijective homomorphism. (The definition of a homomorphism depends on the type of algebraic structure; see, for example: group homomorphism, ring homomorphism, and linear operator).

The identity morphism (identity mapping) is called the **trivial automorphism** in some contexts. Respectively, other (non-identity) automorphisms are called **nontrivial automorphisms**.

48.2 Automorphism group

If the automorphisms of an object X form a set (instead of a proper class), then they form a group under composition of morphisms. This group is called the **automorphism group** of X. That this is indeed a group is simple to see:

- Closure: composition of two endomorphisms is another endomorphism.

- Associativity: composition of morphisms is *always* associative.

- Identity: the identity is the identity morphism from an object to itself, which exists by definition.

- Inverses: by definition every isomorphism has an inverse which is also an isomorphism, and since the inverse is also an endomorphism of the same object it is an automorphism.

The automorphism group of an object X in a category C is denoted $\text{Aut}C(X)$, or simply $\text{Aut}(X)$ if the category is clear from context.

48.3 Examples

- In set theory, an arbitrary permutation of the elements of a set X is an automorphism. The automorphism group of X is also called the symmetric group on X.

- In elementary arithmetic, the set of integers, **Z**, considered as a group under addition, has a unique nontrivial automorphism: negation. Considered as a ring, however, it has only the trivial automorphism. Generally speaking, negation is an automorphism of any abelian group, but not of a ring or field.

- A group automorphism is a group isomorphism from a group to itself. Informally, it is a permutation of the group elements such that the structure remains unchanged. For every group G there is a natural group homomorphism $G \rightarrow \text{Aut}(G)$ whose image is the group $\text{Inn}(G)$ of inner automorphisms and whose kernel is the center of G. Thus, if G has trivial center it can be embedded into its own automorphism group.[1]

- In linear algebra, an endomorphism of a vector space V is a linear operator $V \rightarrow V$. An automorphism is an invertible linear operator on V. When the vector space is finite-dimensional, the automorphism group of V is the same as the general linear group, $\text{GL}(V)$.

- A field automorphism is a bijective ring homomorphism from a field to itself. In the cases of the rational numbers (**Q**) and the real numbers (**R**) there are no nontrivial field automorphisms. Some subfields of **R** have nontrivial field automorphisms, which however do not extend to all of **R** (because they cannot preserve the property of a number having a square root in **R**). In the case of the complex numbers, **C**, there is a unique nontrivial automorphism that sends **R** into **R**: complex conjugation, but there are infinitely (uncountably) many "wild" automorphisms (assuming the axiom of choice).[2][3] Field automorphisms are important to the theory of field extensions, in particular Galois extensions. In the case of a Galois extension L/K the subgroup of all automorphisms of L fixing K pointwise is called the Galois group of the extension.

- The field **Q**p of p-adic numbers has no nontrivial automorphisms.

- In graph theory an automorphism of a graph is a permutation of the nodes that preserves edges and non-edges. In particular, if two nodes are joined by an edge, so are their images under the permutation.

- For relations, see relation-preserving automorphism.

  - In order theory, see order automorphism.

- In geometry, an automorphism may be called a motion of the space. Specialized terminology is also used:

  - In metric geometry an automorphism is a self-isometry. The automorphism group is also called the isometry group.

  - In the category of Riemann surfaces, an automorphism is a bijective biholomorphic map (also called a conformal map), from a surface to itself. For example, the automorphisms of the Riemann sphere are Möbius transformations.

  - An automorphism of a differentiable manifold M is a diffeomorphism from M to itself. The automorphism group is sometimes denoted $\text{Diff}(M)$.

  - In topology, morphisms between topological spaces are called continuous maps, and an automorphism of a topological space is a homeomorphism of the space to itself, or self-homeomorphism (see homeomorphism group). In this example it is *not sufficient* for a morphism to be bijective to be an isomorphism.

48.4   History

One of the earliest group automorphisms (automorphism of a group, not simply a group of automorphisms of points) was given by the Irish mathematician William Rowan Hamilton in 1856, in his icosian calculus, where he discovered an order two automorphism,[4] writing:

> so that μ is a new fifth root of unity, connected with the former fifth root λ by relations of perfect reciprocity.

48.5   Inner and outer automorphisms

In some categories—notably groups, rings, and Lie algebras—it is possible to separate automorphisms into two types, called "inner" and "outer" automorphisms.

In the case of groups, the inner automorphisms are the conjugations by the elements of the group itself. For each element a of a group G, conjugation by a is the operation $\varphi a : G \to G$ given by $\varphi a(g) = aga^{-1}$ (or $a^{-1}ga$; usage varies). One can easily check that conjugation by a is a group automorphism. The inner automorphisms form a normal subgroup of $\mathrm{Aut}(G)$, denoted by $\mathrm{Inn}(G)$; this is called Goursat's lemma.

The other automorphisms are called outer automorphisms. The quotient group $\mathrm{Aut}(G) / \mathrm{Inn}(G)$ is usually denoted by $\mathrm{Out}(G)$; the non-trivial elements are the cosets that contain the outer automorphisms.

The same definition holds in any unital ring or algebra where a is any invertible element. For Lie algebras the definition is slightly different.

48.6   See also

- Endomorphism ring

- Antiautomorphism

- Frobenius automorphism

- Morphism

- Characteristic subgroup

48.7   References

[1] PJ Pahl, R Damrath (2001). "§7.5.5 Automorphisms". *Mathematical foundations of computational engineering* (Felix Pahl translation ed.). Springer. p. 376. ISBN 3-540-67995-2.

[2] Yale, Paul B. (May 1966). "Automorphisms of the Complex Numbers" (PDF). *Mathematics Magazine* **39** (3): 135–141. doi:10.2307/2689301. JSTOR 2689301.

[3] Lounesto, Pertti (2001), *Clifford Algebras and Spinors* (2nd ed.), Cambridge University Press, pp. 22–23, ISBN 0-521-00551-5

[4] Sir William Rowan Hamilton (1856). "Memorandum respecting a new System of Roots of Unity" (PDF). *Philosophical Magazine* **12**: 446.

48.8 External links

- *Automorphism* at Encyclopaedia of Mathematics

- Weisstein, Eric W., "Automorphism", *MathWorld*.

Chapter 49

Isometry

This article is about distance-preserving functions. For other mathematical uses, see isometry (disambiguation). For non-mathematical uses, see Isometric.

In mathematics, an **isometry** is a distance-preserving injective map between metric spaces.

49.1   Introduction

Given a metric space (loosely, a set and a scheme for assigning distances between elements of the set), an isometry is a transformation which maps elements to the same or another metric space such that the distance between the image elements in the new metric space is equal to the distance between the elements in the original metric space. In a two-dimensional or three-dimensional Euclidean space, two geometric figures are congruent if they are related by an isometry: related by either a rigid motion (translation or rotation), or a composition of a rigid motion and a reflection. They are equal, up to an action of a rigid motion, if related by a direct isometry (orientation preserving).

Isometries are often used in constructions where one space is embedded in another space. For instance, the completion of a metric space M involves an isometry from M into M', a quotient set of the space of Cauchy sequences on M. The original space M is thus isometrically isomorphic to a subspace of a complete metric space, and it is usually identified with this subspace. Other embedding constructions show that every metric space is isometrically isomorphic to a closed subset of some normed vector space and that every complete metric space is isometrically isomorphic to a closed subset of some Banach space.

An isometric surjective linear operator on a Hilbert space is called a unitary operator.

49.2   Formal definitions

Let X and Y be metric spaces with metrics dX and dY. A map $f : X \to Y$ is called an **isometry** or **distance preserving** if for any $a, b \in X$ one has

$$d_Y(f(a), f(b)) = d_X(a, b).$$

An isometry is automatically injective. (Otherwise two distinct points, a and b, could be mapped to the same point, which would contradict the coincidence axiom of the metric d.) This proof is similar to the proof that an order embedding between partially ordered sets is injective. Clearly, every isometry between metric spaces is a topological embedding (i.e. a homeomorphism).

A **global isometry**, **isometric isomorphism** or **congruence mapping** is a bijective isometry.

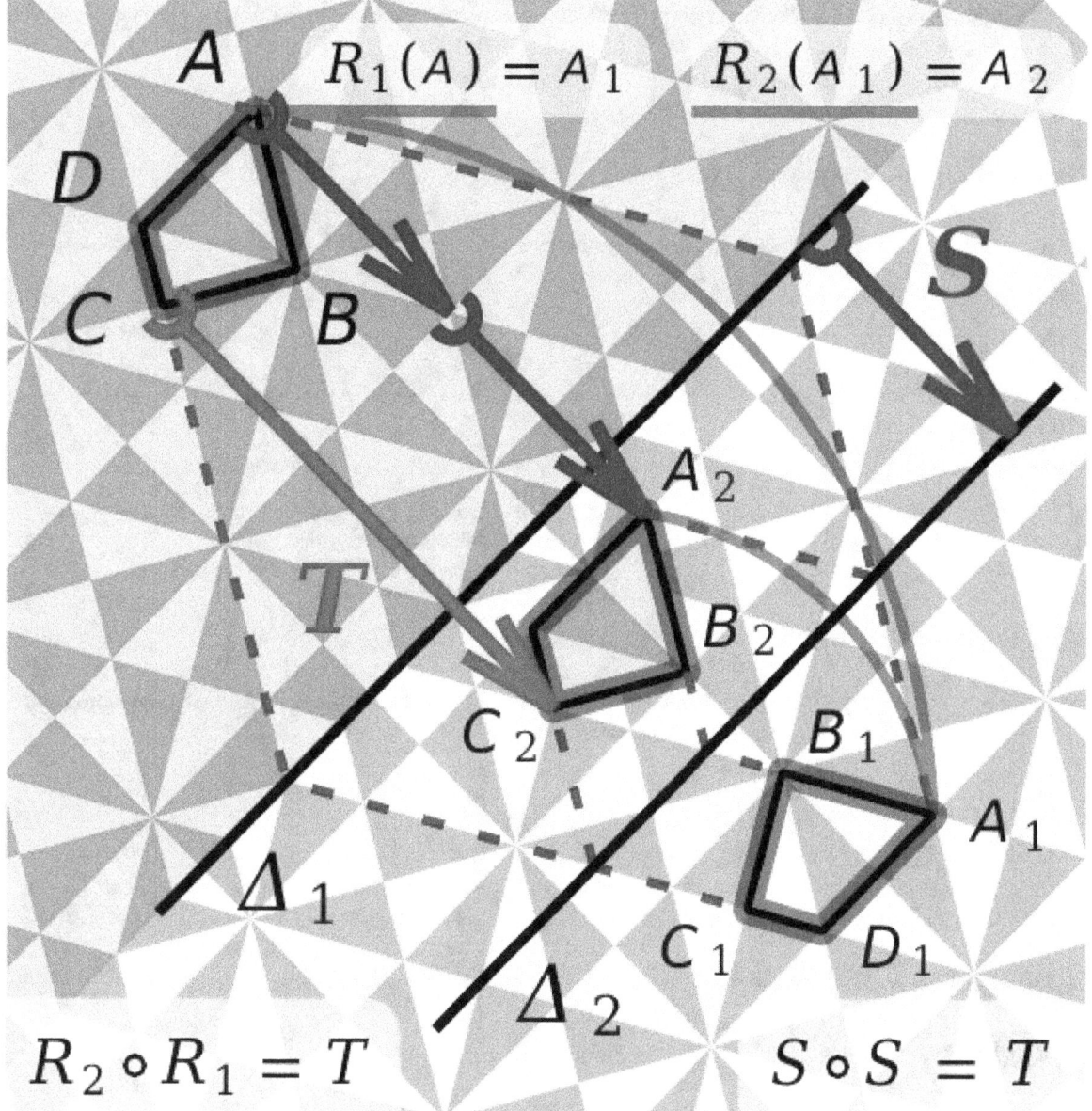

A composition of two indirect isometries is a direct isometry. A reflection in a line is an indirect isometry, like R1 or R2 on the image.
Translation T is a direct isometry: a rigid motion.

Two metric spaces X and Y are called **isometric** if there is a bijective isometry from X to Y. The set of bijective isometries from a metric space to itself forms a group with respect to function composition, called the **isometry group**.

There is also the weaker notion of *path isometry* or *arcwise isometry*:

A **path isometry** or **arcwise isometry** is a map which preserves the lengths of curves; such a map is not necessarily an isometry in the distance preserving sense, and it need not necessarily be bijective, or even injective. This term is often abridged to simply *isometry*, so one should take care to determine from context which type is intended.

49.3 Examples

- Any reflection, translation and rotation is a global isometry on Euclidean spaces. See also Euclidean group.

- The map $x \mapsto |x|$ in $\mathbb{R} \to \mathbb{R}$ is a path isometry but not an isometry. Note that unlike an isometry, it is not injective.

- The isometric linear maps from $\mathbf{C}^n$ to itself are the unitary matrices.

49.4   Linear isometry

Given two normed vector spaces V and W, a **linear isometry** is a linear map $f : V \to W$ that preserves the norms:

$$\|f(v)\| = \|v\|$$

for all v in V. Linear isometries are distance-preserving maps in the above sense. They are global isometries if and only if they are surjective.

By the Mazur-Ulam theorem, any isometry of normed vector spaces over $\mathbf{R}$ is affine.

In an inner product space, the fact that any linear isometry is an orthogonal transformation can be shown by using polarization to prove $<Ax, Ay> = <x, y>$ and then applying the Riesz representation theorem.

49.5   Generalizations

- Given a positive real number ε, an **ε-isometry** or **almost isometry** (also called a **Hausdorff approximation**) is a map $f : X \to Y$ between metric spaces such that

  1. for $x,x' \in X$ one has $|dY(f(x),f(x'))-dX(x,x')| < \varepsilon$, and
  2. for any point $y \in Y$ there exists a point $x \in X$ with $dY(y,f(x)) < \varepsilon$

  That is, an ε-isometry preserves distances to within ε and leaves no element of the codomain further than ε away from the image of an element of the domain. Note that ε-isometries are not assumed to be continuous.

- The **restricted isometry property** characterizes nearly isometric matrices for sparse vectors.

- **Quasi-isometry** is yet another useful generalization.

- One may also define an element in an abstract unital C*-algebra to be an isometry:

  $a \in \mathfrak{A}$ is an isometry if and only if $a^* \cdot a = 1$.

  Note that as mentioned in the introduction this is not necessarily a unitary element because one does not in general have that left inverse is a right inverse.

- On a pseudo-Euclidean space, the term *isometry* means a linear bijection preserving magnitude. See also Quadratic spaces.

49.6   See also

- Motion (geometry)

- Isometric projection

- Euclidean plane isometry

- 3D isometries that leave the origin fixed

- Space group

- Involution

- Isometries in physics

- Homeomorphism group

- Partial isometry

49.7 References

49.8 Further reading

- F. S. Beckman and D. A. Quarles, Jr., *On isometries of Euclidean space*, Proc. Amer. Math. Soc., 4 (1953) 810-815.

Chapter 50

Multiple integral

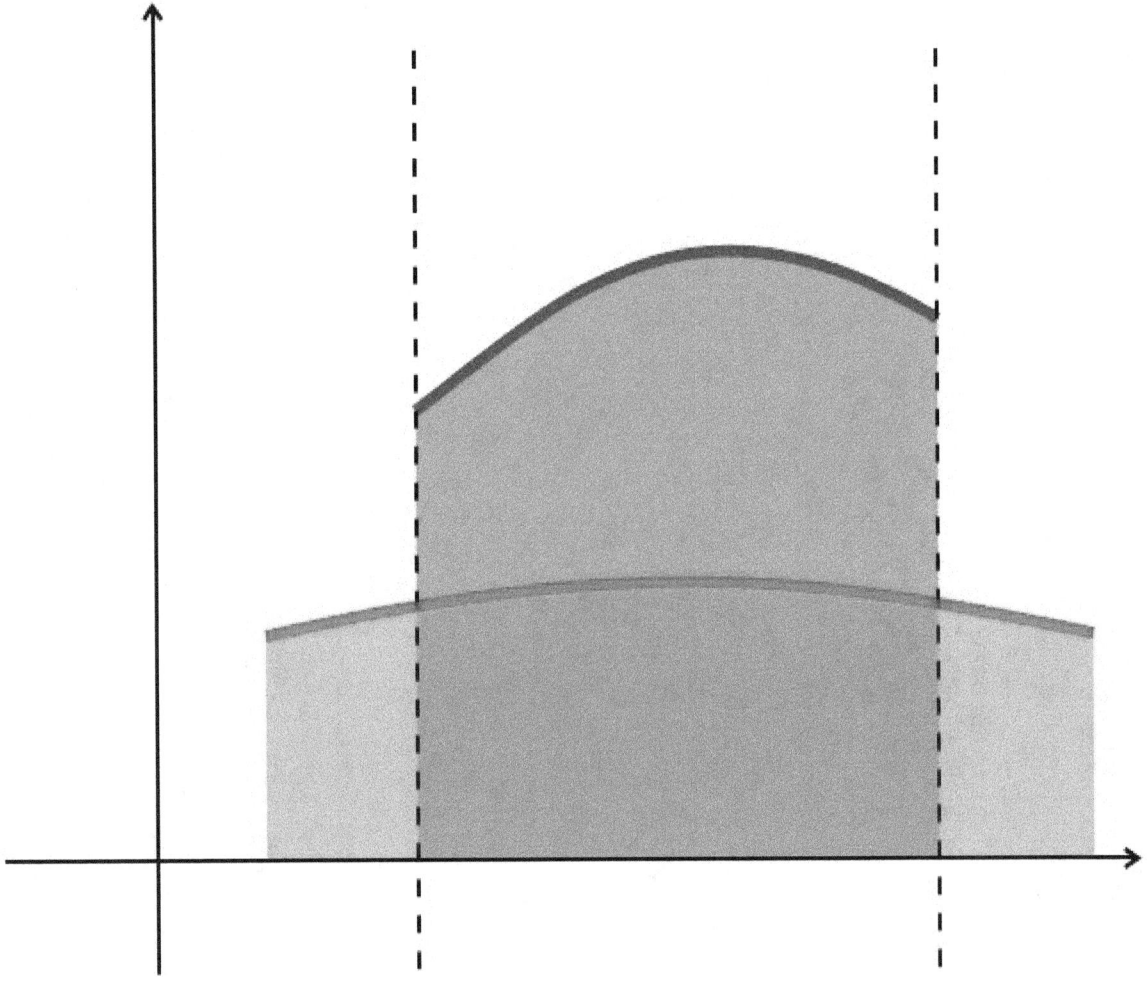

Integral as area between two curves.

The **multiple integral** is a generalization of the definite integral to functions of more than one real variable, for example,

Double integral as volume under a surface $z = 10 - \frac{x^2 - y^2}{8}$ *. The rectangular region at the bottom of the body is the domain of integration, while the surface is the graph of the two-variable function to be integrated.*

$f(x, y)$ or $f(x, y, z)$. Integrals of a function of two variables over a region in $\mathbf{R}^2$ are called double integrals, and integrals of a function of three variables over a region of $\mathbf{R}^3$ are called triple integrals.[1]

50.1   Introduction

Just as the definite integral of a positive function of one variable represents the area of the region between the graph of the function and the *x*-axis, the **double integral** of a positive function of two variables represents the volume of the region between the surface defined by the function (on the three-dimensional Cartesian plane where $z = f(x, y)$) and the plane which contains its domain. (The same volume can be obtained via the **triple integral**—the integral of a function in three variables—of the constant function $f(x, y, z) = 1$ over the above-mentioned region between the surface and the plane.)[1] If there are more variables, a multiple integral will yield hypervolumes of multi-dimensional functions.

Multiple integration of a function in *n* variables: $f(x_1, x_2, ..., xn)$ over a domain *D* is most commonly represented by nested integral signs in the reverse order of execution (the leftmost integral sign is computed last), followed by the function and integrand arguments in proper order (the integral with respect to the rightmost argument is computed last). The domain of integration is either represented symbolically for every argument over each integral sign, or is abbreviated by a variable at the rightmost integral sign:[2]

$$\int \cdots \int_{\mathbf{D}} f(x_1, x_2, \ldots, x_n) \, dx_1 \cdots dx_n$$

Since the concept of an antiderivative is only defined for functions of a single real variable, the usual definition of the indefinite integral does not immediately extend to the multiple integral.

50.2   Mathematical definition

For n > 1, consider a so-called "half-open" *n*-dimensional hyperrectangular domain T, defined as:

$$T = [a_1, b_1) \times [a_2, b_2) \times \cdots \times [a_n, b_n) \subseteq \mathbf{R}^n.$$

Partition each interval [*aj, bj*) into a finite family *Ij* of non-overlapping subintervals *ijα*, with each subinterval closed at the left end, and open at the right end.

Then the finite family of subrectangles *C* given by

$$C = I_1 \times I_2 \times \cdots \times I_n$$

is a partition of *T*; that is, the subrectangles *Ck* are non-overlapping and their union is *T*.

Let $f : T \to \mathbf{R}$ be a function defined on *T*. Consider a partition *C* of *T* as defined above, such that *C* is a family of *m* subrectangles *Cm* and

$$T = C_1 \cup C_2 \cup \cdots \cup C_m$$

We can approximate the total (*n*+1)th-dimensional volume bounded below by the *n*-dimensional hyperrectangle *T* and above by the *n*-dimensional graph of *f* with the following Riemann sum:

$$\sum_{k=1}^{m} f(P_k) \, \mathrm{m}(C_k)$$

where *Pk* is a point in *Ck* and m(*Ck*) is the product of the lengths of the intervals whose Cartesian product is *Ck*, otherwise known as the measure of *Ck*.

The **diameter** of a subrectangle C_k is the largest of the lengths of the intervals whose Cartesian product is C_k. The diameter of a given partition of T is defined as the largest of the diameters of the subrectangles in the partition. Intuitively, as the diameter of the partition C is restricted smaller and smaller, the number of subrectangles m gets larger, and the measure $m(C_k)$ of each subrectangle grows smaller. The function f is said to be **Riemann integrable** if the limit

$$S = \lim_{\delta \to 0} \sum_{k=1}^{m} f(P_k) \, \mathrm{m}(C_k)$$

exists, where the limit is taken over all possible partitions of T of diameter at most δ.[3]

If f is Riemann integrable, S is called the **Riemann integral** of f over T and is denoted

$$\int \cdots \int_T f(x_1, x_2, \ldots, x_n) \, dx_1 \cdots dx_n$$

Frequently this notation is abbreviated as

$$\int_T f(\mathbf{x}) \, d^n\mathbf{x}.$$

where $\mathbf{x}$ represents the n-tuple $(x_1, \ldots x_n)$ and $d^n\mathbf{x}$ is the n-dimensional volume differential.

The Riemann integral of a function defined over an arbitrary bounded n-dimensional set can be defined by extending that function to a function defined over a half-open rectangle whose values are zero outside the domain of the original function. Then the integral of the original function over the original domain is defined to be the integral of the extended function over its rectangular domain, if it exists.

In what follows the Riemann integral in n dimensions will be called the **multiple integral**.

50.2.1 Properties

Multiple integrals have many properties common to those of integrals of functions of one variable (linearity, commutativity, monotonicity, and so on.). One important property of multiple integrals is that the value of an integral is independent of the order of integrands under certain conditions. This property is popularly known as Fubini's theorem.[4]

50.2.2 Particular cases

In the case of $T \subseteq \mathbf{R}^2$, the integral

$$\ell = \iint_T f(x, y) \, dx \, dy$$

is the **double integral** of f on T, and if $T \subseteq \mathbf{R}^3$ the integral

$$\ell = \iiint_T f(x, y, z) \, dx \, dy \, dz$$

is the **triple integral** of f on T.

Notice that, by convention, the double integral has two integral signs, and the triple integral has three; this is a notational convention which is convenient when computing a multiple integral as an iterated integral, as shown later in this article.

50.3   Methods of integration

The resolution of problems with multiple integrals consists, in most of cases, of finding a way to reduce the multiple integral to an iterated integral, a series of integrals of one variable, each being directly solvable. For continuous functions, this is justified by Fubini's theorem. Sometimes, it is possible to obtain the result of the integration by direct examination without any calculations.

The following are some simple methods of integration:[1]

50.3.1   Integrating constant functions

When the integrand is a constant function c, the integral is equal to the product of c and the measure of the domain of integration. If $c = 1$ and the domain is a subregion of $\mathbf{R}^2$, the integral gives the area of the region, while if the domain is a subregion of $\mathbf{R}^3$, the integral gives the volume of the region.

> **Example.** Let $f(x, y) = 2$ and
>
> $$D = \{(x, y) \in \mathbf{R}^2 \;:\; 2 \le x \le 4 \,;\, 3 \le y \le 6\}$$
>
> in which case
>
> $$\int_3^6 \int_2^4 2 \, dx \, dy = 2 \int_3^6 \int_2^4 1 \, dx \, dy = 2 \cdot \text{area}(D) = (2 \cdot 3) \cdot 2 = 12$$
>
> since by definition we have:
>
> $$\int_3^6 \int_2^4 1 \, dx \, dy = \text{area}(D).$$

50.3.2   Use of symmetry

When the domain of integration is symmetric about the origin with respect to at least one of the variables of integration and the integrand is odd with respect to this variable, the integral is equal to zero, as the integrals over the two halves of the domain have the same absolute value but opposite signs. When the integrand is even with respect to this variable, the integral is equal to twice the integral over one half of the domain, as the integrals over the two halves of the domain are equal.

> **Example 1.** Consider the function $f(x, y) = 2\sin(x) - 3y^3 + 5$ integrated over the domain
>
> $$T = \left\{ (x, y) \in \mathbf{R}^2 \;:\; x^2 + y^2 \le 1 \right\},$$
>
> a disc with radius 1 centered at the origin with the boundary included.
>   Using the linearity property, the integral can be decomposed into three pieces:
>
> $$\iint_T (2\sin x - 3y^3 + 5)\, dx\, dy = \iint_T 2\sin x \, dx\, dy - \iint_T 3y^3 \, dx\, dy + \iint_T 5 \, dx\, dy$$
>
> The function $2\sin(x)$ is an odd function in the variable x and the disc T is symmetric with respect to the y-axis, so the value of the first integral is 0. Similarly, the function $3y^3$ is an odd function of y, and T is symmetric with respect to the x-axis, and so the only contribution to the final result is that of the third integral. Therefore the original integral is equal to the area of the disk times 5, or 5π.

Example 2. Consider the function $f(x, y, z) = x \exp(y^2 + z^2)$ and as integration region the sphere with radius 2 centered at the origin,

$$T = \left\{ (x, y, z) \in \mathbf{R}^3 \ : \ x^2 + y^2 + z^2 \leq 4 \right\}.$$

The "ball" is symmetric about all three axes, but it is sufficient to integrate with respect to x-axis to show that the integral is 0, because the function is an odd function of that variable.

50.3.3 Normal domains on $\mathbf{R}^2$

See also: Order of integration (calculus)

This method is applicable to any domain D for which:

- the projection of D onto either the x-axis or the y-axis is bounded by the two values, a and b

- any line perpendicular to this axis that passes between these two values intersects the domain in an interval whose endpoints are given by the graphs of two functions, α and β.

In all cases, the function to be integrated must be continuous on the domain.

x-axis

If the domain D is normal with respect to the x-axis, and $f : D \to \mathbf{R}$ is a continuous function; then $\alpha(x)$ and $\beta(x)$ (defined on the interval $[a, b]$) are the two functions that determine D. Then:

$$\iint_D f(x, y) \, dx \, dy = \int_a^b dx \int_{\alpha(x)}^{\beta(x)} f(x, y) \, dy.$$

y-axis

If D is normal with respect to the y-axis and $f : D \to \mathbf{R}$ is a continuous function; then $\alpha(y)$ and $\beta(y)$ (defined on the interval $[a, b]$) are the two functions that determine D. Then:

$$\iint_D f(x, y) \, dx \, dy = \int_a^b dy \int_{\alpha(y)}^{\beta(y)} f(x, y) \, dx.$$

Example

Consider the region (please see the graphic in the example):

$$D = \left\{ (x, y) \in \mathbf{R}^2 \ : \ x \geq 0, y \leq 1, y \geq x^2 \right\}$$

Calculate

$$\iint_D (x + y) \, dx \, dy.$$

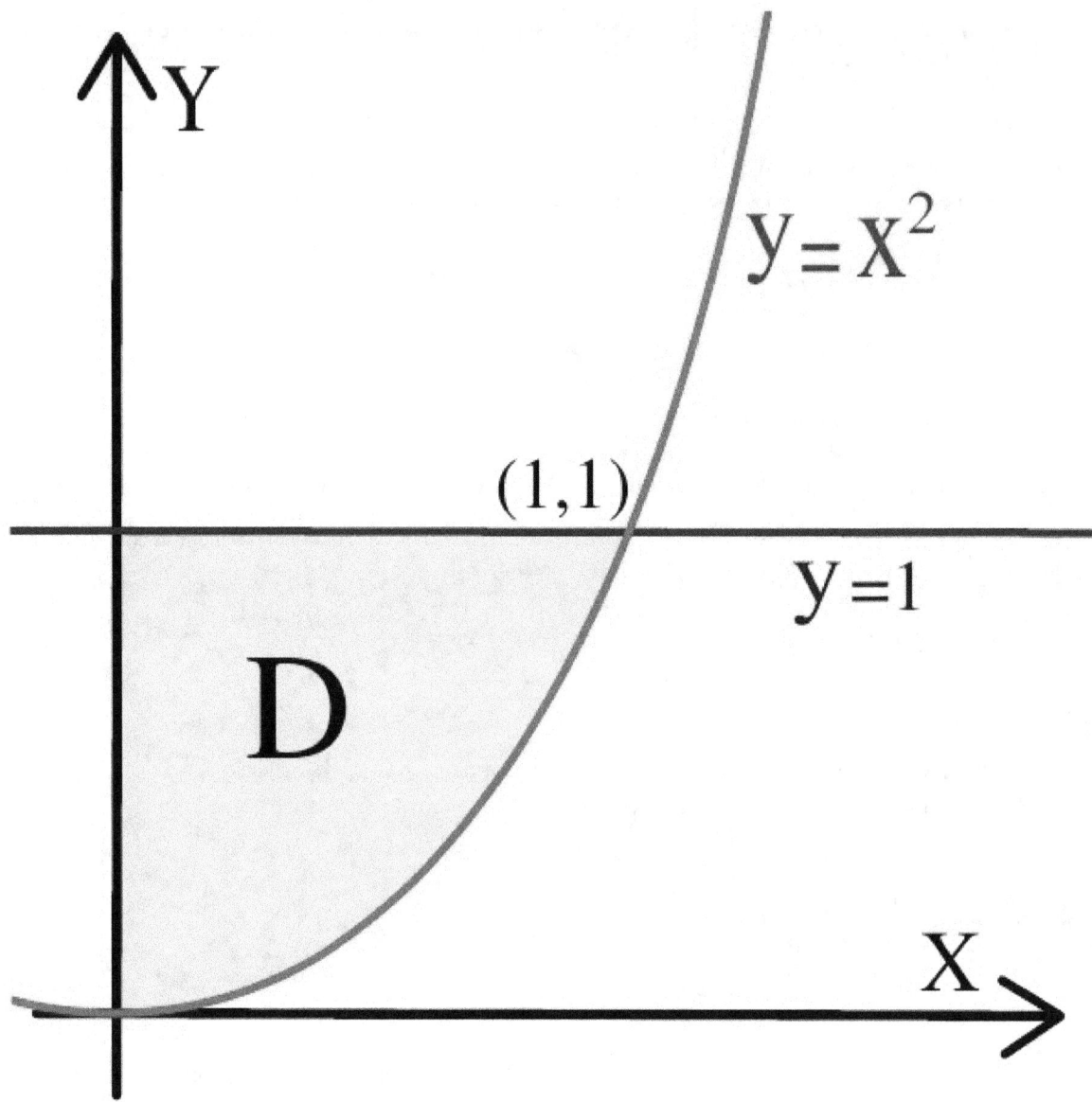

Example: double integral over the normal region D

This domain is normal with respect to both the x- and y-axes. To apply the formulae it is required to find the functions that determine D and the intervals over which these are defined. In this case the two functions are:

$\alpha(x) = x^2$ and $\beta(x) = 1$

while the interval is given by the intersections of the functions with $x = 0$, so the interval is $[a, b] = [0, 1]$ (normality has been chosen with respect to the x-axis for a better visual understanding).

It is now possible to apply the formula:

$$\iint_D (x + y)\, dx\, dy = \int_0^1 dx \int_{x^2}^1 (x + y)\, dy = \int_0^1 dx \left[xy + \frac{y^2}{2} \right]_{x^2}^1$$

(at first the second integral is calculated considering x as a constant). The remaining operations consist of applying the basic techniques of integration:

$$\int_0^1 \left[xy + \frac{y^2}{2} \right]_{x^2}^1 dx = \int_0^1 \left(x + \frac{1}{2} - x^3 - \frac{x^4}{2} \right) dx = \cdots = \frac{13}{20}.$$

If we choose normality with respect to the y-axis we could calculate

$$\int_0^1 dy \int_0^{\sqrt{y}} (x + y)\, dx.$$

and obtain the same value.

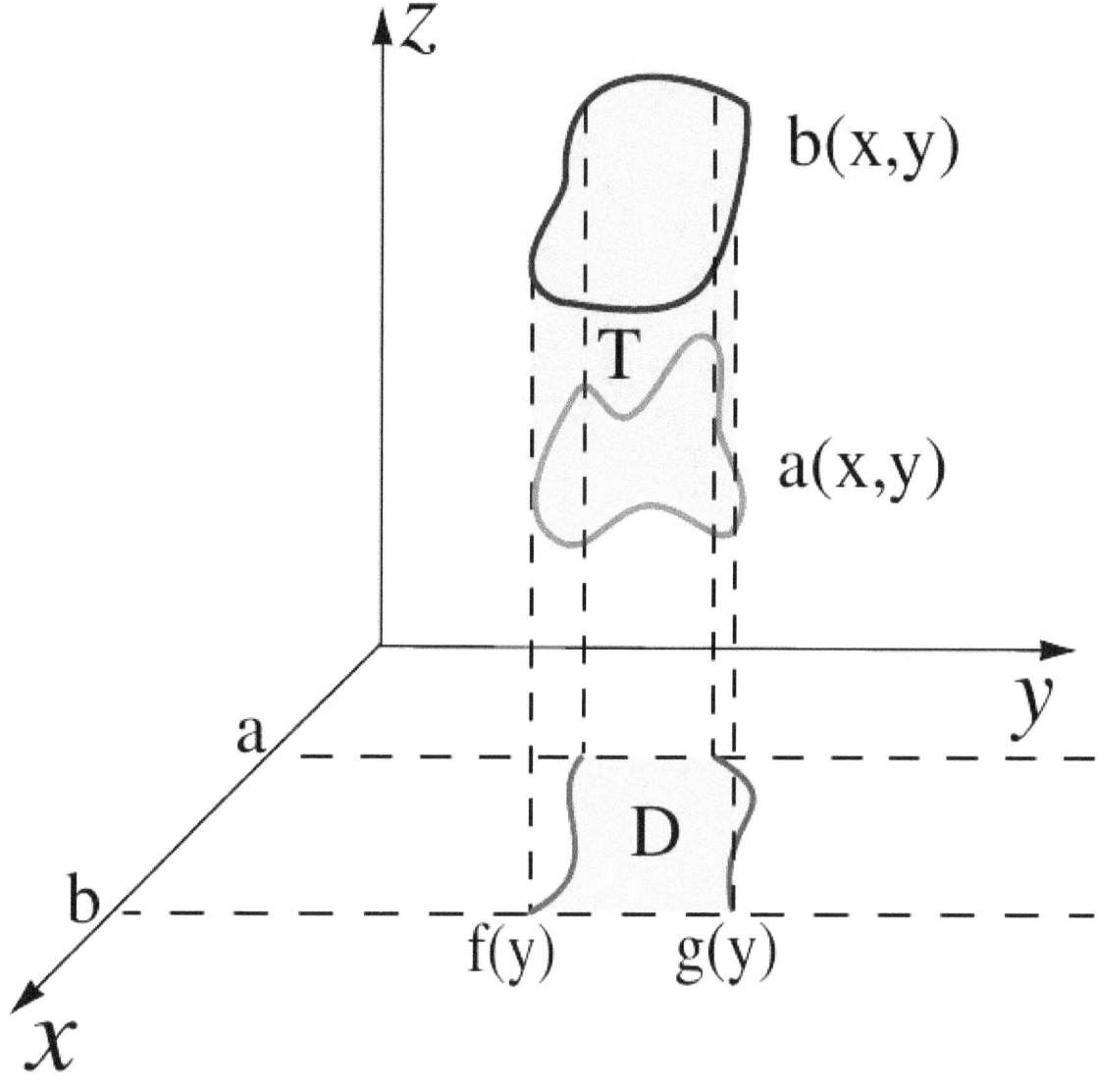

Example of domain in $\mathbf{R}^3$ that is normal with respect to the xy-plane.

Normal domains on $\mathbf{R}^3$

The extension of these formulae to triple integrals should be apparent:

if T is a domain that is normal with respect to the xy-plane and determined by the functions $\alpha(x,y)$ and $\beta(x,y)$, then

$$\iiint_T f(x,y,z)\,dx\,dy\,dz = \iint_D \int_{\alpha(x,y)}^{\beta(x,y)} f(x,y,z)\,dz\,dx\,dy$$

(this definition is the same for the other five normality cases on $\mathbf{R}^3$).

50.3.4   Change of variables

See also: Integration by substitution § Substitution for multiple variables

The limits of integration are often not easily interchangeable (without normality or with complex formulae to integrate). One makes a change of variables to rewrite the integral in a more "comfortable" region, which can be described in simpler formulae. To do so, the function must be adapted to the new coordinates.

> **Example 1a.** The function is $f(x,y) = (x-1)^2 + \sqrt{y}$; if one adopts this substitution $x' = x-1$, $y' = y$ therefore $x = x' + 1$, $y = y'$ one obtains the new function $f_2(x,y) = (x')^2 + \sqrt{y}$.

- Similarly for the domain because it is delimited by the original variables that were transformed before (x and y in example).

- the differentials dx and dy transform via the absolute value of the determinant of the Jacobian matrix containing the partial derivatives of the transformations regarding the new variable (consider, as an example, the differential transformation in polar coordinates).

There exist three main "kinds" of changes of variable (one in $\mathbf{R}^2$, two in $\mathbf{R}^3$); however, more general substitutions can be made using the same principle.

Polar coordinates

See also: Polar coordinate system

In $\mathbf{R}^2$ if the domain has a circular symmetry and the function has some particular characteristics you can apply the *transformation to polar coordinates* (see the example in the picture) which means that the generic points $P(x, y)$ in Cartesian coordinates switch to their respective points in polar coordinates. That allows one to change the shape of the domain and simplify the operations.

The fundamental relation to make the transformation is the following:

$$f(x,y) \to f(\rho\cos\phi, \rho\sin\phi).$$

> **Example 2a.** The function is $f(x,y) = x + y$ and applying the transformation one obtains

$$f(\rho,\phi) = \rho\cos\phi + \rho\sin\phi = \rho(\cos\phi + \sin\phi).$$

> **Example 2b.** The function is $f(x,y) = x^2 + y^2$, in this case one has:

$$f(\rho,\phi) = \rho^2(\cos^2\phi + \sin^2\phi) = \rho^2$$

using the Pythagorean trigonometric identity (very useful to simplify this operation).

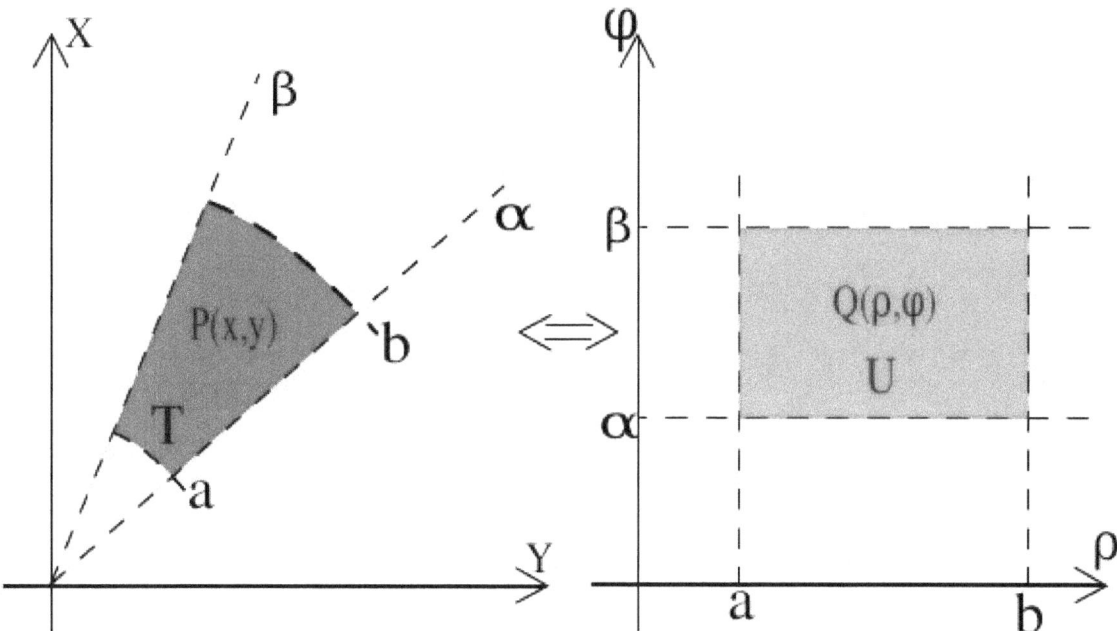

Transformation from cartesian to polar coordinates.

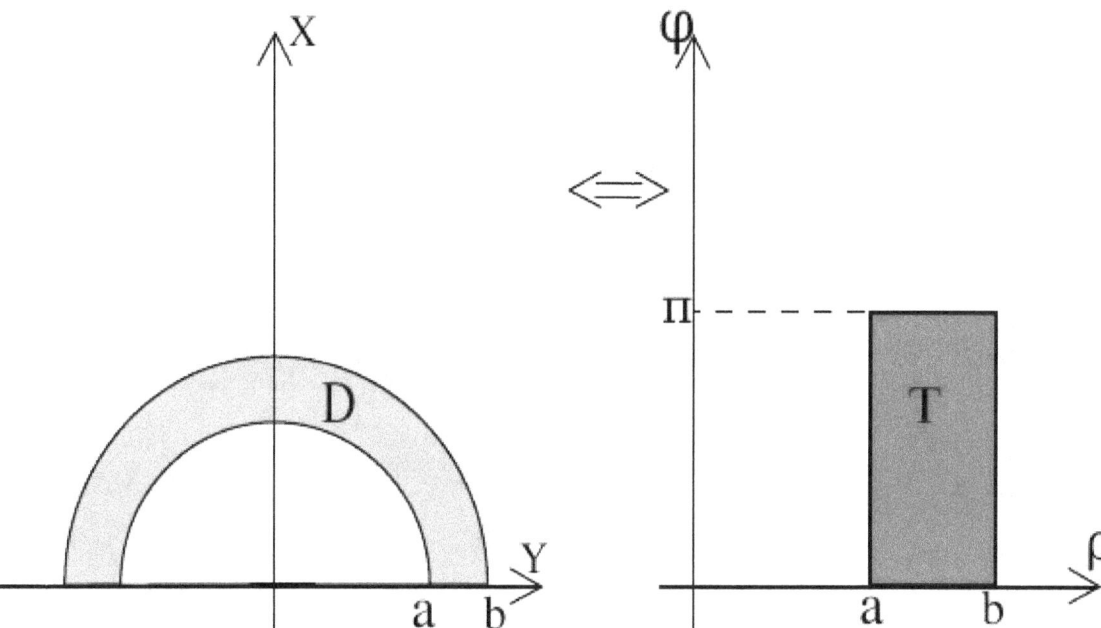

Example of a domain transformation from cartesian to polar.

The transformation of the domain is made by defining the radius' crown length and the amplitude of the described angle to define the ρ, φ intervals starting from x, y.

Example 2c. The domain is $D = \{x^2 + y^2 \leq 4\}$, that is a circumference of radius 2; it's evident that the covered angle is the circle angle, so φ varies from 0 to 2π, while the crown radius varies from 0 to 2 (the crown with the inside radius null is just a circle).

Example 2d. The domain is $D = \{x^2 + y^2 \leq 9,\ x^2 + y^2 \geq 4,\ y \geq 0\}$, that is the circular crown in the positive y half-plane (please see the picture in the example); φ describes a plane angle while ρ varies from 2 to 3. Therefore the transformed domain will be the following rectangle:

$$T = \{2 \leq \rho \leq 3,\ 0 \leq \phi \leq \pi\}.$$

The Jacobian determinant of that transformation is the following:

$$\frac{\partial(x,y)}{\partial(\rho,\phi)} = \begin{vmatrix} \cos\phi & -\rho\sin\phi \\ \sin\phi & \rho\cos\phi \end{vmatrix} = \rho$$

which has been obtained by inserting the partial derivatives of $x = \rho\cos(\varphi)$, $y = \rho\sin(\varphi)$ in the first column respect to ρ and in the second respect to φ, so the $dx\,dy$ differentials in this transformation becomes $\rho\,d\rho\,d\varphi$.

Once the function is transformed and the domain evaluated, it is possible to define the formula for the change of variables in polar coordinates:

$$\iint_D f(x,y)\,dx\,dy = \iint_T f(\rho\cos\phi, \rho\sin\phi)\rho\,d\rho\,d\phi.$$

φ is valid in the $[0, 2\pi]$ interval while ρ, which is a measure of a length, can only have positive values.

Example 2e. The function is $f(x, y) = x$ and the domain is the same as in Example 2d. From the previous analysis of D we know the intervals of ρ (from 2 to 3) and of φ (from 0 to π). Now let's change the function:

$$f(x,y) = x \longrightarrow f(\rho,\phi) = \rho\cos\phi.$$

finally let's apply the integration formula:

$$\iint_D x\,dx\,dy = \iint_T \rho\cos\phi\rho\,d\rho\,d\phi.$$

Once the intervals are known, you have

$$\int_0^\pi \int_2^3 \rho^2\cos\phi\,d\rho\,d\phi = \int_0^\pi \cos\phi\,d\phi \left[\frac{\rho^3}{3}\right]_2^3 = [\sin\phi]_0^\pi \left(9 - \frac{8}{3}\right) = 0.$$

Cylindrical coordinates

In $\mathbf{R}^3$ the integration on domains with a circular base can be made by the *passage in cylindrical coordinates*; the transformation of the function is made by the following relation:

$$f(x,y,z) \to f(\rho\cos\phi, \rho\sin\phi, z)$$

The domain transformation can be graphically attained, because only the shape of the base varies, while the height follows the shape of the starting region.

Example 3a. The region is $D = \{x^2 + y^2 \leq 9,\ x^2 + y^2 \geq 4,\ 0 \leq z \leq 5\}$ (that is the "tube" whose base is the circular crown of Example 2d and whose height is 5); if the transformation is applied, this region is obtained: $T = \{2 \leq \rho \leq 3,\ 0 \leq \phi \leq 2\pi,\ 0 \leq z \leq 5\}$ (that is the parallelepiped whose base is similar to the rectangle in Example 2d and whose height is 5).

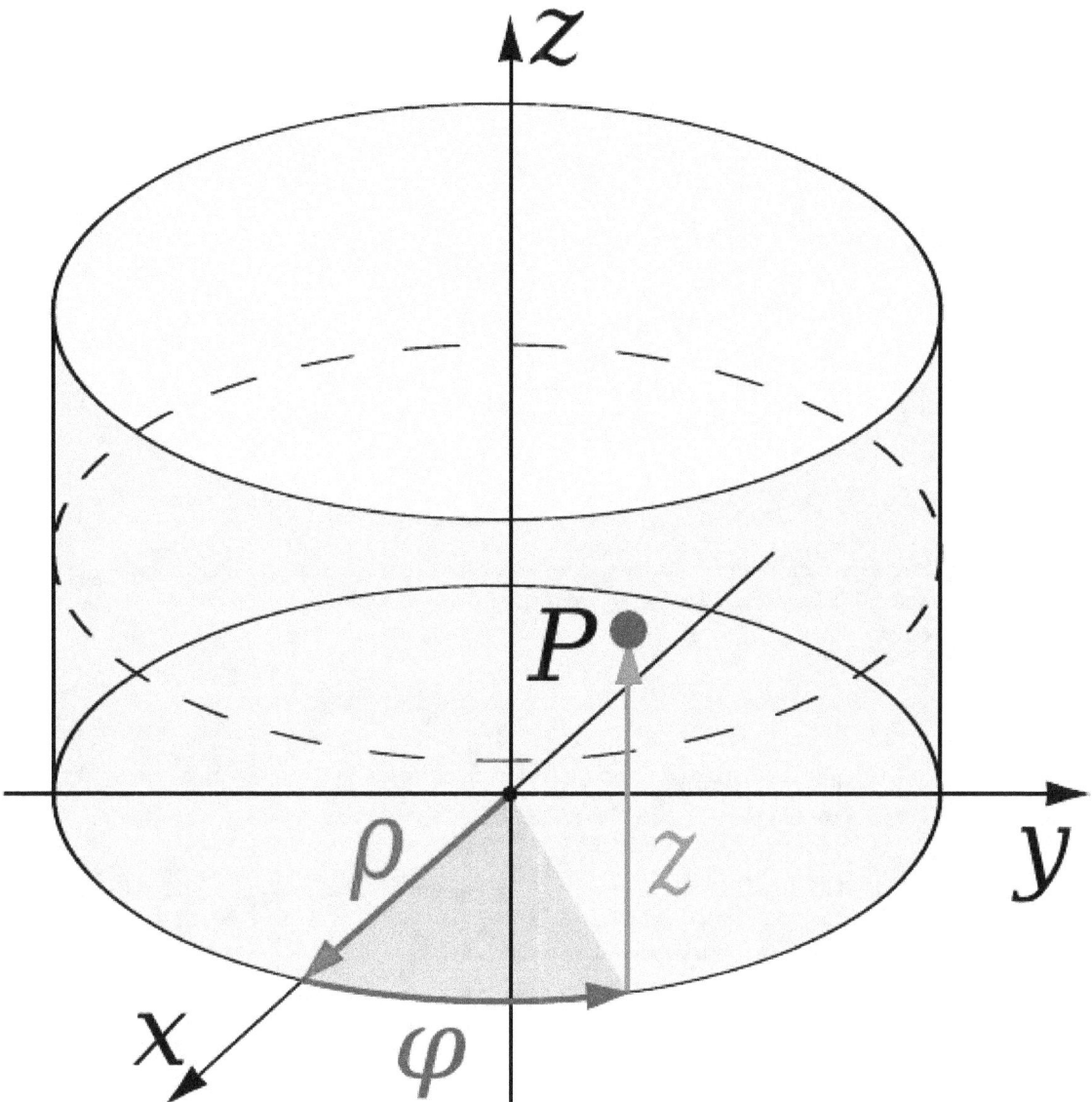

Cylindrical coordinates.

Because the z component is unvaried during the transformation, the $dx\,dy\,dz$ differentials vary as in the passage in polar coordinates: therefore, they become $\rho\,d\rho\,d\varphi\,dz$.

Finally, it is possible to apply the final formula to cylindrical coordinates:

$$\iiint_D f(x,y,z)\,dx\,dy\,dz = \iiint_T f(\rho\cos\phi,\rho\sin\phi,z)\rho\,d\rho\,d\phi\,dz.$$

This method is convenient in case of cylindrical or conical domains or in regions where it is easy to individuate the z interval and even transform the circular base and the function.

Example 3b. The function is $f(x,y,z) = x^2 + y^2 + z$ and as integration domain this cylinder: $D = \{x^2 + y^2 \leq 9,\ -5 \leq z \leq 5\}$. The transformation of D in cylindrical coordinates is the following:

$$T = \{0 \leq \rho \leq 3,\ 0 \leq \phi \leq 2\pi,\ -5 \leq z \leq 5\}.$$

while the function becomes

$$f(\rho \cos \phi, \rho \sin \phi, z) = \rho^2 + z$$

Finally one can apply the integration formula:

$$\iiint_D (x^2 + y^2 + z) \, dx \, dy \, dz = \iiint_T (\rho^2 + z)\rho \, d\rho \, d\phi \, dz;$$

developing the formula you have

$$\int_{-5}^{5} dz \int_0^{2\pi} d\phi \int_0^3 (\rho^3 + \rho z) \, d\rho = 2\pi \int_{-5}^{5} \left[\frac{\rho^4}{4} + \frac{\rho^2 z}{2} \right]_0^3 dz = 2\pi \int_{-5}^{5} \left(\frac{81}{4} + \frac{9}{2} z \right) dz = \cdots = 405\pi.$$

Spherical coordinates

In $\mathbf{R}^3$ some domains have a spherical symmetry, so it's possible to specify the coordinates of every point of the integration region by two angles and one distance. It's possible to use therefore the *passage in spherical coordinates*; the function is transformed by this relation:

$$f(x, y, z) \longrightarrow f(\rho \cos \theta \sin \phi, \rho \sin \theta \sin \phi, \rho \cos \phi)$$

Points on z axis do not have a precise characterization in spherical coordinates, so θ can vary between 0 to 2π.

The better integration domain for this passage is obviously the sphere.

> **Example 4a.** The domain is $D = x^2 + y^2 + z^2 \leq 16$ (sphere with radius 4 and center in the origin); applying the transformation you get this region: $T = \{0 \leq \rho \leq 4, \ 0 \leq \phi \leq \pi, \ 0 \leq \theta \leq 2\pi\}$.
> The Jacobian determinant of this transformation is the following:

$$\frac{\partial(x, y, z)}{\partial(\rho, \theta, \phi)} = \begin{vmatrix} \cos \theta \sin \phi & -\rho \sin \theta \sin \phi & \rho \cos \theta \cos \phi \\ \sin \theta \sin \phi & \rho \cos \theta \sin \phi & \rho \sin \theta \cos \phi \\ \cos \phi & 0 & -\rho \sin \phi \end{vmatrix} = \rho^2 \sin \phi$$

> The $dx \, dy \, dz$ differentials therefore are transformed to $\rho^2 \sin(\varphi) \, d\rho \, d\theta \, d\varphi$.
> This yields the final integration formula:

$$\iiint_D f(x, y, z) \, dx \, dy \, dz = \iiint_T f(\rho \sin \phi \cos \theta, \rho \sin \phi \sin \theta, \rho \cos \phi)\rho^2 \sin \phi \, d\rho \, d\theta \, d\phi.$$

It's better to use this method in case of spherical domains **and** in case of functions that can be easily simplified, by the first fundamental relation of trigonometry, extended in $\mathbf{R}^3$ (please see Example 4b); in other cases it can be better to use cylindrical coordinates (please see Example 4c).

$$\iiint_T f(a, b, c)\rho^2 \sin \phi \, d\rho \, d\theta \, d\phi.$$

The extra ρ^2 and $\sin \phi$ come from the Jacobian.

In the following examples the roles of φ and θ have been reversed.

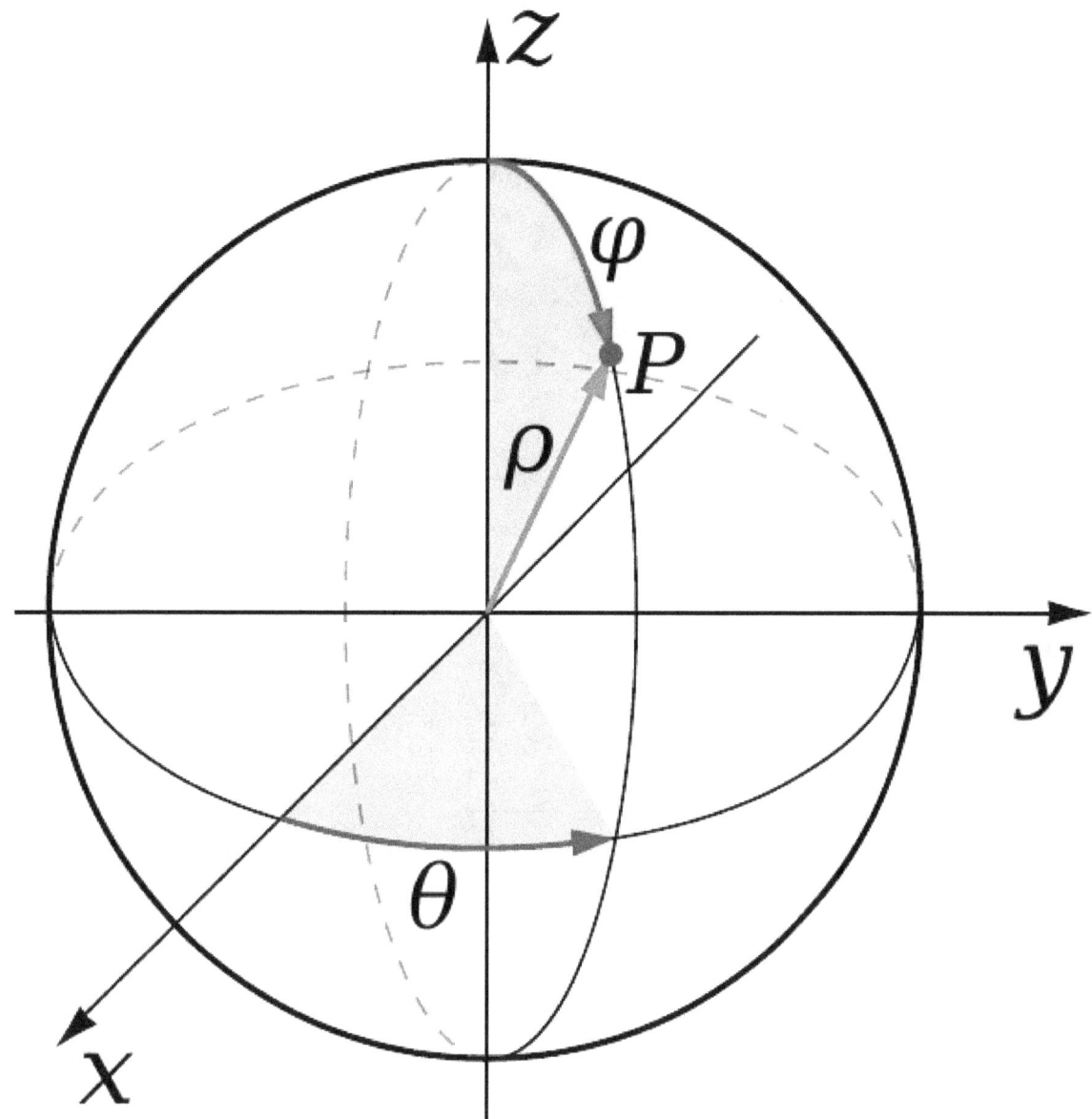

Spherical coordinates.

Example 4b. D is the same region as in Example 4a and $f(x, y, z) = x^2 + y^2 + z^2$ is the function to integrate. Its transformation is very easy:

$$f(\rho \sin \phi \cos \theta, \rho \sin \phi \sin \theta, \rho \cos \phi) = \rho^2,$$

while we know the intervals of the transformed region T from D:

$$(0 \leq \rho \leq 4, \ 0 \leq \phi \leq \pi, \ 0 \leq \theta \leq 2\pi).$$

Let's therefore apply the integration's formula:

$$\iiint_D (x^2 + y^2 + z^2) \, dx \, dy \, dz = \iiint_T \rho^2 \, \rho^2 \sin \theta \, d\rho \, d\theta \, d\phi.$$

and, developing, we get

$$\iiint_T \rho^4 \sin\theta \, d\rho \, d\theta \, d\phi = \int_0^\pi \sin\phi \, d\phi \int_0^4 \rho^4 d\rho \int_0^{2\pi} d\theta = 2\pi \int_0^\pi \sin\phi \left[\frac{\rho^5}{5}\right]_0^4 d\phi = 2\pi \left[\frac{\rho^5}{5}\right]_0^4 \left[-\cos\phi\right]_0^\pi = \frac{4096\pi}{5}.$$

Example 4c. The domain D is the ball with center in the origin and radius $3a$,

$$D = \{x^2 + y^2 + z^2 \leq 9a^2\}$$

and $f(x, y, z) = x^2 + y^2$ is the function to integrate.

Looking at the domain, it seems convenient to adopt the passage in spherical coordinates, in fact, the intervals of the variables that delimit the new T region are obviously:

$$0 \leq \rho \leq 3a, \ 0 \leq \phi \leq 2\pi, \ 0 \leq \theta \leq \pi.$$

However, applying the transformation, we get

$$f(x, y, z) = x^2 + y^2 \longrightarrow \rho^2 \sin^2\theta\cos^2\phi + \rho^2 \sin^2\theta\sin^2\phi = \rho^2 \sin^2\theta$$

Applying the formula for integration we would obtain:

$$\iiint_T \rho^2 \sin^2\theta \rho^2 \sin\theta \, d\rho \, d\theta \, d\phi = \iiint_T \rho^4 \sin^3\theta \, d\rho \, d\theta \, d\phi$$

which is very hard to solve. This problem will be solved by using the passage in cylindrical coordinates. The new T intervals are

$$0 \leq \rho \leq 3a, \ 0 \leq \phi \leq 2\pi, \ -\sqrt{9a^2 - \rho^2} \leq z \leq \sqrt{9a^2 - \rho^2};$$

the z interval has been obtained by dividing the ball in two hemispheres simply by solving the inequality from the formula of D (and then directly transforming $x^2 + y^2$ in ρ^2). The new function is simply ρ^2. Applying the integration formula

$$\iiint_T \rho^2 \rho \, d\rho d\phi dz$$

Then we get

$$\int_0^{2\pi} d\phi \int_0^{3a} \rho^3 d\rho \int_{-\sqrt{9a^2-\rho^2}}^{\sqrt{9a^2-\rho^2}} dz = 2\pi \int_0^{3a} 2\rho^3 \sqrt{9a^2 - \rho^2} \, d\rho$$

$$= -2\pi \int_{9a^2}^0 (9a^2 - t)\sqrt{t} \, dt \qquad\qquad t = 9a^2 - \rho^2$$

$$= 2\pi \int_0^{9a^2} \left(9a^2\sqrt{t} - t\sqrt{t}\right) dt$$

$$= 2\pi \left[\int_0^{9a^2} 9a^2\sqrt{t} \, dt - \int_0^{9a^2} t\sqrt{t} \, dt\right]$$

$$= 2\pi \left[9a^2 \frac{2}{3}t^{\frac{3}{2}} - \frac{2}{5}t^{\frac{5}{2}}\right]_0^{9a^2}$$

$$= 2 \cdot 27\pi a^5 \left(6 - \frac{18}{5}\right)$$

$$= \frac{648\pi}{5}a^5.$$

Thanks to the passage in cylindrical coordinates it was possible to reduce the triple integral to an easier one-variable integral.

See also the differential volume entry in nabla in cylindrical and spherical coordinates.

50.4 Examples

50.4.1 Double integral

Let us assume that we wish to integrate a multivariable function f over a region A.

$$A = \left\{(x, y) \in \mathbf{R}^2 \;:\; 11 \le x \le 14 \;;\; 7 \le y \le 10\right\} \text{ and } f(x, y) = x^2 + 4y$$

From this we formulate the double integral

$$\int_7^{10} \int_{11}^{14} (x^2 + 4y) \, dx \, dy$$

The inner integral is performed first, integrating with respect to x and taking y as a constant, as it is not the variable of integration. The result of this integral, which is a function depending only on y, is then integrated with respect to y.

$$
\begin{aligned}
\int_{11}^{14} (x^2 + 4y) \, dx &= \left(\frac{1}{3}x^3 + 4yx\right)\Big|_{x=11}^{x=14} \\
&= \frac{1}{3}(14)^3 + 4y(14) - \frac{1}{3}(11)^3 - 4y(11) \\
&= 471 + 12y
\end{aligned}
$$

We then integrate the result with respect to y.

$$
\begin{aligned}
\int_7^{10} (471 + 12y) \, dy &= (471y + 6y^2)\big|_{y=7}^{y=10} \\
&= 471(10) + 6(10)^2 - 471(7) - 6(7)^2 \\
&= 1719
\end{aligned}
$$

Observe that the order of integration is sometimes interchangeable:

$$
\begin{aligned}
\int_{11}^{14} \int_7^{10} (x^2 + 4y) \, dy \, dx &= \int_{11}^{14} \left(x^2 y + 2y^2\right)\Big|_{y=7}^{y=10} \, dx \\
&= \int_{11}^{14} (3x^2 + 102) \, dx \\
&= (x^3 + 102x)\Big|_{x=11}^{x=14} \\
&= 1719
\end{aligned}
$$

The instances where the order is interchangeable is determined by Fubini's Theorem.

50.4.2 Computing a volume

Using the methods previously described, it is possible to calculate the volumes of some common solids.

- **Cylinder**: The volume of a cylinder with height h and circular base of radius R can be calculated by integrating the constant function h over the circular base, using polar coordinates.

$$\text{Volume} = \int_0^{2\pi} d\phi \int_0^R h\rho \, d\rho = h2\pi \left[\frac{\rho^2}{2} \right]_0^R = \pi R^2 h$$

This is in agreement with the formula

$$\text{Volume} = \text{area base} \times height$$

- **Sphere**: The volume of a sphere with radius R can be calculated by integrating the constant function 1 over the sphere, using spherical coordinates.

$$
\begin{aligned}
\text{Volume} &= \iiint_D f(x, y, z) \, dx \, dy \, dz \\
&= \iiint_D 1 \, dV \\
&= \iiint_S \rho^2 \sin\phi \, d\rho \, d\theta \, d\phi \\
&= \int_0^{2\pi} d\theta \int_0^\pi \sin\phi \, d\phi \int_0^R \rho^2 \, d\rho \\
&= 2\pi \int_0^\pi \sin\phi \, d\phi \int_0^R \rho^2 \, d\rho \\
&= 2\pi \int_0^\pi \sin\phi \frac{R^3}{3} \, d\phi \\
&= \frac{2}{3}\pi R^3 [-\cos\phi]_0^\pi = \frac{4}{3}\pi R^3.
\end{aligned}
$$

- **Tetrahedron** (triangular pyramid or 3-simplex): The volume of a tetrahedron with its apex at the origin and edges of length l along the x, y and z axes can be calculated by integrating the constant function 1 over the tetrahedron.

$$
\begin{aligned}
\text{Volume} &= \int_0^\ell dx \int_0^{\ell-x} dy \int_0^{\ell-x-y} dz \\
&= \int_0^\ell dx \int_0^{\ell-x} (\ell - x - y) \, dy \\
&= \int_0^\ell \left[\ell^2 - 2\ell x + x^2 - \frac{(\ell-x)^2}{2} \right] dx \\
&= \ell^3 - \ell\ell^2 + \frac{\ell^3}{3} - \left[\frac{\ell^2 x}{2} - \frac{\ell x^2}{2} + \frac{x^3}{6} \right]_0^\ell \\
&= \frac{\ell^3}{3} - \frac{\ell^3}{6} = \frac{\ell^3}{6}
\end{aligned}
$$

This is in agreement with the formula

$$\text{Volume} = \frac{1}{3} \times \text{area base} \times \text{height} = \frac{1}{3} \times \frac{\ell^2}{2} \times \ell = \frac{\ell^3}{6}.$$

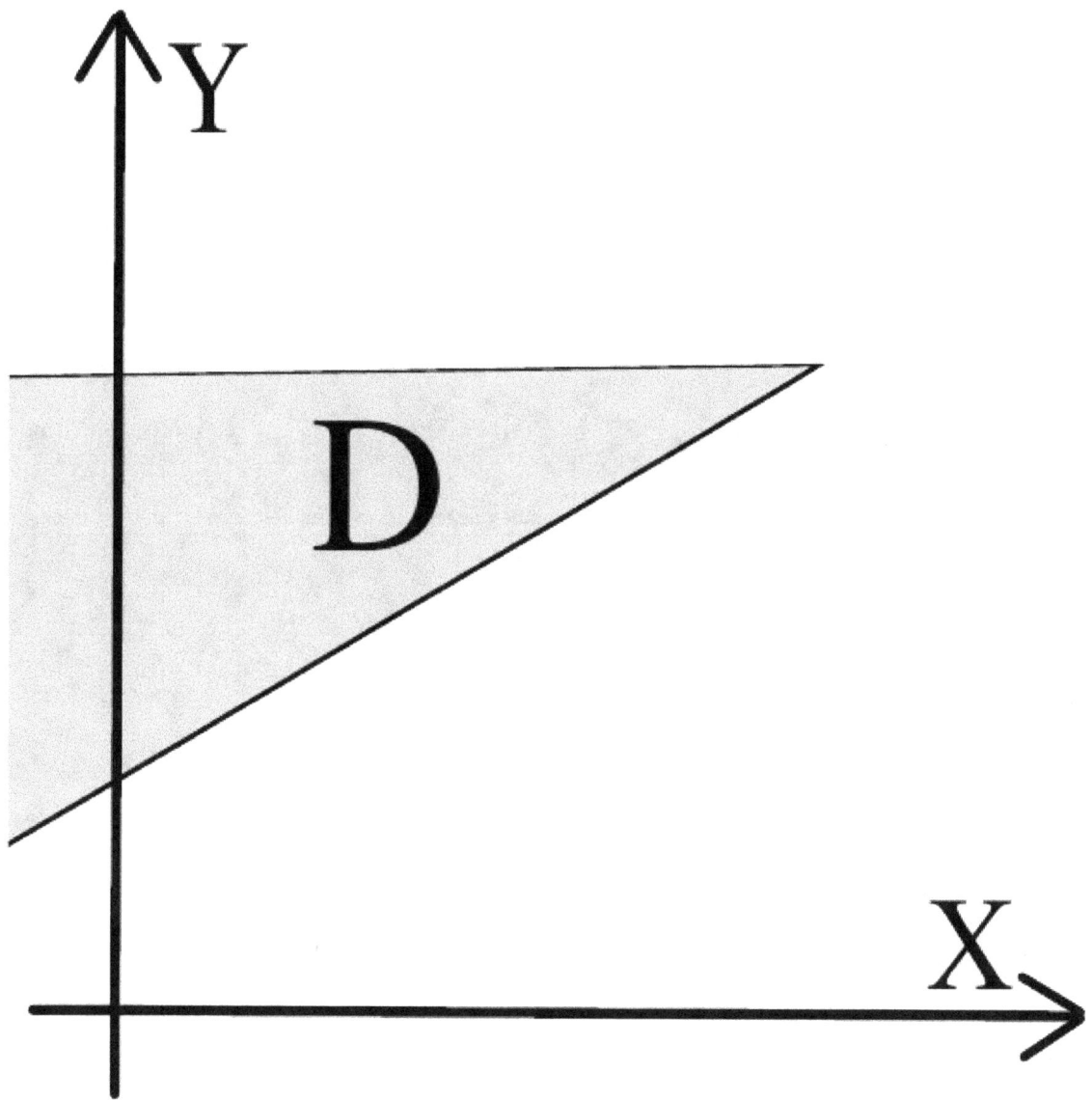

Example of an improper domain.

50.5   Multiple improper integral

In case of unbounded domains or functions not bounded near the boundary of the domain, we have to introduce the double improper integral or the **triple improper integral**.

50.6   Multiple integrals and iterated integrals

See also: Order of integration (calculus)

Fubini's theorem states that if[4]

$$\iint_{A \times B} |f(x,y)|\, d(x,y) < \infty,$$

that is, if the integral is absolutely convergent, then the multiple integral will give the same result as the iterated integral,

$$\iint_{A \times B} f(x,y)\, d(x,y) = \int_A \left(\int_B f(x,y)\, dy \right) dx = \int_B \left(\int_A f(x,y)\, dx \right) dy.$$

In particular this will occur if $|f(x, y)|$ is a bounded function and A and B are bounded sets.

If the integral is not absolutely convergent, care is needed not to confuse the concepts of *multiple integral* and *iterated integral*, especially since the same notation is often used for either concept. The notation

$$\int_0^1 \int_0^1 f(x,y)\, dy\, dx$$

means, in some cases, an iterated integral rather than a true double integral. In an iterated integral, the outer integral

$$\int_0^1 \cdots dx$$

is the integral with respect to x of the following function of x:

$$g(x) = \int_0^1 f(x,y)\, dy.$$

A double integral, on the other hand, is defined with respect to area in the xy-plane. If the double integral exists, then it is equal to each of the two iterated integrals (either "$dy\, dx$" or "$dx\, dy$") and one often computes it by computing either of the iterated integrals. But sometimes the two iterated integrals exist when the double integral does not, and in some such cases the two iterated integrals are different numbers, i.e., one has

$$\int_0^1 \int_0^1 f(x,y)\, dy\, dx \neq \int_0^1 \int_0^1 f(x,y)\, dx\, dy.$$

This is an instance of rearrangement of a conditionally convergent integral.

The notation

$$\int_{[0,1] \times [0,1]} f(x,y)\, dx\, dy$$

may be used if one wishes to be emphatic about intending a double integral rather than an iterated integral.

50.7   Some practical applications

Quite generally, just as in one variable, one can use the multiple integral to find the average of a function over a given set. Given a set $D \subseteq \mathbf{R}^n$ and an integrable function f over D, the average value of f over its domain is given by

$$\bar{f} = \frac{1}{m(D)} \int_D f(x)\,dx.$$

where $m(D)$ is the measure of D.

Additionally, multiple integrals are used in many applications in physics. The examples below also show some variations in the notation.

In mechanics, the moment of inertia is calculated as the volume integral (triple integral) of the density weighed with the square of the distance from the axis:

$$I_z = \iiint_V \rho r^2\,dV.$$

The gravitational potential associated with a mass distribution given by a mass measure dm on three-dimensional Euclidean space $\mathbf{R}^3$ is[5]

$$V(\mathbf{x}) = - \iiint_{\mathbf{R}^3} \frac{G}{|\mathbf{x} - \mathbf{y}|}\,dm(\mathbf{y}).$$

If there is a continuous function $\rho(\mathbf{x})$ representing the density of the distribution at $\mathbf{x}$, so that $dm(\mathbf{x}) = \rho(\mathbf{x})d^3\mathbf{x}$, where $d^3\mathbf{x}$ is the Euclidean volume element, then the gravitational potential is

$$V(\mathbf{x}) = - \iiint_{\mathbf{R}^3} \frac{G}{|\mathbf{x} - \mathbf{y}|}\,\rho(\mathbf{y})\,d^3\mathbf{y}.$$

In electromagnetism, Maxwell's equations can be written using multiple integrals to calculate the total magnetic and electric fields.[6] In the following example, the electric field produced by a distribution of charges given by the volume charge density $\rho(\vec{r})$ is obtained by a *triple integral* of a vector function:

$$\vec{E} = \frac{1}{4\pi\epsilon_0} \iiint \frac{\vec{r} - \vec{r}'}{\|\vec{r} - \vec{r}'\|^3}\,\rho(\vec{r}')\,\mathrm{d}^3 r'.$$

This can also be written as an integral with respect to a signed measure representing the charge distribution.

50.8   See also

- Main analysis theorems that relate multiple integrals:

  - Divergence theorem
  - Stokes' theorem
  - Green's theorem

50.9   References

[1] Stewart, James (2008). *Calculus: Early Transcendentals*, 6th ed., Brooks Cole Cengage Learning. ISBN 978-0-495-01166-8

[2] Larson/Edwards (2014)/ *Multivariable Calculus*, 10th ed., Cengage Learning. ISBN 978-1-285-08575-3

[3] Rudin, Walter. *Principles of Mathematical Analysis*. Walter Rudin Student Series in Advanced Mathematics (3rd ed.). McGraw–Hill. ISBN 978-0-07-054235-8.

[4] Jones, Frank (2001), *Lebesgue Integration on Euclidean Space*, Jones and Bartlett publishers, pp. 527–529.

[5] Kibble, Tom W.B.; Berkshire, Frank H. (2004). *Classical Mechanics* (5th ed.). Imperial College Press. ISBN 978-1-86094-424-6.

[6] Jackson, John D. (1998). *Classical Electrodynamics* (3rd ed.). Wiley. ISBN 0-471-30932-X.

50.10   Further reading

- Robert A. Adams - Calculus: A Complete Course (5th Edition) ISBN 0-201-79131-5.

- R.K.Jain and S.R.K Iyengar- Advanced Engineering Mathematics (Third edition) 2009, Narosa Publishing House ISBN 978-81-7319-730-7

50.11   External links

- Weisstein, Eric W., "Multiple Integral", *MathWorld*.

- L.D. Kudryavtsev (2001), "Multiple integral", in Hazewinkel, Michiel, *Encyclopedia of Mathematics*, Springer, ISBN 978-1-55608-010-4

- Mathematical Assistant on Web online evaluation of double integrals in Cartesian coordinates and polar coordinates (includes intermediate steps in the solution, powered by Maxima (software))

Chapter 51

Skew-symmetric matrix

In mathematics, and in particular linear algebra, a **skew-symmetric** (or **antisymmetric** or **antimetric**[11]) **matrix** is a square matrix A whose transpose is also its negative; that is, it satisfies the condition $-A = A^\mathrm{T}$. If the entry in the i th row and j th column is aij, i.e. $A = (aij)$ then the skew symmetric condition is $aij = -aji$. For example, the following matrix is skew-symmetric:

$$\begin{bmatrix} 0 & 2 & -1 \\ -2 & 0 & -4 \\ 1 & 4 & 0 \end{bmatrix}.$$

51.1   Properties

We assume that the underlying field is not of characteristic 2; that is, that $1 + 1 \neq 0$ where 1 denotes the multiplicative identity and 0 the additive identity of the given field. Otherwise, a skew-symmetric matrix is just the same thing as a symmetric matrix.

Sums and scalar multiples of skew-symmetric matrices are again skew-symmetric. Hence, the skew-symmetric matrices form a vector space. Its dimension is $n(n-1)/2$.

Let Mat$n$ denote the space of $n \times n$ matrices. A skew-symmetric matrix is determined by $n(n-1)/2$ scalars (the number of entries above the main diagonal); a symmetric matrix is determined by $n(n+1)/2$ scalars (the number of entries on or above the main diagonal). Let Skew$n$ denote the space of $n \times n$ skew-symmetric matrices and Sym$n$ denote the space of $n \times n$ symmetric matrices. If $A \in$ Mat$n$ then

$$A = \frac{1}{2}(A - A^\mathrm{T}) + \frac{1}{2}(A + A^\mathrm{T}).$$

Notice that $\frac{1}{2}(A - A^\mathrm{T}) \in$ Skew$n$ and $\frac{1}{2}(A + A^\mathrm{T}) \in$ Sym$n$. This is true for every square matrix A with entries from any field whose characteristic is different from 2. Then, since Mat$n$ = Skew$n$ + Sym$n$ and Skew$n$ ∩ Sym$n$ = {0},

$$\mathrm{Mat}_n = \mathrm{Skew}_n \oplus \mathrm{Sym}_n,$$

where $\oplus$ denotes the direct sum.

Denote with $\langle \cdot, \cdot \rangle$ the standard inner product on $\mathbf{R}^n$. The real n-by-n matrix A is skew-symmetric if and only if

$$\langle Ax, y \rangle = -\langle x, Ay \rangle \quad \forall x, y \in \mathbb{R}^n.$$

415

This is also equivalent to $\langle x, Ax \rangle = 0$ for all x (one implication being obvious, the other a plain consequence of $\langle x + y, A(x + y) \rangle = 0$ for all x and y). Since this definition is independent of the choice of basis, skew-symmetry is a property that depends only on the linear operator A and a choice of inner product.

All main diagonal entries of a skew-symmetric matrix must be zero, so the trace is zero. If $A = (aij)$ is skew-symmetric, $aij = -aji$; hence $aii = 0$.

3x3 skew symmetric matrices can be used to represent cross products as matrix multiplications.

51.1.1   Determinant

Let A be a $n{\times}n$ skew-symmetric matrix. The determinant of A satisfies

$$\det(A) = \det(A^{\mathsf{T}}) = \det(-A) = (-1)^n \det(A).$$

In particular, if n is odd, and since the underlying field is not of characteristic 2, the determinant vanishes. This result is called **Jacobi's theorem**, after Carl Gustav Jacobi (Eves, 1980).

The even-dimensional case is more interesting. It turns out that the determinant of A for n even can be written as the square of a polynomial in the entries of A, which was first proved by Cayley:[2]

$$\det(A) = \text{Pf}(A)^2.$$

This polynomial is called the *Pfaffian* of A and is denoted Pf(A). Thus the determinant of a real skew-symmetric matrix is always non-negative. However this last fact can be proved in an elementary way as follows: the eigenvalues of a real skew-symmetric matrix are purely imaginary (see below) and to every eigenvalue there corresponds the conjugate eigenvalue with the same multiplicity; therefore, as the determinant is the product of the eigenvalues, each one repeated according to its multiplicity, it follows at once that the determinant, if it is not 0, is a positive real number.

The number of distinct terms $s(n)$ in the expansion of the determinant of a skew-symmetric matrix of order n has been considered already by Cayley, Sylvester, and Pfaff. Due to cancellations, this number is quite small as compared the number of terms of a generic matrix of order n, which is $n!$. The sequence $s(n)$ (sequence A002370 in OEIS) is

$$1, 0, 1, 0, 6, 0, 120, 0, 5250, 0, 395010, 0, \dots$$

and it is encoded in the exponential generating function

$$\sum_{n=0}^{\infty} \frac{s(n)}{n!} x^n = (1 - x^2)^{-\frac{1}{4}} \exp\left(\frac{x^2}{4}\right).$$

The latter yields to the asymptotics (for n even)

$$s(n) = \pi^{-\frac{1}{2}} 2^{\frac{3}{4}} \Gamma\left(3/4\right) (n/e)^{n - \frac{1}{4}} \left(1 + O(\frac{1}{n})\right).$$

The number of positive and negative terms are approximatively a half of the total, although their difference takes larger and larger positive and negative values as n increases (sequence A167029 in OEIS).

51.1.2   Spectral theory

Since a matrix is similar to its own transpose, they must have the same eigenvalues. It follows that the eigenvalues of a skew-symmetric matrix always come in pairs $\pm\lambda$ (except in the odd-dimensional case where there is an additional unpaired

0 eigenvalue). From the spectral theorem, for a real skew-symmetric matrix the nonzero eigenvalues are all pure imaginary and thus are of the form $i\lambda_1, -i\lambda_1, i\lambda_2, -i\lambda_2, \ldots$ where each of the λ_k are real.

Real skew-symmetric matrices are normal matrices (they commute with their adjoints) and are thus subject to the spectral theorem, which states that any real skew-symmetric matrix can be diagonalized by a unitary matrix. Since the eigenvalues of a real skew-symmetric matrix are imaginary it is not possible to diagonalize one by a real matrix. However, it is possible to bring every skew-symmetric matrix to a block diagonal form by a special orthogonal transformation.[3] Specifically, every $2n \times 2n$ real skew-symmetric matrix can be written in the form $A = Q \Sigma Q^T$ where Q is orthogonal and

$$\Sigma = \begin{bmatrix} 0 & \lambda_1 & 0 & \cdots & & 0 & \\ -\lambda_1 & 0 & & & & & \\ & & 0 & \lambda_2 & & 0 & \\ & & -\lambda_2 & 0 & & & \\ & \vdots & & & \ddots & \vdots & \\ & 0 & & \cdots & 0 & \lambda_r & \\ & & & & -\lambda_r & 0 & \\ & & & & & & 0 & \\ & & & & & & & \ddots \\ & & & & & & & & 0 \end{bmatrix}$$

for real λ_k. The nonzero eigenvalues of this matrix are $\pm i\lambda_k$. In the odd-dimensional case Σ always has at least one row and column of zeros.

More generally, every complex skew-symmetric matrix can be written in the form $A = U \Sigma U^T$ where U is unitary and Σ has the block-diagonal form given above with complex λ_k. This is an example of the Youla decomposition of a complex square matrix.[4]

51.2   Skew-symmetric and alternating forms

A **skew-symmetric form** φ on a vector space V over a field K of arbitrary characteristic is defined to be a bilinear form

$$\varphi : V \times V \to K$$

such that for all v, w in V,

$$\varphi(v, w) = -\varphi(w, v).$$

This defines a form with desirable properties for vector spaces over fields of characteristic not equal to 2, but in a vector space over a field of characteristic 2, the definition is equivalent to that of a symmetric form, as every element is its own additive inverse.

Where the vector space V is over a field of arbitrary characteristic including characteristic 2, we may define an **alternating form** as a bilinear form φ such that for all vectors v in V

$$\varphi(v, v) = 0.$$

This is equivalent to a skew-symmetric form when the field is not of characteristic 2 as seen from

$$0 = \varphi(v + w, v + w) = \varphi(v, v) + \varphi(v, w) + \varphi(w, v) + \varphi(w, w) = \varphi(v, w) + \varphi(w, v),$$

whence,

$\varphi(v, w) = -\varphi(w, v)$.

A bilinear form φ will be represented by a matrix A such that $\varphi(v, w) = v^{\mathrm{T}} A w$, once a basis of V is chosen, and conversely an $n \times n$ matrix A on K^n gives rise to a form sending (v, w) to $v^{\mathrm{T}} A w$. For each of symmetric, skew-symmetric and alternating forms, the representing matrices are symmetric, skew-symmetric and alternating respectively.

51.3   Infinitesimal rotations

Further information: Generators of rotations
Further information: Infinitesimal rotations
Further information: Infinitesimal rotation tensor

Skew-symmetric matrices over the field of real numbers form the tangent space to the real orthogonal group $O(n)$ at the identity matrix; formally, the special orthogonal Lie algebra. In this sense, then, skew-symmetric matrices can be thought of as *infinitesimal rotations*.

Another way of saying this is that the space of skew-symmetric matrices forms the Lie algebra $o(n)$ of the Lie group $O(n)$. The Lie bracket on this space is given by the commutator:

$$[A, B] = AB - BA.$$

It is easy to check that the commutator of two skew-symmetric matrices is again skew-symmetric:

$$[A, B]^{\mathrm{T}} = B^{\mathrm{T}} A^{\mathrm{T}} - A^{\mathrm{T}} B^{\mathrm{T}} = BA - AB = -[A, B].$$

The matrix exponential of a skew-symmetric matrix A is then an orthogonal matrix R:

$$R = \exp(A) = \sum_{n=0}^{\infty} \frac{A^n}{n!}.$$

The image of the exponential map of a Lie algebra always lies in the connected component of the Lie group that contains the identity element. In the case of the Lie group $O(n)$, this connected component is the special orthogonal group $SO(n)$, consisting of all orthogonal matrices with determinant 1. So $R = \exp(A)$ will have determinant $+1$. Moreover, since the exponential map of a connected compact Lie group is always surjective, it turns out that *every* orthogonal matrix with unit determinant can be written as the exponential of some skew-symmetric matrix. In the particular important case of dimension $n=2$, the exponential representation for an orthogonal matrix reduces to the well-known polar form of a complex number of unit modulus. Indeed, if $n=2$, a special orthogonal matrix has the form

$$\begin{bmatrix} a & -b \\ b & a \end{bmatrix},$$

with $a^2+b^2=1$. Therefore, putting $a=\cos\theta$ and $b=\sin\theta$, it can be written

$$\begin{bmatrix} \cos\theta & -\sin\theta \\ \sin\theta & \cos\theta \end{bmatrix} = \exp\left(\theta \begin{bmatrix} 0 & -1 \\ 1 & 0 \end{bmatrix}\right),$$

which corresponds exactly to the polar form $\cos\theta + i\sin\theta = e^{i\theta}$ of a complex number of unit modulus.

The exponential representation of an orthogonal matrix of order n can also be obtained starting from the fact that in dimension n any special orthogonal matrix R can be written as $R = Q S Q^T$, where Q is orthogonal and S is a block diagonal matrix with $\lfloor n/2 \rfloor$ blocks of order 2, plus one of order 1 if n is odd; since each single block of order 2 is also an orthogonal matrix, it admits an exponential form. Correspondingly, the matrix S writes as exponential of a skew-symmetric block matrix Σ of the form above, $S = \exp(\Sigma)$, so that $R = Q \exp(\Sigma)Q^T = \exp(Q \Sigma Q^T)$, exponential of the skew-symmetric matrix $Q \Sigma Q^T$. Conversely, the surjectivity of the exponential map, together with the above-mentioned block-diagonalization for skew-symmetric matrices, implies the block-diagonalization for orthogonal matrices.

51.4 Coordinate-free

More intrinsically (i.e., without using coordinates), skew-symmetric linear transformations on a vector space V with an inner product may be defined as the bivectors on the space, which are sums of simple bivectors (2-blades) $v \wedge w$. The correspondence is given by the map $v \wedge w \mapsto v^* \otimes w - w^* \otimes v$, where v^* is the covector dual to the vector v; in orthonormal coordinates these are exactly the elementary skew-symmetric matrices. This characterization is used in interpreting the curl of a vector field (naturally a 2-vector) as an infinitesimal rotation or "curl", hence the name.

51.5 Skew-symmetrizable matrix

An n-by-n matrix A is said to be **skew-symmetrizable** if there exist an invertible diagonal matrix D and skew-symmetric matrix S such that $S = DA$. For **real** n-by-n matrices, sometimes the condition for D to have positive entries is added.[5]

51.6 See also

- Symmetric matrix

- Skew-Hermitian matrix

- Symplectic matrix

- Symmetry in mathematics

51.7 References

[1] Richard A. Reyment; K. G. Jöreskog; Leslie F. Marcus (1996). *Applied Factor Analysis in the Natural Sciences*. Cambridge University Press. p. 68. ISBN 0-521-57556-7.

[2] Cayley, Arthur (1847). "Sur les determinants gauches" [On skew determinants]. *Crelle's Journal* **38**: 93–96. Reprintend in Cayley, A. (2009). "Sur les Déterminants Gauches". *The Collected Mathematical Papers* **1**. p. 410. doi:10.1017/CBO9780511 70367-6.ISBN978-0-511-70367-6.

[3] Voronov, Theodore. "Pfaffian." Concise Encyclopedia of Supersymmetry. Springer Netherlands, 2003. 298-298.

[4] Youla, D. C. (1961). "A normal form for a matrix under the unitary congruence group". *Canad. J. Math.* **13**: 694–704. doi:10.4153/CJM-1961-059-8.

[5] Fomin, Sergey; Zelevinsky, Andrei (2001). "Cluster algebras I: Foundations". arXiv:math/0104151v1.

51.8 Further reading

- Eves, Howard (1980). *Elementary Matrix Theory*. Dover Publications. ISBN 978-0-486-63946-8.

- Suprunenko, D. A. (2001), "Skew-symmetric matrix", in Hazewinkel, Michiel, *Encyclopedia of Mathematics*, Springer, ISBN 978-1-55608-010-4

- Aitken, A. C. (1944). "On the number of distinct terms in the expansion of symmetric and skew determinants.". *Edinburgh Math. Notes*.

51.9 External links

- "Antisymmetric matrix". *Wolfram Mathworld*.

- Benner, Peter; Kressner, Daniel. "HAPACK – Software for (Skew-)Hamiltonian Eigenvalue Problems".

- Ward, R. C.; Gray, L. J. (1978). "Algorithm 530: An Algorithm for Computing the Eigensystem of Skew-Symmetric Matrices and a Class of Symmetric Matrices [F2]". *ACM Transactions on Mathematical Software* **4** (3): 286. doi:10.1145/355791.355799. Fortran Fortran90

Chapter 52

Antisymmetric relation

In mathematics, a binary relation R on a set X is **antisymmetric** if there is no pair of distinct elements of X each of which is related by R to the other. More formally, R is antisymmetric precisely if for all a and b in X

if $R(a,b)$ and $R(b,a)$, then $a = b$,

or, equivalently,

if $R(a,b)$ with $a \neq b$, then $R(b,a)$ must not hold.

As a simple example, the divisibility order on the natural numbers is an antisymmetric relation. And what antisymmetry means here is that the only way each of two numbers can be divisible by the other is if the two are, in fact, the same number; equivalently, if n and m are distinct and n is a factor of m, then m cannot be a factor of n.

In mathematical notation, this is:

$$\forall a, b \in X, \ R(a,b) \wedge R(b,a) \ \Rightarrow \ a = b$$

or, equivalently,

$$\forall a, b \in X, \ R(a,b) \wedge a \neq b \Rightarrow \neg R(b,a).$$

The usual order relation $\leq$ on the real numbers is antisymmetric: if for two real numbers x and y both inequalities $x \leq y$ and $y \leq x$ hold then x and y must be equal. Similarly, the subset order $\subseteq$ on the subsets of any given set is antisymmetric: given two sets A and B, if every element in A also is in B and every element in B is also in A, then A and B must contain all the same elements and therefore be equal:

$$A \subseteq B \wedge B \subseteq A \Rightarrow A = B$$

Partial and total orders are antisymmetric by definition. A relation can be both symmetric and antisymmetric (e.g., the equality relation), and there are relations which are neither symmetric nor antisymmetric (e.g., the "preys on" relation on biological species).

Antisymmetry is different from asymmetry, which requires both antisymmetry and irreflexivity.

52.1 Examples

The relation "x is even, y is odd" between a pair (x, y) of integers is antisymmetric:

x is even and y is odd

√ Is true for this case (need not be true for all cases)

Ⓩ Must be false if the check mark with the same number (z) is true for it to be an antisymmetric relation

z√ Is true for this case and requires the circle with the same number (z) to also be true for it to be a symmetric relation

Every asymmetric relation is also an antisymmetric relation.

52.2 See also

- Symmetric relation
- Asymmetric relation
- Symmetry in mathematics

52.3 References

- Weisstein, Eric W., "Antisymmetric Relation", *MathWorld*.

- Lipschutz, Seymour; Marc Lars Lipson (1997). *Theory and Problems of Discrete Mathematics*. McGraw-Hill. p. 33. ISBN 0-07-038045-7.

Chapter 53

Antisymmetry

For the property of matrices, see Skew-symmetric matrix. For the property of mathematical relations, see Antisymmetric relation. For other uses, see Antisymmetric (disambiguation).

In linguistics, **antisymmetry** is a theory of syntactic linearization presented in Richard Kayne's 1994 monograph *The Antisymmetry of Syntax*.[1] The crux of this theory is that hierarchical structure in natural language maps universally onto a particular surface linearization, namely specifier-head-complement branching order. The theory derives a version of X-bar theory. Kayne hypothesizes that all phrases whose surface order is not specifier-head-complement have undergone movements that disrupt this underlying order. Subsequently, there have also been attempts at deriving specifier-complement-head as the basic word order.[2]

Antisymmetry as a principle of word order is reliant on assumptions that many theories of syntax dispute, e.g. constituency structure (as opposed to dependency structure), X-bar notions such as specifier and complement, and the existence of ordering altering mechanisms such as movement and/or copying.

53.1 Asymmetric c-command

The theory is based on a notion of *asymmetric c-command*, c-command being a relation between nodes in a tree originally defined by Tanya Reinhart (1976).[3] Kayne uses a simple definition of c-command based on the "first node up". However, the definition is complicated by his use of a "segment/category distinction". A category is a kind of extended node; if two directly connected nodes in a tree have the same label, these two nodes are both segments of a single category. C-command is defined in terms of categories using the notion of "exclusion". A category excludes all categories not dominated by *both* its segments. A c-commands B if every category that dominates A also dominates B, and A excludes B. The following tree illustrates these concepts:

AP_1 and AP_2 are both segments of a single category. AP does not c-command BP because it does not exclude BP. CP does not c-command BP because both segments of AP do not dominate BP (so it is not the case that every category that dominates CP dominates BP). BP c-commands CP and A. A c-commands C. The definitions above may perhaps be thought to allow BP to c-command AP, but a c-command relation is not usually assumed to hold between two such categories, and for the purposes of antisymmetry, the question of whether BP c-commands AP is in fact moot.

(The above is not an exhaustive list of c-command relations in the tree, but covers all of those that are significant in the following exposition.)

Asymmetric c-command is the relation that holds between two categories, A and B, if A c-commands B but B does not c-command A. This relationship is a primitive in Kayne's theory of linearization, the process that converts a tree structure into a flat (structureless) string of terminal nodes.

53.2   Precedence and asymmetric c-command

Informally, Kayne's theory states that if a nonterminal category A c-commands another nonterminal category B, all the terminal nodes dominated by A must precede all of the terminal nodes dominated by B (this statement is commonly referred to as the "Linear Correspondence Axiom" or LCA). Moreover, this principle must suffice to establish a *complete* and *consistent* ordering of all terminal nodes — if it cannot consistently order all of the terminal nodes in a tree, the tree is illicit. Consider the following tree:

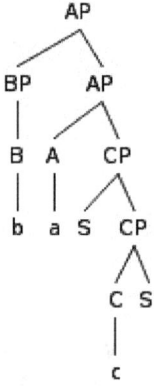

(S and S' may either be simplex structures like BP, or complex structures with specifiers and complements like CP.)

In this tree, the set of pairs of nonterminal categories such that the first member of the pair asymmetrically c-commands the second member is as follows: {<BP, A>, <BP, CP>, <A, CP>}. This gives rise to the total ordering: <b, a, c>.

As a result, there is no right adjunction, and hence in practice no rightward movement either.[4] Furthermore, the underlying order must be specifier-head-complement.

53.3   Derivation of X-bar theory

The example tree in the first section of this article is in accordance with X-bar theory (with the exception that [Spec,CP] is treated as an adjunct). It can be seen that removing any of the structure in the tree (e.g. deleting the C dominating the 'c' terminal, so that the complement of A is [CP c]) will destroy the asymmetric c-command relations necessary for linearly ordering the terminals of the tree.

53.4   The universal order

Kayne notes that his theory permits either a universal specifier-head-complement order or a universal complement-head-specifier order, depending on whether asymmetric c-command establishes precedence or subsequence (S-H-C results from precedence) (pp. 35–36)[1] He argues that there are good empirical grounds for preferring S-H-C as the universal underlying order, since the typologically most widely attested order is for specifiers to precede heads and complements (though the order of heads and complements themselves is relatively free). He further argues that a movement approach to deriving non S-H-C orders is appropriate, since it derives asymmetries in typology (such as the fact that "verb second" languages such as German are not mirrored by any known "verb second-from-last" languages).

53.5   Derived orders: the case of Japanese wh-questions

Perhaps the biggest challenge for antisymmetry is to explain the wide variety of different surface orders across languages. Any deviation from Spec-Head-Comp order (which implies overall Subject-Verb-Object order, if objects are complements) must be explained by movement. Kayne argues that in some cases, the need for extra movements (previously unnecessary because different underlying orders were assumed for different languages) can actually explain some mysterious typological generalizations. His explanation for the lack of wh-movement in Japanese is the most striking example of this. From the mid-1980s onwards, the standard analysis of wh-movement involved the wh-phrase moving leftward to a position on the left edge of the clause called [Spec,CP] (i.e., the specifier of the CP phrase). Thus, a derivation of the English question *What did John buy?* would proceed roughly as follows:

> [CP {Spec,CP position} John did buy what]
>
> *wh-movement* →
>
> [CP What did John buy]

The Japanese equivalent of this sentence is as follows[5] (note the lack of wh-movement):

Japanese has an overt "question particle" (*ka*), which appears at the end of the sentence in questions. It is generally assumed that languages such as English have a "covert" (i.e. phonologically null) equivalent of this particle in the 'C' position of the clause — the position just to the right of [Spec,CP]. This particle is overtly realised in English by movement of an auxiliary to C (in the case of the example above, by movement of *did* to C). Why is it that this particle is on the left edge of the clause in English, but on the right edge in Japanese? Kayne suggests that in Japanese, the *whole of the clause* (apart from the question particle in C) has moved to the [Spec,CP] position. So, the structure for the Japanese example above is something like the following:

> [CP [John-wa nani-o kaimasita] C ka

Now it is clear why Japanese does not have wh-movement — the [Spec,CP] position is already filled, so no wh-phrase can move to it. We therefore predict a seemingly obscure relationship between surface word order and the possibility of wh-movement. A possible alternative to the antisymmetric explanation could be based on the difficulty of parsing languages with rightward movement.[6]

53.6   Dynamic antisymmetry

A weak version of the theory of antisymmetry (Dynamic antisymmetry) has been proposed by Andrea Moro, which allows the generation of non-LCA compatible structures (points of symmetry) before the hierarchical structure is linearized at Phonetic Form. The unwanted structures are then rescued by movement: deleting the phonetic content of the moved element would neutralize the linearization problem.[7] From this perspective, Dynamic Antisymmetry aims at unifying movement and phrase structure, which otherwise would be two independent properties that characterize all human language grammars.

53.7   Antisymmetry and ternary branching

In a recent manuscript, Kayne (2010) has proposed recasting the antisymmetry of natural language as a condition on "Merge", the operation which combines two linguistic elements into one complex linguistic element.[8] Kayne proposes that merging a head H and its complement C yields an ordered pair $< H, C >$ (rather than the standard symmetric set $\{H,C\}$). $< H, C >$ involves immediate temporal precedence (or immediate linear precedence), so that H immediately precedes (i-precedes) C. Kayne proposes furthermore that when a specifier S merges, it forms an ordered pair with the head directly, $< S, H >$, or S i-precedes H. Invoking i-precedence prevents more than two elements from merging with H; only one element can i-precede H (the specifier), and H can i-precede only one element (the complement).

Kayne (2010) notes that $< S, H >, < H, C >$ is not mappable to a tree structure, since H would have two mothers, and that it has the consequence that $< S, H >$ and $< H, C >$ would seem to be constituents. He suggests that $< S, H >, < H, C >$ is replaced by $< S, H, C >$, "with an ordered triple replacing the two ordered pairs and then being mappable to a ternary-branching tree" (pp. 17). Kayne goes on to say, "This would lead to seeing my [(1981)][9] arguments for binary branching to have two subcomponents, the first being the claim that syntax is n-ary branching with n having a single value, the second being that that value is 2. Mapping $[< S, H >, < H, C >$ to $< S, H, C >]$ would retain the first subcomponent and replace 2 by 3 in the second, arguably with no loss in restrictiveness."

53.8   Theoretical arguments

According to Kayne's Antisymmetry theory, there is no head-directionality parameter as such: it is claimed that at an underlying level, all languages are head-initial. In fact, it is argued that all languages have the underlying order Specifier-Head-Complement. Deviations from this order are accounted for by different syntactic movements applied by languages.

Kayne argues that a theory that allows both directionalities would imply an absence of asymmetries between languages, whereas in fact languages are found not to be symmetrical in many respects. Some examples of linguistic asymmetries which may be cited in support of the theory (although they do not concern head direction) are listed below.

- Hanging topics appear at the start of sentences, as in "Henry – I've known that guy for a long time".[10] They are not attested at the end of sentences.[11]

- Number agreement is stronger when the noun phrase precedes the verb (Greenberg's Universal 33). Examples of this are found in English sentences such as *There's books on the table*, where the verb frequently fails to agree with the following plural noun, and in French and Italian compound tenses,[12] where the past participle may agree with a preceding direct object but not with a following one.

- Relative clauses which precede the noun (as in Chinese and Japanese) tend to differ from those that follow the noun: they more often lack complementizers (akin to English *that*) or relative pronouns, and are more likely to be non-finite (this can be found, for example, in Quecha.[13])

- Other areas in which asymmetries are found, according to Kayne, include clitics and clitic dislocation, serial verb constructions, coordination, and forward and backward pronominalization.

In arguing for a universal underlying Head-Complement order, Kayne uses the concept of a probe-goal search (based on the Minimalist program). The idea of probes and goals in syntax is that a head acts as a probe and looks for a goal, namely its complement. Kayne proposes that the direction of the probe-goal search must share the direction of language parsing and production.[14] Parsing and production proceed in a left-to-right direction: the beginning of sentence is heard or spoken first, and the end of the sentence is heard or spoken last. This implies (according to the theory) an ordering whereby probe comes before goal, i.e. head precedes complement.

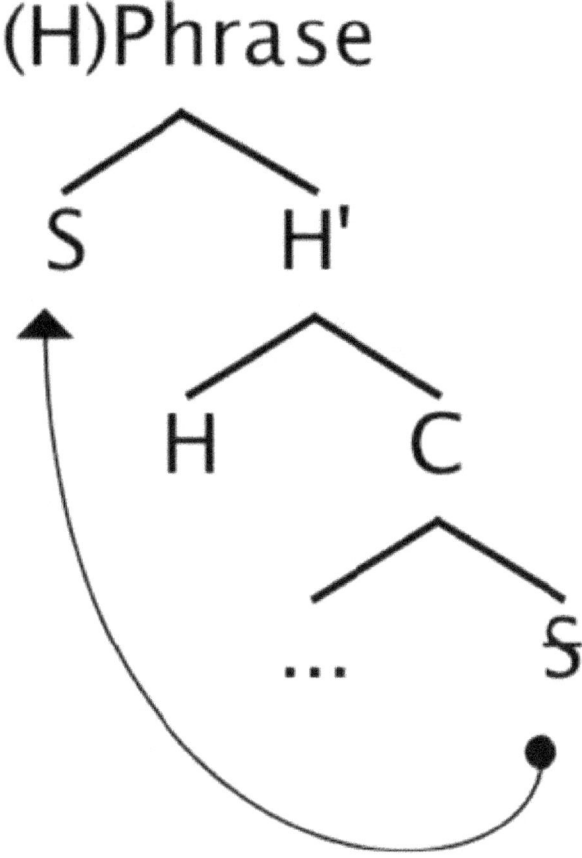

X-bar syntactic tree showing movement of the specifier (S) relative to the head (H) and complement (C)

Kayne's theory also addresses the position of the specifier of a phrase. He represents the relevant scheme as follows:[115]

    S H [$_c$...S...]

The specifier, at first internal to the complement, is moved to the unoccupied position to the left of the head. In terms of merged pairs, this structure can also be represented as:

    <S, H> <H, C>

This process can be mapped onto X-bar syntactic trees as shown in the diagram to the right.

Antisymmetry, then, leads to a universal Specifier-Head-Complement order. The many cases of different ordering found in various languages would have to be explained by syntactic movement away from this underlying base order. It has been pointed out, though, that in predominantly head-final languages such as Japanese and Basque, this would involve complex and massive leftward movement, which is not in accordance with the ideal of grammatical simplicity.[16] An example of the type of movement scheme that would need to be envisaged is provided by Tokizaki:[17]

Here, at each phrasal level in turn, the head of the phrase moves from left to right position relative to its complement. The eventual result reflects the ordering of complex nested phrases found in a language like Japanese.

An attempt to provide evidence for Kayne's scheme is made by Lin,[18] who considered Standard Chinese sentences with the sentence-final particle *le*. This particle is taken to convey perfect aspectual meaning, and thus to be the head of an aspect phrase, having the verb phrase as its complement. If phrases are always to be underlyingly head-initial, then a case like this must entail movement, since the particle comes after the verb phrase. It is proposed that here the complement moves into specifier position, which precedes the head.

As evidence for this, Lin considers (among others) *wh*-adverbials such as *zenmeyang* ("how?"). Based on prior work by Huang,[19] it is postulated that (a) adverbials of this type are subject to movement at Logical Form (LF) level (even though, in Chinese, they do not display *wh*-movement at surface level); and (b) movement is not possible from within a non-complement (Huang's Condition on Extraction Domain or CED). This would imply that *zenmeyang* could not appear in a verb phrase with sentence-final *le*, assuming the above analysis, since that verb phrase has moved into a non-complement (specifier) position, and thus further movement (such as that which *zenmeyang* is required to undergo at LF level) is not possible. Such a restriction on the occurrence of *zenmeyang* is indeed found:[20]

Sentence (b), in which *zenmeyang* co-occurs with sentence-final *le*, is found to be ungrammatical. Lin cites this and other related findings as evidence that the above analysis is correct, supporting the view that Chinese aspect phrases are deeply head-initial.

53.8.1 Surface true approach

According to the "surface true" viewpoint, analysis of head direction must take place at the level of surface derivations, or even the Phonetic Form (PF), i.e. the order in which sentences are pronounced in natural speech. This rejects the idea of an underlying ordering which is then subject to movement, as posited in the Antisymmetry theory and in certain other approaches. In a 2008 article, the linguist Marc Richards argued that a head parameter must only reside at PF, as it is unmaintainable in its original form as a structural parameter.[21] In this approach the relative positions of head and complement that are actually attested at this surface level, which are found to show variation both between and within languages (see above), must be treated as the "true" orderings.

53.8.2 Existence of true head-final languages

Takita (2009) argues against the conclusion of Kayne's Antisymmetry Theory which states that all languages are head-initial at an underlying level. He claims that a language such as Japanese is truly head-final, since the mass movement which would be required to take an underlying head-initial structure to the head-final ones actually found in such languages would violate other constraints. It is implied that such languages are likely following a head-final parameter value, as originally conceived. (For a head-initial/Antisymmetry analysis of Japanese, see Kayne 2003.[22])

Takita's argument is based on the analysis provided by Lin concerning Chinese (see above). Since surface head-final structures are derived from underlying head-initial structures from the act of moving the complements, further extraction from within the moved complement violates CED.

One of the examples of movement which Takita looks at is that of VP-fronting in Japanese. Grammatically, there is not a significant difference between the sentence without VP-fronting (a) and the sentence where the VP moves to the matrix clause (b).[23]

In (b), the fronted VP precedes the matrix subject, confirming that the VP is located in the matrix clause. If Japanese were underlyingly head-initial, (b) should not be grammatical because it allows for extraction of an element (VP2) from the moved complement (CP2).[23]

Thus Takita shows that surface head-final structures in Japanese do not block movement, as they do in Chinese. He concludes that, because Japanese does not block movement as shown in previous sections, it is a genuinely head-final language, and not derived from an underlying, head-initial structure. These results imply that Universal Grammar is equipped with the binary head-directionality, and is not antisymmetric. Takita briefly applies the same tests to Turkish, another seemingly head-final language, and finds similar results.[24]

53.9 References and footnotes

[1] Kayne, Richard S. (1994). *The Antisymmetry of Syntax. Linguistic Inquiry Monograph Twenty-Five*. MIT Press.

[2] Li, Yafei (2005). *A Theory of the Morphology-Syntax Interface*. MIT Press.

[3] Reinhart, Tanya (1979). *The Syntactic Domain of Anaphora*. Doctoral dissertation. M.I.T. Press.

[4] Since any rightward movement must also be downward movement if there are no rightward specifiers or right adjunction, and downward movement is generally assumed to be illicit.

[5] Jamal Ouhalla (1999). *Introducing Transformational Grammar (Second Edition)*. Arnold/Oxford University Press. (See p. 461 for the Japanese example.)

[6] Neeleman, Ad & Peter Ackema (2002). "Effects of Short-Term Storage in Processing Rightward Movement" In S. Nooteboom et al. (eds.) *Storage and Computation in the Language Faculty*. Dordrecht: Kluwer. Pages 219-256.

[7] Moro, A. 2000 Dynamic Antisymmetry. Linguistic Inquiry Monograph Series 38, MIT press. Cambridge, Massachusetts.

[8] Kayne, Richard S. (2010). "Why are there no directionality parameters?" In *WCCFL XXVIII*. Available on http://ling.auf.net/lingBuzz/001100

[9] Kayne, Richard S. (1981) "Unambiguous Paths," in Robert May and Jan Koster (eds.) *Levels of Syntactic* Representation. *Dordrecht: Kluwer. Pages 143-183*

[10] Nolda, Andreas (2004). Topics Detached to the Left: On 'Left Dislocation', 'Hanging Topic', and Related Constructions in German. Berlin: ZAS Papers in Linguistics. pp. 423–448.

[11] Kayne (2010), p. 4.

[12] Kayne (2010), p. 7.

[13] Courtney, Ellen H. (2011). "Learning to produce Quechua relative clauses". Acquisition of Relative Clauses : Processing, typology and function. John Benjamins Publishing Company. p. 150.

[14] Kayne (2010), p. 12.

[15] Kayne (2010), p. 15.

[16] Elordieta, Arantzazu (2014). Biberauer, T.; Sheehan, M., eds. "On the relevance of the Head Parameter in a mixed OV language". Theoretical Approaches to Disharmonic Word Order (Oxford Scholarship Online). p. 5.

[17] Tokizaki, Hisao (2011). "The nature of linear information in the morphosyntax-PF interface". English Linguistics 28 (2), p. 238.

[18] Lin, Tzong-Hong J. (2006). "Complement-to-specifier movement in Mandarin Chinese". Ms., National Tsing Hua University.

[19] Huang, C-T. J. (1982). "Logical relations in Chinese and the theory of grammar". PhD dissertation. MIT.

[20] Takita, Kensuke (2009). "If Chinese is head-initial, Japanese cannot be". Journal of East Asian Linguistics 18 (1), p. 44.

[21] Richards, Marc D. (2008). "Desymmetrization: Parametric variation at the PF-Interface". The Canadian Journal of Linguistics 53 (2-3), p. 283.

[22] Kayne, Richard S. (2003). "Antisymmetry and Japanese". English Linguistics 20: 1–40.

[23] Takita (2009), p. 57.

[24] Takita (2009), p. 59.

Chapter 54

Symmetry in biology

A selection of animals showing the range of possible symmetries, including both radial and bilateral body plans.

Symmetry in biology is the balanced distribution of duplicate body parts or shapes. In nature and biology, symmetry is approximate. For example, plant leaves, while considered symmetrical, rarely match up exactly when folded in half. Symmetry creates a class of patterns in nature, where the near-repetition of the pattern element is by reflection or rotation.

The body plans of most multicellular organisms exhibit some form of symmetry, whether **radial symmetry**, **bilateral symmetry** or "spherical symmetry". A small minority, notably the sponges, exhibit no symmetry (are asymmetric).

54.1 Radial symmetry

Radially symmetric organisms resemble a pie where several cutting planes produce roughly identical pieces. Such an organism exhibits no left or right sides. They have a top and a bottom surface only.

Symmetry has been important historically in the taxonomy of animals; animals with radial symmetry were classified in the taxon Radiata, which is now generally accepted to be a polyphyletic assemblage of different phyla of the Animal kingdom. Most radially symmetric animals are symmetrical about an axis extending from the center of the oral surface, which contains the mouth, to the center of the opposite, or aboral, end. Radial symmetry is especially suitable for sessile animals such as the sea anemone, floating animals such as jellyfish, and slow moving organisms such as starfish. Animals in the phyla cnidaria and echinodermata are radially symmetric,[1] although many sea anemones and some corals have bilateral symmetry defined by a single structure, the siphonoglyph.[2]

Many flowers are radially symmetric or actinomorphic. Roughly identical flower parts – petals, sepals, and stamens occur at regular intervals around the axis of the flower, which is often the female part, with the carpel, style and stigma.[3]

Many viruses have radial symmetries, their coats being composed of a relatively small number of protein molecules arranged in a regular pattern to form polyhedrons, spheres, or ovoids. Most are icosahedrons.[4]

54.1.1 Special forms of radial symmetry

Tetramerism is a variant of radial symmetry found in jellyfish, which have four canals in an otherwise radial body plan.

Pentamerism, another variant of radial symmetry (also called pentaradial and pentagonal symmetry), means the organism is in five parts around a central axis, 72° apart. Among animals, only the echinoderms such as sea stars, sea urchins, and sea lilies are pentamerous as adults, with five arms arranged around the mouth. Being bilaterian animals, however, they initially develop with mirror symmetry as larvae, then gain pentaradial symmetry later.[5]

Flowering plants show fivefold symmetry in many flowers and in various fruits. This is well seen in the arrangement of the five carpels (the botanical fruits containing the seeds) in an apple cut transversely.

Hexamerism is found in the corals and sea anemones (class *Anthozoa*) which are divided into two groups based on their symmetry. The most common corals in the subclass *Hexacorallia* have a hexameric body plan; their polyps have sixfold internal symmetry and the number of their tentacles is a multiple of six.

Octamerism is found in corals of the subclass *Octocorallia*. These have polyps with eight tentacles and octameric radial symmetry. The octopus, however, has bilateral symmetry, despite its eight arms.

54.2 Spherical symmetry

Spherical symmetry occurs in an organism if it is able to be cut into two identical halves through any cut that runs through the organism's center. Organisms which approximate spherical symmetry include the freshwater green alga *Volvox*.[1]

54.3 Bilateral symmetry

"Bilateral symmetry" redirects here. For bilateral symmetry in mathematics, see reflection symmetry.
Main article: Bilateria

In bilateral symmetry (also called plane symmetry), only one plane, called the sagittal plane, will divide an organism into roughly mirror image halves (with respect to external appearance only, see situs solitus). Thus there is approximate reflection symmetry.

Animals that are bilaterally symmetric have mirror symmetry in the sagittal plane, which divides the body vertically into left and right halves, with one of each sense organ and limb pair on either side. The great majority (at least 99%) of animals are bilaterally symmetric, including humans (see also facial symmetry).[6][7][8]

When an organism normally moves in one direction, it inevitably has a front or head end. This end encounters the environment before the rest of the body as the organism moves along, so sensory organs such as eyes tend to be clustered there, and similarly it is the likely site for a mouth as food is encountered.[8] A distinct head, with sense organs connected to a central nervous system, therefore (on this view) tends to develop (cephalization). Given a direction of travel which creates a front/back difference, and gravity which creates a dorsal/ventral difference, left and right are unavoidably distinguished, so a bilaterally symmetric body plan is widespread and found in most animal phyla.[8][9] Bilateral symmetry also permits streamlining and on a traditional view in zoology facilitates locomotion.[8] However, in the Cnidaria, different symmetries exist, and bilateral symmetry is not necessarily aligned with the direction of locomotion, so another mechanism such as internal transport may be needed to explain the origin of bilateral symmetry in animals.[8][10]

The phylum Echinodermata, which includes starfish, sea urchins and sand dollars, is unique among animals in having bilateral symmetry at the larval stage, but fivefold symmetry (pentamerism, a special type of radial symmetry) as adults.[11]

Bilateral symmetry is not easily broken. In experiments using the fruit fly, Drosophila, in contrast to other traits (where laboratory selection experiments always yield a change), right or left-sidedness in eye size, or eye facet number, wing-folding behavior (left over right) show a lack of response.[12]

Females of some species select for symmetry, presumed by biologists to be a mark (technically a "cue") of fitness. Female barn swallows, a species where adults have long tail streamers, prefer to mate with males that have the most symmetrical tails.[13]

Flowers in some families of flowering plants, such as the orchid and pea families, and also most of the figwort family,[14] are bilaterally symmetric (zygomorphic).[15]

54.4   Biradial symmetry

Biradial symmetry is a combination of radial and bilateral symmetry, as in the Ctenophores. Here, the body components are arranged with similar parts on either side of a central axis, and each of the four sides of the body is identical to the opposite side but different from the adjacent side. This may represent a stage in the evolution of bilateral symmetry "from a presumably radially symmetrical ancestor."[10]

54.5   Asymmetry

Further information: List of animals featuring external asymmetry

Not all animals are symmetric. Many members of the phylum Porifera (sponges) have no symmetry (while others are radially symmetric).[16]

It is normal for essentially symmetric animals to show some measure of asymmetry. Usually in humans the left brain is structured differently to the right, the heart is positioned towards the left, and the right hand functions better than the left hand.[17] The scale-eating cichlid Perissodus microlepis develops left or right asymmetries in their mouths and jaws that allow them to be more effective when removing scales from the left or right flank of their prey.[18] The approximately 400 species of flatfish also lack symmetry as adults, though the larvae are bilaterally symmetrical. Adult flatfish rest on one side, and the eye that was on that side has migrated round to the other (top) side of the body.[19]

54.6 References

[1] Chandra, Girish. "Symmetry". IAS. Retrieved 14 June 2014.

[2] Finnerty JR (2003). "The origins of axial patterning in the metazoa: How old is bilateral symmetry?". *The International journal of developmental biology* **47** (7–8): 523–9. PMID 14756328. 14756328 16341006.

[3] Endress, P.K. (February 2001). "Evolution of Floral Symmetry". *Current Opinion Plant Biology* **4** (1): 86–91. doi:10.1016/S1369-5266(00)00140-0. PMID 11163173.

[4] Horne, R. W.; Wildy, P. (1961). "Symmetry in virus architecture". *Virology* **15** (3): 348–373. doi:10.1016/0042-6822(61)90366-X.

[5] Stewart, 2001. pp 64-65.

[6] Valentine, James W. "Bilateria". AccessScience. Retrieved 29 May 2013.

[7] "Bilateral symmetry". Natural History Museum. Retrieved 14 June 2014.

[8] Finnerty, John R. (2005). "Did internal transport, rather than directed locomotion, favor the evolution of bilateral symmetry in animals?" (PDF). *BioEssays* **27**: 1174–1180. doi:10.1002/bies.20299.

[9] "Bilateral (left/right) symmetry". Berkeley. Retrieved 14 June 2014.

[10] Martindale, Mark Q.; Henry, Jonathan Q. (1998). "The Development of Radial and Biradial Symmetry: The Evolution of Bilaterality1" (PDF). *American Zoology* **38** (4): 672–684. doi:10.1093/icb/38.4.672.

[11] Fox, Richard. "*Asterias forbesi*". *Invertebrate Anatomy OnLine*. Lander University. Retrieved 14 June 2014.

[12] Symmetry breaking in fruit flies

[13] Maynard Smith, John; Harper, David (2003). *Animal Signals*. Oxford University Press. pp. 63-65.

[14] "SCROPHULARIACEAE - Figwort or Snapdragon Family". Texas A&M University Bioinformatics Working Group. Retrieved 14 June 2014.

[15] Symmetry, biological, from *The Columbia Electronic Encyclopedia* (2007).

[16] Myers, Phil (2001). "Porifera Sponges". University of Michigan (Animal Diversity Web). Retrieved 14 June 2014.

[17] Lopsided fish show that symmetry is only skin deep *BioMed Central*, 22 January 2010.

[18] Lee HJ, Kusche H, Meyer A (2012) Handed Foraging Behavior in Scale-Eating Cichlid Fish: Its Potential Role in Shaping Morphological Asymmetry. PLoS ONE 7(9): e44670. doi:10.1371/journal.pone.0044670

[19] Friedman, Matt (2008). "The evolutionary origin of flatfish asymmetry". *Nature* **454** (7201): 209–212. doi:10.1038/nature07108. PMID 18615083.

54.7 Bibliography

• Ball, Philip (2009). *Shapes*. Oxford University Press.

• Stewart, Ian (2007). *What Shape is a Snowflake? Magical Numbers in Nature*. Weidenfeld and Nicolson.

These sea anemones have been painted to emphasize their radial symmetry. (Plate from Ernst Haeckel's Kunstformen der Natur*).*

Lilium bulbiferum *displays typical floral symmetry with repeated parts arranged around the axis of the flower.*

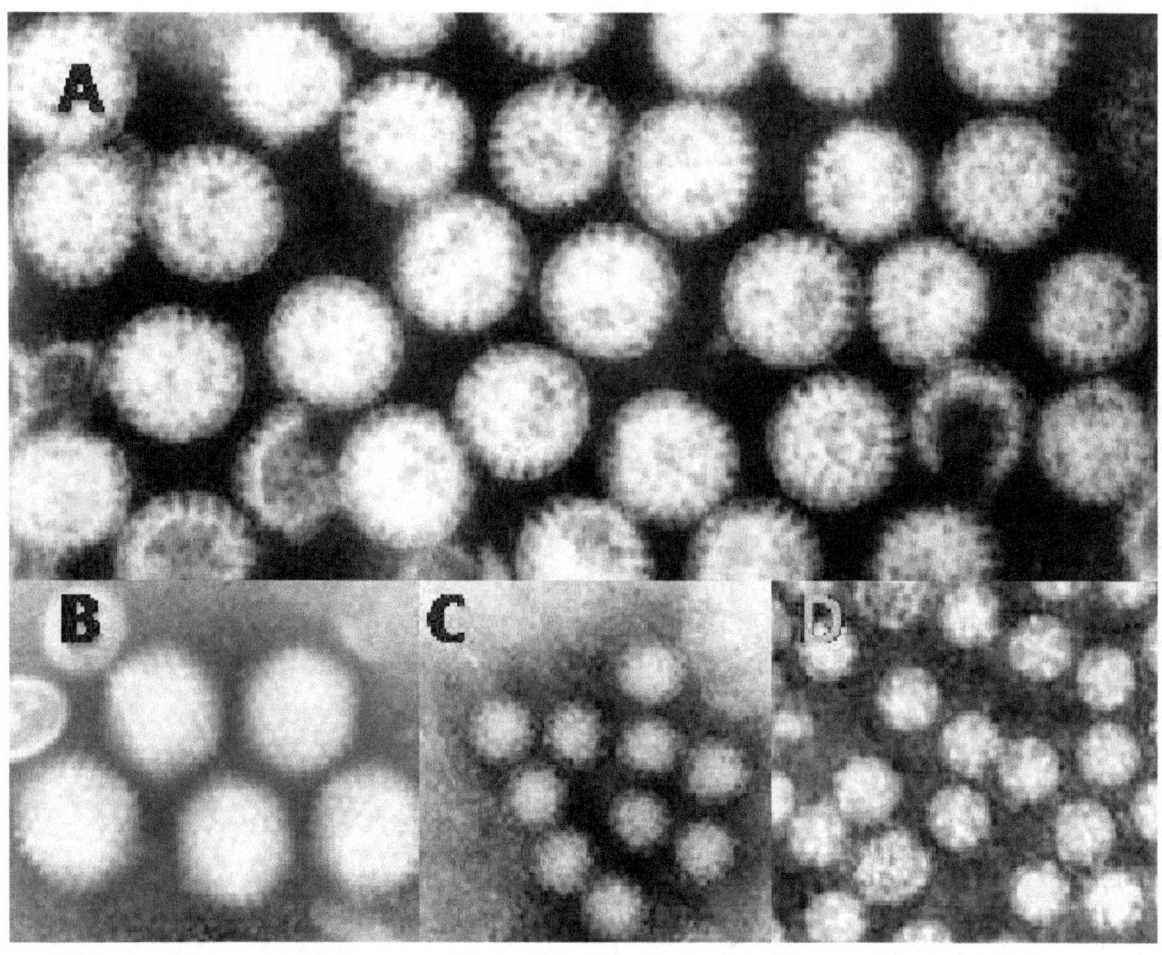

Gastroenteritis viruses have radial symmetry, being icosahedral: A rotavirus, B adenovirus, C norovirus, D astrovirus.

Apple cut horizontally, showing pentamerism

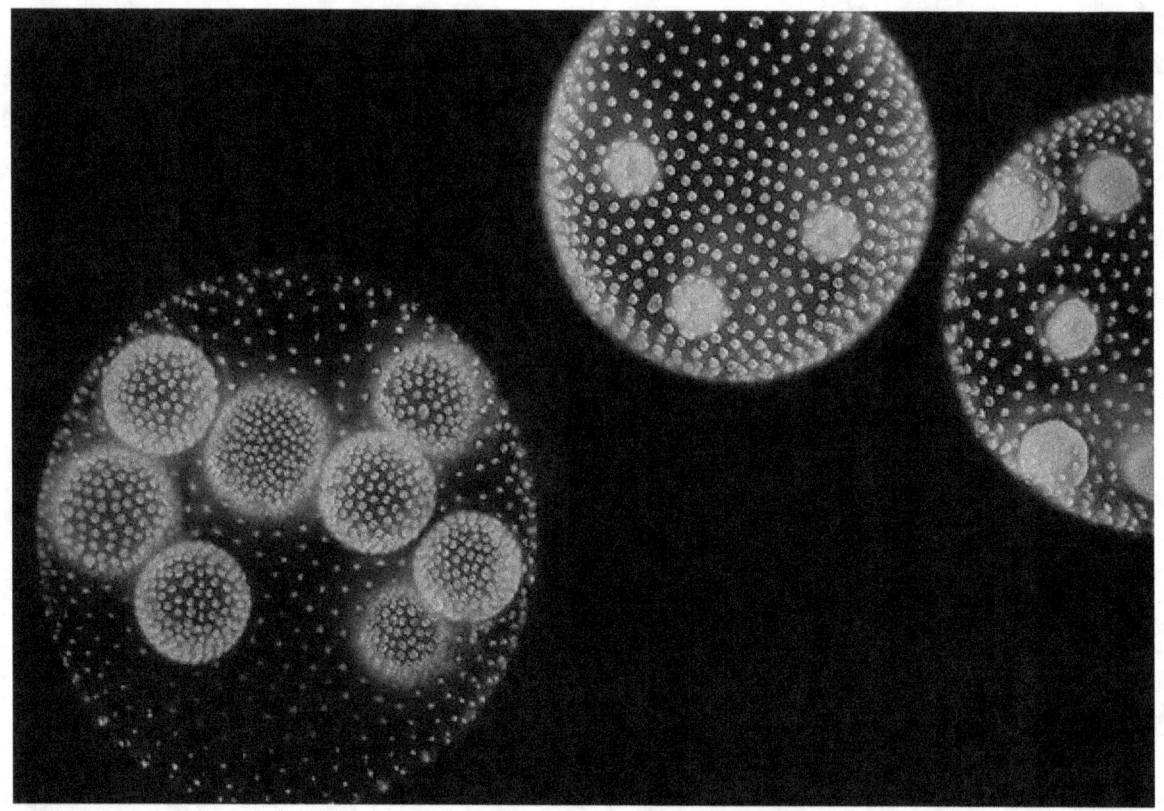

Volvox *is a microscopic green freshwater alga with spherical symmetry. Young colonies can be seen inside the larger ones.*

The small emperor moth, Saturnia pavonia, *displays a deimatic pattern with bilateral symmetry.*

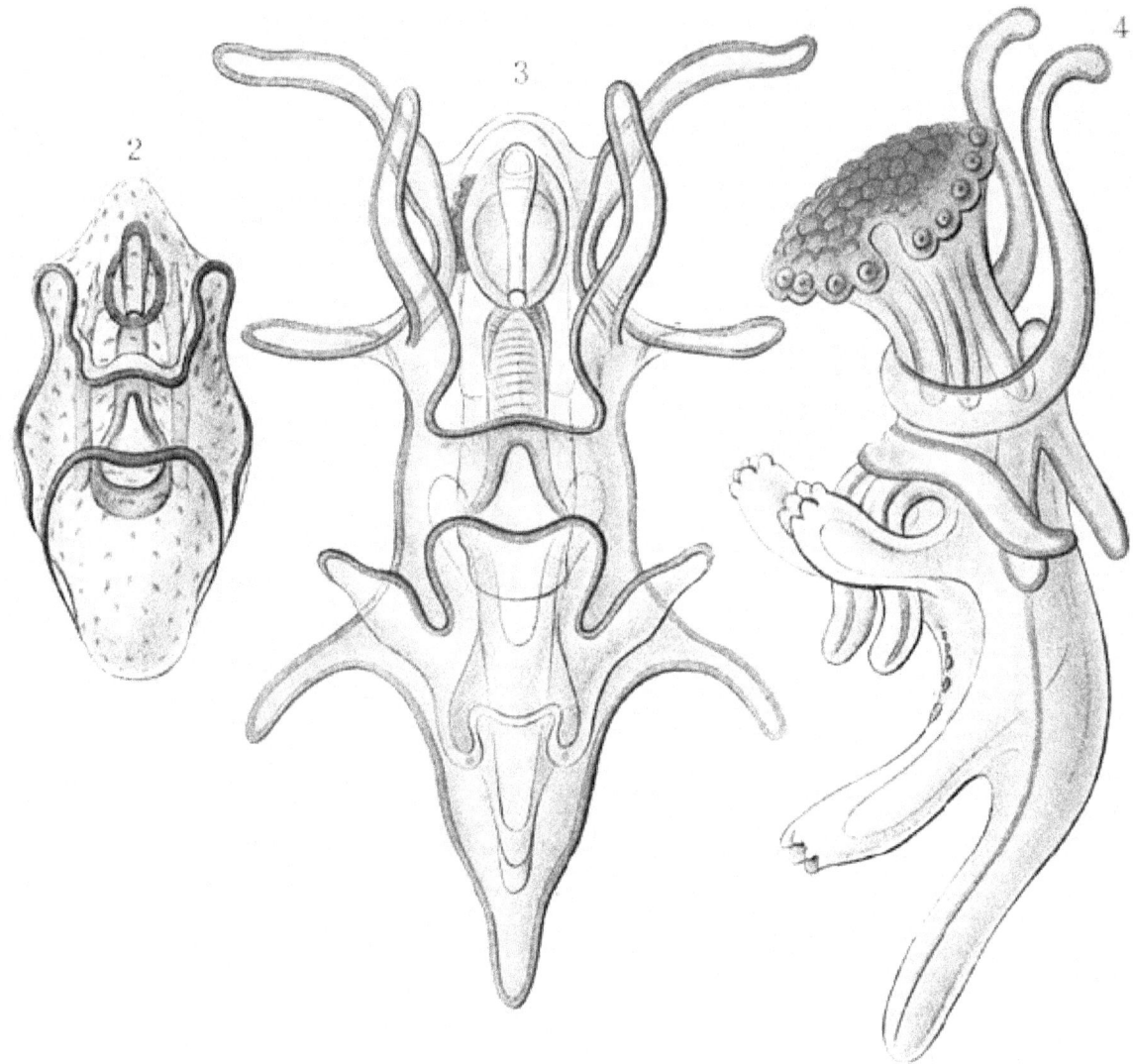

Starfish larvae are bilaterally symmetric, whereas the adults have fivefold symmetry.

Flower of bee orchid, Ophrys apifera *is bilaterally symmetrical (zygomorphic). The lip of the flower resembles the (bilaterally symmetric) abdomen of a female bee; pollination occurs when a male bee attempts to mate with it.*

Chapter 55

Patterns in nature

Patterns in nature are visible regularities of form found in the natural world. These patterns recur in different contexts and can sometimes be modelled mathematically. Natural patterns include symmetries, trees, spirals, meanders, waves, foams, tessellations, cracks and stripes.[1] Early Greek philosophers studied pattern, with Plato, Pythagoras and Empedocles attempting to explain order in nature. The modern understanding of visible patterns developed gradually over time.

In the 19th century, Belgian physicist Joseph Plateau examined soap films, leading him to formulate the concept of a minimal surface. German biologist and artist Ernst Haeckel painted hundreds of marine organisms to emphasise their symmetry. Scottish biologist D'Arcy Thompson pioneered the study of growth patterns in both plants and animals, showing that simple equations could explain spiral growth. In the 20th century, British mathematician Alan Turing predicted mechanisms of morphogenesis which give rise to patterns of spots and stripes. Hungarian biologist Aristid Lindenmayer and French American mathematician Benoît Mandelbrot showed how the mathematics of fractals could create plant growth patterns.

Mathematics, physics and chemistry can explain patterns in nature at different levels. Patterns in living things are explained by the biological processes of natural selection and sexual selection. Studies of pattern formation make use of computer models to simulate a wide range of patterns.

55.1   History

Early Greek philosophers attempted to explain order in nature, anticipating modern concepts. Plato (c 427 – c 347 BC) — looking only at his work on natural patterns — argued for the existence of universals. He considered these to consist of ideal forms (εἶδος *eidos*: "form") of which physical objects are never more than imperfect copies. Thus, a flower may be roughly circular, but it is never a perfect mathematical circle.[2] Pythagoras explained patterns in nature like the harmonies of music as arising from number, which he took to be the basic constituent of existence.[3] Empedocles to an extent anticipated Darwin's evolutionary explanation for the structures of organisms.[4]

In 1202, Leonardo Fibonacci (c 1170 – c 1250) introduced the Fibonacci number sequence to the western world with his book *Liber Abaci*.[5] Fibonacci gave an (unrealistic) biological example, on the growth in numbers of a theoretical rabbit population.[6] In 1917, D'Arcy Wentworth Thompson (1860–1948) published his book *On Growth and Form*. His description of phyllotaxis and the Fibonacci sequence, the mathematical relationships in the spiral growth patterns of plants, is classic. He showed that simple equations could describe all the apparently complex spiral growth patterns of animal horns and mollusc shells.[7]

The Belgian physicist Joseph Plateau (1801–1883) formulated the mathematical problem of the existence of a minimal surface with a given boundary, which is now named after him. He studied soap films intensively, formulating Plateau's laws which describe the structures formed by films in foams.[8]

The German psychologist Adolf Zeising (1810–1876) claimed that the golden ratio was expressed in the arrangement of plant parts, in the skeletons of animals and the branching patterns of their veins and nerves, as well as in the geometry of crystals.[9][10][11]

Ernst Haeckel (1834–1919) painted beautiful illustrations of marine organisms, in particular Radiolaria, emphasising their symmetry to support his faux-Darwinian theories of evolution.[12]

The American photographer Wilson Bentley (1865–1931) took the first micrograph of a snowflake in 1885.[13]

In 1952, Alan Turing (1912–1954), better known for his work on computing and codebreaking, wrote *The Chemical Basis of Morphogenesis*, an analysis of the mechanisms that would be needed to create patterns in living organisms, in the process called morphogenesis.[14] He predicted oscillating chemical reactions, in particular the Belousov–Zhabotinsky reaction. These activator-inhibitor mechanisms can, Turing suggested, generate patterns of stripes and spots in animals, and contribute to the spiral patterns seen in plant phyllotaxis.[15]

In 1968, Hungarian theoretical biologist Aristid Lindenmayer (1925–1989) developed the L-system, a formal grammar which can be used to model plant growth patterns in the style of fractals.[16] L-systems have an alphabet of symbols that can be combined using production rules to build larger strings of symbols, and a mechanism for translating the generated strings into geometric structures. In 1975, after centuries of slow development of the mathematics of patterns by Gottfried Leibniz, Georg Cantor, Helge von Koch, Wacław Sierpiński and others, Benoît Mandelbrot wrote a famous paper, *How Long Is the Coast of Britain? Statistical Self-Similarity and Fractional Dimension*, crystallising mathematical thought into the concept of the fractal.[17]

55.2 Causes

Living things like orchids, hummingbirds, and the peacock's tail have abstract designs with a beauty of form, pattern and colour that artists struggle to match.[18] The beauty that people perceive in nature has causes at different levels, notably in the mathematics that governs what patterns can physically form, and among living things in the effects of natural selection, that govern how patterns evolve.[19]

Mathematics seeks to discover and explain abstract patterns or regularities of all kinds.[20][21] Visual patterns in nature find explanations in chaos theory, fractals, logarithmic spirals, topology and other mathematical patterns. For example, L-systems form convincing models of different patterns of tree growth.[16]

The laws of **physics** apply the abstractions of mathematics to the real world, often as if it were perfect. For example a crystal is perfect when it has no structural defects such as dislocations and is fully symmetric. Exact mathematical perfection can only approximate real objects.[22] Visible patterns in nature are governed by physical laws; for example, meanders can be explained using fluid dynamics.

In biology, **natural selection** can cause the development of patterns in living things for several reasons, including camouflage,[23] sexual selection,[23] and different kinds of signalling, including mimicry[24] and cleaning symbiosis.[25] In plants, the shapes, colours, and patterns of flowers like the lily have evolved to optimise insect pollination (other plants may be pollinated by wind, birds, or bats). European honey bees and other pollinating insects are attracted to flowers by a radial pattern of colours and stripes (some visible only in ultraviolet light) that serve as nectar guides that can be seen at a distance; by scent; and by rewards of sugar-rich nectar and edible pollen.[26]

55.3 Types of pattern

55.3.1 Symmetry

Further information: Symmetry in biology, Floral symmetry and Crystal symmetry

Symmetry is pervasive in living things. Animals mainly have bilateral or mirror symmetry, as do the leaves of plants and some flowers such as orchids.[27] Plants often have radial or rotational symmetry, as do many flowers and some groups of animals such as sea anemones. Fivefold symmetry is found in the echinoderms, the group that includes starfish, sea urchins, and sea lilies.[28]

Among non-living things, snowflakes have striking sixfold symmetry: each flake is unique, its structure forming a record of the varying conditions during its crystallisation, with nearly the same pattern of growth on each of its six arms.[29] Crystals in general have a variety of symmetries and crystal habits; they can be cubic or octahedral, but true crystals cannot have fivefold symmetry (unlike quasicrystals).[30] Rotational symmetry is found at different scales among non-living things including the crown-shaped splash pattern formed when a drop falls into a pond,[31] and both the spheroidal shape and rings of a planet like Saturn.[32]

Symmetry has a variety of causes. Radial symmetry suits organisms like sea anemones whose adults do not move: food and threats may arrive from any direction. But animals that move in one direction necessarily have upper and lower sides, head and tail ends, and therefore a left and a right. The head becomes specialised with a mouth and sense organs (cephalisation), and the body becomes bilaterally symmetric (though internal organs need not be).[33] More puzzling is the reason for the fivefold (pentaradiate) symmetry of the echinoderms. Early echinoderms were bilaterally symmetrical, as their larvae still are. Sumrall and Wray argue that the loss of the old symmetry had both developmental and ecological causes.[34]

- Animals often show mirror or bilateral symmetry, like this tiger.

- Echinoderms like this starfish have fivefold symmetry.

- Fivefold symmetry can be seen in many flowers and some fruits like this medlar.

- Snowflakes have sixfold symmetry.

- Each snowflake is unique but symmetrical.

- Fluorite showing cubic crystal habit

- Water splash approximates radial symmetry.

- Garnet showing rhombic dodecahedral crystal habit

- Volvox has spherical symmetry.

- Sea anemones have rotational symmetry.

55.3.2 Trees, fractals

Fractals are infinitely self-similar, iterated mathematical constructs having fractal dimension.[17][35][36] Infinite iteration is not possible in nature so all 'fractal' patterns are only approximate. For example, the leaves of ferns and umbellifers (Apiaceae) are only self-similar (pinnate) to 2, 3 or 4 levels. Fern-like growth patterns occur in plants and in animals including bryozoa, corals, hydrozoa like the air fern, *Sertularia argentea*, and in non-living things, notably electrical discharges. Lindenmayer system fractals can model different patterns of tree growth by varying a small number of parameters including branching angle, distance between nodes or branch points (internode length), and number of branches per branch point.[16]

Fractal-like patterns occur widely in nature, in phenomena as diverse as clouds, river networks, geologic fault lines, mountains, coastlines,[37] animal coloration, snow flakes,[38] crystals,[39] blood vessel branching,[40] and ocean waves.[41]

- Leaf of Cow Parsley, *Anthriscus sylvestris*, is 2- or 3-pinnate, not infinite

- Fractal spirals: Romanesco broccoli showing self-similar form

- Angelica flowerhead, a sphere made of spheres (self-similar)

- Trees: Lichtenberg figure: high voltage dielectric breakdown in an acrylic polymer block

- Trees: dendritic Copper crystals (in microscope)

55.3.3   Spirals

Further information: phyllotaxis

Spirals are common in plants and in some animals, notably molluscs. For example, in the nautilus, a cephalopod mollusc, each chamber of its shell is an approximate copy of the next one, scaled by a constant factor and arranged in a logarithmic spiral.[42] Given a modern understanding of fractals, a growth spiral can be seen as a special case of self-similarity.[43]

Plant spirals can be seen in phyllotaxis, the arrangement of leaves on a stem, and in the arrangement (parastichy[44]) of other parts as in composite flower heads and seed heads like the sunflower or fruit structures like the pineapple[45]:337 and snake fruit, as well as in the pattern of scales in pine cones, where multiple spirals run both clockwise and anticlockwise. These arrangements have explanations at different levels – mathematics, physics, chemistry, biology – each individually correct, but all necessary together.[46] Phyllotaxis spirals can be generated mathematically from Fibonacci ratios: the Fibonacci sequence runs 1, 1, 2, 3, 5, 8, 13... (each subsequent number being the sum of the two preceding ones). For example, when leaves alternate up a stem, one rotation of the spiral touches two leaves, so the pattern or ratio is 1/2. In hazel the ratio is 1/3; in apricot it is 2/5; in pear it is 3/8; in almond it is 5/13.[47] In disc phyllotaxis as in the sunflower and daisy, the florets are arranged in Fermat's spiral with Fibonacci numbering, at least when the flowerhead is mature so all the elements are the same size. Fibonacci ratios approximate the golden angle, $137.508°$, which governs the curvature of Fermat's spiral.[48]

From the point of view of physics, spirals are lowest-energy configurations[49] which emerge spontaneously through self-organizing processes in dynamic systems.[50] From the point of view of chemistry, a spiral can be generated by a reaction-diffusion process, involving both activation and inhibition. Phyllotaxis is controlled by proteins that manipulate the concentration of the plant hormone auxin, which activates meristem growth, alongside other mechanisms to control the relative angle of buds around the stem.[51] From a biological perspective, arranging leaves as far apart as possible in any given space is favoured by natural selection as it maximises access to resources, especially sunlight for photosynthesis.[45]

- Fibonacci spiral

- Bighorn sheep, *Ovis canadensis*

- Spirals: phyllotaxis of spiral aloe, *Aloe polyphylla*

- *Nautilus* shell's logarithmic growth spiral

- Fermat's spiral: seed head of sunflower, *Helianthus annuus*

- Multiple Fibonacci spirals: red cabbage in cross section

- Gastropod mollusc shell, *Trochoidea liebetruti*, showing how opening moves around, outward, and downwards as it grows

55.3.4   Chaos, flow, meanders

In mathematics, a dynamical system is chaotic if it is (highly) sensitive to initial conditions (the so-called "butterfly effect"[52]), which requires the mathematical properties of topological mixing and dense periodic orbits.[53]

Alongside fractals, chaos theory ranks as an essentially universal influence on patterns in nature. There is a relationship between chaos and fractals—the *strange attractors* in chaotic systems have a fractal dimension.[54] Some cellular automata, simple sets of mathematical rules that generate patterns, have chaotic behaviour, notably Stephen Wolfram's Rule 30.[55]

Vortex streets are zigzagging patterns of whirling vortices created by the unsteady separation of flow of a fluid, most often air or water, over obstructing objects.[56] Smooth (laminar) flow starts to break up when the size of the obstruction or the velocity of the flow become large enough compared to the viscosity of the fluid.

Meanders are sinuous bends in rivers or other channels, which form as a fluid, most often water, flows around bends. As soon as the path is slightly curved, the size and curvature of each loop increases as helical flow drags material like sand and gravel across the river to the inside of the bend. The outside of the loop is left clean and unprotected, so erosion accelerates, further increasing the meandering in a powerful positive feedback loop.[57]

- Chaos: shell of gastropod mollusc the cloth of gold cone, *Conus textile*, resembles Rule 30 cellular automaton

- Chaos: vortex street of clouds

- Meanders: dramatic meander scars and oxbow lakes in the broad flood plain of the Rio Negro, seen from space

- Meanders: sinuous path of Rio Cauto, Cuba

- Meanders: sinuous snake crawling

- Meanders: symmetrical Brain Coral, *Diploria strigosa*

55.3.5   Waves, dunes

Waves are disturbances that carry energy as they move. Mechanical waves propagate through a medium – air or water, making it oscillate as they pass by.[58] Wind waves are sea surface waves that create the characteristic chaotic pattern of any large body of water, though their statistical behaviour can be predicted with wind wave models.[59] As waves in water or wind pass over sand, they create patterns of ripples. When winds blow over large bodies of sand, they create dunes, sometimes in extensive dune fields as in the Taklamakan desert. Dunes may form a range of patterns including crescents, very long straight lines, stars, domes, parabolas, and longitudinal or Seif ('sword') shapes.[60]

Barchans or crescent dunes are produced by wind acting on desert sand; the two horns of the crescent and the slip face point downwind. Sand blows over the upwind face, which stands at about 15 degrees from the horizontal, and falls on to the slip face, where it accumulates up to the angle of repose of the sand, which is about 35 degrees. When the slip face exceeds the angle of repose, the sand avalanches, which is a nonlinear behaviour: the addition of many small amounts of sand causes nothing much to happen, but then the addition of a further small amount suddenly causes a large amount to avalanche.[61] Apart from this nonlinearity, barchans behave rather like solitary waves.[62]

- Waves: breaking wave in a ship's wake

- Dunes: sand dunes in Taklamakan desert, from space

- Dunes: barchan crescent sand dune

- Wind ripples with dislocations in Sistan, Afghanistan

55.3.6   Bubbles, foam

A soap bubble forms a sphere, a surface with minimal area — the smallest possible surface area for the volume enclosed. Two bubbles together form a more complex shape: the outer surfaces of both bubbles are spherical; these surfaces are joined by a third spherical surface as the smaller bubble bulges slightly into the larger one.[8]

A foam is a mass of bubbles; foams of different materials occur in nature. Foams composed of soap films obey Plateau's laws, which require three soap films to meet at each edge at 120° and four soap edges to meet at each vertex at the tetrahedral angle of about 109.5°. Plateau's laws further require films to be smooth and continuous, and to have a constant average curvature at every point. For example, a film may remain nearly flat on average by being curved up in one direction (say, left to right) while being curved downwards in another direction (say, front to back).[63][64] Structures with minimal surfaces can be used as tents. Lord Kelvin identified the problem of the most efficient way to pack cells of equal volume as a foam in 1887; his solution uses just one solid, the bitruncated cubic honeycomb with very slightly curved faces to meet Plateau's laws. No better solution was found until 1993 when Denis Weaire and Robert Phelan proposed the Weaire–Phelan structure; the Beijing National Aquatics Center adapted the structure for their outer wall in the 2008 Summer Olympics.[65]

At the scale of living cells, foam patterns are common; radiolarians, sponge spicules, silicoflagellate exoskeletons and the calcite skeleton of a sea urchin, *Cidaris rugosa*, all resemble mineral casts of Plateau foam boundaries.[66][67] The skeleton of the Radiolarian, *Aulonia hexagona*, a beautiful marine form drawn by Haeckel, looks as if it is a sphere composed wholly of hexagons, but this is mathematically impossible. The Euler characteristic states that for any convex polyhedron, the number of faces plus the number of vertices (corners) equals the number of edges plus two. A result of this formula is that any closed polyhedron of hexagons has to include exactly 12 pentagons, like a soccer ball, Buckminster Fuller geodesic dome, or fullerene molecule. This can be visualised by noting that a mesh of hexagons is flat like a sheet of chicken wire, but each pentagon that is added forces the mesh to bend (there are fewer corners, so the mesh is pulled in).[68]

- Foam of soap bubbles: 4 edges meet at each vertex, at angles close to 109.5°, as in two C-H bonds in methane.

- Radiolaria drawn by Haeckel in his *Kunstformen der Natur* (1904).

- Haeckel's Spumellaria; the skeletons of these Radiolaria have foam-like forms.

- Buckminsterfullerene C_{60}: Richard Smalley and colleagues synthesised the fullerene molecule in 1985.

- Brochosomes (secretory microparticles produced by leafhoppers) often approximate fullerene geometry.

- Circus tent approximates a minimal surface.

- Beijing's National Aquatics Center for the 2008 Olympic games has a Weaire-Phelan surface.

- Equal spheres (gas bubbles) in a surface foam

55.3.7   Tessellations

Main article: tessellation

Tessellations are patterns formed by repeating tiles all over a flat surface. There are 17 wallpaper groups of tilings.[69] While common in art and design, exactly repeating tilings are less easy to find in living things. The cells in the paper nests of social wasps, and the wax cells in honeycomb built by honey bees are well-known examples. Among animals, bony fish, reptiles or the pangolin, or fruits like the Salak are protected by overlapping scales or osteoderms, these form more-or-less exactly repeating units, though often the scales in fact vary continuously in size. Among flowers, the Snake's Head Fritillary, *Fritillaria meleagris*, have a tessellated chequerboard pattern on their petals. The structures of minerals provide good examples of regularly repeating three-dimensional arrays. Despite the hundreds of thousands of known minerals, there are rather few possible types of arrangement of atoms in a crystal, defined by crystal structure, crystal system, and point group; for example, there are exactly 14 Bravais lattices for the 7 lattice systems in three-dimensional space.[70]

- Crystals: cube-shaped crystals of halite (rock salt); cubic crystal system, Isometric hexoctahedral crystal symmetry

- Arrays: honeycomb is a natural tessellation

- Bismuth hopper crystal illustrating the stairstep crystal habit.

- Tilings: tessellated flower of Snake's Head Fritillary, *Fritillaria meleagris*

- Tilings: overlapping scales of Common Roach, *Rutilus rutilus*

- Tilings: overlapping scales of snakefruit or salak, *Salacca zalacca*

- Tessellated pavement: a rare rock formation on the Tasman Peninsula

55.3.8 Cracks

Cracks are linear openings that form in materials to relieve stress. When an elastic material stretches or shrinks uniformly, it eventually reaches its breaking strength and then fails suddenly in all directions, creating cracks with 120 degree joints, so three cracks meet at a node. Conversely, when an inelastic material fails, straight cracks form to relieve the stress. Further stress in the same direction would then simply open the existing cracks; stress at right angles can create new cracks, at 90 degrees to the old ones. Thus the pattern of cracks indicates whether the material is elastic or not.[71] In a tough fibrous material like oak tree bark, cracks form to relieve stress as usual, but they do not grow long as their growth is interrupted by bundles of strong elastic fibres. Since each species of tree has its own structure at the levels of cell and of molecules, each has its own pattern of splitting in its bark.[72]

- Old pottery surface, white glaze with mainly 90° cracks

- Drying inelastic mud in the Rann of Kutch with mainly 90° cracks

- Veined gabbro with 90° cracks, near Sgurr na Stri, Skye

- Drying elastic mud in Sicily with mainly 120° cracks

- Cooled basalt at Giant's Causeway. Vertical mainly 120° cracks giving hexagonal columns

- Palm trunk with branching vertical cracks (and horizontal leaf scars)

55.3.9 Spots, stripes

Leopards and ladybirds are spotted; angelfish and zebras are striped.[73] These patterns have an evolutionary explanation: they have functions which increase the chances that the offspring of the patterned animal will survive to reproduce. One function of animal patterns is camouflage;[23] for instance, a leopard that is harder to see catches more prey. Another function is signalling[24] — for instance, a ladybird is less likely to be attacked by predatory birds that hunt by sight, if it has bold warning colours, and is also distastefully bitter or poisonous, or mimics other distasteful insects. A young bird may see a warning patterned insect like a ladybird and try to eat it, but it will only do this once; very soon it will spit out the bitter insect; the other ladybirds in the area will remain unmolested. The young leopards and ladybirds, inheriting genes that somehow create spottedness, survive. But while these evolutionary and functional arguments explain why these animals need their patterns, they do not explain how the patterns are formed.[73]

- Dirce Beauty butterfly, *Colobura dirce*

- Grevy's Zebra, *Equus grevyi*

- Royal Angelfish, *Pygoplites diacanthus*

- Leopard, *Panthera pardus pardus*

- Array of Ladybirds by G.G. Jacobson

- Breeding pattern of Cuttlefish, *Sepia officinalis*

55.3.10 Pattern formation

Main article: Pattern formation

Alan Turing,[14] and later the mathematical biologist James Murray, described a mechanism that spontaneously creates spotted or striped patterns: a reaction-diffusion system.[74] The cells of a young organism have genes that can be switched on by a chemical signal, a morphogen, resulting in the growth of a certain type of structure, say a darkly pigmented patch of skin. If the morphogen is present everywhere, the result is an even pigmentation, as in a black leopard. But if it is unevenly distributed, spots or stripes can result. Turing suggested that there could be feedback control of the production of the morphogen itself. This could cause continuous fluctuations in the amount of morphogen as it diffused around the body. A second mechanism is needed to create standing wave patterns (to result in spots or stripes): an inhibitor chemical that switches off production of the morphogen, and that itself diffuses through the body more quickly than the morphogen, resulting in an activator-inhibitor scheme. The Belousov–Zhabotinsky reaction is a non-biological example of this kind of scheme, a chemical oscillator.[74]

Later research has managed to create convincing models of patterns as diverse as zebra stripes, giraffe blotches, jaguar spots (medium-dark patches surrounded by dark broken rings) and ladybird shell patterns (different geometrical layouts of spots and stripes, see illustrations).[75] Richard Prum's activation-inhibition models, developed from Turing's work, use six variables to account for the observed range of nine basic within-feather pigmentation patterns, from the simplest, a central pigment patch, via concentric patches, bars, chevrons, eye spot, pair of central spots, rows of paired spots and an array of dots.[76][77]:6 More elaborate models simulate complex feather patterns in the Guinea fowl, *Numida meleagris*, in which the individual feathers feature transitions from bars at the base to an array of dots at the far (distal) end. These require an oscillation created by two inhibiting signals, with interactions in both space and time.[77]:7-8

Patterns can form for other reasons in the vegetated landscape of tiger bush[78] and fir waves.[79] Tiger bush stripes occur on arid slopes where plant growth is limited by rainfall. Each roughly horizontal stripe of vegetation effectively collects the rainwater from the bare zone immediately above it.[78] Fir waves occur in forests on mountain slopes after wind disturbance, during regeneration. When trees fall, the trees that they had sheltered become exposed and are in turn more likely to be damaged, so gaps tend to expand downwind. Meanwhile, on the windward side, young trees grow, protected by the wind shadow of the remaining tall trees.[79] Natural patterns are sometimes formed by animals, as in the Mima mounds of the Northwestern United States and some other areas, which appear to be created over many years by the burrowing activities of pocket gophers.[80]

In permafrost soils with an active upper layer subject to annual freeze and thaw, patterned ground can form, creating circles, nets, ice wedge polygons, steps, and stripes. Thermal contraction causes shrinkage cracks to form; in a thaw, water fills the cracks, expanding to form ice when next frozen, and widening the cracks into wedges. These cracks may join up to form polygons and other shapes.[81]

- Giant pufferfish, *Tetraodon mbu*

- Detail of Giant pufferfish skin pattern

- Snapshot of simulation of Belousov-Zhabotinsky reaction

- Guinea fowl, *Numida meleagris* feathers transition from barred to spotted, both in-feather and across the bird

- Aerial view of a tiger bush plateau in Niger

- Fir waves in White Mountains, New Hampshire

- Patterned ground: a melting pingo with surrounding ice wedge polygons near Tuktoyaktuk, Canada

55.4   See also

- Emergence

- Evolutionary history of plants

- Mathematics and art

55.5   References

[1] Stevens, Peter. *Patterns in Nature*, 1974. Page 3.

[2] Balaguer, Mark (7 April 2009) [2004]. "Stanford Encyclopedia of Philosophy". *Platonism in Metaphysics*. Stanford University. Retrieved 4 May 2012.

[3] The so-called Pythagoreans, who were the first to take up mathematics, not only advanced this subject, but saturated with it, they fancied that the principles of mathematics were the principles of all things. Aristotle, *Metaphysics 1–5* , cc. 350 BC

[4] Aristotle reports Empedocles arguing that, "[w]herever, then, everything turned out as it would have if it were happening for a purpose, there the creatures survived, being accidentally compounded in a suitable way; but where this did not happen, the creatures perished." *The Physics*, B8, 198b29 in Kirk, et al., 304).

[5] Singh, Parmanand. *Acharya Hemachandra and the (so called) Fibonacci Numbers*. Math. Ed. Siwan, 20(1):28–30, 1986. ISSN 0047-6269

[6] Knott, Ron. "Fibonacci's Rabbits". University of Surrey Faculty of Engineering and Physical Sciences.

[7] About D'Arcy. D' Arcy 150. University of Dundee and the University of St Andrews. Retrieved 16 October 2012.

[8] Stewart, Ian. 2001. Pages 108–109.

[9] Padovan, Richard (1999). *Proportion*. Taylor & Francis. pp. 305–306. ISBN 978-0-419-22780-9.

[10] Padovan, Richard (2002). "Proportion: Science, Philosophy, Architecture". *Nexus Network Journal* 4(1): 113–1224-001-0008-7.

[11] Zeising, Adolf (1854). *Neue Lehre van den Proportionen des meschlischen Körpers*. preface.

[12] Ball, Philip. *Shapes*. 2009. Page 41.

[13] Hannavy, John (2007). *Encyclopedia of Nineteenth-Century Photography* 1. CRC Press. p. 149. ISBN 0-415-97235-3.

[14] Turing, A. M. (1952). "The Chemical Basis of Morphogenesis". *Philosophical Transactions of the Royal Society B* **237** (641): 37–72. Bibcode:1952RSPTB.237...37T. doi:10.1098/rstb.1952.0012.

[15] Ball, Philip. *Shapes*. 2009. Pages 163, 247–250.

[16] Rozenberg, Grzegorz; Salomaa, Arto. *The mathematical theory of L systems*. Academic Press, New York, 1980. ISBN 0-12-597140-0

[17] Mandelbrot, Benoît B. (1983). *The fractal geometry of nature*. Macmillan.

[18] Forbes, Peter. *All that useless beauty*. The Guardian. Review: Non-fiction. 11 February 2012.

[19] Stevens, Peter. 1994. Page 222.

[20] Steen, L.A.. *The Science of Patterns*. Science, 240: 611–616, 1998. Summary at ascd.org

[21] Devlin, Keith. *Mathematics: The Science of Patterns: The Search for Order in Life, Mind and the Universe* (Scientific American Paperback Library) 1996

[22] Tatarkiewicz, Władysław. *Perfection in the Sciences. II. Perfection in Physics and Chemistry*, Dialectics and Humanism, vol. VII, no. 2 (spring 1980), p. 139.

[23] Darwin, Charles. *On the Origin of Species*. 1859, chapter 4.

[24] Wickler, W. (1968). *Mimicry in plants and animals*. New York: McGraw-Hill.

[25] Poulin, R.; Grutter, A.S. (1996) "Cleaning symbiosis: proximate and adaptive explanations". *Bioscience* 46(7): 512–517. (subscription required)

[26] Koning, Ross (1994). "Plant Physiology Information Website". *Pollination Adaptations*. Ross Koning. Retrieved May 2, 2012.

[27] Stewart, Ian. 2001. Pages 48-49.

[28] Stewart, Ian. 2001. Pages 64-65.

[29] Stewart, Ian. 2001. Page 52.

[30] Stewart, Ian. 2001. Pages 82-84.

[31] Stewart, Ian. 2001. Page 60.

[32] Stewart, Ian. 2001. Page 71.

[33] Hickman, Cleveland P.; Roberts, Larry S.; Larson, Allan (2002). "Animal Diversity" (PDF). *Chapter 8: Acoelomate Bilateral Animals* (Third ed.). McGraw-Hill. p. 139. Retrieved October 25, 2012.

[34] Sumrall, Colin D.; Wray, Gregory A. (January 2007). "Ontogeny in the fossil record: diversification of body plans and the evolution of "aberrant" symmetry in Paleozoic echinoderms". *Paleobiology* **33** (1): 149–163. doi:10.1666/06053.1.

[35] Falconer, Kenneth (2003). *Fractal Geometry: Mathematical Foundations and Applications*. John Wiley.

[36] Briggs, John (1992). *Fractals:The Patterns of Chaos*. Thames and Hudson. p. 148.

[37] Batty, Michael (1985-04-04). "Fractals – Geometry Between Dimensions". *New Scientist* (Holborn Publishing Group) **105** (1450): 31.

[38] Meyer, Yves; Roques, Sylvie (1993). *Progress in wavelet analysis and applications: proceedings of the International Conference "Wavelets and Applications," Toulouse, France – June 1992*. Atlantica Séguier Frontières. p. 25.

[39] Carbone, Alessandra; Gromov, Mikhael; Prusinkiewicz, Przemyslaw (2000). *Pattern formation in biology, vision and dynamics*. World Scientific. p. 78. ISBN 9789810237929.

[40] Hahn, Horst K.; Georg, Manfred; Peitgen, Heinz-Otto (2005). "Fractal aspects of three-dimensional vascular constructive optimization". In Losa, Gabriele A.; Nonnenmacher, Theo F. *Fractals in biology and medicine*. Springer. pp. 55–66.

[41] Addison, Paul S. (1997). *Fractals and chaos: an illustrated course*. CRC Press. pp. 44–46.

[42] Maor, Eli. *e: The Story of a Number*. Princeton University Press, 2009. Page 135.

[43] Ball, 2009. *Shapes* pp 29-32.

[44] "Spiral Lattices & Parastichy". Smith College. Retrieved 24 September 2013.

[45] Kappraff, Jay (2004). "Growth in Plants: A Study in Number" (PDF). *Forma* **19**: 335–354.

[46] Ball, Philip. *Shapes*. 2009. Page 13.

[47] Coxeter, H. S. M. (1961). *Introduction to geometry*. Wiley. p. 169.

[48] Prusinkiewicz, Przemyslaw; Lindenmayer, Aristid (1990). *The Algorithmic Beauty of Plants*. Springer-Verlag. pp. 101–107. ISBN 978-0-387-97297-8.

[49] Levitov LS (15 March 1991). "Energetic Approach to Phyllotaxis" (PDF). *Europhys. Lett.* **14** (6): 533–9. Bibcode:1991EL533L. doi:10.1209/0295-5075/14/6/006. (subscription required)

[50] Douady, S; Couder, Y. (March 1992). "Phyllotaxis as a physical self-organized growth process". *Physical Review Letters* **68** (13): 2098–2101. Bibcode:1992PhRvL..68.2098D. doi:10.1103/PhysRevLett.68.2098. PMID 10045303. (subscription required)

[51] Ball, Philip. *Shapes*. 2009. Pages 163, 249–250.

[52] Lorenz, Edward N. (March 1963). "Deterministic Nonperiodic Flow". *Journal of the Atmospheric Sciences* **20** (2): 130–141. Bibcode:1963JAtS...20..130L. doi:10.1175/1520-0469(1963)020<0130:DNF>2.0.CO;2. ISSN 1520-0469. Retrieved 3 June 2010.

[53] Elaydi, Saber N. (1999). *Discrete Chaos*. Chapman & Hall/CRC. p. 117.

[54] Ruelle, David. *Chance and Chaos*. Princeton University Press, 1991.

[55] Wolfram, Stephen. *A New Kind of Science*. Wolfram Media, 2002.

[56] von Kármán, Theodore. *Aerodynamics*. McGraw-Hill (1963): ISBN 978-0070676022. Dover (1994): ISBN 978-0486434858.

[57] Lewalle, Jacques (2006). "Flow Separation and Secondary Flow: Section 9.1". *Lecture Notes in Incompressible Fluid Dynamics: Phenomenology, Concepts and Analytical Tools* (PDF). Syracuse, NY: Syracuse University..

[58] French, A.P. *Vibrations and Waves.* Nelson Thornes, 1971.

[59] Tolman, H.L. (2008), "Practical wind wave modeling", in Mahmood, M.F., *CBMS Conference Proceedings on Water Waves: Theory and Experiment* (PDF), Howard University, USA, 13–18 May 2008: World Scientific Publ.

[60] "Types of Dunes". USGS. 29 October 1997. Retrieved May 2, 2012.

[61] Strahler, A. & Archibold, O.W. *Physical Geography: Science and Systems of the Human Environment.* John Wiley, 4th edition 2008. Page 442.

[62] Schwämmle, V.; Herrman, H.J. (2003). "Solitary wave behaviour of sand dunes". *Nature* **426** (Dec. 11): 619–620 Abstract. Bibcode:2003Natur.426..619S. doi:10.1038/426619a. PMID 14668849.(subscription required)

[63] Philip Ball. *Shapes.* 2009. Page 68.

[64] Frederick J. Almgren, Jr. and Jean E. Taylor, *The geometry of soap films and soap bubbles,* Scientific American, vol. 235, pp. 82–93, July 1976.

[65] Ball, Philip. *Shapes.* 2009. Pages 73–76.

[66] Ball, Philip. *Shapes.* 2009. Pages 96–101.

[67] Brodie, Christina (February 2005). "Geometry and Pattern in Nature 3: The holes in radiolarian and diatom tests". Microscopy-UK. Retrieved May 28, 2012.

[68] Ball, Philip. *Shapes.* 2009. Pages 51–54.

[69] Armstrong, M.A. (1988). *Groups and Symmetry.* New York: Springer-Verlag.

[70] Hook, J.R.; Hall, H.E. *Solid State Physics* (2nd Edition). Manchester Physics Series, John Wiley & Sons, 2010. ISBN 978-0-471-92804-1

[71] Stevens, Peter. 1974. Page 207.

[72] Stevens, Peter. 1974. Page 208.

[73] Ball, Philip. *Shapes.* 2009. Pages 156–158.

[74] Ball, Philip. *Shapes.* 2009. Pages 159–167.

[75] Ball, Philip. *Shapes.* 2009. Pages 168–180.

[76] Rothenburg, David. 2011. Pages 93–95.

[77] Prum, Richard O.; Williamson, Scott (2002). "Reaction–diffusion models of within-feather pigmentation patterning" (PDF). *Proceedings Royal Society London B* **269**: 781–792. doi:10.1098/rspb.2001.1896.

[78] Tongway, D.J.; Valentin, C. & Seghieri, J. (2001). *Banded vegetation patterning in arid and semiarid environments.* New York: Springer-Verlag.

[79] D'Avanzo, C. (22 February 2004). "Fir Waves: Regeneration in New England Conifer Forests". TIEE. Retrieved 26 May 2012.

[80] Morelle, Rebecca. "'Digital gophers' solve Mima mound mystery". BBC News. Retrieved 9 December 2013.

[81] "Permafrost: Patterned Ground". US Army Corps of Engineers. Retrieved 17 February 2015.

55.6 Bibliography

55.6.1 Pioneering authors

- Fibonacci, Leonardo. *Liber Abaci*, 1202.
  - ----- translated by Sigler, Laurence E. *Fibonacci's Liber Abaci.* Springer, 2002.
- Haeckel, Ernst. *Kunstformen der Natur* (Art Forms in Nature), 1899–1904.
- Thompson, D'Arcy Wentworth. *On Growth and Form.* Cambridge, 1917.

55.6.2   General books

- Adam, John A. *Mathematics in Nature: Modeling Patterns in the Natural World.* Princeton University Press, 2006.

- Ball, Philip. *Nature's Patterns: a tapestry in three parts. 1:Shapes. 2:Flow. 3:Branches.* Oxford, 2009.

- Murphy, Pat and Neill, William. *By Nature's Design.* Chronicle Books, 1993.

- Rothenburg, David. *Survival of the Beautiful: Art, Science and Evolution.* Bloomsbury Press, 2011.

- Stevens, Peter S. *Patterns in Nature.* Little, Brown & Co. 1974.

- Stewart, Ian. *What Shape is a Snowflake? Magical Numbers in Nature.* Weidenfeld & Nicolson, 2001.

55.6.3   Patterns from nature (as art)

- Edmaier, Bernard. *Patterns of the Earth.* Phaidon Press, 2007.

- Macnab, Maggie. *Design by Nature: Using Universal Forms and Principles in Design.* New Riders, 2012.

- Nakamura, Shigeki. *Pattern Sourcebook: 250 Patterns Inspired by Nature..* Books 1 and 2. Rockport, 2009.

- O'Neill, Polly. *Surfaces and Textures: A Visual Sourcebook.* Black, 2008.

- Porter, Eliot, and Gleick, James. *Nature's Chaos.* Viking Penguin, 1990.

55.7   External links

- Fibonacci Numbers and the Golden Section

- Phyllotaxis: an Interactive Site for the Mathematical Study of Plant Pattern Formation

Natural patterns form as wind blows sand in the dunes of the Namib Desert. The crescent shaped dunes and the ripples on their surfaces repeat wherever there are suitable conditions.

Patterns of the veiled chameleon, Chamaeleo calyptratus, *evolved for camouflage and to signal mood and breeding condition.*

Fibonacci patterns occur widely in plant structures including this cone of Queen sago, Cycas circinalis

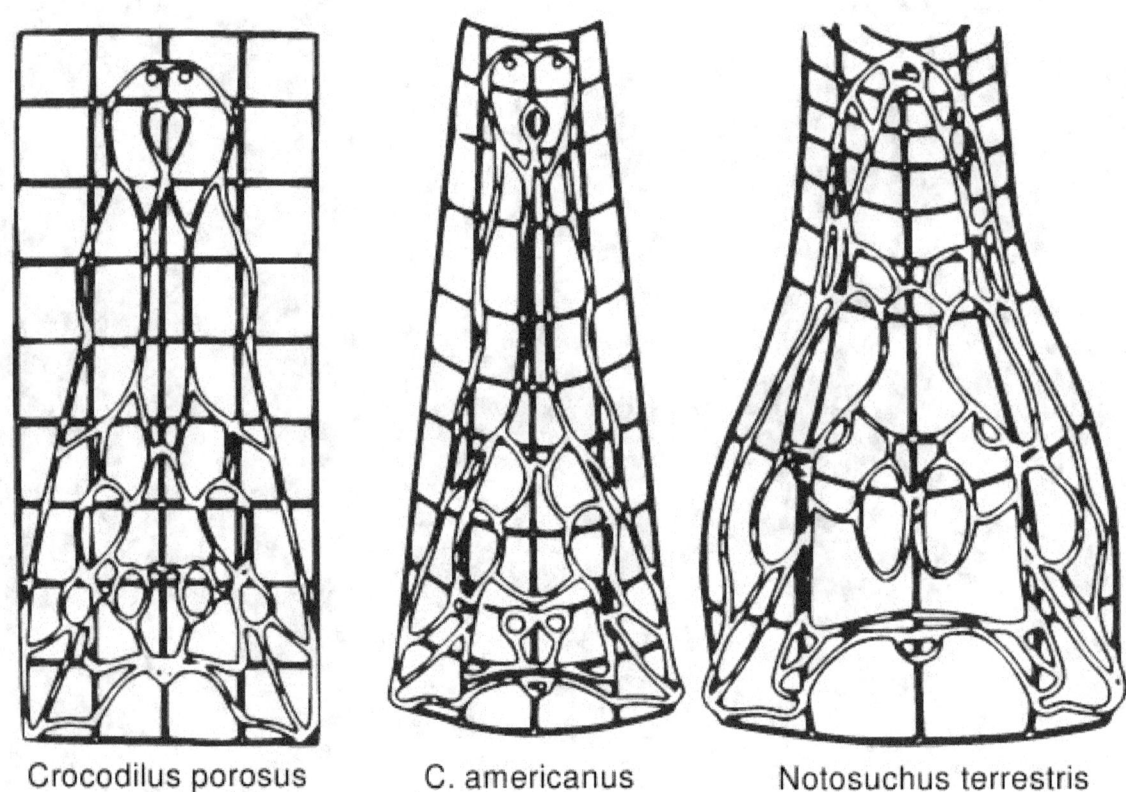

Crocodilus porosus   C. americanus   Notosuchus terrestris

D'Arcy Thompson pioneered the study of growth and form in his 1917 book

Composite patterns: aphids and newly born young in arraylike clusters on Sycamore leaf, divided into polygons by veins, which are avoided by the young aphids

The growth patterns of certain trees resemble these Lindenmayer system fractals.

Chapter 56

List of animals featuring external asymmetry

This is a list of animals that markedly feature external asymmetry in some form. They are exceptions to the general pattern of symmetry in biology. In particular, these animals do not exhibit bilateral symmetry which permits streamlining and is common in animals.[1]

This list is incomplete; you can help by expanding it.

56.1   Birds

The crossbill has an unusual beak in which the upper and lower tips cross each other.[2]

The wrybill is the only species of bird in the world with a beak that is bent sideways (always to the right).[2]

Many owl species, such as the barn owl have asymmetrically positioned ears to enhance sound positioning.

56.2   Fish

Many flatfish, such as flounders, have eyes placed asymmetrically in the adult fish. The fish has the usual symmetrical body structure when it is young, but as it matures and moves to living close to the sea bed, the fish lies on its side, and the head twists so that both eyes are on the top.[4]

The jaws of the scale-eating cichlid *Perissodus microlepis* occur in two distinct morphological forms. One morph has its jaw twisted to the left, allowing it to eat scales more readily on its victim's right flank. The other morph has its jaw twisted to the right, which makes it easier to eat scales on its victim's left flank. The relative abundance of the two morphs in populations is regulated by frequency-dependent selection.[3][5][6]

56.3   Mammals

The narwhal has a helical tusk on its upper left jaw.

The sperm whale has a single nostril on its upper left head. The right nostril forms a phonic lip. The source of the air forced through the phonic lips is the right nasal passage. While the left nasal passage opens to the blow hole, the right nasal passage has evolved to supply air to the phonic lips. It is thought that the nostrils of the land-based ancestor of the

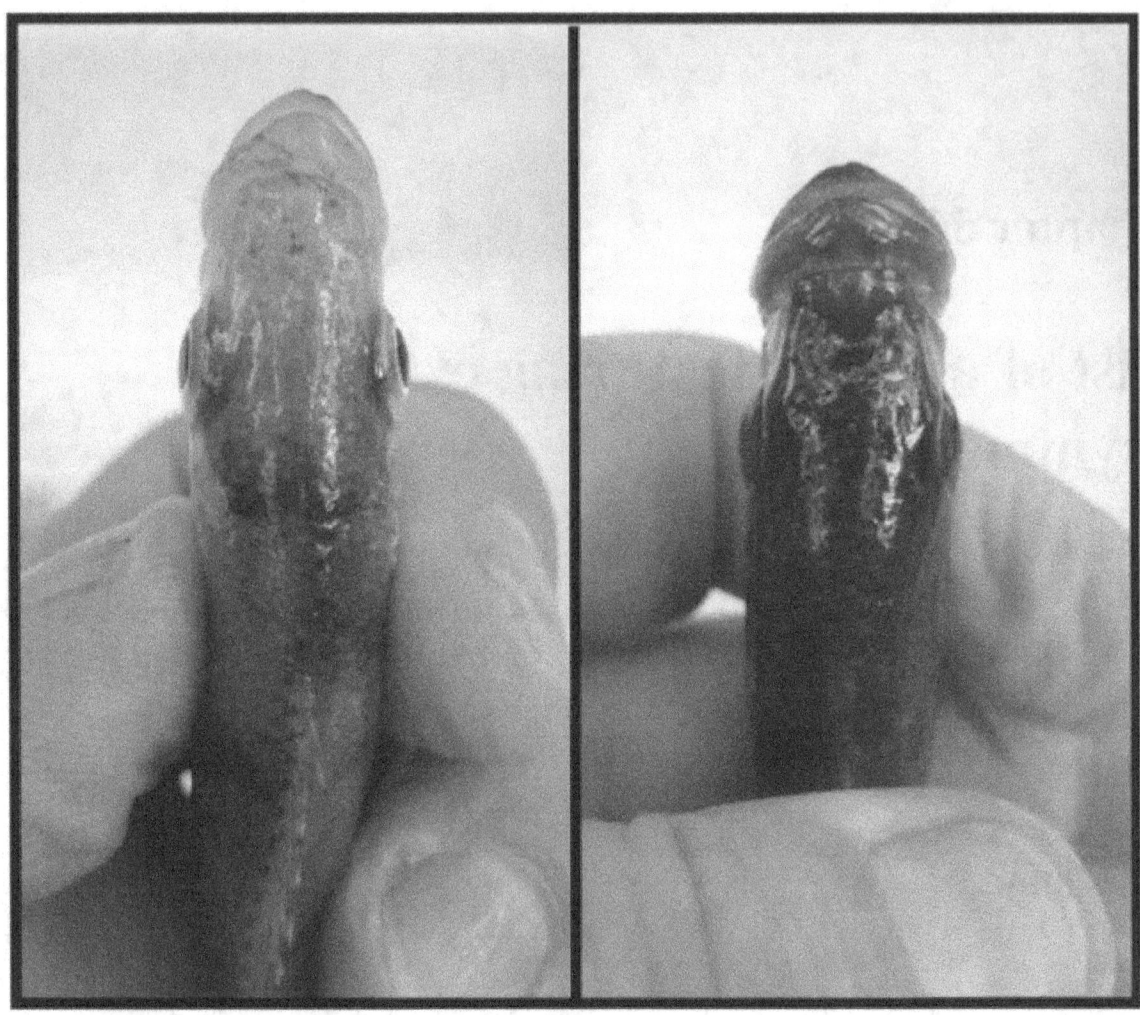

Dorsal view of right-bending (left) and left-bending (right) jaw morphs[3]

sperm whale migrated through evolution to their current functions, the left nostril becoming the blowhole and the right nostril becoming the phonic lips.

Honey badgers of the subspecies *signata* have a second lower molar on the left side of their jaws, but not the right.[7]

56.4   Invertebrates

Fiddler crabs and hermit crabs have one claw much larger than the other. If a male fiddler loses its large claw, it will grow another on the opposite side after moulting. A soft abdomen is also present in a hermit crab as an asymmetrical modification.

All gastropods are asymmetrical. This is easily seen in snails and sea snails, which have helical shells. At first glance slugs appear externally symmetrical, but their pneumostome (breathing hole) is always on the right side. The origin of asymmetry in gastropods is a subject of scientific debate.[8] Other gastropods develop external asymmetry, such as Glaucus atlanticus that develops asymmetrical cerata as they mature.

Sessile animals such as sponges are asymmetrical.[1]

Corals build colonies that are not symmetrical, but the individual polyps exhibit radial symmetry.

Alpheidae feature asymmetrical claws that lack pincers.

Certain polyopisthocotylean monogeneans are asymmetrical, as an adaptation to their attachment to the gill of their fish hosts. Certain parasitic copepods which live inside the gill chamber of their fish hosts are asymmetrical.

- Head of a male crossbill

- Side and top view of a wrybill's beak

- A winter flounder

- A Fiddler crab

- Hermit crabs also have different sized claws

- A Roman snail

- Chicoreus palmarosae, a sea snail

- A red slug, clearly showing the pneumostome

- A sponge

- Coral

- *Protocotyle euzetmaillardi*, a polyopisthocotylean monogenean

- *Heteromicrocotyloides megaspinosus*, a polyopisthocotylean monogenean

- Silhouettes of bodies of 8 species of polyopisthocotylean monogeneans, all asymmetrical

56.5 See also

- Symmetry in biology

56.6 References

[1] Symmetry, biological, cited at FactMonster.com from *The Columbia Electronic Encyclopedia* (2007).

[2] Ngutuparore, the wrybill. Narena Olliver, New Zealand Birds Limited. Retrieved 8 February 2012.

[3] Lee, H. J.; Kusche, H.; Meyer, A. (2012). "Handed Foraging Behavior in Scale-Eating Cichlid Fish: Its Potential Role in Shaping Morphological Asymmetry". *PLoS ONE* **7** (9): e44670. doi:10.1371/journal.pone.0044670.

[4] American Plaice. Canadian Fisheries. Retrieved 8 February 2012.

[5] Hori, M. (1993). "Frequency-dependent natural selection in the handedness of scale-eating cichlid fish". *Science* **260**: 216–219.

[6] Stewart, T. A.; Albertson, R. C. (2010). "Evolution of a unique predatory feeding apparatus: functional anatomy, development and a genetic locus for jaw laterality in Lake Tanganyika scale-eating cichlids". *BMC Biology* **8** (1): 8. doi:10.1186/1741-7007-8-8.

[7] Rosevear 1974, pp. 114–16

[8] Louise R. Page (2006). "Modern insights on gastropod development: Reevaluation of the evolution of a novel body plan". *Integrative and Comparative Biology* **46** (2): 134–143. doi:10.1093/icb/icj018. PMID 21672730. Retrieved 8 February 2012.

56.7 Text and image sources, contributors, and licenses

56.7.1 Text

- **Symmetry** *Source:* https://en.wikipedia.org/wiki/Symmetry?oldid=684614541 *Contributors:* AxelBoldt, Bryan Derksen, The Anome, Tarquin, XJaM, Stevertigo, Patrick, Michael Hardy, Earth, HarmonicSphere, Haakon, CatherineMunro, Cyan, AugPi, Marksuppes, Rossami, Jeandré du Toit, Smack, RodC, Charles Matthews, Jitse Niesen, Wik, Zoicon5, Maximus Rex, Jeffrey Smith, Hyacinth, Nv8200pa, Phys, Omegatron, Bevo, Olathe, Finlay McWalter, Jeffq, Robbot, Altenmann, Romanm, Mayooranathan, Gandalf61, Rursus, Burtonator, JeffC, Paul Murray, ElBenevolente, Alanyst, Marc Venot, Giftlite, Pandammonium, Tom harrison, Herbee, Dratman, DO'Neil, Andycjp, Geni, Yarnover, Quadell, Antandrus, Joseph Myers, Tomruen, C4-enwiki, Porges, Venu62, Imroy, Discospinster, Paul August, Tompw, Rgdboer, Hayabusa future, Uieoa, Mpulier, Benbread, Alansohn, BadSanta, CuriousOne, Jheald, Jon Cates, Oleg Alexandrov, Feezo, Stemonitis, Mel Etitis, Woohookitty, Linas, David Haslam, Guy M, Ruud Koot, Tabletop, GregorB, Pfalstad, Marudubshinki, Graham87, Magister Mathematicae, V8rik, BD2412, Martin von Gagern, Pmj, KYPark, Stardust8212, Salix alba, SchuminWeb, Mathbot, Margosbot~enwiki, RexNL, Ben Babcock, SpectrumDT, King of Hearts, Scoops, Eric B, PointedEars, YurikBot, Wavelength, RussBot, Splash, Grubber, CambridgeBayWeather, NawlinWiki, 0waldo, Wiki alf, Anetode, Mysid, Trojanavenger, Tomabbott, Enormousdude, Cullinane, Arthur Rubin, MathsIsFun, Willtron, Eugee, GrinBot~enwiki, MelRip, Segv11, AceVentura, Sbyrnes321, RonnieBrown, Brentt, Lviatour, SmackBot, Incnis Mrsi, Melchoir, Pavlović, Pgk, KocjoBot~enwiki, Delldot, Atomota, Xaosflux, Gilliam, Ennorehling, Bluebot, Fplay, MalafayaBot, Silly rabbit, CSWarren, Ikiroid, Octahedron80, Kostmo, John Reaves, Antabus, Tamfang, OrphanBot, Rrburke, Cybercobra, TheLimbicOne, Akriasas, Sadi Carnot, ArglebargleIV, Ybact, Lakinekaki, Terry Bollinger, Bjankuloski06en~enwiki, Firefox13, 16@r, Violncello, Abel Cavaşi, Dreftymac, MlckStephenson, Joseph Solis in Australia, Theone00, S0me l0ser, 'Ff'lo, Debanjum, JForget, CmdrObot, Iced Kola, The Font, CBM, MarsRover, Shadow121, Cydebot, Rifleman 82, JFreeman, Jgbeldock, Xminivann, Vanished User jdksfajlasd, Thijs!bot, Epbr123, Kilva, Mojo Hand, Miesling, Yaragn, TimVickers, Coyets, Darvasg, Steelpillow, JAnDbot, Txomin, Struthious Bandersnatch, Dcooper, Bongwarrior, Dekimasu, Soulbot, Baccyak4H, SparrowsWing, Johnbibby, Seberle, Justanother, Japo, David Eppstein, Cpl Syx, DerHexer, JaGa, Wdflake, Khalid Mahmood, Pax:Vobiscum, Falcor84, Monurkar~enwiki, Gwern, Jtir, MartinBot, Schmloof, R'n'B, Pomte, Wlodzimierz, J.delanoy, BigrTex, Nigholith, Chiswick Chap, Lbeaumont, Vanished user 39948282, DWPittelli, VolkovBot, Thisisborin9, JohnBlackburne, Soliloquial, AllS33ing1, Philip Trueman, TXiKiBoT, Bbik, CosmonautLaunchPad, Mercurywoodrose, Zamphuor, A4bot, Weena Eloi, John Ellsworth, LeaveSleaves, Falcon8765, Brianga, Hrafn, SieBot, Nubiatech, Malcolmxl5, Caltas, Keilana, Bentogoa, Flyer22 Reborn, Oxymoron83, JerroldPease-Atlanta, JackSchmidt, Onopearls, Termer, Nn123645, Denisarona, ImageRemovalBot, Tanvir Ahmmed, ClueBot, Snigbrook, The Thing That Should Not Be, BenWillard, Hal8999, Hafspajen, DragonBot, Watchduck, Ottre, Sun Creator, Brews ohare, Promethean, Hans Adler, SchreiberBike, Taranet, Vybr8, Qwfp, TimothyRias, XLinkBot, Dthomsen8, Petitjeanmichel, HOOTmag, Addbot, Jpinacheeto, MrOllie, 5 albert square, Xev lexx, PjOfAustralia, Wytenus208, Tide rolls, Zorrobot, Snaily, Legobot, Cote d'Azur, Luckas-bot, Yobot, Pink!Teen, NotARusski, Gobbleswoggler, AnomieBOT, Jim1138, Materialscientist, Citation bot, ArthurBot, Xqbot, Joshua.mccall, The Evil IP address, GrouchoBot, Lillebi, Omnipaedista, RibotBOT, Shadowjams, Hersfold tool account, FrescoBot, Finalius, Pinethicket, Thinking of England, RobinK, TobeBot, Imaebn, Lotje, Vancouver Outlaw, Jave7784, Bea.miau, Onel5969, TjBot, DexDor, John of Reading, Kpufferfish, RA0808, ZxxZxxZ, ZéroBot, Fæ, Bollyjeff, Parsonscat, Aknicholas, Wayne Slam, Ben Tamari, Jacobisq, Jay-Sebastos, Donner60, Scientific29, ChuispastonBot, Weimer, ClueBot NG, Malleus Felonius, Scalelore, Graythos1, ساجد امجد ساجد, Darian25, Helpful Pixie Bot, BG19bot, Snaevar-bot, Qx2020, Канеюку, Cyberpower678, MusikAnimal, Krupasindhu Muduli, Klilidiplomus, Wannabemodel, Huntlj88, Tonusamuel, Harshul Ravindran, GoShow, Khazar2, EuroCarGT, Kelvinsong, Volvens, Mogism, Brirush, Leprof 7272, Mark viking, Tentinator, Oj.jain, Ginsuloft, Argent2, Monkbot, Crystallizedcarbon, Rakshith12Kiran, Loraof, Malc9141, Yolo pizza pocket, New User Person and Anonymous: 381

- **Symmetry (physics)** *Source:* https://en.wikipedia.org/wiki/Symmetry_(physics)?oldid=684906899 *Contributors:* The Anome, Heron, Stevertigo, Patrick, Michael Hardy, Bloodshedder, Rorro, Giftlite, BenFrantzDale, Netoholic, NetBot, I9Q79oL78KiL0QTFHgyc, Physicistjedi, Danski14, Mattpickman, Reaverdrop, Oleg Alexandrov, Woohookitty, StradivariusTV, Commander Keane, Mpatel, BD2412, Eyu100, Mathbot, Nihiltres, Jrtayloriv, X42bn6, Bhny, Archelon, Paul D. Anderson, Sbyrnes321, SmackBot, Incnis Mrsi, Complexica, Mets501, Quodfui, JRSpriggs, Myasuda, AndrewHowse, Cydebot, Michael C Price, Christian75, Divey, YK Times, PhilKnight, Email4mobile, Homunq, CodeCat, Janus Shadowsong, YohanN7, Paradoctor, Fratrep, ClueBot, Ottre, Bob108, Brews ohare, SchreiberBike, Thamuzino, Rror, Bradv, Addbot, Fyrael, Stylus881, PV=nRT, Luckas-bot, Yobot, Hotbody, Manganite, Rubinbot, Point-set topologist, Locobot, A. di M., FrescoBot, HRoestBot, MastiBot, 8af4bf06611c, ZéroBot, Maschen, Shivsagardharam, Vkpd11, Muskid, Dexbot, Hemlisp, Augustus Leonhardus Cartesius, Rabbitflyer, KHEname, BioticPixels, Matli, Ryanexler and Anonymous: 40

- **Physical system** *Source:* https://en.wikipedia.org/wiki/Physical_system?oldid=678857570 *Contributors:* Charles Matthews, Reddi, Patrick0Moran, Ratjed, Siroxo, Andycjp, Blazotron, Karol Langner, Discospinster, Bobo192, Nigelj, Maureen, Mdd, Shimeru, BD2412, Ian Pitchford, Mar-gosbot~enwiki, Dhollm, Jpbowen, WMarsh, SmackBot, Chris the speller, Stevenmitchell, JHunterJ, Pierre cb, Levineps, Iridescent, CapitalR, Courcelles, Cydebot, Christian75, PamD, Thijs!bot, Berto, Headbomb, Lfstevens, JAnDbot, Grimlock, Tercer, Maurice Carbonaro, 5Q5, Peko2, Lantonov, Kenneth M Burke, VolkovBot, TXiKiBoT, Falcon8765, PaddyLeahy, Phe-bot, Washdivad, CristianCantoro, XLinkBot, Addbot, Vejvančický, LaaknorBot, Fryed-peach, TaBOT-zerem, Piano non troppo, Codicorumus, Xqbot, Quixotec, Banhtrung1, Pierre5018, Hhhippo, ZéroBot, Netknowle, ChuispastonBot, RockMagnetist, ClueBot NG, Jorgenev, Le Comte, AleksandrKönig, KennethH5, Desiree-1d5sauce, Integic, KasparBot, Abirag, AlexCramer, Kremlin00 and Anonymous: 52

- **Symmetry group** *Source:* https://en.wikipedia.org/wiki/Symmetry_group?oldid=669914051 *Contributors:* AxelBoldt, Tarquin, Jkominek, Josh Grosse, Nonenmac, Stevertigo, Edward, Patrick, Dominus, Stevenj, Charles Matthews, Dysprosia, AndrewKepert, RedWolf, Sverdrup, Tobias Bergemann, Giftlite, Snags, BenFrantzDale, Fropuff, Jason Quinn, Auximines, Beland, Bornintheguz, Rich Farmbrough, Qutezuce, Fadereu, Oleg Alexandrov, MFH, Tokek, BD2412, Martin von Gagern, Eubot, Mathbot, Debivort, Siddhant, YurikBot, Wavelength, Reverendgraham, KSmrq, Raven4x4x, LarryLACa, Cullinane, Modify, Wikipedist~enwiki, IstvanWolf, TimBentley, Silly rabbit, Tsca.bot, Tamfang, TheLimbicOne, Pcgomes, Vina-iwbot~enwiki, Maverick starstrider, Mets501, Sir Vicious, Thijs!bot, Kilva, Hannes Eder, R'n'B, JohnBlackburne, LokiClock, TXiKiBoT, Anonymous Dissident, Eubulides, Anchor Link Bot, Addbot, Romaioi, Luckas-bot, Xqbot, GrouchoBot, Lillebi, RobinK, EmausBot, Minimac's Clone, Quondum, Maschen, ClueBot NG, Helpful Pixie Bot, Russell157 and Anonymous: 30

- **Circular symmetry** *Source:* https://en.wikipedia.org/wiki/Circular_symmetry?oldid=609106121 *Contributors:* Patrick, Charles Matthews, BryanD, Dirac1933, Incnis Mrsi, Bluebot, Chetvorno, Atart, Jim.henderson, SieBot, Addbot, LaaknorBot, Amirobot, Handle Dan Down 43-1, 뭐, FrescoBot and Anonymous: 6

- **Rotational symmetry** *Source:* https://en.wikipedia.org/wiki/Rotational_symmetry?oldid=680634850 *Contributors:* Damian Yerrick, The Anome, Nonenmac, Tedernst, Edward, Patrick, Charles Matthews, Fibonacci, MathMartin, Giftlite, Peruvianllama, Anythingyouwant, Tomruen, Lumidek, Joyous!, CALR, Discospinster, Dbachmann, Zscout370, Laurascudder, Stesmo, Orzetto, Wtmitchell, Ceyockey, A D Monroe III, Zanaq, Linas, MONGO, GregorB, Phillipedison1891, Nightscream, Nivix, Gurch, Srleffler, Roboto de Ajvol, CambridgeBayWeather, Voyevoda, Mysid, PyroGamer, Enormousdude, MathsIsFun, That Guy, From That Show!, Akrabbim, Bigcheesegs, David Kernow, Inenis Mrsi, KnowledgeOfSelf, Melchoir, Bluebot, Alan smithee, Baa, Dreadstar, TenPoundHammer, Ninnnu-enwiki, A. Parrot, KokomoNYC, Fairuse-Bot, Pontificake, Falconus, Epbr123, Jed, Dfrg.msc, Nick Number, Mentifisto, Abu-Fool Danyal ibn Amir al-Makhiri, Lifanixi, Shtorman, Farrin, Klapsin, Rappon, JAnDbot, Magioladitis, VoABot II, Infovarius, MartinBot, CommonsDelinker, Acalamari, KylieTastic, RJASE1, JohnBlackburne, Zarcusian, Maximillion Pegasus, Nburoojy, Martin451, CreatureOfEbil, Yintan, Proud Ho, Decoratrix, JerroldPease-Atlanta, Demonic soldier, ClueBot, Zack wadghiri, Fred Fnord, 7&6=thirteen, Polly, Aitias, Qwfp, Crowsnest, Danowen, Dark Mage, RP459, Addbot, Tide rolls, Quantumobserver, Ptbotgourou, PowerUserPCDude, Zad68, Erik9bot, BoomerAB, FrescoBot, MGA73bot, Jen_carrol, EmausBot, WikitanvirBot, Njm6679, A930913, Wayne Slam, Alborzagros, RockMagnetist, ClueBot NG, HMSSolent, Wikih101, Pizzalover87, Rinkle gorge, Epicgenius, Ormistonrocks433, Filedelinkerbot, Verdana Bold, 泰泰泰泰 and Anonymous: 112

- **Global symmetry** *Source:* https://en.wikipedia.org/wiki/Global_symmetry?oldid=606799358 *Contributors:* TakuyaMurata, Charles Matthews, Phys, Mpatel, BD2412, Conscious, SmackBot, Michael C Price, Pamputt, AlleborgoBot, Addbot, GrouchoBot, Mcoupal, Erik9bot, Hauntedpz, ThaeliosActual and Anonymous: 2

- **Local symmetry** *Source:* https://en.wikipedia.org/wiki/Local_symmetry?oldid=674751475 *Contributors:* Physicistjedi, Mpatel, BD2412, BradBeattie, SmackBot, Maksim-e-enwiki, Myasuda, Michael C Price, Headbomb, Dougher, R'n'B, Jeepday, Cuzkatzimhut, Niceguyedc, JavierReynaldo, Jaime Saldarriaga, SchreiberBike, AnomieBOT, FrescoBot, Citation bot 1, AlexUT, GoingBatty, Maschen, MerllwBot, Helpful Pixie Bot, Brad7777 and Anonymous: 10

- **Gauge theory** *Source:* https://en.wikipedia.org/wiki/Gauge_theory?oldid=677681547 *Contributors:* The Anome, Michael Hardy, Tobias Bergemann, Ancheta Wis, TedPavlic, Xezbeth, MuDavid, Bender235, Pt, Phils, BD2412, Rjwilmsi, JocK, Modify, Teply, SmackBot, RDBury, Henning Makholm, Byelf2007, Michael C Price, Biblbroks, Headbomb, Nick Number, Fashionslide, VectorPosse, Magioladitis, Bakken, Email4mobile, JaGa, Policron, Squids and Chips, Cuzkatzimhut, VolkovBot, Red Act, Michael H 34, Setreset, Jwpitts, Tcamps42, Moonriddengirl, ClueBot, Mastertek, TimothyRias, XLinkBot, Addbot, Mortense, Eric Drexler, Bte99, Zorrobot, Luckas-bot, AnomieBOT, Christopher.Gordon3, Citation bot, Northryde, Xqbot, Pra1998, Gsard, A. di M., Erik9bot, FrescoBot, Fortdj33, Citation bot 1, Ganondolf, RedBot, RobinK, Mary at CERN, EmausBot, Brent Perreault, Slawekb, Cogiati, Maschen, Isocliff, ClueBot NG, Helpful Pixie Bot, Bibcode Bot, Dzustin, Brendan.Oz, ChrisGualtieri, SD5bot, Dexbot, Enyokoyama, Joeinwiki, Dath Thou Even Lift, Dhm4444, Dbw1976, KasparBot and Anonymous: 42

- **Continuous symmetry** *Source:* https://en.wikipedia.org/wiki/Continuous_symmetry?oldid=635098217 *Contributors:* Patrick, Charles, BD2412, SmackBot, Ealdent, Cronholm144, Special-T, Kilva, Addbot, Erik9bot, Brirush and Anonymous: 2

- **Spacetime symmetries** *Source:* https://en.wikipedia.org/wiki/Spacetime_symmetries?oldid=627801392 *Contributors:* Edward, Charles Matthews, Hooperbloob, Oleg Alexandrov, Mpatel, BD2412, Salix alba, Ligulem, BradBeattie, Bornhj, Hillman, Grafen, That Guy, From That Show!, SmackBot, Mgiganteus1, Myasuda, Michael H 34, Legobot, Yobot, Citation bot, RockSolidCosmo, Maschen, AHusain314 and Anonymous:1

- **Continuous function** *Source:* https://en.wikipedia.org/wiki/Continuous_function?oldid=685561131 *Contributors:* AxelBoldt, Zundark, Ap, Toby-enwiki, Edemaine, Youandme, Michael Hardy, Wshun, Isomorphic, Ellywa, Ams80, Iulianu, Stevenj, Glenn, BenKovitz, Pizza Puzzle, Schneelocke, Charles Matthews, Stan Lioubomoudrov, Dcoetzee, Dysprosia, Jitse Niesen, Zoicon5, Hyacinth, Sabbut, Joseaperez, Bloodshedder, Phil Boswell, Robbot, MathMartin, Yacht, Bkell, Intangir, Aetheling, Tobias Bergemann, Tosha, Giftlite, Markus Krötzsch, Mikez, Lupin, MSGJ, Jason Quinn, Nayuki, Mormegil, Felix Wiemann, Rich Farmbrough, TedPavlic, Guanabot, Harriv, Paul August, Bender235, BenjBot, Kwamikagami, Rsmelt, Army1987, NetBot, Robotje, Andywall, Delius, Monkey 32606, Mdd, Msh210, Dallashan-enwiki, Arthena, ABCD, Sligocki, Fiedorow, Ultramarine, Oleg Alexandrov, Linas, StradivariusTV, Pdn-enwiki, Smmurphy, Jacj, Graham87, Jshadias, Rjwilmsi, Penumbra2000, FlaBot, Splarka, Ian Pitchford, Mathbot, Jrtayloriv, Fresheneesz, Kri, Chobot, Krishnavedala, YurikBot, Jimp, Fabartus, Musicpvm, NawlinWiki, Rick Norwood, Seb35, Twin Bird, Cheeser1, Klutzy, DomenicDenicola, Tlevine, Igiffin, Kompik, Arthur Rubin, Netrapt, JahJah, Sardanaphalus, SmackBot, RDBury, Thierry Caro, Inenis Mrsi, Melchoir, K-UNIT, Eskimbot, MalafayaBot, Nbarth, ZyMOS, Darth Panda, AdamSmithee, T00h00, Vina-iwbot-enwiki, Igrant, SashatoBot, Lambiam, Jim.belk, EdC-enwiki, Dr.K., Noleander, Ashted, Newone, Domitori, Rhetth, Tawkerbot2, CRGreathouse, Sniffnoy, WLior, Gregbard, MC10, Xantharius, Thijs!bot, LachlanA, Lee Larson, Salgueiro-enwiki, Rbb l181, JAnDbot, Thenub314, 01001, Transcendence, Jakob.scholbach, Cic, Tiagofassoni, Sullivan.t.j, David Eppstein, Error792, R'n'B, Gombang, Policron, STBotD, HyDeckar, DorganBot, Izno, Quiet Silent Bob, VolkovBot, Pasixxxx, Larryisgood, Mplourde-enwiki, Leoremy, A4bot, Hqb, Ctmt, Don4of4, Wolfrock, Sapphic, AlleborgoBot, Katzmik, GirasoleDE, SieBot, Stca74, Craigy90, Henry Delforn (old), Iameukarya, Thehotelambush, Svick, Anchor Link Bot, Rinconsoleao, Sbacle, UKoch, Ramzzhakim, Mspraveen, Timhoooey, PixelBot, Johnuniq, Crowsnest, QYV, SilvonenBot, Jujubot-enwiki, D.M. from Ukraine, Addbot, Fgnievinski, Leszek Jańczuk, Forich, Favonian, LinkFA-Bot, Kisbesbot, Lightbot, PV=nRT, Zorrobot, Yobot, AnomieBOT, Evilchicken1234, Materialscientist, ArthurBot, Xqbot, Bdmy, RibotBOT, Raulshe, Pillcrow, Grinevitski, Scibuff, Citation bot 1, Tkuvho, JumpDiscont, Reach Out to the Truth, Grumpfel, Faolin42, Vanjka-ivanych, Slawekb, Bethnim, Tuxedo junction, Mobius Bot, Roman3, Quondum, D.Lazard, EWikist, Ulipaul, ChuispastonBot, ClueBot NG, Wcherowi, Frietjes, Koertefa, BG19bot, Walrus068, HGK745, Solomon7968, Deltasun, Felidofractals, IkamusumeFan, Freeze S, APerson, Dexbot, ספרטא, Mark viking, CsDix, Ginsuloft, Paul2520, Whikie, SindHind and Anonymous: 166

- **Smoothness** *Source:* https://en.wikipedia.org/wiki/Smoothness?oldid=682390339 *Contributors:* Michael Hardy, TakuyaMurata, Charles Matthews, Lowellian, Tobias Bergemann, Eric Kvaalen, Neelix, Bonadea, Fgnievinski, Fflittle, Josve05a, ClueBot NG, Frietjes, Abitslow, Arash.yazdani59 and Anonymous: 6

- **Discrete symmetry** *Source:* https://en.wikipedia.org/wiki/Discrete_symmetry?oldid=678186111 *Contributors:* Edward, Michael Hardy, Phys, Lumidek, Jag123, Gary, SmackBot, David Eppstein, Addbot, Jeff Muscato, Dc987, Yellow octopus, Reak spoughly, Azzifeldman and Anonymous: 2

- **T-symmetry** *Source:* https://en.wikipedia.org/wiki/T-symmetry?oldid=671700931 *Contributors:* Zundark, Roadrunner, Stevertigo, Patrick, JohnOwens, Michael Hardy, Tim Starling, Albertplanck, Chinju, Karada, SebastianHelm, Stevenj, Angela, Ehn, Phys, Mp-enwiki, SJRubenstein, Giftlite, Lee J Haywood, Pcarbonn, Carandol–enwiki, AmarChandra, Hidaspal, Pjacobi, Martpol, ReallyNiceGuy, Dataphile, Roy-Boy, I9Q79oL78KiL0QTFHgyc, Cherlin, Bucephalus, Jheald, Reaverdrop, Mpatel, Marudubshinki, Strait, Lionelbrits, Moskvax, Wavelength, Mushin, Bambaiah, Yamara, Ihope127, SCZenz, 2over0, StuRat, Reyk, Tim R, SmackBot, QFT, Daqu, Richard L. Peterson, Zarniwoot, Hypnosifl, Mets501, Dan Gluck, JRSpriggs, CmdrObot, Galo1969X, Thijs!bot, Mbell, Headbomb, Pjvpjv, Infophile, Timeron–enwiki, Laksman–enwiki, JayJung, Magioladitis, Thasaidon, Grimlock, Bjheiden, Slash, Maurice Carbonaro, VolkovBot, Red Act, Someguy1221, Eubulides, SieBot, Likebox, Mr. Stradivarius, PhysicsGrad2013, Dave.bradi, EverettYou, Agor153, Forbes72, Addbot, LaaknorBot, Legobot, Luckas-bot, Manganite, Citation bot, Companicus, Xqbot, Minibikini, Rurigok, FrescoBot, Tom.Reding, Dude1818, Al8217, Justpoppingintosayhi, EmausBot, Ebrambot, Pandeist, Bibcode Bot, Shuikouhw, Adamb76, Mark viking, Kfitzell29, Wish vishal and Anonymous: 45

• **Parity (physics)** *Source:* https://en.wikipedia.org/wiki/Parity_(physics)?oldid=685934580 *Contributors:* Patrick, TakuyaMurata, Charles Matthews, Phys, SolLando, Tobias Bergemann, Giftlite, Xerxes314, Beland, Karol Langner, Lumidek, CALR, Pak21, Nvj, Cmdrjameson, Eruantalon, Sergio Macías, Wtmitchell, Knowledge Seeker, Count Iblis, Oleg Alexandrov, Joriki, Marudubshinki, Ae77, Nihiltres, Thecurran, Wave-length, Bambaiah, Archelon, Pseudomonas, Kabirramola, E2mb0t–enwiki, Elkman, GrinBot–enwiki, SmackBot, Incnis Mrsi, Tom Lougheed, Leifisme, QFT, Wiki me, Akriasas, WhiteHatLurker, Erwin, JarahE, JRSpriggs, Raghunathan, Usgnus, Cydebot, Michael C Price, Thijs!bot, Barticus88, Mbell, Headbomb, Pjvpjv, Dougher, Magioladitis, Thasaidon, Dirac66, HEL, Tarotcards, Idioma-bot, Gerrit C. Groenenboom, Cuzkatzimhut, Red Act, Pamputt, Antixt, SieBot, BotMultichill, Paolo.dL, Anchor Link Bot, ClueBot, Sun Creator, DumZiBoT, Lazyrussian, Rror, TravisAF, Addbot, Luckas-bot, Yobot, Tonyrex, PianoDan, Citation bot, ArthurBot, Omnipaedista, Theaucitron, Sławomir Biały, CraigPemberton, Merongb10, RedBot, TobeBot, Heurisko, Linguisticgeek, Queller69, RjwilmsiBot, EmausBot, Albear-And, ZéroBot, Quondum, Kmva, ClueBot NG, Greedohun, Tamila Shalumova, Helpful Pixie Bot, Bibcode Bot, Vkpd11, Slumdog2011, Goodbear3, MuonRay, Abitslow, JellyPatotie, Are you freaking kidding me and Anonymous: 64

• **Glide reflection** *Source:* https://en.wikipedia.org/wiki/Glide_reflection?oldid=679497811 *Contributors:* Patrick, Michael Hardy, Glenn, CharlesMatthews, Henrygb, Tosha, Tomruen, Wgw4, Alexb@cut-the-knot.com, Aholtman, Mathbot, CiaPan, Roboto de Ajvol, SmackBot, Smith609, Ezrakilty, Cydebot, Thijs!bot, .anacondabot, David Eppstein, Katalaveno, Aboluay, ClueBot, Addbot, Ronhjones, Leszek Jańczuk, Luckas-bot, AnomieBOT, MathsPoetry, N4m3, Akerans, Solomon7968, Trevayne08, Brad7777, Kelvinsong, Dexbot and Anonymous: 7

• **Wallpaper group** *Source:* https://en.wikipedia.org/wiki/Wallpaper_group?oldid=683812022 *Contributors:* Edward, Patrick, Michael Hardy, SGBailey, CesarB, Ahoerstemeier, Cimon Avaro, Schneelocke, Charles Matthews, Hyacinth, Fibonacci, Phys, Mordomo, Hajor, Lowellian, Mayooranathan, P0lyglut, Timrollpickering, AnomalousArtemis, Pengo, Tosha, Giftlite, Noe, Joseph Myers, Tomruen, Rich Farmbrough, Pak21, Kjoonlee, C S, 99of9, Johntinker, Keenan Pepper, Kwikwag, Greg Kuperberg, HenkvD, Oleg Alexandrov, Woohookitty, Jeff3000, Trapolator, Mpatel, Lasunncty, Martin von Gagern, Salix alba, R.e.b., John Baez, Mathbot, Tedder, Chobot, Karlscherer3, Spacepotato, Dmharvey, KSmrq, Gaius Cornelius, Anomalocaris, Magicmonster, Ruhrfisch, Zwobot, Gadget850, SilentC, Paul D. Anderson, Krótki, SmackBot, RDBury, Nihonjoe, Prodego, Bluebot, TimBentley, Mhym, Dogears, Breno, Jim.belk, CmdrObot, Tkircher, Asmeurer, Madmarigold, VoABot II, Lenschulwitz, Seberle, David Eppstein, JaGa, Nevit, CommonsDelinker, BigrTex, Maproom, LesPaul75, Chiswick Chap, The enemies of god, Saibod, Cwkmail, Paolo.dL, JackSchmidt, Titian1962, ImageRemovalBot, ClueBot, Plastikspork, JuPitEer, C G Strauss, Snacks, Addbot, Calculuslover, Yobot, Nallimbot, Xqbot, Lpetrich, RjwilmsiBot, Sculleyjp, Oldmanjank, Aughost, SporkBot, Jaspervdg, Helpful Pixie Bot, Toni 001, Winslowa and Anonymous: 65

• **C-symmetry** *Source:* https://en.wikipedia.org/wiki/C-symmetry?oldid=548565045 *Contributors:* Tim Starling, Albertplanck, SebastianHelm, Charles Matthews, The Anomebot, Phys, Jmabel, Xerxes314, Hidaspal, Pjacobi, Bender235, Linas, Mike Peel, Mathbot, YurikBot, Mushin, RussBot, Huatulco–enwiki, Spike Wilbury, SmackBot, Tom Lougheed, Gyrobo, QFT, Erwin, Alan.ca, Thijs!bot, Headbomb, Davidhorman, Alphachimpbot, Magioladitis, Thasaidon, Grimlock, Maliz, A4bot, Venny85, AlleborgoBot, WikHead, Addbot, Mpfiz, Zorrobot, Legobot, Luckas-bot, Citation bot, Maxis ftw, Xqbot, Trebauchet1986, Thinking of England, RobinK, ZéroBot, Ego White Tray, QuantumSquirrel, Steve Bz, Gabobaby and Anonymous: 14

• **CPT symmetry** *Source:* https://en.wikipedia.org/wiki/CPT_symmetry?oldid=684553583 *Contributors:* AxelBoldt, Bryan Derksen, MadSurgeon, Hfastedge, Albertplanck, Charles Matthews, The Anomebot, Phys, Secretlondon, Chuunen Baka, Ary29, Brianjd, Nicobn–enwiki, Hidaspal, Pjacobi, Bender235, Omnifarious, Keenan Pepper, Axl, Qcomp, Joke137, Christopher Thomas, BD2412, Rjwilmsi, Strait, Mushin, Bambaiah, JabberWok, Schlafly, SamuelRiv, Yonir, Finell, SmackBot, Tom Lougheed, The Monster, Unyoyega, QFT, Jmnbatista, Ligulembot, Cronholm144, CapitalR, Van helsing, Myasuda, Michael C Price, ChKa, Robsinden, Headbomb, Magioladitis, WolfmanSF, Thasaidon, Grimlock, Homunq, VolkovBot, Red Act, Mihaip, Synthebot, Likebox, WurmWoode, Razimantv, MystBot, Addbot, FiriBot, Luckas-bot, Yobot, AnomieBOT, Citation bot, Omnipaedista, Mark Schierbecker, FrescoBot, Citation bot 1, Etincelles, NotAnonymous0, UncertaintyPrinciples, ZéroBot, NuclearDuckie, Suslindisambiguator, Quondum, Zueignung, ClueBot NG, Bibcode Bot, Wisconsinitee, Ultra snozbarg and Anonymous: 28

• **CP violation** *Source:* https://en.wikipedia.org/wiki/CP_violation?oldid=678895835 *Contributors:* Roadrunner, Stevertigo, Michael Hardy, Albertplanck, TakuyaMurata, Angela, Julesd, Netsnipe, Palfrey, Raven in Orbit, Coren, Charles Matthews, The Anomebot, Phys, Donarreiskoffer, Pigsonthewing, COGDEN, Ruakh, Giftlite, Jmnbpt, Harp, Xerxes314, Gracefool, ConradPino, HorsePunchKid, Mako098765, WhiteDragon, Karol Langner, Pmanderson, Lumidek, Rich Farmbrough, Hidaspal, Bobo192, Davidruben, Elipongo, Foobaz, I9Q79oL78KiL0QTFHgyc, M0rph, MPerel, Pearle, Sligocki, Evil Monkey, Dirac1933, BlastOButter42, Kusma, Kay Dekker, Oleg Alexandrov, Linas, Nopherox, Marudubshinki, Strait, Tawker, Ligulem, Mathbot, Lmatt, Goudzovski, Chobot, Gdrbot, Bhny, Limulus, JabberWok, NawlinWiki, Grafen, Crasshopper, Tonywalton, Square87–enwiki, Fram, ArielGold, GrinBot–enwiki, MacsBug, SmackBot, HalfShadow, PeterSymonds, Dauto, Chris the speller, Tigerhawkvok, Can't sleep, clown will eat me, QFT, Voyajer, Wen D House, Pwjb, Ligulembot, Drunken Pirate, GTFleming, Erwin, Ryulong, Dan Gluck, Lottamiata, Qqs83, IRevLinas, CRGreathouse, CmdrObot, Wafulz, Vyznev Xnebara, Friendofthehose, Vanished user vjhsduheuiui4t5hjri, Simon Brady, DumbBOT, Gimmetrow, Raoul NK, Mbell, Cosmi, Applecore91, Headbomb, WilliamH, Insane99, Dinagling, Leevclarke, Txomin, Igodard, Kaonslau–enwiki, Thasaidon, Parsecboy, Kevinmon, Homunq, Tonyfaull, DerHexer, Dr. Morbius, MartinBot, Rettetast, Warrickball, Felixbecker2, The dark lord trombonator, Extransit, 1310342–enwiki, Larryisgood, Rich Janis, Jackfork, Venny85, Rknasc, PaddyLeahy, SieBot, Nintendostar, Wing gundam, Scasa–enwiki, LonelyMarble, ClueBot, Likebreakfe, Rotational, Chimesmonster, Yakrami, NuclearWarfare, DumZiBoT, Saeed.Veradi, MystBot, Airplaneman, Addbot, Cxz111, Landon1980, Leszek Jańczuk, Debresser, Tide rolls, Lightbot, Zorrobot, Micko.hjort–enwiki, LuK3, Legobot, Luckas-bot, Yobot, Amirobot, AnomieBOT, Killiondude, Citation bot, Quebec99, Xqbot, Spidern, False vacuum, FrescoBot, Paine Ellsworth, Citation bot 1, Sunandclouds, Elockid, Mutinus, Thinking of England, Sanomi, The Perfection, Dizanl, Ofercomay, John of Reading, Bphyswiki, Zerkroz, Fæ, Arbnos, Ebrambot, AManWithNoPlan, Kweckzilber,

John of Reading, Superlaser1, Slawekb, Somethingcompletelydifferent, Quondum, D.Lazard, Zephyrus Tavvier, Jjenkins5123, Helpful Pixie Bot, Solomon7968, Deltahedron, Mark L MacDonald, Makecat-bot, Spectral sequence, Umberto Lupo, CsDix, Cyrapas, Promise her a definition, Zimboras, Dhm4444, WillemienH and Anonymous: 54

- **Lorentz group** *Source:* https://en.wikipedia.org/wiki/Lorentz_group?oldid=683852993 *Contributors:* AxelBoldt, Ram-Man, Patrick, Michael Hardy, Stupidmoron, Charles Matthews, Timwi, Jitse Niesen, Phys, PuzzletChung, Josh Cherry, Giftlite, Lethe, Fropuff, Pjacobi, Paul August, Rgdboer, Teorth, Msh210, Yurivict, Woohookitty, Linas, Hfarmer, Rjwilmsi, Ligulem, R.e.b., FlaBot, Mathbot, Chobot, Roboto de Ajvol, Wavelength, Hillman, KSmrq, SEWilcoBot, Welsh, SmackBot, Eskimbot, Bluebot, Nbarth, QFT, Sammy1339, Ligulembot, Ulner, Mwj-enwiki, Dan Gluck, RobHar, Atreyu81, JAnDbot, David Eppstein, JaGa, Paulnwatts, T.Needham, Lantonov, Cuzkatzimhut, Camrn86, JohnBlackburne, Geometry guy, Neparis, Drschawrz, YohanN7, Count Truthstein, TimothyRias, Addbot, Cesiumfrog, Niout, MichalKotowski, Rubinbot, Citation bot, DSisyphBot, FrescoBot, Netheril96, Quondum, Maschen, Bstoica, Pashilkar, T.seppelt, A.entropy, Mark viking, CsDix, Al'Beroya, Café Bene, Michael Lee Baker and Anonymous: 23

- **Poincaré group** *Source:* https://en.wikipedia.org/wiki/Poincar%C3%A9_group?oldid=685453530 *Contributors:* AxelBoldt, Zundark, The Anome, XJaM, Mbecker, Stevertigo, Patrick, Michael Hardy, Marco Krohn, AugPi, Stupidmoron, Charles Matthews, Phys, Anupamsr, Giftlite, Lethe, Fropuff, DefLog-enwiki, Jossi, Rich Farmbrough, Bender235, Rgdboer, Aronbeekman, Danski14, Keenan Pepper, Gene Nygaard, Oleg Alexandrov, JFG, Mpatel, Allen3, Rjwilmsi, DVdm, YurikBot, That Guy, From That Show!, SmackBot, Incnis Mrsi, Nbarth, Tsca.bot, Kcordina, Cybercobra, JRSpriggs, Cydebot, Headbomb, Nearyan, JAnDbot, Fetchcomms, Yill577, SHCarter, Sullivan.t.j, Cuzkatzimhut, XCelam, Drschawrz, YohanN7, VVVBot, Phe-bot, Addbot, Luckas-bot, Ptbotgourou, AnomieBOT, Omnipaedista, Thinking of England, Meaghan, Jowa fan, Skater00, ZéroBot, Quondum, Git2010, Maschen, JFB80, Dexbot, CsDix, Prokaryotes, Kfitzell29, Ryanexler and Anonymous: 32

- **Standard Model (mathematical formulation)** *Source:* https://en.wikipedia.org/wiki/Standard_Model_(mathematical_formulation)?oldid=676455116 *Contributors:* The Anome, Phys, BenRG, Giftlite, Alison, Wmahan, Beland, Setokaiba, Kaldari, Masudr, Bender235, Mykhal, Pt, Cmdrjameson, Kuratowski's Ghost, Wdyoung, Mandarax, BD2412, Qwertyus, Mathbot, Tone, Whosasking, Bambaiah, Dna-webmaster, Closedmouth, Caco de vidro, SmackBot, Maksim-e-enwiki, Robotbeat, Incnis Mrsi, Baad, Dauto, Chris the speller, Acipsen, QFT, Grover cleveland, Akriasas, Yevgeny Kats, K.enevoldsen, Xxanthippe, Headbomb, Marwie, Gnixon, Christopher Cooper, Magioladitis, DinoBot, HEL, Lseixas, Schucker, Themel, Flyte35, Ptrslv72, Wing gundam, Bamkin, General Epitaph, Kitsunegami, Brews ohare, NuclearWarfare, SkyLined, Truthnlove, Jeffrey.Yepez, Lightbot, Yobot, Mateusz.Kwasnicki, Bci2, Omnipaedista, John S. Peterson, Ernsts, A. di M., FrescoBot, Ganondolf, Wikipelli, Maschen, Zueignung, Isocliff, Clearlyfakeusername, Asadsphotogremlin, Dsperlich, MerllwBot, Qwerty12345, Randomnonsense, Cjean42, Epicgenius, Euan Richard, Dimension10, Impsswoon, Lathamboyle and Anonymous: 63

- **Spontaneous symmetry breaking** *Source:* https://en.wikipedia.org/wiki/Spontaneous_symmetry_breaking?oldid=678321175 *Contributors:* AxelBoldt, Bryan Derksen, XJaM, Edward, Michael Hardy, Lexor, Charles Matthews, Timwi, Reddi, Phys, Bevo, Dusik, Nagelfar, Giftlite, Lethe, Alison, JeffBobFrank, Jcobb, Gotanda, Gadfium, DragonflySixtyseven, Lumidek, FT2, Hidaspal, Ascánder, Mal-enwiki, Bender235, Clement Cherlin, PhilHibbs, MPS, Shenme, Physicistjedi, Kocio, StuTheSheep, Linas, Jmhodges, Dennis Estenson II, Salix alba, Jehochman, BjKa, Chobot, YurikBot, Ugha, Bambaiah, Archelon, Zzuuzz, Reyk, Roques, RupertMillard, SmackBot, Maksim-e-enwiki, Complexica, Colonies Chris, Jmnbatista, Lagrangian, Akriasas, P199, JarahE, Hetar, Dan Gluck, JMK, Harej bot, Ezrakilty, Thijs!bot, Barticus88, Headbomb, Arcresu, Hillarryous, Dougher, Gökhan, JAnDbot, Yuksing, Attarparn, Jpod2, R'n'B, Natsirtguy, Lseixas, BernardZ, Cuzkatzimhut, Holme053, TXiKiBoT, Red Act, Michael H 34, Pamputt, Moose-32, SieBot, Wing gundam, Renatops, Denisarona, Mastertek, BlueDevil, MelonBot, Truthnlove, Addbot, Yakiv Gluck, Zahd, LaaknorBot, SpBot, OlEnglish, Luckas-bot, Yobot, Yotcmdr, Christopher Pritchard, Zimboz Montizawooba, Obersachsebot, False vacuum, Waleswatcher, Gsard, A. di M., 永澤, CES1596, Freddy78, Pmokeefe, RobinK, Mary at CERN, Marie Poise, Slightsmile, Quondum, Shovkovy, Maschen, Boris Breuer, Vatsal19, Helpful Pixie Bot, Bibcode Bot, Ahhaha, Kalmiopsiskid, Fraulein451, Dexbot, Lugia2453, CMTdrew, Mparisi90, LudicrousTripe and Anonymous: 85

- **Lorentz covariance** *Source:* https://en.wikipedia.org/wiki/Lorentz_covariance?oldid=679857908 *Contributors:* Roadrunner, Stevertigo, Patrick, Michael Hardy, Charles Matthews, Reddi, Doradus, Phys, Drxenocide, Tobias Bergemann, Greyengine5, Herbee, Fropuff, Yath, Lumidek, Kate, Chris Howard, Masudr, Hidaspal, Teorth, Oleg Alexandrov, Ruud Koot, Mpatel, Rjwilmsi, YurikBot, Bhny, Archelon, Grafen, Schlafly, Modify, GrinBot-enwiki, Incnis Mrsi, Gilliam, AlexDitto, Cybercobra, Canadianshoper, JRSpriggs, CmdrObot, Saintrain, Thijs!bot, Omooney, Headbomb, D.H, Escarbot, Tim Shuba, Pkoppenb, WolfmanSF, Maurice Carbonaro, Lantonov, Thurth, TXiKiBoT, Kawakameha, Antixt, Neparis, YohanN7, Henry Delforn (old), Curtdbz, Ajoykt, Djr32, SchreiberBike, MystBot, Addbot, Michele.allegra, DOI bot, Aboctok, SpBot, Luckas-bot, Yobot, AnomieBOT, Citation bot, J04n, Pandamonia, FrescoBot, Nunc aut numquam, Citation bot 1, Haael, UncertaintyPrinci-ples, Fizz2010, Quondum, AManWithNoPlan, SDLEECY, Raidr, Kevin Gorman, Bibcode Bot, Mr. viktor.stepanov, Dilaton, Andyhowlett, Monkbot and Anonymous: 42

- **Transformation(function)** *Source:* https://en.wikipedia.org/wiki/Transformation_(function)?oldid=665037827 *Contributors:* Stevertigo, MichaelHardy, Glenn, Charles Matthews, Aleph4, Tobias Bergemann, Piotrus, PhotoBox, Brian0918, Rgdboer, Bookofjude, Arthena, Joris Gillis, Oleg Alexandrov, BD2412, RxS, NeonMerlin, Ewlyahoocom, Fresheneesz, Siddhant, YurikBot, Wavelength, RussBot, Zwobot, CLW, Finell, Brentt, SmackBot, Mouse Nightshirt, Kirbytime, Neelix, Cydebot, Konradek, Jazzam, David Eppstein, 0612, Chromega, Rohan Ghatak, R'n'B, J.delanoy, Lantonov, Christophre, Dependent Variable, Geometry guy, Synthebot, Portalian, This, that and the other, Yintan, Oxymoron83, Aboluay, ClueBot, Carriearchdale, Good Olfactory, Addbot, Kisbesbot, Teles, Jarble, Ht686rg90, NotARusski, Alexkin, AnomieBOT, IRP, Erik9bot, Thehelpfulbot, Alarics, Quondum, Aughost, Lorem Ip, Peter Karlsen, DASHBotAV, ClueBot NG, Hyiltiz, Weherowi, Mesoderm, محلا, سراجدامحدس, Helpful Pixie Bot, Gorthian, Brad7777, Saehry.Coolestkidyouknow, Zeiimer, JMP EAX, Loraof and Anonymous: 73

- **Circle group** *Source:* https://en.wikipedia.org/wiki/Circle_group?oldid=683179952 *Contributors:* Zundark, Michael Hardy, TakuyaMurata, Karada, Eric119, Revolver, Charles Matthews, Giftlite, Fropuff, HorsePunchKid, Eep², ZeroOne, Rgdboer, Keenan Pepper, Oleg Alexandrov, Linas, Juan Marquez, Mathbot, Elpaw, Dmharvey, Archelon, Netrapt, SmackBot, Incnis Mrsi, Melchoir, Bluebot, Richard L. Peterson, Jim.belk, JoeBot, Freelance Intellectual, Tac-Tics, WISo, Kilva, RobHar, Dogru144, LokiClock, Hesam7, Arcfrk, JackSchmidt, Mr. Stradivarius, Addbot, Fgnievinski, CanadianLinuxUser, HerculeBot, Yobot, Jgmoxness, AnomieBOT, Ciphers, Sławomir Biały, Chricho, Quantum-Squirrel, Bulldog73, The Anonymouse, CsDix, Blackbombchu, Loraof and Anonymous: 16

- **Special unitary group** *Source:* https://en.wikipedia.org/wiki/Special_unitary_group?oldid=676056824 *Contributors:* AxelBoldt, Taral, Stevertigo, Michael Hardy, Looxix-enwiki, Charles Matthews, Dysprosia, Rudminjd, Phys, Robbot, Robinh, Tobias Bergemann, Giftlite, BenFrantzDale, Lethe, Fropuff, Jason Quinn, Eequor, Lumidek, Vivacissamamente, 4pq1injbok, Xezbeth, MuDavid, Paul August, Spoon!, Giraffedata, Eric Kvaalen, Fourthords, Joriki, Simetrical, JATerg, GregorB, BD2412, Rjwilmsi, HappyCamper, Marozols, Nowhither, Itinerant1,

Roboto de Ajvol, YurikBot, StuffOfInterest, RussBot, JabberWok, Archelon, Crasshopper, Tetracube, Reyk, Banus, KnightRider–enwiki, SmackBot, Incnis Mrsi, Tom Lougheed, Movementarian, Silly rabbit, Nbarth, Tamfang, Chlewbot, Speedplane, Harryboyles, Ryulong, Vaughan Pratt, Vyznev Xnebara, MatthewMain, WISo, Dr.enh, Michael C Price, Quibik, Thijs!bot, Koeplinger, Headbomb, Savant13, Magioladitis, VoABot II, Jlenthe, Etale, David Eppstein, Haseldon, Cuzkatzimhut, JohnBlackburne, LokiClock, Anonymous Dissident, StevenJohnston, Yartsa, Drsehawrz, YohanN7, Jasondet, JackSchmidt, OKBot, Mr. Stradivarius, Ideal gas equation, Gigacephalus, Cacadril, Count Truthstein, Addbot, Eric Drexler, Morriswa, Shender, Luckas-bot, Yobot, Niout, Dickdock, AnomieBOT, Collieuk, Citation bot, Waltruda, Tkuvho, LittleWink, Jonesey95, Tim1357, EmausBot, Groemaer, NN22, Zueignung, Bomazi, David C Bailey, 汉语维基, Lemingue, BG19bot, Trodemaster, Hillbillyholiday, CsDix, Prokaryotes, Zimboras, Impsswoon, MarkovianStumble and Anonymous: 95

- **Symmetry in quantum mechanics** *Source:* https://en.wikipedia.org/wiki/Symmetry_in_quantum_mechanics?oldid=683045133 *Contributors:* Bearcat, BD2412, Wavelength, Colonies Chris, Cesium 133, Headbomb, Camrn86, YohanN7, AnomieBOT, FrescoBot, Lonaowna, Quondum, Maschen, Stefan Neumeier, AdventurousSquirrel, Khazar2, Dexbot, Mark viking, Mathphysman, Sumeruhazra, Henderson Duff and Anonymous: 3

- **Noether's theorem** *Source:* https://en.wikipedia.org/wiki/Noether'{}s_theorem?oldid=674709031 *Contributors:* Zundark, The Anome, Tarquin, Heron, Stevertigo, Michael Hardy, Albertplanck, Looxix–enwiki, Stevan White, AugPi, Hollgor, Charles Matthews, Dysprosia, Jitse Niesen, Phys, Tobias Bergemann, David Gerard, Wejalawaga–enwiki, Ancheta Wis, Giftlite, Sj, Harp, Tseller, Lupin, Dratman, Jcobb, DefLog–enwiki, HorsePunchKid, Karol Langner, AmarChandra, Lumidek, Zowie, CALR, Nathan Penton, Luqui, Dbachmann, MuDavid, Bender235, Jnestorius, Ntmatter, Teorth, ..Ajvol.., Brim, Rmz, Babajobu, Osmodiar, 汉语维基, Kusma, Oleg Alexandrov, Linas, David Haslam, Ae-a, Pfalstad, Driftwoodzebulin, Marudubshinki, BD2412, Rjwilmsi, MarSch, R.e.b., Ems57fcva, Rangek, Gseryakov, Alfred Centauri, Chobot, DVdm, WriterHound, Nahmad–enwiki, RussBot, Archelon, PaulGarner, Tong–enwiki, Długosz, Jpowell, Light current, Enormousdude, 2over0, Reyk, CharlesHBennett, Fourohfour, Paul D. Anderson, SmackBot, BeteNoir, Tom Lougheed, InverseHypercube, Melchoir, JC-Santos, RDBrown, Silly rabbit, Complexica, Modest Genius, Chlewbot, Berland, QFT, Jwy, Yevgeny Kats, Lambiam, Kuru, Atoll, Pathosbot, JRSpriggs, Chetvorno, ZICO, CBM, Vyznev Xnebara, Ksoileau, Myasuda, Cydebot, WillowW, Michael C Price, Thijs!bot, Headbomb, Oreo Priest, JAnDbot, .anacondabot, Magioladitis, David Eppstein, Church of emacs, Andejons, Quantling, Lisagosselin, Missphysics, Cuzkatzimhut, AlnoktaBOT, Barbacana, Red Act, Imadeitmyself, Arcfrk, Veshapa, SteakNShake, Thehotelambush, Lisatwo, StewartMH, ClueBot, Mallodi, Twolves14, Jjauregui, Brews ohare, Bbbeard, 1ForTheMoney, AnonyScientist, Pointillist blur, Ziofil–enwiki, Addbot, Substar, Download, Favonian, Deamon138, Luckas-bot, TaBOT-zerem, Niout, Helena srilowa, Materialscientist, Citation bot, Qiushi, Xqbot, Omnipaedista, RibotBOT, Gsard, GliderMaven, FrescoBot, Sae1962, Craig Pemberton, Citation bot 1, Night Jaguar, RjwilmsiBot, Wikeithpedia, Vincent Semeria, Slawekb, Quondum, Zephyrus Tavvier, Maschen, Zfeinst, RockMagnetist, Starshipenterprise, Wgolf, Helpful Pixie Bot, Mlhalpern, Bibcode Bot, Petermahlzahn, F=q(E+v^B), Mark viking, Jb1944, Airwoz, Epic Wink, Kfitzell29, Theoretical wormhole, Ayegbayo and Anonymous: 125

- **Electroweak interaction** *Source:* https://en.wikipedia.org/wiki/Electroweak_interaction?oldid=684180440 *Contributors:* Tobias Hoevekamp, Marj Tiefert, Tarquin, AstroNomer–enwiki, Stevertigo, Michael Hardy, Looxix–enwiki, Schneelocke, Charles Matthews, Phys, Robbot, Post-dlf, Wikibot, Paul Murray, Ancheta Wis, Giftlite, Dbenbenn, Harp, Herbee, Fropuff, Xerxes314, JeffBobFrank, Jcobb, Varlaam, HorsePunchKid, Beland, Kaldari, Icairns, Lumidek, Jørgen Friis Bak, Rich Farmbrough, Pjacobi, Paul August, Pt, Fwb22, Rudchenko, Superstring, Mpa-tel, Christopher Thomas, Rjwilmsi, ElKevbo, Drrngrvy, Naraht, Chobot, Jaraalbe, Roboto de Ajvol, YurikBot, Ugha, Bambaiah, Caco devidro, SmackBot, Tom Lougheed, KocjoBot–enwiki, Jagged 85, Betacommand, Dauto, Colonies Chris, Neutronium, InnocentMind, Jg-wacker, Jaganath, Fangfufu, EmreDuran, JRSpriggs, Drinibot, Michael C Price, Thijs!bot, Headbomb, Escarbot, Certain, Yellowdesk, Yill577,StudierMalMarburg, St.Geoluca Hadge, Duendeverde, HEL, Maurice Carbonaro, Reibot, Antixt, Moose-32, Ptrslv72, MikeRumex, SieBot,Graham Beards, Likebox, Toddst1, Iknowyourider, ClueBot, Brews ohare, TimothyRias, Lightbot, Legobot, Luckas-bot, Yobot, Amirobot,AnomieBOT, Bsimmons666, Girl Scout cookie, Materialscientist, Citation bot, ArthurBot, LilHelpa, Xqbot, Omnipaedista, Jmahl42, Fruit-lord, Citation bot 1, Tom.Reding, MastiBot, کاشف عقیل, ZéroBot, AbigwikiFan, Quondum, Maschen, Cerlbar, ClueBot NG, Asi013, Stalpotaten, Bibcode Bot, Rwbest, Gravitoweak, Ownedroad9, ChrisGualtieri, KJHealey, Cropgood, Cjean42, ElŞahin, Jwratner1, KasparBot andAnonymous: 58

- **Higgs mechanism** *Source:* https://en.wikipedia.org/wiki/Higgs_mechanism?oldid=685001216 *Contributors:* CYD, Roadrunner, Ubiquity, Michael Hardy, Julesd, Palfrey, Charles Matthews, Doradus, Tpbradbury, Phys, Shisolo, David Gerard, Ancheta Wis, Giftlite, Herbee, Edcolins, Lumidek, Ukexpat, Benzh–enwiki, Chris Howard, FT2, Mat cross, David Schaich, Saintswithin, Mal–enwiki, Bender235, Viriditas, BDD, Oleg Alexandrov, Kelly Martin, Linas, BoLingua, Duncan.france, Christopher Thomas, BD2412, Rjwilmsi, Koavf, Strait, R.e.b., Jehochman, FlaBot, Goudzovski, Markdroberts, Gareth E Kegg, Chobot, Algebraist, YurikBot, Bambaiah, Wester, Darsie, AVM, Bhny, Stephenb, Długosz, Dna-webmaster, Tetracube, Caco de vidro, Finell, Triple333, SmackBot, Maksim-e–enwiki, ZerodEgo, Chris the speller, Sbharris, Jmnbatista, Lambiam, JorisvS, Ckatz, Meco, Newone, Benabik, MarsRover, Myasuda, Xxanthippe, Michael C Price, Quibik, Ldussan, Difty, Thijs!bot, Epbr123, Headbomb, West Brom 4ever, Mattfiller, D.H, RogierBrussee, VoABot II, Bakken, Jpod2, RickyCayley, JohnWilliams, Hekerui, Rif Winfield, MartinBot, Haydarhan, Gilleke, Cuzkatzimhut, VolkovBot, Off-shell, LokiClock, TXiKiBoT, Calwiki, Moose-32, Ptrslv72, Coffee, Gerakibot, Likebox, JacquesPHI, Henry Delforn (old), Pac72, Mr. Stradivarius, LoserJoke, ClueBot, General Epitaph, Wwheaton, Drmies, Auntof6, Brews ohare, M.O.X, Crowsnest, XLinkBot, Scvblwxq, Addbot, Eric Drexler, SpBot, Bob K31416, Barak Sh, Tide rolls, Yoavd, Luckas-bot, Yobot, Ptbotgourou, Fraggle81, Galaxydraem, AnomieBOT, Ciphers, Rubinbot, ArthurBot, LilHelpa, Xqbot, TheAMmollusc, Capricorn42, DSisyphBot, RibotBOT, Waleswatcher, Benzen, FrescoBot, BenzolBot, XeBot, Citation bot 1, Benji1986, O.anatinus, RedBot, MastiBot, Aknochel, Beth Ann Lindstrom, Felix0411, Meier99, Mary at CERN, WildBot, EmausBot, WikitanvirBot, Japs 88, LHC Tommy, Slawekb, JSquish, Quondum, L Kensington, Cerlbar, Zueignung, BabbaQ, CBuiltother, PhysicsAboveAll, Giuseppe Vitiello, Jj1236, Parthdu, Curb Chain, Bibcode Bot, Tirebiter78, Ownedroad9, ChrisGualtieri, Dexbot, Abits52, Konbini, CuriousMind01, Ajsal.ea, Itchmean, Cjean42, Crigeos, Crbeals, Jwratner1, Atotalstranger, Jzampardi, KasparBot and Anonymous: 146

- **Symmetry in mathematics** *Source:* https://en.wikipedia.org/wiki/Symmetry_in_mathematics?oldid=680387149 *Contributors:* Patrick, Michael Hardy, Ixfd64, Charles Matthews, Markhurd, Giftlite, AdamJacobMuller, PhotoBox, Pgimeno–enwiki, Woohookitty, Dzine, DePiep, Math-bot, Wavelength, DeadEyeArrow, Cullinane, Chris the speller, Ohconfucius, Hbachus, Konradek, HappyInGeneral, Magioladitis, MartinBot,Eliko, Rgoodermote, Gill110951, Mrh30, Anonymous Zebra, Arcfrk, Calabraxthis, Budhen, Ottre, Xodarap00, HOOTmag, Addbot, Yobot,Ht686rg90, J04n, Shadowjams, Rhalah, I dream of horses, Nickanc, Duoduoduo, Racerx11, ClueBot NG, Mrt3366, Kelvinsong, Brirush,Jumpow, Mark viking, Monkbot and Anonymous: 33

• **Lie algebra** *Source:* https://en.wikipedia.org/wiki/Lie_algebra?oldid=683314667 *Contributors:* AxelBoldt, Zundark, Miguel~enwiki, Michael Hardy, Wshun, Joel Koerwer, TakuyaMurata, Suisui, Kragen, Rossami, Iorsh, Loren Rosen, Charles Matthews, Dysprosia, Michael Larsen, Grendelkhan, Phys, Tobias Bergemann, David Gerard, Weialawaga~enwiki, Tosha, Giftlite, BenFrantzDale, Lethe, Fropuff, Curps, Jeremy Henty, Jason Quinn, Python eggs, Chameleon, DefLog~enwiki, CryptoDerk, CSTAR, Pyrop, Guanabot, Pj.de.bruin, Vsmith, Gauge, Pt, Kwamikagami, Wood Thrush, Reinyday, Foobaz, Msh210, Arthena, Spangineer, Dirac1933, Drbreznjev, Oleg Alexandrov, Linas, Isnow, BD2412, NatusRoma, MarSch, Mathbot, Margosbot~enwiki, RexNL, Masnevets, YurikBot, Wavelength, Hairy Dude, Michael Slone, Lenthe, Stephenb, Grubber, Trovatore, Asimy, Crasshopper, Curpsbot-unicodify, Sbyrnes321, SmackBot, Incnis Mrsi, Grokmoo, Kmarinas86, Bluebot, Silly rabbit, Nbarth, Thomas Bliem, Chlewbot, BlackFingolfin, Noegenesis, Rschwieb, AlainD, Harold f, CmdrObot, Shirulashem, Headbomb, Second Quantization, Dachande, RobHar, B-80, Jrw@pobox.com, Deflective, Englebert, Vanish2, R'n'B, Bogey97, Maurice Carbonaro, Supermanifold, Policron, Fylwind, Cuzkatzimhut, VolkovBot, JohnBlackburne, LokiClock, Ndbrian1, Hesam7, Geometry guy, Drorata, Arcfrk, StevenJohnston, YohanN7, SieBot, Stca74, Jenny Lam, Paolo.dL, JackSchmidt, Mr. Stradivarius, Fatchat, Veromies, JP.Martin-Flatin, Count Truthstein, Addbot, Roentgenium111, Lightbot, Legobot, Luckas-bot, Yobot, Niout, Jason Recliner, Esq., DutchCanadian, Delilahblue, AnomieBOT, Twri, SassoBot, Kaoru Itou, D'ohBot, Darij, Juniuswikiae, Prtmrz, Rausch, Jkock, Adam cohenus, TobeBot, Lotje, Doctor Zook, Slawekb, Quondum, Mikhail Ryazanov, ClueBot NG, Dd314, BG19bot, Teika kazura, Walterpfeifer, Pfeiferwalter, IkamusumeFan, Flbsimas, Deltahedron, Saung Tadashi, Mark L MacDonald, Danielbrice, Enyokoyama, CsDix, 314Username, Forgetfulfunctor00, CaptainLama, KasparBot, Texnico, Douga137 and Anonymous: 92

• **Diffeomorphism** *Source:* https://en.wikipedia.org/wiki/Diffeomorphism?oldid=678510305 *Contributors:* JeLuF, Maury Markowitz, Michael Hardy, TakuyaMurata, AugPi, Poor Yorick, Med, Charles Matthews, Dysprosia, Kuszi, MathMartin, Pascalromon, Tosha, Connelly, Giftlite, Lethe, Fropuff, CryptoDerk, Paul August, Rgdboer, Physicistjedi, Msh210, BRW, Oleg Alexandrov, R.e.b., Mathbot, Lmatt, Bgwhite, Mhwu, YurikBot, Wavelength, RussBot, Woseph, Gaius Cornelius, SmackBot, Silly rabbit, Nakon, Dreadstar, Mathsci, Myasuda, Dharma6662000, Thijs!bot, Headbomb, LachlanA, Nosirrom, Ensign beedrill, Policron, LokiClock, Geometry guy, Rybu, BotMultichill, JerroldPease-Atlanta, He7d3r, Topology Expert, PV=nRT, Legobot, Yobot, AnomieBOT, Point-set topologist, RibotBOT, Sławomir Biały, Citation bot 1, Åkebråke, Redrose64, MondalorBot, Fly by Night, Slawekb, ZéroBot, Quondum, Chester Markel, Uni.Liu, Helpful Pixie Bot, BG19bot, Muses' house, Herve.lombaert, Jeremy112233, Hillbillyholiday, CsDix, Pwm86 and Anonymous: 44

• **Symmetry breaking** *Source:* https://en.wikipedia.org/wiki/Symmetry_breaking?oldid=687075275 *Contributors:* Michael Hardy, Kku, Charles Matthews, Phys, Aleron235, Giftlite, C8to, Thincat, FT2, Xezbeth, Gauge, TheParanoidOne, BD2412, Rjwilmsi, Nihiltres, Debivort, Hairy Dude, Splash, Dan Gluck, Thermochap, Ben MacDui, Parthasarathy.kr, R'n'B, DadaNeem, NigelHarris, Moose-32, Brews ohare, Torage, Stephen Poppitt, Addbot, DOI bot, Dranorter, Download, Ptbotgourou, AnomieBOT, RibotBOT, 沈澄心, AllCluesKey, Enredanrestos, Citation bot 1, DrilBot, Mary at CERN, RjwilmsiBot, EmausBot, WikitanvirBot, Nyxhadanielle, Shivsagardharam, Qx2020, HMman, Mark viking and Anonymous: 20

• **Wheeler–Feynman absorber theory** *Source:* https://en.wikipedia.org/wiki/Wheeler%E2%80%93Feynman_absorber_theory?oldid= 687265654 *Contributors:* Kku, Intangir, Giftlite, Dratman, Bender235, Pt, RJFJR, GregorB, Altman, MarSch, Hydrargyrum, Salsb, Larsobrien, DRosen-bach, SmackBot, Fairley, Chris the speller, Colonies Chris, Pegua, QFT, Eliyak, Agencius, Dicklyon, Myasuda, Michael a lowry, Geostar1024, Hoffes2, Maurice Carbonaro, Yone Fernandes, Dave.bradi, Unica111, Addbot, Yobot, Skirpichev, AnomieBOT, Omnipaedista, Canned Soul, FrescoBot, Goodbye Galaxy, RandomDSdevel, Christopher1968, Fcy, EmausBot, ZéroBot, TonyMath, Maschen, Helpful Pixie Bot, BibcodeBot, BG19bot, BattyBot, Wadsworth baxter, Mogism, Andyhowlett, Laramasama, Mohakhe and Anonymous: 48

• **Invariant (mathematics)** *Source:* https://en.wikipedia.org/wiki/Invariant_(mathematics)?oldid=684918295 *Contributors:* Patrick, Michael Hardy, Charles Matthews, Robbot, MathMartin, Tobias Bergemann, Giftlite, BenFrantzDale, Rgdboer, Krakhan, OoberMick, Mattpickman, Poppafuze, BD2412, FlaBot, Alexb@cut-the-knot.com, YurikBot, Michael Slone, Gaius Cornelius, Gwaihir, Grafen, Mindthief, Cullinane, Ghazer~enwiki, Zvika, RDBury, Melchoir, Optikos, Nbarth, Jackzhp, Thijs!bot, Paxinum, JAnDbot, Magioladitis, Powerinthelines, David Eppstein, TXiKiBoT, Arcfrk, -Midorihana-, Rulerofutumno, Addbot, Atharris, Luckas-bot, Vini 17bot5, AnomieBOT, Twri, Almabot, Charvest, David Nemati, FrescoBot, Cerebral paul c, Magmalex, KHamsun, Anita5192, ClueBot NG, Ignacitum, Nathanielfirst, Makecat-bot, Das O2, Some1Redirects4You, The Quixotic Potato, Nishantgupta9 and Anonymous: 20

• **Symmetric relation** *Source:* https://en.wikipedia.org/wiki/Symmetric_relation?oldid=676529262 *Contributors:* Patrick, Looxix~enwiki, William M. Connolley, Charles Matthews, MathMartin, Tobias Bergemann, Giftlite, Elektron, Ascánder, Paul August, Syp, Joriki, Isnow, Salix alba, Margosbot~enwiki, Fresheneesz, Roboto de Ajvol, Laurentius, Bota47, Arthur Rubin, Incnis Mrsi, Unyoyega, Jdthood, Vina-iwbot~enwiki, Gregbard, Thijs!bot, Mouchoir le Souris, David Eppstein, Jamelan, Henry Delforn (old), ClueBot, Libcub, Addbot, Luckas-bot, ArthurBot, Xqbot, Adavis444, RedBot, EmausBot, ZéroBot, Zap Rowsdower, EdoBot, DASHBotAV, 28bot, Kasirbot, Nosuchforever, Aryan5496 and Anonymous: 18

• **Symmetry (geometry)** *Source:* https://en.wikipedia.org/wiki/Symmetry_(geometry)?oldid=675934995 *Contributors:* Tomruen, BD2412, Solomon7968 and Brirush

• **Symmetric polynomial** *Source:* https://en.wikipedia.org/wiki/Symmetric_polynomial?oldid=676765195 *Contributors:* Patrick, Michael Hardy, Charles Matthews, Giftlite, Bob.v.R, Icairns, Bender235, Zaslav, Oleg Alexandrov, Linas, Small potato, Gwaihir, Arthur Rubin, KnightRiderenwiki, SmackBot, Polyade, JCSantos, Nbarth, Stephen B Streater, Gco, CBM, Mon4, Dogaroon, RobHar, Stellmach, JoergenB, Warut, STBotD, YohanN7, Ali 24789, J.Gowers, DumZiBoT, Marc van Leeuwen, Addbot, Zdaugherty, Firoja, Luckas-bot, Yobot, GrouchoBot, D'ohBot, Orenburg1, EmausBot, Basemaze, Wikfr, Joel B. Lewis, SoSivr, KasparBot and Anonymous: 22

• **Symmetric tensor** *Source:* https://en.wikipedia.org/wiki/Symmetric_tensor?oldid=678435599 *Contributors:* Wesley, Michael Hardy, Looxix~enwiki, Cyp, Charles Matthews, Giftlite, Sj, BenFrantzDale, Billlion, Kundor, Pearle, Msh210, Keenan Pepper, Oleg Alexandrov, Linas, Mpatel, MarSch, Wavelength, Hillman, CambridgeBayWeather, Alex Bakharev, Deville, SmackBot, Chris the speller, JRSpriggs, Eric Lengyel, Vector-Posse, CosineKitty, Pomte, Lantonov, Daniele.tampieri, TXiKiBoT, Thric3, Mbroshi, SchreiberBike, Addbot, LaaknorBot, Yobot, Justpasha, Sławomir Biały, Trappist the monk, Slawekb, Quondum, Jaspervdg, Maschen, Ricardopaleari, BG19bot, Pulga0907, F=q(E+v^B), Hoobster, Dexbot, Theoretical wormhole, Ntheazk and Anonymous: 12

• **Automorphism** *Source:* https://en.wikipedia.org/wiki/Automorphism?oldid=639114467 *Contributors:* AxelBoldt, LC~enwiki, Tarquin, Jan Hidders, Youssefsan, Hephaestos, Edward, Bdesham, Patrick, Chas zzz brown, Michael Hardy, William M. Connolley, AugPi, MatrixFrog, Dysprosia, Phys, SirJective, Robbot, Huppybanny, Altenmann, Tosha, Giftlite, MSGJ, Fropuff, Peruvianllama, Kaldari, Rdsmith4, Sam

Hocevar, Noisy, Rich Farmbrough, Xezbeth, Zaslav, Elwikipedista~enwiki, Gauge, Rgdboer, Crisófilax, Obradovic Goran, Keenan Pepper, Mcmillin24, Oleg Alexandrov, Pixeltoo, Qwertyus, Amire80, VKokielov, Mathbot, Algebraist, YurikBot, Archelon, TechnoGuyRob, BOT-Superzerocool, Cbogart2, Reyk, Pred, Nbarth, Henning Makholm, Mets501, Dreftymac, Happy-melon, Gregbard, Gogo Dodo, Dogaroon, MishaMisha, Hannes Eder, Salgueiro~enwiki, Ksanyi, David Eppstein, Policron, Caiodnh, TXiKiBoT, Mskalak13, Thehotelambush, JackSchmidt, Mild Bill Hiccup, Brews ohare, MelonBot, TimothyRias, Marc van Leeuwen, DOI bot, Topology Expert, PV=nRT, Legobot, JRB-Europe, Citation bot, ArthurBot, Nishantjr, Omnipaedista, SassoBot, Shadowjams, FrescoBot, Citation bot 1, Orenburg1, Quondum, Tommy Jantarek, Helpful Pixie Bot, Vagobot, MathKnight-at-TAU, Valentinovna, Bg9989, Monkbot, Dyott and Anonymous: 24

- **Isometry** Source: https://en.wikipedia.org/wiki/Isometry?oldid=682534821 Contributors: Zundark, Patrick, Michael Hardy, TakuyaMurata, Glenn, Charles Matthews, KRS, Hyacinth, Alembert~enwiki, Robbot, MathMartin, Lupo, Tosha, Giftlite, Lupin, Fropuff, Mike Rosoft, Reuben, Snowolf, Jopxton, Oleg Alexandrov, Peya, Isnow, Marudubshinki, Grammarbot, Salix alba, FlaBot, Alexb@cut-the-knot.com, Mathbot, Siddhant, YurikBot, RussBot, Gaius Cornelius, Crasshopper, ManoaChild, Bota47, Silly rabbit, Jtabbsvt, UKER, LaMenta3, Jackzhp, Sniffnoy, TheTito, Thijs!bot, B-80, Turgidson, Vanish2, Albmont, Trioculite, Sullivan.t.j, TomyDuby, Trumpet marietta 45750, VolkovBot, Matematico~enwiki, SieBot, Harry~enwiki, Marino-slo, Niceguyedc, DragonBot, Addbot, AkhtaBot, Topology Expert, TutterMouse, Uscitizenjason, Peti610botH, KamikazeBot, 4th-otaku, Omnipaedista, Point-set topologist, RibotBOT, EmausBot, WikitanvirBot, Quondum, Aughost, Mfluch, ClueBot NG, Wcherowi, Vagobot, Brad7777, Brirush, Limit-theorem, Noix07, GeoffreyT2000, Hmeaun and Anonymous: 51

- **Multiple integral** Source: https://en.wikipedia.org/wiki/Multiple_integral?oldid=676871019 Contributors: Manning Bartlett, Patrick, Michael Hardy, BenRG, Alan Liefting, Giftlite, Mboverload, Supaluminal, Abdull, D6, Andros 1337, Chizu, Mineralogy, Cmprince, Oleg Alexandrov, Shreevatsa, LOL, Ilario, Salix alba, Mathbot, NekoDaemon, Chobot, YurikBot, Wavelength, KSmrq, Gaius Cornelius, PoorLeno, Tong~enwiki, Saric, Fram, JLaTondre, Matsoftware, Segv11, Med-, Ankurdave, SmackBot, Maksim-e~enwiki, AnOddName, RuudVisser, Jjbeard~enwiki, Dream out loud, Andrei Stroe, Mental Blank, Lambiam, Karakal, Cronholm144, Hvn0413, Igfm2, TestUser001, Hannibal MD, FilipeS, AndrewHowse, Quinnculver, Jameboy, Xantharius, Thijs!bot, Dmhrown00, Einsidler, Hannes Eder, User A1, R'n'B, Pomte, Fylwind, Homo logos, S, VolkovBot, LokiClock, Philip Trueman, Anonymous Dissident, Geometry guy, Gamesguru2, DspDoubleE, Yintan, Wikiarm, ClueBot, Justin W Smith, Mostargue, ChandlerMapBot, F-402, Brews ohare, BOTarate, Franklin.vp, 1ForTheMoney, StevenDH, Kal-El-Bot, Addbot, EconoPhysicist, TStein, LinkFA-Bot, Jasper Deng, Vini 17bot5, Kingpin13, Rtyq2, The High Fin Sperm Whale, Xqbot, Drilnoth, Absolutelegend, DrilBot, MastiBot, Tcnuk, Hager e, Trappist the monk, Ganeshsashank, Hornlitz, Alph Bot, EmausBot, AvicBot, ZéroBot, Raxod502, In base 4, Amiruchka, IznoRepeat, ClueBot NG, Snotbot, Prakash goraniya, Nanna garu, Joel B. Lewis, Euty, Ojha.iiitm, Helpful Pixie Bot, Shashank rathore, BattyBot, Dexbot, חובֵרִים, Brirush, CsDix and Anonymous: 100

- **Skew-symmetric matrix** Source: https://en.wikipedia.org/wiki/Skew-symmetric_matrix?oldid=677108349 Contributors: AxelBoldt, Tarquin, Tbackstr, XJaM, Patrick, Michael Hardy, TakuyaMurata, Kevin Baas, Andres, Charles Matthews, Jitse Niesen, Zero0000, Robbot, Josh Cherry, Tobias Bergemann, Giftlite, BenFrantzDale, Herbee, Fropuff, Syp, Iyerkri, Msh210, Burn, Oleg Alexandrov, Mattfister, LOL, Username314, Jshadias, Rjwilmsi, Godzatswing, FlaBot, Algebraist, YurikBot, KSmrq, Lunch, SmackBot, RDBury, Maksim-e~enwiki, Melchoir, Bluebot, Oli Filth, Octahedron80, Nbarth, Brienanni, Juansempere, Haseldon, Doniminico, DorganBot, Kyap, Partha lal, PMajer, Mcbeth50, TXiKiBoT, A4bot, Ocolon, Neparis, Quietbritishjim, Paolo.dL, Alexbot, Bender2k14, Addbot, Luckas-bot, Yobot, Calle, 9258fahsflkh917fas, Citation bot, ArthurBot, LilHelpa, Lizard86, Xqbot, Adam Dent, RibotBOT, Grinevitski, Kiefer.Wolfowitz, Jonesey95, RedBot, Bourbakista, Trappist the monk, ErikvanB, Bobmath, RjwilmsiBot, The tree stump, WikitanvirBot, Quondum, AManWithNoPlan, Zblumz, Wcherowi, Vacation9, Helpful Pixie Bot, BG19bot, BattyBot, TwoTwoHello and Anonymous: 42

- **Antisymmetric relation** Source: https://en.wikipedia.org/wiki/Antisymmetric_relation?oldid=653233446 Contributors: AxelBoldt, Patrick, Michael Hardy, Charles Matthews, Dcoetzee, MathMartin, Rholton, Tobias Bergemann, Giftlite, Sam Hocevar, Lumidek, Nparikh, Ascánder, Paul August, El C, Spoon!, Rpresser, Jumbuck, EvenT, Adrian.benko, Isnow, Nivaca, Fresheneesz, Chobot, Roboto de Ajvol, Kelovy, Arthur Rubin, Wasseralm, SmackBot, RDBury, Incnis Mrsi, InverseHypercube, Jcarroll, Bluebot, Jdthood, Lambiam, DabMachine, Mike Fikes, Gregbard, WillowW, JAnDbot, TAnthony, Catskineater, Mark lee stillwell, Semmelweiss, PaulTanenbaum, Jackfork, Henry Delforn (old), DuaneL.Anderson, P30Carl, Hakuku, Addbot, LaaknorBot, CarsracBot, Verbal, Legobot, Luckas-bot, Yobot, Xqbot, Adavis444, Erik9bot, De bezige bij, EmausBot, Theophil789, Joel B. Lewis, AvocatoBot, MadGuy7023, Alexjbest, Nbrader, Boga159 and Anonymous: 19

- **Antisymmetry** Source: https://en.wikipedia.org/wiki/Antisymmetry?oldid=665991333 Contributors: SimonP, Michael Hardy, AugPi, Cadr, Beland, N-k, Matve, Mpatel, Rjwilmsi, Quuxplusone, Hairy Dude, Straughn, Bluebot, Saxbryn, CmdrObot, Thijs!bot, Svenonius, Andrea moro, Semmelweiss, Addbot, Yobot, AnomieBOT, Txebixev, RibotBOT, Tjo3ya, EmausBot, John of Reading, Rachel.dicerbo, Catawampus122, BG19bot, W. P. Uzer, Crom daba and Anonymous: 14

- **Symmetry in biology** Source: https://en.wikipedia.org/wiki/Symmetry_in_biology?oldid=673222049 Contributors: Patrick, Ahoerstemeier, Darkwind, Charles Matthews, Haukurth, Grendelkhan, Nv8200pa, Xanzzibar, Giftlite, Tom harrison, Frencheigh, Toytoy, LucasVB, Ukexpat, Trevor MacInnes, Revision17, Stepp-Wulf, HCA, Ivan Bajlo, Violetriga, Syp, Kwamikagami, Bobo192, BrokenSegue, Pearle, Alansohn, TracyRenee, Wouterstomp, Python kiss, Knowledge Seeker, BlastOButter42, Rubidius, Balderai, BD2412, FreplySpang, DePiep, Rjwilmsi, Yamamoto Ichiro, Bmicomp, Cactus.man, Dj Capricorn, Roboto de Ajvol, YurikBot, Wavelength, Rsrikanth05, NawlinWiki, Nineteenthly, VinceBowdren, Anetode, Brandon, Epipelagic, Silverchemist, Enormousdude, ASmartKid, Nil Einne, Chris the speller, Bluebot, Baa, Alphathon, OrphanBot, Nakon, TheLimbicOne, Pissant, Richard001, Cephal-odd, TenPoundHammer, MrDarwin, MattHucke, Kuru, Artman40, Freederick, Kaarel, Ewulp, Tawkerbot2, Sashag, Neelix, Ark-pl, Luigifan, NERIUM, AntiVandalBot, Quintote, TimVickers, Fayenatic london, Spencer, Kerotan, Magioladitis, VoABot II, JamesBWatson, SineWave, Catgut, David Eppstein, The cattr, R'n'B, J.delanoy, Nigholith, Chiswick Chap, DadaNeem, Touch Of Light, WJBscribe, Idioma-bot, Philip Trueman, TheDPRK, Davwillev, Jfryan, Decoratrix, Anchor Link Bot, ClueBot, Excirial, Ottre, Promethean, Tnxman307, Versus22, Vybr8, Alexius08, Vianello, Egor2b, Addbot, Fyrael, Fluffernutter, Tide rolls, Jarble, Halaster, Legobot, Luckas-bot, Bunnyhop11, II MusLiM HyBRiD II, Axelarater, Eric-Wester, AnomieBOT, KDS4444, Rubinbot, Jim1138, Henrik Buschmann, Bluerasberry, Materialscientist, ArthurBot, Thegodofbigthings, Frosted14, SassoBot, Smallman12q, Eugene-elgato, FrescoBot, Shades97, Pinethicket, Zetifree, Felipe lord, Onel5969, Noommos, Carmichael, Mentibot, ClueBot NG, Naldo 911, A520, Jdoggggg, Snotbot, Mesoderm, Shadowcamel, Widr, Helpful Pixie Bot, Calabe1992, BG19bot, SounakSPHS, ChrisGualtieri, Atreyiu, Sminthopsis84, Swagrunner, BioByte, Kenanwang, Blackbombchu, ZAIDZAID12 and Anonymous: 244

- **Patterns in nature** Source: https://en.wikipedia.org/wiki/Patterns_in_nature?oldid=686910778 Contributors: Kku, Alan Liefting, Viriditas, BD2412, Bruce1ee, Sonitus, Wavelength, Gaius Cornelius, Gadget850, Tamfang, Fmindlin, Cybercobra, Valenciano, RichardF, Novangelis, Neoking, Mirrormundo, Tillman, Lopkiol, David Eppstein, Ampy1, Chiswick Chap, DadaNeem, Paradoctor, Yintan, ImageRemovalBot, Elassint, Hafspajen, Piledhigheranddeeper, CharlieRCD, Excirial, MightySaiyan, MrOllie, Jarble, VanishedUser sdu9aya9fasdsopa, Materialscientist, 3family6, Omnipaedista, Edgars2007, FrescoBot, Patchy1, Pinethicket, Tom.Reding, Tomcat7, Animalparty, Dustynyfeathers, GA bot, John Cline, Medeis, ClueBot NG, Gilderien, Mesoderm, Widr, KLBot2, Bibcode Bot, Kempf EK, M0rphzone, Lucquessoy, Dariusz wozniak, Spectral sequence, Saleh Masoumi, Faroffthunder, WikiEnthusiastNumberTwenty-Two, Anrnusna, Monkbot, Fench and Anonymous: 26

- **List of animals featuring external asymmetry** *Source:* https://en.wikipedia.org/wiki/List_of_animals_featuring_external_asymmetry?oldid= 685015964 *Contributors:* Benlisquare, Epipelagic, Cydebot, Fayenatic london, Invertzoo, Roxy the dog, Jarble, FrescoBot, Pinethicket, Ykvach, Helpful Pixie Bot, Stjep, Jeanloujustine and Anonymous: 14

56.7.2   Images

- **File:15crossings-decorative-knot.svg** *Source:* https://upload.wikimedia.org/wikipedia/commons/8/89/15crossings-decorative-knot.svg *License:* Public domain *Contributors:* Self-made image, converted from a version of the following vector PostScript source code: %! 300 396 translate 1.5 dup scale 18 rotate 5{/y 1.1 def/z .75 def -15.1572 51.1194 moveto -33.0803 106.281 lineto -28.0356 107.92 -22.5461 107.488 -17.8201 105.08 curveto 17.8201 86.9199 lineto 22.5461 84.512 28.0356 84.0798 33.0803 85.719 curveto 20.7196 123.761 lineto 20.7196 -12.3607 y mul add 123.761 38.042 y mul add -40 z mul -111.301 add 56.5684 z mul 57.9495 add -111.301 57.9495 curveto stroke/x .3 def 20.7196 123.761 moveto -20.7196 68.2385 lineto -41.4392 x mul -20.7196 add -55.5225 x mul 68.2385 add -53.3013 27.3811 -53.3013 1.3811 curveto stroke 72 rotate} repeat showpage %EOF *Original artist:* AnonMoos

- **File:20_petit_paon_de_nuit.jpg** *Source:* https://upload.wikimedia.org/wikipedia/commons/a/af/20_petit_paon_de_nuit.jpg *License:* CC-BY-SA-3.0 *Contributors:* Own work *Original artist:* jean-pierre Hamon

- **File:2C_3_1979.JPG** *Source:* https://upload.wikimedia.org/wikipedia/commons/4/4c/2C_3_1979.JPG *License:* CC BY-SA 3.0 *Contributors:* Gerard Caris *Original artist:* Gerard Caris

- **File:AIP-Sakurai-best.JPG** *Source:* https://upload.wikimedia.org/wikipedia/commons/2/2b/AIP-Sakurai-best.JPG *License:* Public domain *Contributors:* Own work *Original artist:* self

- **File:A_code_snippet_for_a_rhombic_repetitive_pattern.svg** *Source:* https://upload.wikimedia.org/wikipedia/commons/9/9c/A_code_ for_a_rhombic_repetitive_pattern.svg *License:* CC BY-SA 3.0 *Contributors:* Own work *Original artist:* Baelde

- **File:Academ_Reflections_with_parallel_axis_on_wallpaper.svg** *Source:* https://upload.wikimedia.org/wikipedia/commons/9/9c/Acad Reflections_with_parallel_axis_on_wallpaper.svg *License:* CC BY-SA 3.0 *Contributors:* Own work *Original artist:* Yves Baelde

- **File:Ambox_important.svg** *Source:* https://upload.wikimedia.org/wikipedia/commons/b/b4/Ambox_important.svg *License:* Public domain *Contributors:* Own work, based off of Image:Ambox scales.svg *Original artist:* Dsmurat (talk - contribs)

- **File:Andersonphoto.jpg** *Source:* https://upload.wikimedia.org/wikipedia/commons/8/8d/Andersonphoto.jpg *License:* Copyrighted free use *Contributors:* ? *Original artist:* ?

- **File:Antisymmetry_php_basic_tree_structure.png** *Source:* https://upload.wikimedia.org/wikipedia/commons/a/a8/Antisymmetry_php_ tree_structure.png*License:* Public domain*Contributors:* Created usinghttp://www.ironcreek.net/phpsyntaxtree/*Original artist:* CadratEnglish Wikipedia

- **File:Antisymmetry_segment_category_distinction.png** *Source:* https://upload.wikimedia.org/wikipedia/commons/f/f0/Antisymmetry_ segment_category_distinction.png*License:* Public domain*Contributors:* Created usinghttp://www.ironcreek.net/phpsyntaxtree/*Original artist:* CadratEnglish Wikipedia

- **File:Aphids_and_live_young_under_Sycamore_leaf.JPG** *Source:* https://upload.wikimedia.org/wikipedia/commons/d/d1/Aphids_and_ young_under_Sycamore_leaf.JPG *License:* CC BY-SA 3.0 *Contributors:* Own work *Original artist:* Chiswick Chap

- **File:Areabetweentwographs.svg** *Source:* https://upload.wikimedia.org/wikipedia/commons/f/f9/Areabetweentwographs.svg *License:* Public domain *Contributors:* ? *Original artist:* ?

- **File:Asymmetric_(PSF).svg** *Source:* https://upload.wikimedia.org/wikipedia/commons/f/f8/Asymmetric_%28PSF%29.svg *License:* Public domain *Contributors:* Archives of Pearson Scott Foresman, donated to the Wikimedia Foundation *Original artist:* Pearson Scott Foresman

- **File:BigPlatoBig.png** *Source:* https://upload.wikimedia.org/wikipedia/commons/3/3a/BigPlatoBig.png *License:* CC BY-SA 3.0 *Contributors:* Own work *Original artist:* Brirush

- **File:Bothodd.png** *Source:* https://upload.wikimedia.org/wikipedia/commons/d/de/Bothodd.png *License:* Public domain *Contributors:* Own work *Original artist:* Fresheneesz (talk) (Uploads)

- **File:Brent_method_example.svg** *Source:* https://upload.wikimedia.org/wikipedia/commons/2/2d/Brent_method_example.svg *License:* CC0 *Contributors:* Own work *Original artist:* Krishnavedala

- **File:Brillouin_Zone_(1st,_FCC).svg** *Source:* https://upload.wikimedia.org/wikipedia/commons/c/c1/Brillouin_Zone_%281st%2C_FCC% 29.svg *License:* Public domain *Contributors:* Own work *Original artist:* Inductiveload

- **File:Bump2D_illustration.png** *Source:* https://upload.wikimedia.org/wikipedia/commons/2/24/Bump2D_illustration.png *License:* Public domain *Contributors:* Own work *Original artist:* Oleg Alexandrov

- **File:C0_function.svg** *Source:* https://upload.wikimedia.org/wikipedia/commons/c/c5/C0_function.svg *License:* Public domain *Contributors:* This file was derived from: C0 function.png *Original artist:* Dega180

- **File:Edit-clear.svg** *Source:* https://upload.wikimedia.org/wikipedia/en/f/f2/Edit-clear.svg *License:* Public domain *Contributors:* The *Tango! Desktop Project. Original artist:*

  The people from the Tango! project. And according to the meta-data in the file, specifically: "Andreas Nilsson, and Jakub Steiner (although minimally)."

- **File:Electroweak.svg** *Source:* https://upload.wikimedia.org/wikipedia/commons/d/dc/Electroweak.svg *License:* CC BY-SA 3.0 *Contributors:* Own work *Original artist:* Cjean42

- **File:Elementary_particle_interactions.svg** *Source:* https://upload.wikimedia.org/wikipedia/commons/4/4c/Elementary_particle_.svg *License:* Public domain *Contributors:* en:Image:Interactions.png *Original artist:* en:User:TriTertButoxy, User:Stannered

- **File:Esempio-formulediriduzione-r2.svg** *Source:* https://upload.wikimedia.org/wikipedia/commons/1/1c/Esempio-formulediriduzione-r2.svg *License:* CC-BY-SA-3.0 *Contributors:* No machine-readable source provided. Own work assumed (based on copyright claims). *Original artist:* No machine-readable author provided. Cronholm144 assumed (based on copyright claims).

- **File:Esempio_trasformazione_dominio_da_cartesiano_polare.svg** *Source:* https://upload.wikimedia.org/wikipedia/commons/c/c2/Esempio_trasformazione_dominio_da_cartesiano_polare.svg *License:* CC-BY-SA-3.0 *Contributors:* No machine-readable source provided. Own work assumed (based on copyright claims). *Original artist:* No machine-readable author provided. Cronholm144 assumed (based on copyright claims).

- **File:Even_and_odd_antisymmetric_relation.png** *Source:* https://upload.wikimedia.org/wikipedia/commons/c/cf/Even_and_odd_relation.png *License:* Public domain *Contributors:* From the very same page I am now uploading it to. *Original artist:* Fresheneesz

- **File:Example_of_continuous_function.svg** *Source:* https://upload.wikimedia.org/wikipedia/commons/7/7f/Example_of_continuous_function.svg *License:* Public domain *Contributors:* This file was derived from Example of continuous function.png: <a href='//commons.wikimedia.org/wiki/File:Example_of_continuous_function.png' class='image'><img alt='Example of continuous function.png' src='https://upload.wikimedia.org/wikipedia/commons/thumb/8/87/Example_of_continuous_function.png/50px-Example_of_continuous_function.png' width='50' height='50' srcset='https://upload.wikimedia.org/wikipedia/commons/thumb/8/87/Example_of_continuous_function.png/75px-Example_of_continuous_function.png 1.5x, https://upload.wikimedia.org/wikipedia/commons/thumb/8/87/Example_of_continuous_function.png/100px-Example_of_continuous_function.png 2x' data-file-width='217' data-file-height='219' /></a>*Original artist:* Example_of_continuous_function.png: User:Pasixxxx

- **File:ExponentialMap-01.png** *Source:* https://upload.wikimedia.org/wikipedia/commons/0/06/ExponentialMap-01.png *License:* Public domain *Contributors:* ? *Original artist:* ?

- **File:FWF_Samuel_Monnier_détail.jpg** *Source:* https://upload.wikimedia.org/wikipedia/commons/9/99/FWF_Samuel_Monnier_d%C3%A9tail.jpg *License:* CC BY-SA 3.0 *Contributors:* Own work (low res file) *Original artist:* Samuel Monnier

- **File:Farsh1.jpg** *Source:* https://upload.wikimedia.org/wikipedia/commons/4/40/Farsh1.jpg *License:* CC-BY-SA-3.0 *Contributors:* Transferred from en.wikipedia to Commons. *Original artist:* Zereshk at English Wikipedia. [1]

- **File:Feynman-Diagram.svg** *Source:* https://upload.wikimedia.org/wikipedia/commons/e/e3/Feynman-Diagram.svg *License:* Public domain *Contributors:* own work, based on Image:Feynman-Diagram.jpg *Original artist:* helix84

- **File:Finland_road_sign_166.svg** *Source:* https://upload.wikimedia.org/wikipedia/commons/5/58/Finland_road_sign_166.svg *License:* Public domain *Contributors:* Finnish Transport Agency *Original artist:* Unknown

- **File:Folder_Hexagonal_Icon.svg** *Source:* https://upload.wikimedia.org/wikipedia/en/4/48/Folder_Hexagonal_Icon.svg *License:* Cc-by-sa-3.0 *Contributors:* ? *Original artist:* ?

- **File:Frieze_example_p1.png** *Source:* https://upload.wikimedia.org/wikipedia/commons/1/10/Frieze_example_p1.png *License:* GFDL *Contributors:* File:Frieze2b.png *Original artist:* AndrewKepert

- **File:Frieze_example_p11g.png** *Source:* https://upload.wikimedia.org/wikipedia/commons/7/76/Frieze_example_p11g.png *License:* GFDL *Contributors:* File:Frieze2b.png *Original artist:* AndrewKepert

- **File:Frieze_example_p2mg.png** *Source:* https://upload.wikimedia.org/wikipedia/commons/7/7b/Frieze_example_p2mg.png *License:* GFDL *Contributors:* File:Frieze2b.png *Original artist:* AndrewKepert

- **File:Function-x3.svg** *Source:* https://upload.wikimedia.org/wikipedia/commons/f/f5/Function_x%5E3.svg *License:* CC-BY-SA-3.0 *Contributors:* self-made, based on Image:Function x^2.svg by Qualc1 *Original artist:* Oleg Alexandrov

- **File:Function_x\char"005E\relax{}2.svg** *Source:* https://upload.wikimedia.org/wikipedia/commons/d/dd/Function_x%5E2.svg *License:* BY-SA-3.0 *Contributors:* Self-made using gnuplot *Original artist:* Qualc1

- **File:Gastroenteritis_viruses.jpg** *Source:* https://upload.wikimedia.org/wikipedia/commons/7/71/Gastroenteritis_viruses.jpg *License:* CC BY 3.0 *Contributors:* Transferred from en.wikipedia by Ronhjones *Original artist:* en:User:Graham Beards at en.wikipedia

- **File:Glide_reflection.png** *Source:* https://upload.wikimedia.org/wikipedia/commons/5/5f/Glide_reflection.png *License:* CC-BY-SA-3.0 *Contributors:* ? *Original artist:* ?

- **File:Glide_reflection.svg** *Source:* https://upload.wikimedia.org/wikipedia/commons/e/e8/Glide_reflection.svg *License:* CC0 *Contributors:* Own work *Original artist:* Kelvinsong

- **File:Great_Mosque_of_Kairouan,_west_portico_of_the_courtyard.jpg** *Source:* https://upload.wikimedia.org/wikipedia/commons/4/42/Great_Mosque_of_Kairouan%2C_west_portico_of_the_courtyard.jpg *License:* CC BY-SA 2.0 *Contributors:* Flickr: marble arch *Original artist:* James Rose

- **File:Haeckel_Actiniae.jpg** *Source:* https://upload.wikimedia.org/wikipedia/commons/a/a9/Haeckel_Actiniae.jpg *License:* Public domain *Contributors:* Kunstformen der Natur (1904), plate/planche 49: Actiniae (see here, here, here and here) *Original artist:* Ernst Haeckel

- **File:Wallpaper_group_diagram_p4_square.svg**Source: https://upload.wikimedia.org/wikipedia/commons/0/08/Wallpaper_group_diagram_p4_square.svgLicense: Public domainContributors: generated by self written XSLT available from thecategory overviewOriginal artist: Martin von Gagern

- **File:Wallpaper_group_diagram_p4g_square.svg**Source: https://upload.wikimedia.org/wikipedia/commons/0/0c/Wallpaper_group_diagram_p4g_square.svgLicense: Public domainContributors: generated by self written XSLT available from the category overviewOriginal artist:Martin von Gagern

- **File:Wallpaper_group_diagram_p4m_square.svg**Source: https://upload.wikimedia.org/wikipedia/commons/4/49/Wallpaper_group_diagram_p4m_square.svgLicense: Public domainContributors: generated by self written XSLT available from the category overviewOriginal artist:Martin von Gagern

- **File:Wallpaper_group_diagram_p6.png** Source: https://upload.wikimedia.org/wikipedia/commons/4/49/Wallpaper_group_diagram_p6.png License: Public domain Contributors: generated by inkscape from SVG generated by self written XSLT Original artist: Martin von Gagern

- **File:Wallpaper_group_diagram_p6.svg** Source: https://upload.wikimedia.org/wikipedia/commons/a/a3/Wallpaper_group_diagram_p6.svg License: Public domain Contributors: generated by self written XSLT available from the category overview Original artist: Martin von Gagern

- **File:Wallpaper_group_diagram_p6m.svg** Source: https://upload.wikimedia.org/wikipedia/commons/b/b1/Wallpaper_group_diagram_p6m.svg License: Public domain Contributors: generated by self written XSLT available from the category overview Original artist: Martin von Gagern

- **File:Wallpaper_group_diagram_pg.svg** Source: https://upload.wikimedia.org/wikipedia/commons/e/ee/Wallpaper_group_diagram_pg.svg License: Public domain Contributors: generated by self written XSLT available from the category overview Original artist: Martin von Gagern

- **File:Wallpaper_group_diagram_pg_rotated.svg**Source: https://upload.wikimedia.org/wikipedia/commons/f/f4/Wallpaper_group_pg_rotated.svg License: Public domain Contributors: Derived from unrotated version Original artist: User:Tomruen

- **File:Wallpaper_group_diagram_pgg.svg** Source: https://upload.wikimedia.org/wikipedia/commons/c/e7/Wallpaper_group_diagram_pgg.svg License: Public domain Contributors: generated by self written XSLT available from the category overview Original artist: Martin von Gagern

- **File:Wallpaper_group_diagram_pgg_square.svg** Source: https://upload.wikimedia.org/wikipedia/commons/b/b3/Wallpaper_group_pgg_square.svgLicense: Public domainContributors: Derived from file:Wallpaper group diagram pgg square.svgOriginal artist: User:Tomruen

- **File:Wallpaper_group_diagram_pm.svg** Source: https://upload.wikimedia.org/wikipedia/commons/7/7e/Wallpaper_group_diagram_pm.svg License: Public domain Contributors: generated by self written XSLT available from the category overview Original artist: Martin von Gagern

- **File:Wallpaper_group_diagram_pm_rotated.svg**Source: https://upload.wikimedia.org/wikipedia/commons/f/f8/Wallpaper_group_pm_rotated.svg License: Public domain Contributors: Derived from unrotated version Original artist: User:Tomruen

- **File:Wallpaper_group_diagram_pmg.svg** Source: https://upload.wikimedia.org/wikipedia/commons/b/b5/Wallpaper_group_diagram_pmg.svg License: Public domain Contributors: generated by self written XSLT available from the category overview Original artist: Martin von Gagern

- **File:Wallpaper_group_diagram_pmg_rotated.svg**Source: https://upload.wikimedia.org/wikipedia/commons/7/72/Wallpaper_group_pmg_rotated.svg License: Public domain Contributors: Derived from unrotated version Original artist: User:Tomruen

- **File:Wallpaper_group_diagram_pmm.svg**Source: https://upload.wikimedia.org/wikipedia/commons/6/6b/Wallpaper_group_diagram_pmm.svgLicense: Public domainContributors: generated by self written XSLT available from thecategory overviewOriginal artist: Martin von Gagern

- **File:Wallpaper_group_diagram_pmm_square.svg**Source: https://upload.wikimedia.org/wikipedia/commons/d/d8/Wallpaper_group_diagram_pmm_square.svgLicense: Public domainContributors: Derived from file:Wallpaper group diagram pmm square.svgOriginal artist: User:Tomruen

- **File:Wiki_letter_w_cropped.svg** Source: https://upload.wikimedia.org/wikipedia/commons/1/1c/Wiki_letter_w_cropped.svg License: CC-BY-SA-3.0 Contributors:

- Wiki_letter_w.svg Original artist: Wiki_letter_w.svg: Jarkko Piiroinen

- **File:Wikinews-logo.svg** Source: https://upload.wikimedia.org/wikipedia/commons/2/24/Wikinews-logo.svg License: CC BY-SA 3.0 Contributors: This is a cropped version of Image:Wikinews-logo-en.png. Original artist: Vectorized by Simon 01:05, 2 August 2006 (UTC) Updated by Time3000 17 April 2007 to use official Wikinews colours and appear correctly on dark backgrounds. Originally uploaded by Simon.

- **File:Wikisource-logo.svg** Source: https://upload.wikimedia.org/wikipedia/commons/4/4c/Wikisource-logo.svg License: CC BY-SA 3.0 Contributors: Rei-artur Original artist: Nicholas Moreau

- **File:Wiktionary-logo-en.svg** Source: https://upload.wikimedia.org/wikipedia/commons/f/f8/Wiktionary-logo-en.svg License: Public domain Contributors: Vector version of Image:Wiktionary-logo-en.png. Original artist: Vectorized by Fvasconcellos (talk · contribs), based on original logo tossed together by Brion Vibber

- **File:Wooden_hourglass_3.jpg** Source: https://upload.wikimedia.org/wikipedia/commons/7/70/Wooden_hourglass_3.jpg License: CC-BY-SA-3.0 Contributors: Own work Original artist: User:S Sepp

- **File:World_line.svg** Source: https://upload.wikimedia.org/wikipedia/commons/1/16/World_line.svg License: CC-BY-SA-3.0 Contributors: Transferred from en.wikipedia.
Original artist: SVG version: K. Aainsqatsi at en.wikipedia

- **File:X-Bar_movement_example_(Antisymmetry).png** Source: https://upload.wikimedia.org/wikipedia/commons/3/34/X-Bar_movement_example_%28Antisymmetry%29.png License: CC BY-SA 4.0 Contributors: Own work Original artist: FulcoE

- **File:X\char"005E\relax{}2sin(x\char"005E\relax{}$-$1).svg** *Source:* https://upload.wikimedia.org/wikipedia/commons/d/db/X%5E2sin%28x%5E-1%29.svg *License:* Attribution *Contributors:*
- TV_pie3.png *Original artist:*
- derivative work: Pbroks13 (talk)
- **File:Yemen_Chameleon_(cropped).jpg** *Source:* https://upload.wikimedia.org/wikipedia/commons/f/fc/Yemen_Chameleon_%28cropped%29.jpg *License:* CC BY-SA 3.0 *Contributors:* Own work *Original artist:* Chiswick Chap

56.7.3   Content license

- Creative Commons Attribution-Share Alike 3.0